Lecture Notes in Computer Science 15671

The series Lecture Notes in Computer Science (LNCS), including its subseries Lecture Notes in Artificial Intelligence (LNAI) and Lecture Notes in Bioinformatics (LNBI), has established itself as a medium for the publication of new developments in computer science and information technology research, teaching, and education.

LNCS enjoys close cooperation with the computer science R & D community, the series counts many renowned academics among its volume editors and paper authors, and collaborates with prestigious societies. Its mission is to serve this international community by providing an invaluable service, mainly focused on the publication of conference and workshop proceedings and postproceedings. LNCS commenced publication in 1973.

Ulrich Schmid · Roman Kuznets
Editors

Structural Information and Communication Complexity

32nd International Colloquium, SIROCCO 2025
Delphi, Greece, June 2–4, 2025
Proceedings

Springer

Editors
Ulrich Schmid
TU Wien
Vienna, Austria

Roman Kuznets
Institute of Computer Science of the Czech Academy of Sciences
Prague, Czech Republic

ISSN 0302-9743 ISSN 1611-3349 (electronic)
Lecture Notes in Computer Science
ISBN 978-3-031-91735-6 ISBN 978-3-031-91736-3 (eBook)
https://doi.org/10.1007/978-3-031-91736-3

This Springer imprint is published by the registered company Springer Nature Switzerland AG
The registered company address is: Gewerbestrasse 11, 6330 Cham, Switzerland

Preface

SIROCCO, the *International Colloquium On Structural Information and Communication Complexity*, is devoted to the study of the interplay between structural knowledge, communication, and computing in decentralized systems of multiple communicating entities. Special emphasis is given to innovative methodological and solution approaches leading to a better understanding of the relationship between computing and communication. SIROCCO has a tradition of interesting and productive scientific meetings in a relaxed and pleasant atmosphere, attracting leading researchers in a variety of fields in which communication and knowledge play a significant role.

This volume contains the papers presented at the 32nd instance of SIROCCO, which was held on June 2–4, 2025 in the famous historic town of Delphi, Greece, at the European Cultural Centre of Delphi. It also includes the citation for the 2025 Prize for Innovation in Distributed Computing, which was awarded to *Sergio Rajsbaum* for his pioneering advancements in the theory of distributed computing. SIROCCO 2025 also featured three keynote lectures, presented by

– *Hagit Attiya* (Technion, Israel) on
  "A Gold Standard, but Not a Silver Bullet"

– *Joseph Y. Halpern* (Cornell University, USA) on
  "Actual Causality: A Survey"

– *Christos Papadimitriou* (Columbia University, USA) on
  "Computing with Dynamical Systems: Game Theory and the Brain"

An abstract of each keynote lecture is included in the proceedings.

SIROCCO 2025 received 57 submissions, each of which was read and evaluated by at least three members of the PC, assisted by 47 external reviewers, using the single-blind reviewing process outlined on page xi. The Program Committee finally selected 24 regular papers and 4 brief announcements for inclusion in the conference program and in the proceedings. The brief announcements were the result of inviting the authors of regular submissions that could not be accepted as a full paper, but were considered promising by the PC members, to provide a 6-page brief announcement version of their work.

The Best Paper Award of SIROCCO 2025 went to

Trisha Chakraborty, *Abir Islam*, *Valerie King*,
Daniel Rayborn, *Jared Saia*, *Maxwell Young*

for their paper "Bankrupting DoS Attackers." The Best Student Paper Award of SIROCCO 2025 was given to

Christian Cachin, *David Lehnherr*, *Thomas Studer*,

for their paper "Simplicial Belief." Revised and expanded versions of several additional selected regular papers will be considered for publication in a special issue of the journal *Distributed Computing*.

We wish to thank the many contributors to SIROCCO 2025: the authors of the submitted papers, the PC members and the reviewers, the three keynote speakers, the conference general chairs and local organizers *Konstantinos Georgiou* and *Euripides Markou*, and the proceedings chair and co-editor *Roman Kuznets*.

SIROCCO 2025 also acknowledges the use of the EasyChair system for handling submissions and is indebted to Springer LNCS for publishing the proceedings.

Special thanks to the supporters of SIROCCO 2025 listed below, particularly the co-organizer, ATHENA Archimedes Unit, for their generous support of our keynote speakers and the Prize for Innovation in Distributed Computing awardee; Toronto Metropolitan University, Faculty of Science, and the Department of Mathematics for covering substantial conference expenses; and Springer LNCS for supporting the Best Student Paper Award.

March 2025 Ulrich Schmid

SIROCCO 2025 Co-organizers

SIROCCO 2025 Supporting Organizations

2025 Prize for Innovation in Distributed Computing

We are pleased to announce that the 2025 Prize for Innovation in Distributed Computing is awarded to **Sergio Rajsbaum (National Autonomous University of Mexico)** for his pioneering advancements in the theory of distributed computing.

Sergio has played a central role in developing and using the topological approach as described in his "ACM Notable Books of the Year"-listed book [16] with Maurice Herlihy and Dmitry Kozlov. Together with more than 100 different collaborators, he has co-authored more than 250 scientific publications in journals and conferences. In particular, Sergio has been a leading figure in advancing the basic wait-free asynchronous read/write shared memory model to t-resilient models [3], more powerful shared memory objects [17], synchronous or partially synchronous message-passing models [22, 19], and dependent or Byzantine failures [21]. He also introduced novel tasks like loop agreement [20] and symmetry breaking [23, 6] and developed fundamental undecidability results on task solvability [18].

A remarkable feature of Sergio's research work is its breadth and, in particular, the various bridges to other areas and application domains. Notable examples of the former are a suite of papers devoted to runtime verification (culminating in the "Notable Publications of JACM"-listed paper [2]), his work on interval-linearizability [7] recently surveyed in *Communications of the ACM* [8], and his well-cited contributions to condition-based consensus [25] and layered analysis [24]. Among Sergio's interdisciplinary work are ground-breaking publications [13, 15, 14] on simplicial epistemic multi-agent logic, which have been developed in a collaboration with leading experts in this field, like Hans van Ditmarsch and Éric Goubault, and contributions to robot coordination [1].

Sergio has a long-established relation to SIROCCO, and a rich history of SIROCCO papers, co-authored by prominent researchers like Carole Delporte-Gallet, Hugues Fauconnier, Pierre Fraigniaud, and Michel Raynal. In particular, the body of work on symmetry breaking tasks was started in a SIROCCO 2011 paper [23]. Similarly, the structure of immediate snapshots was studied in [9], resource coordination in [4], and anonymous t-resilient task solvability in [12]. A topological perspective of algorithms for dynamic networks was developed in [5]. A paper that fitted the scope of SIROCCO particularly well is [11], which studied, for the first time, the communication complexity of wait-free task solvability. And last but not least, Sergio's SIROCCO 2021 paper [26] on unconditionally secure information transmission and the SIROCCO 2024 paper [10] on non-negotiating distributed computing attest to his still active relation to this conference, which is also confirmed by his role as the SIROCCO 2023 PC chair [27].

The 2025 Award Committee:

Keren Censor-Hillel, Chair (Technion, Israel)
Yuval Emek (Technion, Israel)
Paola Flocchini (University of Ottawa, Canada)
Merav Parter (Weizmann Institute, Israel)
Christian Scheideler (University of Paderborn, Germany)

References

1. M. Alcántara, A. Castañeda, D. Flores-Peñaloza, and S. Rajsbaum: "The topology of look-compute-move robot wait-free algorithms with hard termination" *Distributed Computing* **32**:235–255 (2019) https://doi.org/10.1007/s00446-018-0345-3
2. B. Bonakdarpour, P. Fraigniaud, S. Rajsbaum, D. Rosenblueth, and C. Travers: "Decentralized asynchronous crash-resilient runtime verification" *Journal of the ACM* **69**, Art. 34 (2022) https://doi.org/10.1145/3550483
3. E. Borowsky, E. Gafni, N. Lynch, and S. Rajsbaum: "The BG distributed simulation algorithm" *Distributed Computing* **14**:127–146 (2001) https://doi.org/10.1007/pl00008933
4. A. Castañeda, P. Fraigniaud, E. Gafni, S. Rajsbaum, and M. Roy: "Asynchronous coordination under preferences and constraints" *SIROCCO 2016, LNCS* **9988**:111–126 https://doi.org/10.1007/978-3-319-48314-6_8
5. A. Castañeda, P. Fraigniaud, A. Paz, S. Rajsbaum, M. Roy, and C. Travers: "A topological perspective on distributed network algorithms" *SIROCCO 2019, LNCS* **11639**:3–18 https://doi.org/10.1007/978-3-030-24922-9_1
6. A. Castañeda and S. Rajsbaum: "New combinatorial topology bounds for renaming: The upper bound" *Journal of the ACM* **59**, Art. 3 (2012) https://doi.org/10.1145/2108242.2108245
7. A. Castañeda, S. Rajsbaum, and M. Raynal: "Unifying concurrent objects and distributed tasks: Interval-linearizability" *Journal of the ACM* **65**, Art. 45 (2018) https://doi.org/10.1145/3266457
8. A. Castañeda, S. Rajsbaum, and M. Raynal: "A linearizability-based hierarchy for concurrent specifications" *Communications of the ACM* **66**:86–97 (2023) https://doi.org/10.1145/3546826
9. C. Delporte, H. Fauconnier, S. Rajsbaum, and M. Raynal: "t-Resilient immediate snapshot is impossible" *SIROCCO 2016, LNCS* **9988**:177–191 https://doi.org/10.1007/978-3-319-48314-6_12
10. C. Delporte-Gallet, H. Fauconnier, P. Fraigniaud, S. Rajsbaum, and C. Travers: "Non-negotiating distributed computing" *SIROCCO 2024, LNCS* **14662**:208–225 https://doi.org/10.1007/978-3-031-60603-8_12
11. C. Delporte-Gallet, H. Fauconnier, and S. Rajsbaum: "Communication complexity of wait-free computability in dynamic networks" *SIROCCO 2020, LNCS* **12156**:291–309 https://doi.org/10.1007/978-3-030-54921-3_17
12. C. Delporte-Gallet, H. Fauconnier, S. Rajsbaum, and N. Yanagisawa: "A characterization of t-resilient colorless task anonymous solvability" *SIROCCO 2018, LNCS* **11085**:178–192 https://doi.org/10.1007/978-3-030-01325-7_18

13. H. van Ditmarsch, É. Goubault, M. Lazić, J. Ledent, and S. Rajsbaum: "A dynamic epistemic logic analysis of equality negation and other epistemic covering tasks" *Journal of Logical and Algebraic Methods in Programming* **121**, Art. 100662 (2021) https://doi.org/10.1016/j.jlamp.2021.100662
14. É. Goubault, R. Kniazev, J. Ledent, and S. Rajsbaum: "Semi-simplicial set models for distributed knowledge" *LICS 2023* https://doi.org/10.1109/lics56636.2023.10175737
15. É. Goubault, J. Ledent, and S. Rajsbaum: "A simplicial complex model for dynamic epistemic logic to study distributed task computability" *Information and Computation* **278**, Art. 104597 (2021) https://doi.org/10.1016/j.ic.2020.104597
16. M. Herlihy, D. Kozlov, and S. Rajsbaum: *Distributed Computing through Combinatorial Topology*, Morgan Kaufmann (2013) https://doi.org/10.1016/c2011-0-07032-1
17. M. Herlihy and S. Rajsbaum: "Set consensus using arbitrary objects" *PODC 1994*, pp. 324–333 https://doi.org/10.1145/197917.198119
18. M. Herlihy and S. Rajsbaum: "The decidability of distributed decision tasks" *STOC 1997*, pp. 589–598 https://doi.org/10.1145/258533.258652
19. M. Herlihy and S. Rajsbaum: "Algebraic spans" *Mathematical Structures in Computer Science* **10**:549–573 (2000) https://doi.org/10.1017/s0960129500003170
20. M. Herlihy and S. Rajsbaum: "A classification of wait-free loop agreement tasks" *Theoretical Computer Science* **291**:55–77 (2003) https://doi.org/10.1016/s0304-3975(01)00396-6
21. M. Herlihy and S. Rajsbaum: "Simulations and reductions for colorless tasks" *PODC 2012*, pp. 253–260 https://doi.org/10.1145/2332432.2332483
22. M. Herlihy, S. Rajsbaum, and M.R. Tuttle: "Unifying synchronous and asynchronous message-passing models" *PODC 1998*, pp. 133–142 https://doi.org/10.1145/277697.277722
23. D. Imbs, S. Rajsbaum, and M. Raynal: "The universe of symmetry breaking tasks" *SIROCCO 2011, LNCS* **6796**:66–77 https://doi.org/10.1007/978-3-642-22212-2_7
24. Y. Moses and S. Rajsbaum: "A layered analysis of consensus" *SIAM Journal on Computing* **31**:989–1021 (2002) https://doi.org/10.1137/s0097539799364006
25. A. Mostefaoui, S. Rajsbaum, and M. Raynal: "Conditions on input vectors for consensus solvability in asynchronous distributed systems" *Journal of the ACM* **50**:922–954 (2003) https://doi.org/10.1145/950620.950624
26. S. Rajsbaum: "A distributed computing perspective of unconditionally secure information transmission in Russian cards problems" *SIROCCO 2021, LNCS* **12810**:277–295 https://doi.org/10.1007/978-3-030-79527-6_16
27. S. Rajsbaum et al. (eds.): *SIROCCO 2023. LNCS* **13892** https://doi.org/10.1007/978-3-031-32733-9

Details of the SIROCCO 2025 Reviewing Process

Since the topics list of SIROCCO 2025 was extended by additional topics like topology and epistemic logic in distributed computing, we expected a larger number of and more diverse submissions compared to previous years. To keep the load per PC member reasonably small, a relatively large PC, consisting of 32 distinguished members of the scientific community, was hence formed. Special emphasis was put on ensuring that each of the overall 16 conference topics was covered by at least three expert PC members.

In addition, a number of quality-enhancing measures were foreseen in the single-blind reviewing process for SIROCCO 2025:

(i) To facilitate effective paper bidding, EasyChair's ability to match the selected topics of the submissions with the selected topics of expertise of the PC members was used to generate an initial bidding proposal for every PC member, which could be modified during the actual paper bidding phase. The final result of the bidding phase allowed EasyChair to find an optimal paper assignment (3 reviewers per submission) in a single assignment run, in negligible time.

(ii) In order not to rule out the most competent reviewers for a submission by an overly restrictive conflict of interest policy, prohibitive CoI (like supervisor or personal relations, to be declared during bidding as usual) that forbid any access to the reviewing process and milder forms of CoI (like occasional co-authorship, to be declared in the "comments to the PC section" of the reviews) were distinguished. PC members who appeared as co-authors of a submission were automatically excluded from the entire reviewing and discussion process of their paper. Apart from this, PC-co-authored submissions were handled like other submissions. From the overall 10 PC co-authored submissions, 7 made it into the program, and no PC member was a co-author of more than 2 submissions.

(iii) A reviewing process with two intermediate reviews before the final review was enforced. The first intermediate review just asked for the reviewers' actual expertise for reviewing the assigned papers [1 week after paper assignment]; the second intermediate review asked for an estimate of the overall merit figure (and optionally major strengths and weaknesses) [3 weeks after paper assignment]. The intermediate reviews were used to assign additional PC members or external experts to submissions that either did not have at least 2 reviewers with expertise 3 ("knowledgeable") or 4 ("expert") or suffered from controversial merit figure estimates (a difference larger than or equal to 2, from knowledgeable reviewers). Fortunately, only 7 submissions ended up with the need to assign additional reviewers.

(iv) The full reviews were due 6 weeks after paper assignment, which allowed two discussion phases before notification. During the first discussion phase, the reviewers of a submission were supposed to either (i) resolve controversial merit figures or (ii) determine both a proponent and an opponent who summarized the pro- and contra-arguments in a comment, to allow other PC members in the second discussion phase to also contribute efficiently. The few submissions falling under (ii) could hence be resolved without much effort in the latter.

As a final result, 21 submissions ended up with an average score of at least 1.3 (at least one "accept" and no rejects), which was set as the threshold for a "safe accept," whereas 20 submissions with an average score below 0.0 were considered a "safe reject." From the remaining 16 submissions with an average score between 0.0 and 1.3, the second discussion phase ended up with accepting three additional submissions (one after shepherding) and sending out six invitations for brief announcements, of which four were accepted by the authors.

Organization

Program Committee Chair

Ulrich Schmid	TU Wien, Austria

Program Committee

Ittai Abraham	Intel Labs, Israel
Hagit Attiya	Technion, Israel
Armando Castañeda	National Autonomous University of Mexico, Mexico
Bernadette Charron-Bost	École Normale Supérieure Paris, France
Stéphane Devismes	Université de Picardie, France
Hans van Ditmarsch	University of Toulouse, CNRS, IRIT, France
Faith Ellen	University of Toronto, Canada
Yuval Emek	Technion, Israel
Paola Flocchini	University of Ottawa, Canada
Pierre Fraigniaud	Université Paris Cité, CNRS, France
Matthias Függer	ENS Paris-Saclay, CNRS, France
Eric Goubault	École Polytechnique Paris, France
Magnús M. Halldórsson	Reykjavik University, Iceland
Joseph Y. Halpern	Cornell University, USA
Colette Johnen	University of Bordeaux, France
Eyal Kushilevitz	Technion, Israel
Roman Kuznets	Czech Academy of Sciences, Czechia
Petr Kuznetsov	Télécom Paris, Institut Polytechnique de Paris, France
Christoph Lenzen	CISPA Helmholtz Center for Information Security, Germany
Michael Mendler	University of Bamberg, Germany
Yoram Moses	Technion, Israel
Thomas Nowak	ENS Paris-Saclay, France
Ami Paz	Université Paris-Saclay, CNRS, France
Rajmohan Rajaraman	Northeastern University, USA
Andréa W. Richa	Arizona State University, USA
Christian Scheideler	University of Paderborn, Germany
Stefan Schmid	TU Berlin, Germany

Jukka Suomela	Aalto University, Finland
Gadi Taubenfeld	Reichman University, Israel
Roger Wattenhofer	ETH Zurich, Switzerland
Jennifer Welch	Texas A&M University, USA

Organizing Committee Chairs

Konstantinos Georgiou	Toronto Metropolitan University, Canada
Euripides Markou	University of Ioannina, Greece

Proceedings Chair

Roman Kuznets	Czech Academy of Sciences, Czechia

Steering Committee

Keren Censor-Hillel (Chair)	Technion, Israel
Yuval Emek	Technion, Israel
Merav Parter	Weizmann Institute of Science, Israel
Andrzej Pelc	University of Quebec, Canada
Sergio Rajsbaum	National Autonomous University of Mexico, Mexico

Additional Reviewers

Altisen, Karine
Angelopoulos, Spyros
Artmann, Matthias
Bankhamer, Gregor
Bhattacharya, Adri
Bienkowski, Marcin
Bramas, Quentin
Cosson, Romain
D'Amore, Francesco
Di Stefano, Gabriele
Dobrev, Stefan
Durand, Anaïs
Dürr, Christoph
Equi, Massimo
Esperet, Louis
Feletti, Caterina
Hanusse, Nicolas
Harbig, Jonas
Hummes Flores, Bernardo
Ilcinkas, David
Klasing, Ralf
Kuijer, Louwe B.
Kulkarni, Sandeep
Lamani, Anissa
Le Bouder, Gabriel
Liedtke, David

Liu, Quanquan
Modanese, Augusto
Newman, Ilan
Padalkin, Andreas
Pagourtzis, Aris
Paramonov, Anton
Pattanayak, Debasish
Piselli, Francesco
Raskin, Michael
Rincon-Galeana, Hugo
Robinson, Peter
Sela, Gal
Skretas, Georgios
Srinivas, Shreyas
Steffen, Martin
Sudo, Yuichi
Šimić, Borna
Travers, Corentin
Uitto, Jara
Werthmann, Julian
Zhu, Leqi

Abstracts of Invited Talks

A Gold Standard, but Not a Silver Bullet

Hagit Attiya

Department of Computer Science, Technion, Israel

Abstract. Linearizability is the gold standard for libraries of concurrent data structures. Indeed, it is exactly the minimal condition that ensures an abstract data structure can be substituted by its concurrent implementation.

Nevertheless, linearizability is no silver bullet. First, linearizability is based on refinement, which ensures only inclusion of behaviours and does not preserve performance or security properties. Second, it only preserves the interface behaviours of sequential objects and does not capture additional properties like history independence.

This talk will propose augmented sequential objects as a way to capture properties that go beyond input/output relations.

H. Attiya—Supported by the Israel Science Foundation (grant number 22/1425).

Actual Causality: A Survey

Joseph Y. Halpern

Cornell University, USA

Abstract. What does it mean that an event C "actually caused" event E? The problem of defining actual causation goes beyond mere philosophical speculation. For example, in many legal arguments, it is precisely what needs to be established in order to determine responsibility. (What exactly was the actual cause of the car accident or the medical problem?) The philosophy literature has been struggling with the problem of defining causality since the days of Hume, in the 1700s. Many of the definitions have been couched in terms of counterfactuals. (C is a cause of E if, had C not happened, then E would not have happened.) In 2001, Judea Pearl and I introduced a new definition of actual cause, using Pearl's notion of structural equations to model counterfactuals. The definition has been revised twice since then, extended to deal with notions like "responsibility" and "blame", and applied in databases and program verification. I survey the last 15 years of work here, including joint work with Judea Pearl, Hana Chockler, and Chris Hitchcock. The talk will be completely self-contained.

Computing with Dynamical Systems: Game Theory and the Brain

Christos Papadimitriou

Columbia University, USA

Abstract. Dynamical systems is a deep and elegant branch of Mathematics developed at the dawn of the 20th century to understand ordinary differential equations, and which has since powerfully informed Physics, Neuroscience and Economics among other fields — and yet it has not touched CS much. This is strange because dynamical systems are coterminous with computation — and not just because of quantum computers: A chip is a complex dynamical system of atoms and electrons engineered to have computationally meaningful attractors we call bit vectors. After a brief and targeted introduction to the subject, I will recount how dynamical systems are currently informing my work on two important research problems: (a) A reformulation of what a game means in view of the many problematic aspects of the incumbent Nash equilibrium; and (b) Understanding the computational processes whereby our brain creates cognition, intelligence, and language.

Contents

Regular Papers

Regular Papers

Support + Belief = Decision Trust

Alessandro Aldini[1], Agata Ciabattoni[2], Dominik Pichler[2(✉)], and Mirko Tagliaferri[1(✉)]

[1] University of Urbino, Urbino, Italy
{alessandro.aldini,mirko.tagliaferri}@uniurb.it
[2] TU Wien, Vienna, Austria
{agata,dominik}@logic.at

Abstract. We present SBTrust, a logical framework designed to formalize decision trust. Our logic integrates a doxastic modality with a novel non-monotonic conditional operator that establishes a positive support relation between statements and is closely related to a known dyadic deontic modality. For SBTrust, we provide semantics, proof theory and complexity results, as well as motivating examples. Compared to existing approaches, our framework seamlessly accommodates the integration of multiple factors in the emergence of trust.

Keywords: Decision trust · Non-monotonic conditional · Beliefs · Deontic logic

1 Introduction

Decision trust is defined as the willingness to depend on something (or somebody) with a feeling of relative security, although negative consequences are possible. This notion plays a central role in computer-mediated interactions. For instance, in e-commerce, when there is an abundance of vendors in a marketplace offering nearly identical products, customers use trust to decide whom to buy from [42]. Similarly, in the next generation Internet of Things, smart sensors, edge computing nodes, and cloud computing data centres rely on trust to share services such as data routing and analytics and to assess who to communicate with [19]. Trust is becoming increasingly important also in the interaction between AI-enhanced technologies and their users. In particular, a variety of design methodologies and policies are required to foster trust toward artificial decision-making processes [26]. In spite of their differences, in all scenarios, interactions are governed by trust evaluations that depend on various conditions, e.g., reputation scores, Quality of Service (QoS), transparency, perceived accuracy, ease of use, and the trustee's ability to behave as expected by the trustor.

Those facts drove the development of various models for assessing trust, see, e.g., [3, 12]. Yet, each existing model relies on specific conditions for the emergence of trust, thus failing in environments where multifaceted aspects play a role in establishing trust relations (see, e.g., [41] for trust in e-commerce, or [6], where it is shown that computer science students use multiple sources of information to determine their trust in AI systems). This calls for Trust models combining multiple aspects to evaluate the presence

U. Schmid and R. Kuznets (Eds.): SIROCCO 2025, LNCS 15671, pp. 3–22, 2025.
https://doi.org/10.1007/978-3-031-91736-3_1

or lack of trust in the environment [43]. To address this need, we introduce SBTrust, a logic that allows reasoning about decision trust relying on varied enabling conditions.

In our logic, trusting a formula φ means that the trustor is willing to accept the formula as being true, although it might be false. This acceptance-based interpretation of trust is compatible with influential conceptual analyses of the notion of trust that show that trusting a proposition boils down to using the proposition as a premise in one's reasoning, even though the proposition might be false [20].

In SBTrust, Trust is a derived operator whose constituents are a support connective and a belief operator (hence the logic's name). Whenever it is both believed that a formula φ supports a formula ψ, and that φ is true, then ψ is φ-trusted.

The notion of support, establishing a form of positive influence between two statements, is modeled through a novel dyadic operator $\rightsquigarrow$, where $\varphi \rightsquigarrow \psi$ is read as: in the most likely φ-scenarios ψ holds. The operator $\rightsquigarrow$ yields a non-monotonic conditional sharing properties with the KLM logic $\mathbb{P}$ of preferential reasoning [29] although without validating cautious monotonicity. We characterize $\rightsquigarrow$ with semantics and proof theory; its axioms and rules turn out to axiomatize the flat (i.e., non-nested) fragment of Åqvist system $\mathbb{F}$ [7] – a foundational preference-based logic for normative reasoning. The notion of belief (what is considered to be true from a subjective standpoint) is modeled through a belief operator B, obtained through the normal modal logic $\mathbb{KD}4$. Hence, our Trust operator ($T_\varphi\psi$) is built using those ingredients - $B\varphi \wedge B(\varphi \rightsquigarrow \psi) \rightarrow T_\varphi\psi$.

In the following, we provide motivations for introducing yet a new logical framework for decision trust. The key ingredient is the support operator $\rightsquigarrow$, for which we discuss in Sect. 2 the (undesired and) required properties. For our logic we present syntax (Sect. 3), semantics (Sect. 3.2), and establish the connection between $\rightsquigarrow$ and Åqvist system $\mathbb{F}$. Soundness, completeness, and complexity (for the satisfiability and the model checking problem) for SBTrust are established in Sect. 4.

1.1 Decision Trust: State of the Art

Decision trust logical formalisms can be classified into the following paradigms [8]:

- **Policy-based models**: trust is obtained by implementing hard-security mechanisms based on cryptographic protocols and access control, e.g., [39]. Logical frameworks for policy-based mechanisms are defined in various papers, including [1].
- **Reputation-based models**: trust is obtained through indications of past interactions that are evaluated by gathering and manipulating performance scores for those interactions, see, e.g., [9]. Logical approaches in this setting include [2].
- **Cognitive models**: trust derives from the combination of various factors, including the agent's disposition, the importance/utility of a situation [33], and the agent's expectation and willingness [13]; several logics formalize such aspects [4, 17].

Although models within a given paradigm are employed in real-world applications (e.g., [28]), they often rely on partial features of trust or assume very specific conditions, making them limited. Policy-based models flatten trust on the use of (cryptographic) protocols and regulations that fail whenever they circularly rely on some trust conditions - *the problem of trusting the policy-makers* [27]. Reputation-based models flatten trust on scores that often represent only a proxy for trust - *the problem of the insufficiency of reputation for trust* [12]. Unlike other paradigms, cognitive models capture more nuanced notions of trust, but rely on representing agents' cognitive features, which may not always be possible and can lead to complexity issues. This creates a trade-off between effectively modeling various aspects of trust and the complexity of estimating all its constituting elements in real-world settings. The following example illustrates how the three paradigms model agent trust and highlights their shortcomings.

Example 1. [Amazon] As a leading online retailer, Amazon prioritizes consumer trust to drive transactions, enforcing protocols and vendor rules. Imagine a customer assessing whether to trust the proposition "Amazon vendor V_i is reliable" ($\boxed{GoodV_i}$). In a *policy-based model* of trust, the customer would trust $GoodV_i$ only if V_i meets Amazon's internal policies (e.g., V_i is a registered company). Yet, this approach has drawbacks: (i) requirements could be manipulated, giving the customer a false sense of security; (ii) trust extends beyond regulations, as customers' trust is not solely based on vendors complying with policies; (iii) evaluating trust would be shifted from whether a customer trusts V_i to whether it trusts Amazon (as a policy maker) and its policies - trust that would depend on other policies Amazon has to abide by, *ad infinitum* (the problem of establishing primitive trusted policy-makers). In a *reputation-based model*, trust in $GoodV_i$ depends solely on V_i's positive reviews. However, this has two limitations: (i) new vendors lack reviews, making it difficult to establish trust; (ii) reviews can be manipulated, (e.g., in 2017, *The Shed at Dulwich* restaurant became London's top restaurant on Tripadvisor, although serving fake food). A *cognitive model* of trust allows trust estimations based on agents' cognitive features (e.g., the intention of the vendor to provide a good service). However, the trust triggers are limited to the cognitive features representable within the model, restricting its application.

2 Support Operator

We introduce the support operator, $\rightsquigarrow$, explaining our choice of a non-monotonic operator and its key properties with examples. Henceforth, we will shorten the reading of $\varphi \rightsquigarrow \psi$ to: given φ, then ψ is most likely. Throughout this paper, we restrict to non-nested applications of $\rightsquigarrow$ as it simplifies the technical aspects and adequately models the scenarios of interest without unnecessary complexity.

2.1 Why yet Another Notion of Support

Various notions of support and axiomatizations as a conditional operator appear in the literature; see, e.g. [15] for three potential readings as an evidence operator. What all authors agree upon is that the operator should be non-monotonic, i.e., given $\varphi \rightsquigarrow \psi$, there is no reason why $\varphi \land \xi \rightsquigarrow \psi$ should be the case. This is because additional information (ξ) may undermine the previously established supporting statement. We also assume that support is a non-monotonic operator, but our rationale distinguish our view from existing ones. In particular, contraposition $((\phi \rightsquigarrow \psi) \rightarrow (\neg\psi \rightsquigarrow \neg\phi))$, combined with right weakening (if $\phi \models \psi$, then $\chi \rightsquigarrow \phi \models \chi \rightsquigarrow \psi$), leads to monotonicity for the $\rightsquigarrow$ operator. Differently from [15], to avoid monotonicity, we give up contraposition (as motivated in Example 2) rather than right weakening, which is a reasonable assumption.

Example 2 (Contraposition and Modus Ponens). Assume that $GoodV_i$ supports that V_i's products are delivered fast ($\boxed{FastV_i}$), i.e., $GoodV_i \rightsquigarrow FastV_i$. This should not imply that if the delivery is slow, then it is most likely that the vendor is not a good one ($\neg FastV_i \rightsquigarrow \neg GoodV_i$), as the delay may depend on other reasons. For analogous reasons, we do not have that $GoodV_i$ and $GoodV_i \rightsquigarrow FastV_i$ imply $FastV_i$.

To proceed methodically, we draw upon the axiomatizations of non-monotonic conditionals from [29], known as KLM systems, as they serve as cornerstones for non-monotonic reasoning. We begin with an example demonstrating why the principle

$$\textbf{CM} \quad (\varphi \rightsquigarrow \psi) \wedge (\varphi \rightsquigarrow \chi) \rightarrow ((\varphi \wedge \psi) \rightsquigarrow \chi)$$

of cautious monotonicity is unfit for formalizing our notion of support (see also Remark 5).

Example 3. Let $\boxed{DefCV_i}$ denote that customer C receives a defective item from vendor V_i, $\boxed{CompCV_i}$ that C submits a formal complaint regarding V_i, and $\boxed{RefCV_i}$ that C is refunded by V_i. It is most likely that C will complain after receiving a defective item, $DefCV_i \rightsquigarrow CompCV_i$. In the Amazon marketplace, we also have that $DefCV_i \rightsquigarrow RefCV_i$. However, having $DefCV_i \wedge RefCV_i$ does NOT mean that $CompCV_i$ is most likely, and this invalidates **CM**.

Remark 1. **CM** is tightly connected to transitivity of the underlying semantic models. In contrast to KLM logics, we do not assume transitive models, as discussed in Sect. 3.2.

2.2 Intuitive Properties

We introduce the properties that we envision for the concept of support, illustrating their rationale by refining the scenario in Example 1. As will be shown in Theorem 1, many of the properties discussed below are inter-derivable, leading to a more concise axiomatization for $\rightsquigarrow$. Henceforth, by axioms, we mean axiom schemata. The naming conventions for the considered properties are taken from the KLM systems [29] and $\mathbb{F}$ [7,34].

As the support operator $\rightsquigarrow$ applies to boolean formulas we expect all classical tautologies to be provable. Moreover, since any fact intuitively supports itself, the axiomatization of $\rightsquigarrow$ should be able to derive the following axiom

$$\textbf{ID} : \varphi \rightsquigarrow \varphi$$

The presence of this axiom highlights that $\rightsquigarrow$ does not establish a causal relation, see Remark 2. Moreover, we want our support system to not support contradictions. In essence, anything supporting a contradiction must be dismissed:

$$\textbf{ST} : (\varphi \rightsquigarrow \bot) \rightarrow \neg\varphi$$

Example 4. Let $\boxed{\mathit{CompliantV_i}}$ mean that vendor V_i is compliant with *"Amazon Seller Terms and Conditions"* and let V_i be a vendor with an average rating of 4.5 stars for her main product j ($\boxed{\mathit{TopRatingV_{i,j}}}$). Assume that $\mathit{CompliantV_i}$ and $\mathit{TopRatingV_{i,j}}$ support that V_i is a good vendor, i.e., $(\mathit{CompliantV_i} \wedge \mathit{TopRatingV_{i,j}}) \rightsquigarrow \mathit{GoodV_i}$. This implies that being compliant supports the connection between having good reviews and being a good vendor, as expected by Amazon and their customers, i.e., $\mathit{CompliantV_i} \rightsquigarrow (\mathit{TopRatingV_{i,j}} \to \mathit{GoodV_i})$, which leads to the following axiom.

$$\mathbf{SH} : ((\varphi \wedge \psi) \rightsquigarrow \chi) \to (\varphi \rightsquigarrow (\psi \to \chi))$$

This axiom expresses that deductions performed under strong assumptions may be useful even if the assumptions are not known facts.

It is quite natural to assume that if a statement supports two other statements, it supports their conjunction, as expressed by the axiom

$$\mathbf{AND} : (\varphi \rightsquigarrow \psi) \wedge (\varphi \rightsquigarrow \chi) \to (\varphi \rightsquigarrow (\psi \wedge \chi))$$

Due to **ST**, this axiom can never be used to derive that a non-contradictory statement supports a contradiction. We illustrate this with Example 5, involving the lottery paradox [30], a stumbling block for default reasoning systems (see [35]).

Example 5. The paradox states that in a fair lottery, it is rational to assume that each individual ticket is likely not to win. By allowing to infer from two statements being likely that their conjunction is also likely, one concludes that two tickets are likely not to win. By iterating this reasoning, we can infer that it is likely that no ticket will win, which contradicts the fact that a winning ticket exists. The paradox does not apply to $\rightsquigarrow$. Indeed, if we assume that every ticket is most likely not to win, $\top \rightsquigarrow \neg T_i$, we can infer $\top \rightsquigarrow \bigwedge \neg T_i$ by **AND**. $\bigwedge \neg T_i$ being a contradiction, using **ST** we could derive $\neg\top$, which is impossible. This implies that our original assumption was wrong. Therefore, it cannot be assumed that $\top \rightsquigarrow \neg T_i$ for every ticket, thus stopping the paradox.

A support operator should also satisfy the **CUT** axiom, as illustrated by Example 6:

$$\mathbf{CUT} : (\varphi \rightsquigarrow \psi) \wedge ((\varphi \wedge \psi) \rightsquigarrow \chi) \to (\varphi \rightsquigarrow \chi)$$

Example 6. Let $\boxed{\mathit{AuthV_i}}$ stand for V_i is authenticated on the Amazon marketplace. Obviously, an authenticated and compliant vendor is most likely to be a legitimate business ($\boxed{\mathit{LegitV_i}}$), i.e., $(\mathit{AuthV_i} \wedge \mathit{CompliantV_i}) \rightsquigarrow \mathit{LegitV_i}$. Moreover, due to Amazon's policies, $\mathit{AuthV_i} \rightsquigarrow \mathit{CompliantV_i}$. This implies that given $\mathit{AuthV_i}$ it is already most likely that V_i is legitimate, $\mathit{AuthV_i} \rightsquigarrow \mathit{LegitV_i}$.

Example 7. Let $\boxed{\mathit{FairV_i}}$ mean that V_i abides by the *"Acting Fairly"* policy of the "Amazon's Code of Conduct". Assume we have both $\mathit{CompliantV_i} \rightsquigarrow \mathit{GoodV_i}$ and $\mathit{FairV_i} \rightsquigarrow \mathit{GoodV_i}$. These two facts imply that it should be sufficient to satisfy $\mathit{CompliantV_i}$ or $\mathit{FairV_i}$ to be considered a good vendor, i.e., $(\mathit{CompliantV_i} \vee \mathit{FairV_i}) \rightsquigarrow \mathit{GoodV_i}$. This example leads to the following axiom.

$$\mathbf{OR} : (\varphi \rightsquigarrow \psi) \wedge (\chi \rightsquigarrow \psi) \to ((\varphi \vee \chi) \rightsquigarrow \psi)$$

Example 8. Let $\boxed{GoodQoSV_{i,j}}$ mean that vendor V_i offers high QoS for product j. Assume that V_i sells two distinct products a and b, using the same commercial infrastructure (logistics, customer care, and so on). Hence, it would be absurd that V_i offers high QoS only for one of the two products, i.e., $\neg(GoodQoSV_{i,a} \leftrightarrow GoodQoSV_{i,b}) \rightsquigarrow \bot$. Hence, whatever $GoodQoSV_{i,a}$ supports, it should also be supported by $GoodQoSV_{i,b}$, and vice versa. This example leads to the following axiom.

$$\textbf{LL+} : (\neg(\varphi \leftrightarrow \psi) \rightsquigarrow \bot) \to ((\varphi \rightsquigarrow \chi) \leftrightarrow (\psi \rightsquigarrow \chi))$$

Example 9. Let $\boxed{GoodPriceV_i}$ and $\boxed{AmazonChoice_{i,j}}$ mean that V_i uses competitive prices, and that the product j sold by V_i is labeled as "Amazon's Choice". By Amazon's policy, $(GoodPriceV_i \wedge FastV_i \wedge TopRatingV_{i,j}) \to AmazonChoice_{i,j}$. Assume $GoodV_i \rightsquigarrow GoodPriceV_i$ (a good vendor is most likely to price well), $GoodV_i \rightsquigarrow FastV_i$ (a good vendor is most likely to deliver quickly), and $GoodV_i \rightsquigarrow TopRatingV_{i,j}$ (a good vendor is most likely to have good reviews). These supports together should imply that $GoodV_i$ supports $AmazonChoice_{i,j}$. This motivates the following rule.

$$\textbf{RCK} : \frac{\varphi_1 \wedge \cdots \wedge \varphi_n \to \varphi_{n+1}}{(\psi \rightsquigarrow \varphi_1) \wedge \cdots \wedge (\psi \rightsquigarrow \varphi_n) \to (\psi \rightsquigarrow \varphi_{n+1})}$$

of which Right Weakening (**RW**, see Remark 2), represents the particular case $n = 1$.

Example 10. Let V_i be a vendor selling product j and assume that $GoodQoSV_{i,j} \rightsquigarrow FastV_i$ and $\neg(AmazonChoice_{i,j} \rightsquigarrow DefCV_i)$. If these conditions hold, then customer C will have a good purchasing experience. Given that this implication holds, under the same hypothesis, it follows that C having a negative purchasing experience supports a contradiction. This example leads to the following "$\mathbb{S}5$-like" rule.

$$\mathbf{S5_F} : \frac{[\neg](\varphi_1 \rightsquigarrow \psi_1) \wedge \cdots \wedge [\neg](\varphi_n \rightsquigarrow \psi_n) \to \chi}{[\neg](\varphi_1 \rightsquigarrow \psi_1) \wedge \cdots \wedge [\neg](\varphi_n \rightsquigarrow \psi_n) \to (\neg\chi \rightsquigarrow \bot)}$$

where $[\neg](\varphi_i \rightsquigarrow \psi_i)$ stands for either $(\varphi_i \rightsquigarrow \psi_i)$ or its negated version $\neg(\varphi_i \rightsquigarrow \psi_i)$ and the same formulas are negated in the premise as in the conclusion of the rule. The rule is named because, when considered alongside other axioms and rules, $\mathbf{S5_F}$ grants the operator $\rightsquigarrow$ all the properties of an $\mathbb{S}5$-modality for the shallow fragment (see Theorem 3). As shown in Sect. 4, $\mathbf{S5_F}$ lets the operator $\rightsquigarrow$ behave locally like an absolute operator, playing a crucial role in the completeness proof.

Remark 2. (Most of) The axioms and rules discussed above are present in well-known systems. For instance, the KLM logic $\mathbb{P}$ of preferential reasoning, which interprets the dyadic operator $\varphi \mid\!\sim \psi$ as "*φ typically implies ψ*", contains the rule **RW** (see below) and axioms **ID**, **CUT**, **AND** and **OR**. I/O logics [32] and their causal versions [11], whose dyadic operator is interpreted as a dyadic obligation and a causal relation, respectively, share **RW**, **CUT**, **AND** and **OR** (but not **ID**). Note that KLM and (deontic and causal) I/O logics also contain the rule **LLE** (weaker than axiom **LL+**) below:

$$\textbf{RW} : \frac{\varphi_1 \to \varphi_2}{(\psi \rightsquigarrow \varphi_1) \to (\psi \rightsquigarrow \varphi_2)} \qquad \textbf{LLE} : \frac{\varphi \leftrightarrow \psi}{(\varphi \rightsquigarrow \chi) \leftrightarrow (\psi \rightsquigarrow \chi)}$$

An important difference between these logics and our $\rightsquigarrow$ operator is the direct interaction of support formulas and propositional formulas due to the rule $\mathbf{S5_F}$.

3 A Logical Framework for Decision Trust

Our logic SBTrust combines the $\rightsquigarrow$ operator with a belief operator B. For the former, we use a (subset of) the discussed axioms and rules and for the latter a $\mathbb{KD}4$ modality.[1]

3.1 Syntax and Axiomatization

The language $\mathcal{L}$ of SBTrust consists of a countable set of propositional variables or atoms (ranging over $p, q, \ldots$), the connectives $\wedge$ and $\neg$ of classical logic (the other propositional connectives are defined as usual), the binary support operator $\rightsquigarrow$, and the unary belief operator B. $\mathcal{L}$ is defined by the following two layers grammar:

$$\varphi := \bot \mid p \mid \varphi \wedge \varphi \mid \neg\varphi \qquad \alpha := \varphi \mid \varphi \rightsquigarrow \varphi \mid B\alpha \mid \alpha \wedge \alpha \mid \neg\alpha$$

We use $\varphi, \psi, \chi, \delta$, and π for formulas of classical logic CL, and α and β for general formulas in $\mathcal{L}$. $\mathcal{L}_T$ and $\mathcal{L}_{CL}$ will denote the set of formulas of SBTrust and of CL, respectively. We identify theoremhood in SBTrust with derivability in its Hilbert system.

Definition 1. *SBTrust is obtained by extending any axiom system for propositional classical logic: (indicating its axioms by)* (**CL**) *and the Modus Ponens rule* **MP**, *together with the following axioms and rules.*

For the support operator *the axiom schemata are:*

$$\textbf{(ID)}\ \varphi \rightsquigarrow \varphi \qquad \textbf{(SH)}\ ((\psi \wedge \chi) \rightsquigarrow \varphi) \rightarrow (\psi \rightsquigarrow (\chi \rightarrow \varphi))$$
$$\textbf{(ST)}\ (\varphi \rightsquigarrow \bot) \rightarrow \neg\varphi \quad \textbf{(LL+)}\ (\neg(\varphi \leftrightarrow \psi) \rightsquigarrow \bot) \rightarrow ((\varphi \rightsquigarrow \chi) \leftrightarrow (\psi \rightsquigarrow \chi))$$

and the corresponding inference rules are **RCK** *and* $\mathbf{S5_F}$, *with the restriction that their applications yield formulas within the language* $\mathcal{L}$.

For the belief operator *the axiom schemata are:*

$$\textbf{(KB)}\ B(\alpha \rightarrow \beta) \rightarrow (B\alpha \rightarrow B\beta) \quad \textbf{(DB)}\ B\alpha \rightarrow \neg B\neg\alpha \quad \textbf{(4B)}\ B\alpha \rightarrow BB\alpha$$

and Necessitation for B (**NB**) *which derives* $B(\varphi)$ *whenever* φ *has been proven.*

Trust arises as a combination of support and belief, i.e., $T_\varphi\psi := B\varphi \wedge B(\varphi \rightsquigarrow \psi)$.

Definition 2. *A derivation of* φ_n *is a sequence* $\varphi_1, \ldots, \varphi_n$ *where each* φ_i *is either an axiom instance or follows from the previous ones by applying one of the rules. We write* $\Gamma \vdash \varphi$ *if there is a derivation of* φ *or if* $\psi_1 \wedge \ldots \wedge \psi_m \rightarrow \varphi$ *for some* $\psi_1, \ldots, \psi_m \in \Gamma$.

Remark 3. The newly defined trust operator inherits numerous properties from the support and belief axioms and rules. While a more comprehensive study of this operator on its own is planned for future work, here we present a specific property as an initial example. By definition, $T_\varphi\psi \rightarrow B\varphi$. Using axiom **DB** and rule **NB**, we derive $\neg B\bot$

[1] Other normal modal logics could be used to formalize the epistemic/doxastic notion, e.g., a $\mathbb{KD}45$ modality (see [25] for a general discussion), without major effects on the logic.

which implies $\neg T_{\bot}\psi$, meaning it is impossible to trust something based on a contradiction. Now, assuming $T_{\varphi}\bot$, we derive $B\varphi$ by definition. However, from the axioms **ST** and **KB**, we can also derive $B\neg\varphi$, leading to a contradiction by axiom **DB**. Thus, it follows that $\neg T_{\varphi}\psi$ is derivable whenever either φ or ψ (or both) are contradictions. This property ensures that the trust operator does not permit trusting a contradiction or trusting something based on a contradiction-a desirable feature in this framework.

We prove that all the axioms and rules stated in Sect. 2.2 are derivable in SBTrust.

Theorem 1. *The rules* **RW** *and* **LLE***, as well as the axioms* **AND**, **CUT**, *and* **OR** *are derivable in the system for* $\rightsquigarrow$.

Proof. We will show the case for **LLE**. We assume that the formula $\varphi \leftrightarrow \psi$ is provable, then we can derive $\varphi \rightsquigarrow \chi \leftrightarrow \psi \rightsquigarrow \chi$ via the following:

$$\begin{array}{lr}
(1)\ \neg(\varphi \leftrightarrow \psi) \to \bot & \textit{Hyp.} + (\mathbf{CL}) \\
(2)\ (\neg(\varphi \leftrightarrow \psi) \rightsquigarrow \neg(\varphi \leftrightarrow \psi)) \to (\neg(\varphi \leftrightarrow \psi) \rightsquigarrow \bot) & (\mathbf{RCK}) \\
(3)\ \neg(\varphi \leftrightarrow \psi) \rightsquigarrow \bot & (\mathbf{ID} + \mathbf{MP}) \\
(4)\ (\varphi \rightsquigarrow \chi) \leftrightarrow (\psi \rightsquigarrow \chi) & (\mathbf{LL+} + \mathbf{MP})
\end{array}$$

Remark 4. From **RW** and **LLE**, we can derive the replacement of equivalent formulas within the support operator in compliance with the syntactic restrictions of the language.

Remark 5. Another reason for rejecting **CM** is that, in conjunction with **CUT** and **RW**, it permits to derive **REC** - $((\varphi \rightsquigarrow \psi) \wedge (\psi \rightsquigarrow \varphi)) \to ((\varphi \rightsquigarrow \chi) \leftrightarrow (\psi \rightsquigarrow \chi))$, which is too strong for a support operator since two statements that support each other do not necessarily support the same statements.

A strong connection holds between $\rightsquigarrow$ and $\mathbb{F}$, the dyadic deontic logic introduced in [7] and axiomatized in [34] using **MP** and the necessitation rule for $\Box$, and the axioms

CL All truth-functional tautologies	$\mathbf{K}_{\Box}$ $\Box(\varphi \to \psi) \to \Box\varphi \to \Box\psi$
T $\Box\varphi \to \varphi$	**5** $\Diamond\varphi \to \Box\Diamond\varphi$
COK $\bigcirc(\psi \to \chi/\varphi) \to (\bigcirc(\psi/\varphi) \to \bigcirc(\chi/\varphi))$	**Abs** $\bigcirc(\varphi/\psi) \to \Box\bigcirc(\varphi/\psi)$
Nec $\Box\varphi \to \bigcirc(\varphi/\psi)$	**Ext** $\Box(\varphi \leftrightarrow \psi) \to (\bigcirc(\chi/\varphi) \leftrightarrow \bigcirc(\chi/\psi))$
ID $\bigcirc(\varphi/\varphi)$	**SH** $\bigcirc(\varphi/\psi \wedge \chi) \to \bigcirc(\chi \to \varphi/\psi)$
$\mathbf{D}^*$ $\Diamond\psi \to (\bigcirc(\varphi/\psi) \to \neg\bigcirc(\neg\varphi/\psi))$	

where $\bigcirc(\psi/\varphi)$ stands for "ψ *is obligatory under the condition* φ". The flat fragment of the language of $\mathbb{F}$ –where $\Box$ and $\bigcirc$ apply only to formulas of $\mathcal{L}_{CL}$– can be translated into our language as follows:

Definition 3. *Let* χ *be any formula in the flat fragment of* $\mathbb{F}$. *The translation* χ^* *is*

$$\varphi^* \mapsto \varphi \qquad (\Box\varphi)^* \mapsto \neg\varphi \rightsquigarrow \bot \qquad (\bigcirc(\psi/\varphi))^* \mapsto \varphi \rightsquigarrow \psi$$

Remark 6. The dual of $\Box$ gets translated to $(\Diamond\varphi)^* := \neg(\varphi \rightsquigarrow \bot)$.

The theorem below establishes a first link between $\mathbb{F}$ and $\rightsquigarrow$.

Theorem 2. *The translation * of all axioms and rules of* $\mathbb{F}$ *– but* **5** *and* **Abs** *– are derivable in the axiomatization for* $\rightsquigarrow$.

Proof. The claim for **T**, **Ext**, **ID**, and **SH** follows directly from the translation. The translation of axiom **D*** is $\neg(\varphi \rightsquigarrow \bot) \rightarrow \neg((\varphi \rightsquigarrow \psi) \wedge (\varphi \rightsquigarrow \neg\psi))$. Its contraposition $((\varphi \rightsquigarrow \psi) \wedge (\varphi \rightsquigarrow \neg\psi)) \rightarrow (\varphi \rightsquigarrow \bot)$ is an instance of **AND**. The case $\mathbf{K}_\Box$ follows by:

$$(1)\ ((\neg(\varphi \rightarrow \psi) \rightsquigarrow \bot) \wedge (\neg\varphi \rightsquigarrow \bot)) \rightarrow \psi \qquad (\mathbf{ST} + \mathbf{ST} + \mathbf{CL})$$
$$(2)\ ((\neg(\varphi \rightarrow \psi) \rightsquigarrow \bot) \wedge (\neg\varphi \rightsquigarrow \bot)) \rightarrow (\neg\psi \rightsquigarrow \bot) \qquad (\mathbf{S5_F})$$

The translations of the axioms **Nec** and **COK** can be derived in a similar way and are omitted.

Remark 7. The translation of axioms **Abs** and **5** from $\mathbb{F}$ results in formulas containing nested applications of $\rightsquigarrow$. In Sect. 3.2, we will see that the axioms and rules for $\rightsquigarrow$ axiomatize the shallow fragment of $\mathbb{F}$. In this regard, the rule $\mathbf{S5_F}$, which does not correspond to any rule known in the literature, does not follow from the remaining axioms and rules for $\rightsquigarrow$, and it is needed to derive (some) flat formulas holding in $\mathbb{F}$.

Example 11. Assume that a vendor V_1 believes the two formulas supported by $DefCV_i$ discussed in Example 3. If V_1 sends a defective item to customer C ($DefCV_1$), from $B(DefCV_1)$ and $B(DefCV_1 \rightsquigarrow CompCV_1)$ we derive $T_{DefCV_1}(CompCV_1)$. Similarly, it also holds that $T_{DefCV_1}(RefCV_1)$. Now, assume that C does indeed receive the refund, thus $B(DefCV_1 \wedge RefCV_1)$. We show that V cannot trust that a complaint will not be submitted, $\neg T_{DefCV_1 \wedge RefCV_1}(\neg CompCV_1)$. We use the following abbreviations to write a concise derivation: Let $d := DefCV_1$, $r := RefCV_1$, and $c := CompCV_1$, and, by hypothesis, $T_d(c) \wedge T_d(r)$, i.e., $B(d) \wedge B(d \rightsquigarrow c) \wedge B(d \rightsquigarrow r)$. Hence:

$$(1)\ (d \wedge (d \rightsquigarrow c)) \rightarrow \neg(d \rightsquigarrow \neg c) \qquad (\mathbf{ST}+\mathbf{D}^*)$$
$$(2)\ (d \rightsquigarrow r) \wedge \neg(d \rightsquigarrow \neg c) \rightarrow \neg((d \wedge r) \rightsquigarrow \neg c) \quad (\mathbf{CUT}+\mathbf{CL})$$
$$(3)\ d \wedge (d \rightsquigarrow c) \wedge (d \rightsquigarrow r) \rightarrow \neg((d \wedge r) \rightsquigarrow \neg c) \qquad (1 \wedge 2)$$

Then, applying rule **NB** to (3) and using the hypothesis together with axiom **KB**, we derive $B(\neg((d \wedge r) \rightsquigarrow \neg c))$. This formula, by axiom **DB**, finally implies $\neg T_{d \wedge r}(\neg c)$.

3.2 Semantics

For evaluating a formula of the form $\varphi \rightsquigarrow \psi$, we intuitively consider only the most likely φ-scenarios and check whether ψ holds in those scenarios. This approach is inspired by preference-based logics [22], in which a conditional statement "If φ then ψ" is interpreted as among the "best" possible scenarios in which φ is true, ψ is true as well. Notice that in this approach transitivity of the relation is not assumed, in contrast to the KLM systems. The semantics for SBTrust is built on preference-based models [40] (for the support statements) and relational models (for the belief operator). Also used in KLM logics and in Åqvist system $\mathbb{F}$, preference-based models are triples $\langle S, \succeq, V \rangle$, where S denotes a set of states, V a valuation function, and the preference relation $\succeq \subseteq S \times S$ compares pairwise the states in S. In our context, $w \succeq v$ represents a pairwise comparison stating that w is at least as likely as v ($\mathbb{F}$ uses a *better than* interpretation, while the KLM logic $\mathbb{P}$ uses a *preferred to* interpretation).[2] Our "as likely as" interpretation has to be read as comparing two *specific* elements. Like in $\mathbb{F}$, the relations apply to pairs of worlds without imposing a global ordering, and therefore, they do not necessarily have to be transitive; for an illustration, see Example 12. Although close to $\mathbb{F}$, our approach differs in two important aspects. First, in our semantics, we include a component for belief formulas. Second, instead of having a unique preference frame, in a *Trust frame*, the set of states S is partitioned into multiple preference frames $\langle S_i, \succeq_i \rangle$; this allows us to consider different support systems within the same model. Note that, without partitioning, the unique given support system would necessarily have to be believed. In our interpretation, a statement $\varphi \rightsquigarrow \psi$ is true in a state $w \in S_i$ if among all the φ-states in S_i ($||\varphi||_i$) the most likely φ-states in S_i ($most(||\varphi||_i)$) satisfy ψ; $most(||\varphi||_i)$ consists of the maximal elements in the set of all φ-states of S_i according to $\succeq_i$.

In line with systems $\mathbb{F}$ and $\mathbb{P}$, our semantics include a limitedness condition: $||\varphi||_i \neq \emptyset \Rightarrow most(||\varphi||_i) \neq \emptyset$. Limitedness allows us to express when a formula φ is impossible in a partition S_i using $\varphi \rightsquigarrow \bot$. Hence, limitedness restricts SBTrust to support only non-contradictory options (with the exception of $\bot \rightsquigarrow \bot$). Consequently, an agent will never place trust in a blatant contradiction, nor will she place φ-based trust in contradicting statements, invalidating $T_\varphi \psi \wedge T_\varphi \neg\psi$. The definitions in this section characterize the frames and models on which our logic is based.

Definition 4 (Trust frame). *Let* $\mathcal{F} := \langle S, (S_i)_{i\in I}, (\succeq_i)_{i\in I}, R\rangle$ *where*

- $\langle S, R\rangle$ *is a serial and transitive Kripke frame;*
- $(S_i)_{i\in I}$ *is a partition of* S*, i.e.,* $\bigcup_{i\in I} S_i = S$ *and* $\forall i, j \in I:\ S_i \cap S_j = \emptyset$*;*
- *For each* $i \in I$*:* $\langle S_i, (\succeq_i)\rangle$ *is a preference frame (therefore* $\succeq_i \subseteq S_i \times S_i$*).*

Definition 5 (Trust model, truth conditions). *Let* $\mathcal{M} := \langle S, (S_i)_{i\in I}, (\succeq_i)_{i\in I}, R, V\rangle$ *where*

- $\mathcal{F} := \langle S, (S_i)_{i\in I}, (\succeq_i)_{i\in I}, R\rangle$ *is a Trust frame;*
- $V : Prop \mapsto 2^S$ *is a Valuation function;*
- *For each* $i \in I$*,* $\langle S_i, (\succeq_i), V\rangle$ *fulfills the limitedness condition: for every propositional formula* φ*, it holds that* $||\varphi||_i \neq \emptyset \Rightarrow most(||\varphi||_i) \neq \emptyset$*;*
- $\mathcal{M}, s \models p$ *iff* $s \in V(p)$*;*
- $\mathcal{M}, s \models \neg\alpha$ *iff* $\mathcal{M}, s \not\models \alpha$*;*
- $\mathcal{M}, s \models \alpha \wedge \beta$ *iff* $\mathcal{M}, s \models \alpha$ *and* $\mathcal{M}, s \models \beta$*;*
- $\mathcal{M}, s \models \varphi \rightsquigarrow \psi$ *iff* $most(||\varphi||_i) \subseteq ||\psi||_i$ *for* $s \in S_i$*;*
- $\mathcal{M}, s \models B(\varphi)$ *iff* $\forall v : (sRv \rightarrow \mathcal{M}, v \models \varphi)$*;*

  $||\varphi||_i := \{v \in S_i : \mathcal{M}, v \models \varphi\}$*;* $most(||\varphi||_i) := \{s \in ||\varphi||_i : \forall v[(v \in ||\varphi||_i \wedge v \succeq_i s) \rightarrow s \succeq_i v]\}$*.*

---

[2] A preference frame consists of a set of worlds and a binary relation, exactly like a Kripke frame. The difference lies in their different usages in the model.

Semantic consequence ($\Phi \models \alpha$) and validity ($\models \alpha$) are defined as usual.

Remark 8. Two observations are in order. Firstly, we do not assume any property on the relations $\succeq_i$ to keep the model as general as possible, and in particular, we do not require the relations to be transitive. This is related to the idea that the pairwise ordering between the states subsumes considerations about different contexts of evaluation, i.e., the sets $most(\_)$, which are tied to the antecedents of support formulas. The use of non-transitive models is consistent with the discussion presented in [18] about the non-transitivity of descriptive theories of preferences (see also [10] and [38] about intransitive preferences within social choice theories). Note that assuming transitivity, would closely connect $most(||\varphi \wedge \psi||_i)$ and $most(||\varphi||_i)$, forcing cautious monotonicity of the framework, which is an unwanted property for our use cases (see Examples 3 and 12). Secondly, $\mathcal{M}, w \models \varphi \rightsquigarrow \psi$ holds if w is part of a set of states in which φ supports ψ is true. An agent may or may not believe $\varphi \rightsquigarrow \psi$, independently from the fact that $\varphi \rightsquigarrow \psi$ holds or not. This gives us the possibility to capture an agent's misinformation.

The example below, inspired by [37], illustrates the reason we do not assume transitivity in the preference relation.

Example 12. Let V_1, V_2, and V_3 be three vendors with competing offers for the same product. Consider a scenario where the offers of V_1 and V_2 are similar, apart from a key attribute (e.g., delivery time), for which V_2 presents a significantly better offer. Hence, it is more likely that V_2 is preferred to V_1. For analogous reasons but for a different attribute (e.g., quality of reviews), it is possible that V_3 is preferred to V_2. However, it could be that all the attributes of V_1 and V_3 are similar enough to be incomparable. This situation is represented in the following model.

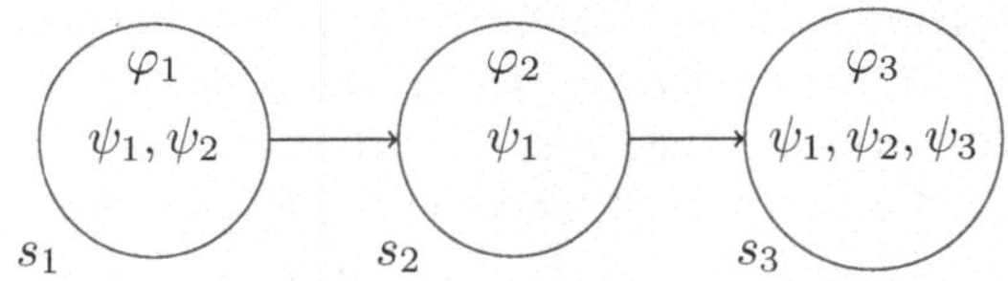

Notice that, for simplicity, we use a single-element partition $\{s_1, s_2, s_3\}$ and omit the accessibility relation R. Moreover, each formula φ_i represents that the vendor V_i is chosen (ignore, for the moment, the other formulas), and the solid arrows express the preference relations. Thus, for example, s_2 is more likely than s_1, s_3 is more likely than s_2, but it is not the case that s_3 is more likely than s_1 or vice versa. We use this example to emphasize the relation between the lack of transitivity and the rejection of **CM**. Let ψ_1 denote that V_1, V_2, and V_3 are Amazon vendors, ψ_2 denotes that Amazon offers a discount on products from long-standing vendors, and ψ_3 that Amazon offers a discount on products from new vendors. Assume that V_1 and V_3 are new vendors. Based on these assumptions, which do not affect the preference relations shown above, the following supports can be demonstrated easily: $\psi_1 \rightsquigarrow \psi_2$ and $\psi_1 \rightsquigarrow \psi_3$ (Amazon frequently supports discounts), but $(\psi_1 \wedge \psi_2) \not\rightsquigarrow \psi_3$ (it is not usual that Amazon promotes different discounts at the same time), which represent a violation of **CM**. Formally, given that $most(||\psi_1||) = \{s_3\}$, the support statements $\psi_1 \rightsquigarrow \psi_2$ and $\psi_1 \rightsquigarrow \psi_3$ are satisfied; however, $most(||\psi_1 \wedge \psi_2||) = \{s_1, s_3\}$ and s_1 violates the formula $(\psi_1 \wedge \psi_2) \rightsquigarrow \psi_3$, implying that the formula is not satisfied. Notice that if we imposed transitivity, thus relating s_1 to s_3, we would recover **CM** because in this case, $most(||\psi_1 \wedge \psi_2||)$ becomes $\{s_3\}$. Moreover, with **CM**, we would also get **REC** (see Remark 5), which, given that $\psi_1 \rightsquigarrow \psi_2$ and $\psi_2 \rightsquigarrow \psi_1$ hold, implies $(\psi_1 \rightsquigarrow \psi_3) \leftrightarrow (\psi_2 \rightsquigarrow \psi_3)$. Since $\psi_1 \rightsquigarrow \psi_3$ holds, we would also have $\psi_2 \rightsquigarrow \psi_3$, which is unreasonable since Amazon usually offers discounts on different types of vendors at different times.

We now examine the relation between $\mathbb{F}$ and $\rightsquigarrow$. Theorem 2 has already highlighted a syntactic connection (from $\mathbb{F}$ to $\rightsquigarrow$ via the translation * in Definition 3). Here, by using their semantics, we uncover a stronger tie. Recall that $\mathbb{F}$ is sound and complete w.r.t. all preference models $\langle S, \succeq, V \rangle$ which fulfill the limitedness condition, see [34]. We denote by $\models^{\mathbb{F}}$ the semantical consequence relation in $\mathbb{F}$.

Theorem 3. *For any set of formulas Γ and formula α in the language of $\mathbb{F}$ that do not contain nested modal operators, we have: $\Gamma \models^{\mathbb{F}} \alpha \Leftrightarrow \Gamma^* \models \alpha^*$.*

Proof. Both directions proceed by contraposition.

($\Rightarrow$) We show that given a Trust model invalidating the semantical consequence for support we can find a preference model invalidating the semantical consequence for $\mathbb{F}$. Assume that $\Gamma^* \not\models \alpha^*$. Hence, there exists a Trust model $\mathcal{M} = \langle S, (S_i)_{i \in I}, (\succeq_i)_{i \in I}, R, V \rangle$ and a state $s \in S_i$ such that $\forall \beta^* \in \Gamma^* : \mathcal{M}, s \models \beta^*$ and $\mathcal{M}, s \not\models \alpha^*$. We cut down the Trust model into a preference model as follows $\mathcal{M}' := \langle S_i, \succeq_i, V \rangle$. By definition, $\mathcal{M}'$ is a preference model fulfilling the limitedness condition.[3] Observe that no formula in $\Gamma^* \cup \{\alpha^*\}$ contains the operator B. Hence, the evaluation of the formulas in $\Gamma^* \cup \{\alpha^*\}$ at the state $s \in S_i$ coincides with the evaluation of the formulas in $\Gamma \cup \{\alpha\}$ in $\mathcal{M}$. We can therefore conclude $\forall \beta \in \Gamma : \mathcal{M}', s \models \beta$ and $\mathcal{M}', s \not\models \alpha$.

($\Leftarrow$) Given a preference model invalidating $\Gamma \models^{\mathbb{F}} \alpha$, we provide a Trust model invalidating $\Gamma^* \models \alpha^*$. Assume to have the preference model $\mathcal{M} = \langle S, \succeq, V \rangle$, fulfilling the limitedness condition and a state $s \in S$ such that $\forall \gamma \in \Gamma : \mathcal{M}, s \models \gamma$ and $\mathcal{M}, s \not\models \alpha$. We extend it into a Trust model with only one element in the partition, as follows $\mathcal{M}' := \langle S, (S_i)_{i \in I}, (\succeq_i)_{i \in I}, V, R \rangle$, with $I := \{1\}$, $S_1 := S$, $\succeq_1 := \succeq$ and $R := S \times S$. By definition, $\mathcal{M}'$ is a Trust model. Again no formula in $\Gamma^* \cup \{\alpha^*\}$ contains the operator B. Hence the evaluation of the formulas in $\Gamma^* \cup \{\alpha^*\}$ at the state $s \in S$ coincides with the evaluation of the formulas in $\Gamma \cup \{\alpha\}$ in $\mathcal{M}$. Hence $\forall \beta^* \in \Gamma^* : \mathcal{M}', s \models \beta^*$ and $\mathcal{M}', s \not\models \alpha^*$.

The Soundness and Completeness of SBTrust w.r.t. Trust models, proved in the next section, implies that the axioms and rules of $\rightsquigarrow$ axiomatize the flat fragment of $\mathbb{F}$.

4 Soundness, Completeness and Computational Complexity

We start with the soundness of SBTrust w.r.t. Trust models.

Theorem 4 (Strong Soundness). $\Phi \vdash \alpha \Rightarrow \Phi \models \alpha$ *for any* $\Phi \subseteq \mathcal{L}_T$ *and* $\alpha \in \mathcal{L}_T$.

[3] In [34], limitedness is stated for every formula of $\mathbb{F}$. Since the truth sets of obligations and modalities are those of $\top$ or $\bot$, our limitedness condition is equivalent to the one given in $\mathbb{F}$.

Proof. Proceed by induction on the derivation length, distinguishing cases according to the last rule applied. We show below the details for axiom **LL+** and rule $\mathbf{S5_F}$.

LL+: Given a Trust model $\mathcal{M} = \langle S, (S_i)_{i\in I}, (\succeq_i)_{i\in I}, R, V\rangle$ and a state $s \in S_i$ such that $\mathcal{M}, s \models \neg(\varphi \leftrightarrow \psi) \rightsquigarrow \bot$, then we get $most(||\neg(\varphi \leftrightarrow \psi)||_i) \subseteq \emptyset$. Given the limitedness assumption, this is equivalent to $||\neg(\varphi \leftrightarrow \psi)||_i = \emptyset$ and furthermore to $||\varphi \leftrightarrow \psi||_i = S_i$. Hence, φ and ψ are equivalent in every state of S_i. Therefore the sets $most(||\varphi||_i)$ and $most(||\psi||_i)$ coincide, i.e., $\mathcal{M}, s \models (\varphi \rightsquigarrow \chi) \leftrightarrow (\psi \rightsquigarrow \chi)$.
$\mathbf{S5_F}$: Given a Trust model $\mathcal{M} = \langle S, (S_i)_{i\in I}, (\succeq_i)_{i\in I}, R, V\rangle$, we assume $((\neg)(\varphi_1 \rightsquigarrow \psi_1) \wedge \cdots \wedge (\neg)(\varphi_n \rightsquigarrow \psi_n)) \rightarrow \chi$ to be true in every state of $\mathcal{M}$. Given a state $s \in S_i$ such that $\mathcal{M}, s \models ((\neg)(\varphi_1 \rightsquigarrow \psi_1) \wedge \cdots \wedge (\neg)(\varphi_n \rightsquigarrow \psi_n))$ holds, it follows that $\forall w \in S_i$ $\mathcal{M}, w \models ((\neg)(\varphi_1 \rightsquigarrow \psi_1) \wedge \cdots \wedge (\neg)(\varphi_n \rightsquigarrow \psi_n))$ because by virtue of the semantics of $\rightsquigarrow$, all the states of a given partition class satisfy the same (negated) support formulas. Therefore, we have that $\forall w \in S_i$ $\mathcal{M}, w \models \chi$, which means $||\neg\chi||_i = \emptyset$ and finally $\mathcal{M}, s \models \neg\chi \rightsquigarrow \bot$.

Completeness is shown via the canonical model construction, adapted to our framework from [22]. The needed modifications are the following. First, we have to ensure that SBTrust allows us to derive all the axioms and rules required for the construction to proceed. Furthermore, unlike the models considered in [22], Trust models incorporate multiple preference frames, the belief operator B, and include the limitedness condition.

The modifications are implemented as follows. The required axioms and rules for their proof to go through are those of $\mathbb{F}$ without $\mathbf{D}^*$ (which corresponds to limitedness). In Sect. 3.1 we have shown that with the exception of **5** and **Abs** the axioms of $\mathbb{F}$ are derivable in SBTrust. We prove that we can derive all the necessary properties of the canonical model even in the absence of axioms **5** and **Abs**, by relying on other rules of SBTrust, primarily on $\mathbf{S5_F}$. The multiple preference frames are handled by partitioning the maximal consistent sets used in the canonical model construction into equivalence classes containing the same support formulas. The addition of belief is easy: We equip the canonical model with the accessibility relation in the usual Kripke fashion. Incorporating the limitedness condition poses the challenge of guaranteeing that, in every preference model of our canonical model, each non-empty set $||\varphi||_i$ contains a maximal element according to the preference relation. We address this by using the axiom **ST**.

Definition 6. *A set $\Gamma \subseteq \mathcal{L}_T$ is called a maximal consistent set (MCS for short) if (a) $\Gamma \nvdash \bot$, and (b) For every $\alpha \in \mathcal{L}_T$ either $\alpha \in \Gamma$ or $\neg\alpha \in \Gamma$.*

Although not all the states in a model validate the same support formulas, we still need to make sure that all the states inside the same preference frame S_i do. For that reason, we partition the maximal consistent sets into equivalence classes, containing the same support formulas. We also define a set $\rightsquigarrow_\varphi (\Gamma)$ containing all formulas which are supported in a MCS Γ by a formula φ.

Definition 7. *Given $\Gamma, \Delta \subseteq \mathcal{L}_T$ and $\varphi \in \mathcal{L}_{CL}$ we define:*

- $\Gamma^{\rightsquigarrow} := \{\chi \rightsquigarrow \psi \in \Gamma\}$ *(We write $\Gamma \leftrightsquigarrow \Delta$ if $\Gamma^{\rightsquigarrow} = \Delta^{\rightsquigarrow}$);*

- $\rightsquigarrow_\varphi(\Gamma) := \{\psi : \varphi \rightsquigarrow \psi \in \Gamma\}$ *(We call Δ φ-likely for Γ if $\rightsquigarrow_\varphi(\Gamma) \subseteq \Delta$).*

Fact 1 *$\leftrightsquigarrow$ is an equivalence relation on the set of all MCSs. We write $[\Gamma]_{\leftrightsquigarrow}$ for the equivalence class containing Γ.*

Each equivalence class serves as a basis for a preference frame in our canonical model. The maximal consistent sets which are φ-likely for Γ are our candidates for the most likely φ states in the preference frame based on the equivalence class $[\Gamma]_{\leftrightsquigarrow}$, as they contain all formulas supported by φ.

Before proceeding, we need a result stating that if ψ is not supported by φ in Γ, then we can construct a MCS Δ with the same support formulas as Γ, including the negation of ψ and all the propositions supported by φ.

Lemma 1. *Given a MCS Γ and a propositional formula ψ with $\psi \notin \rightsquigarrow_\varphi(\Gamma)$, then there exists $\Delta \in [\Gamma]_{\leftrightsquigarrow}$ such that $\{\neg\psi\} \cup \rightsquigarrow_\varphi(\Gamma) \subseteq \Delta$.*

Proof. We prove the consistency of the set $A := \{\neg\psi\} \cup \rightsquigarrow_\varphi(\Gamma) \cup \Gamma^{\rightsquigarrow} \cup \{\neg(\chi \rightsquigarrow \gamma) : \chi \rightsquigarrow \gamma \notin \Gamma\}$. If consistent, we can extend it to an MCS $\Delta \subseteq [\Gamma]_{\leftrightsquigarrow}$. Assuming $A \vdash \bot$, there exist $\varphi_1, ..., \varphi_n \in \rightsquigarrow_\varphi(\Gamma)$, $\pi_1 \rightsquigarrow \psi_1, ..., \pi_m \rightsquigarrow \psi_m \in \Gamma^{\rightsquigarrow}$, and $\neg(\chi_1 \rightsquigarrow \gamma_1), ..., \neg(\chi_k \rightsquigarrow \gamma_k) \in \Gamma$ such that $\vdash (\alpha \wedge \varphi_1 \wedge ... \wedge \varphi_n \wedge \neg\psi) \rightarrow \bot$, where $\alpha := (\pi_1 \rightsquigarrow \psi_1) \wedge ... \wedge (\pi_m \rightsquigarrow \psi_m) \wedge \neg(\chi_1 \rightsquigarrow \gamma_1) \wedge ... \wedge \neg(\chi_k \rightsquigarrow \gamma_k)$. Using **CL** axioms, we obtain $\vdash \alpha \rightarrow ((\varphi_1 \wedge ... \wedge \varphi_n) \rightarrow \psi)$. From $\mathbf{S5_F}$, we derive $\vdash \alpha \rightarrow (\neg(\varphi_1 \wedge ... \wedge \varphi_n \rightarrow \psi) \rightsquigarrow \bot)$, and by **Nec**, we get $\vdash \alpha \rightarrow (\varphi \rightsquigarrow (\varphi_1 \wedge ... \wedge \varphi_n \rightarrow \psi))$. By **RCK** and the maximal consistency of the set Γ, this implies $\psi \in \rightsquigarrow_\varphi(\Gamma)$, contradicting our assumption. Thus, A is consistent.

We define the states and preference relations $\succeq_\Gamma$ for our canonical model (the index Γ is a representative of an equivalence class of the equivalence relation $\leftrightsquigarrow$). The states are (Δ, φ, i) where Δ is a MCS in the equivalence class $[\Gamma]_{\leftrightsquigarrow}$, $i \in \{0, 1, 2\}$, and φ is a propositional formula. φ and i are used to pinpoint the maximal φ-states according to the relation $\succeq_\Gamma$ and to ensure that the maximal elements of $\succeq_\Gamma$ coincide with the states satisfying the supported formulas within Γ, see Corollary 2.

For a MCS Γ we use the following notation:

- $S_\Gamma := [\Gamma]_{\leftrightsquigarrow} \times \mathcal{L}_{CL} \times \{0, 1, 2\}$ and $[\delta]_\Gamma := \{(\Delta, \varphi, i) \in S_\Gamma : \delta \in \Delta\}$.

Definition 8. *The preference relation $\succeq_\Gamma \subseteq S_\Gamma \times S_\Gamma$ is such that $(\Delta, \varphi, i) \succeq_\Gamma (\Omega, \psi, j)$ holds iff at least one of the following conditions holds:*

- *Δ is φ-likely for Γ and $\varphi \in \Omega$;*
- *($i = 1$ and $j = 0$) or ($i = 2$ and $j = 1$) or ($i = 0$ and $j = 2$).*

After having defined our preference relations we show that the maximal elements in $[\delta]_\Gamma$ are δ-likely for Γ. This is what we are aiming for as we want all the elements in $max([\delta]_\Gamma)$ to fulfill every formula supported by δ in Γ. Furthermore, if an MCS $\Delta \in S_\Gamma$ is δ-likely for Γ then (Δ, δ, i) is a maximal element in $[\delta]_\Gamma$. These results will be the core of our completeness proof since they can be used to show that a support formula is true in a state of S_Γ if and only if the formula appears in Γ. To prove this, we start proving the following technical lemma that establishes a connection between an (Δ, φ, i) appearing in $max([\delta]_\Gamma)$ and the MCS Γ.

Lemma 2. *Given* $(\Delta, \varphi, i) \in max([\delta]_\Gamma)$ *then:*
(a) Δ *is* φ*-likely for* Γ*;* (b) $\neg(\delta \to \varphi) \rightsquigarrow \bot \in \Gamma$.

Proof. We assume $i = 0$. The other cases are similar.

(a) Given $(\Delta, \varphi, 0) \in max([\delta]_\Gamma)$, we have $(\Delta, \varphi, 1) \succeq_\Gamma (\Delta, \varphi, 0)$ by construction of $\succeq_\Gamma$. This implies $(\Delta, \varphi, 0) \succeq_\Gamma (\Delta, \varphi, 1)$ by maximality. The latter only holds if Δ is φ-likely for Γ, and $\varphi \in \Delta$ since no other condition applies.

(b) Assume to have a MCS $\Omega \in S_\Gamma$ s.t. $\delta \in \Omega$ but $\varphi \notin \Omega$. For Ω we have $(\Delta, \varphi, 0) \not\succeq_\Gamma (\Omega, \delta, 1)$, but $(\Omega, \delta, 1) \succeq_\Gamma (\Delta, \varphi, 0)$ which contradicts $(\Delta, \varphi, 0) \in max([\delta]_\Gamma)$. Hence, it follows that such an MCS $\Omega \in S_\Gamma$ does not exist, which means $[\delta]_\Gamma \subseteq [\varphi]_\Gamma$. In other words, every MCS in S_Γ contains the formula $\delta \to \varphi$. In particular, each MCS Π with $\Pi \in [\Gamma]_{\leftrightsquigarrow}$ contains the formula $\delta \to \varphi$, which means $[\Gamma]^{\rightsquigarrow} \cup \{\neg(\chi \rightsquigarrow \gamma) : \chi \rightsquigarrow \gamma \notin \Gamma\} \cup \{\neg(\delta \to \varphi)\}$ is inconsistent. This lets us infer $[\Gamma]^{\rightsquigarrow} \cup \{\neg(\chi \rightsquigarrow \gamma) : \chi \rightsquigarrow \gamma \notin \Gamma\} \vdash \delta \to \varphi$, hence we can find finitely many support formulas in $\Gamma^{\rightsquigarrow}$ and finitely many negated support formulas in $\{\neg(\chi \rightsquigarrow \gamma) : \chi \rightsquigarrow \gamma \notin \Gamma\}$ s.t. $\vdash (\varphi_1 \rightsquigarrow \psi_1) \wedge \cdots \wedge (\varphi_n \rightsquigarrow \psi_n) \wedge \neg(\chi_1 \rightsquigarrow \gamma_1) \wedge \ldots \wedge \neg(\chi_k \rightsquigarrow \gamma_k) \to (\delta \to \varphi)$. By the use of $\mathbf{S5_F}$ we get $\vdash (\varphi_1 \rightsquigarrow \psi_1) \wedge \cdots \wedge (\varphi_n \rightsquigarrow \psi_n) \wedge \neg(\chi_1 \rightsquigarrow \gamma_1) \wedge \ldots \wedge \neg(\chi_k \rightsquigarrow \gamma_k) \to (\neg(\delta \to \varphi) \rightsquigarrow \bot)$ and finally $\neg(\delta \to \varphi) \rightsquigarrow \bot \in \Gamma$.

Corollary 1. *Given* $(\Delta, \varphi, i) \in S_\Gamma$ *then:*
(a) $(\Delta, \varphi, i) \in max([\delta]_\Gamma)$ *implies* Δ *is* δ*-likely for* Γ*;*
(b) Δ *being* δ*-likely for* Γ *implies* $(\Delta, \delta, i) \in max([\delta]_\Gamma)$.

Proof. (a) By $(\Delta, \varphi, i) \in max([\delta]_\Gamma)$ and Lemma 2 we derive $\neg(\delta \to \varphi) \rightsquigarrow \bot \in \Gamma$. Using the derivable rule **LLE** we get $\neg(\delta \leftrightarrow (\delta \wedge \varphi)) \rightsquigarrow \bot \in \Gamma$. Let us take an arbitrary $\gamma \in \rightsquigarrow_\delta (\Gamma)$; this means $\delta \rightsquigarrow \gamma \in \Gamma$. Since $\neg(\delta \leftrightarrow (\delta \wedge \varphi)) \rightsquigarrow \bot \in \Gamma$ we can apply **LL+** to derive $(\delta \wedge \varphi) \rightsquigarrow \gamma \in \Gamma$. Furthermore, by applying **SH**, we get $\varphi \rightsquigarrow \delta \to \gamma \in \Gamma$. By Lemma 2 we know that Δ is φ-likely for Γ, therefore $\delta \to \gamma \in \Delta$. By assumption, we have $\delta \in \Delta$, which lets us conclude $\gamma \in \Delta$. Since γ was arbitrary we get $\rightsquigarrow_\delta (\Gamma) \subseteq \Delta$.

(b) As Δ is δ-likely for Γ, by axiom **ID** we have $\delta \in \rightsquigarrow_\delta (\Gamma) \subseteq \Delta$ and hence $\Delta \in [\delta]_\Gamma$. Taking an arbitrary MCS $\Omega \in S_\Gamma$ with $\Omega \in [\delta]_\Gamma$ and an arbitrary propositional formula π, we end up with $(\Delta, \delta, i) \succeq_\Gamma (\Omega, \pi, j)$ by Definition 8.

We can finally show that our construction works as intended, namely that every formula ψ supported by a formula δ according to a MCS Γ is contained in all the maximal δ states in the equivalence set of Γ.

Corollary 2. *Given a MCS* Γ *and two propositional formulas* δ *and* ψ*, it holds that* $\delta \rightsquigarrow \psi \in \Gamma$ *if and only if for all* $(\Delta, \varphi, i) \in max([\delta]_\Gamma) : \psi \in \Delta$.

Proof. Given $(\Delta, \varphi, i) \in max([\delta]_\Gamma)$, Corollary 1 ensures that Δ is δ-likely for Γ. Since $\psi \in \rightsquigarrow_\delta (\Gamma) \subseteq \Delta$, the claim follows. Now, by contraposition, we assume $\psi \notin \rightsquigarrow_\delta (\Gamma)$. By Lemma 1, there is an MCS $\Delta \in [\Gamma]_{\leftrightsquigarrow}$ such that $\{\neg\psi\} \cup \rightsquigarrow_\delta (\Gamma) \subseteq \Delta$. Then, Δ is δ-likely for Γ, and Corollary 1 gives $(\Delta, \varphi, i) \in max([\delta]_\Gamma)$. We conclude $\psi \notin \Delta$ by consistency.

Let us fix a set I consisting of one representative of each equivalent class of $\leftrightsquigarrow$.

Definition 9 (Canonical model). *Let* $\mathcal{M}^{Can} := \langle S, (S_\Gamma)_{\Gamma \in I}, (\succeq_\Gamma)_{\Gamma \in I}, R, V \rangle$ *where:*

- $S := \bigcup_{\Gamma \in I} S_\Gamma$;
- $V(p) := \{(\Delta, \varphi, i) \in S : p \in \Delta\}$;
- *the preference relation* $\succeq_\Gamma \subseteq S_\Gamma \times S_\Gamma$ *is as in Definition 8;*
- $R \subseteq S \times S$ *is defined as* $(\Delta, \varphi, i) R (\Omega, \psi, j)$ *if for all* $\alpha \in \mathcal{L}_T$: $(B(\alpha) \in \Delta \Rightarrow \alpha \in \Omega)$.

We begin the final steps of our completeness proof: the truth lemma and the proof that the canonical model $\mathcal{M}^{Can}$ (from Definition 9) is a Trust model, starting with the truth lemma. Let (Δ, π, i) be a state of the canonical model $\mathcal{M}^{Can}$, s.t. Δ is a MCS (Definition 6), and (π, i) are used to pinpoint arbitrary maximal π-states. This lemma says that a formula α is satisfied by the pointed-canonical model $\mathcal{M}^{Can}, (\Delta, \pi, i)$ iff $\alpha \in \Delta$.

Lemma 3 (Truth lemma). $\mathcal{M}^{Can}, (\Delta, \pi, i) \models \alpha$ *iff* $\alpha \in \Delta$.

Proof. Proceeds by structural induction on α. If $\alpha \in \mathcal{L}_{CL}$, the claim follows from **CL**. By i.h., we derive $||\varphi||_\Gamma = [\varphi]_\Gamma$ and $most(||\varphi||_\Gamma) = max([\varphi]_\Gamma)$ for all $\varphi \in \mathcal{L}_{CL}$.

Let α be $\varphi \rightsquigarrow \psi$. We start with the case $\alpha \in \Delta$. Take Γ with $\Delta \in [\Gamma]_{\leftrightsquigarrow}$. Given an arbitrary $(\Omega, \gamma, j) \in S_\Gamma$ with $(\Omega, \gamma, j) \in max(||\varphi||_\Gamma)$ we get $(\Omega, \gamma, j) \in max([\varphi]_\Gamma)$ by the i.h. Because $\psi \in \rightsquigarrow_\varphi (\Gamma)$ Corollary 2 implies that $\psi \in \Omega$. By the induction hypothesis, we conclude $(\Omega, \gamma, j) \models \psi$, implying $(\Delta, \pi, i) \models \varphi \rightsquigarrow \psi$. For the other direction, assume $\alpha \notin \Delta$. In this case, we have to find a maximal φ state which does not satisfy ψ. The assumption $\psi \notin \rightsquigarrow_\varphi (\Delta)$ lets us derive $\psi \notin \rightsquigarrow_\varphi (\Gamma)$. By Corollary 2, we obtain a $(\Pi, \chi, i) \in max([\varphi]_\Gamma)$ with $\psi \notin \Pi$. By induction hypothesis we get $(\Pi, \chi, i) \in max(||\varphi||_\Gamma)$ and $(\Pi, \chi, i) \models \neg\psi$. This means $(\Delta, \pi, i) \not\models \varphi \rightsquigarrow \psi$.

If α is of the form $B(\varphi)$ both directions of the claim follow directly from the induction hypothesis and the construction of R in Definition 9.

Lemma 4. $\mathcal{M}^{Can}$ *is a Trust model.*

Proof. First, we prove that for each $\Gamma \in I$ $\langle S_\Gamma, (\succeq_\Gamma), V \rangle$ fulfils the limitedness condition. Let $\varphi \in \mathcal{L}_{CL}$ and $\Gamma \in I$ with $[\varphi]_\Gamma \neq \emptyset$. Hence there exists a MCS $\Delta \in [\Gamma]_{\leftrightsquigarrow}$ with $\varphi \in \Delta$. **ST** tells us that $\neg(\varphi \rightsquigarrow \bot) \in \Delta$. This implies $\varphi \rightsquigarrow \bot \notin \Delta$ and finally $\bot \notin \rightsquigarrow_\varphi (\Gamma)$. Given that the set $\rightsquigarrow_\varphi (\Gamma)$ is closed under consequences because of **RCK**, we conclude that it is consistent. By Lemma 1, we can extend $\rightsquigarrow_\varphi (\Gamma)$ to a MCS $\Pi \in [\Gamma]_{\leftrightsquigarrow}$. By construction Π is φ-likely for Γ. By Corollary 1(b) we obtain $(\Pi, \varphi, i) \in max([\varphi]_\Gamma)$, which makes $max([\varphi]_\Gamma)$ non-empty. By Lemma 3 it follows $||\varphi||_\Gamma \neq \emptyset \Rightarrow most(||\varphi||_\Gamma) \neq \emptyset$. Since φ and Γ were arbitrary we are done. Finally, the fact that the relation R is transitive and serial, follows from the axioms for B as in standard Kripke semantics.

Theorem 5 (Strong Completeness). $\Phi \models \alpha \Rightarrow \Phi \vdash \alpha$, *for all* $\Phi \subseteq \mathcal{L}_T$ *and* $\alpha \in \mathcal{L}_T$.

Proof. By contraposition. From $\Phi \nvdash \alpha$ it follows that $\Phi \cup \{\neg\alpha\}$ is consistent and can therefore be extended to a MCS Δ. By Lemma 3 every formula in Δ holds in the canonical model in a state of the form (Δ, γ, i). Hence $\forall \beta \in \Phi : (\Delta, \gamma, i) \models \beta$ and $(\Delta, \gamma, i) \not\models \alpha$, which gives us $\Phi \not\models \alpha$ by Lemma 4.

Remark 9. Since the deduction theorem holds in SBTrust, Theorem 5 implies compactness.

We discuss the computational complexity of SBTrust, splitting the problem into two parts: (i) reducing SBTrust by ignoring its support part and focusing on the Boolean and belief parts, and then (ii) reinstating the support part, to complete the proof.

Definition 10. *The SBTrust-reduction is obtained by reducing $\mathcal{L}_T$ to $\mathcal{L}'_T$ and transforming a model $\mathcal{M} := \langle S, \_, \_, R, V \rangle$ into a model $\mathcal{M}' := \langle S, R, V' \rangle$ as follows:*

- *φ formulas of $\mathcal{L}_T$ remain unchanged in $\mathcal{L}'_T$;*
- *All $\alpha \in \mathcal{L}_T$ of the form $\varphi \rightsquigarrow \varphi$, are mapped to fresh atoms taken from a set Prop′, where Prop′ $\bigcap$ Prop $= \emptyset$; all the other α formulas are adjusted accordingly;*
- *V' extends V to include in its domain Prop′, according to the following rule: if $\mathcal{M}, s \models \varphi \rightsquigarrow \psi$ and p' is the atom corresponding to $\varphi \rightsquigarrow \psi$, then $s \in V'(p')$.*

Theorem 6. *The decision problem for SBTrust is PSPACE-complete.*

Proof. The result for a SBTrust-reduction follows from the fact that $\mathcal{L}'_T$ is a set of $\mathbb{KD}4$ formulas and the $\mathcal{M}'$s are serial and transitive models [23]. Now, take a conjunction of the support formulas corresponding to the atoms of *Prop′* that are mapped to true. The SAT problem for the support formulas is PSPACE: in [36], it is shown that the satisfiability problem for the logic $\mathbb{F}$ is NP-complete. The result follows by Theorem 3.

Finally, we state the complexity of the model checking problem for SBTrust.

Theorem 7. *Given $\mathcal{M} := \langle S, \_, \_, R, V \rangle$ and $\alpha \in \mathcal{L}_T$, let n be the number of states in S and r the number of pairs sRv. Let k be the number of support formulas, and k' the number of belief modalities, plus the number of atomic propositions and of connectives in α. The complexity of the model checking problem is $O(k \cdot n^2 + (k + k') \cdot (n + r))$.*

Proof. We use the splitting methodology: we first consider the modal part of the formula α and then the support part. For the first stage, apply a SBTrust-reduction to $\mathcal{M}$ and α. This takes at most $k + k'$-steps. What is left is a model-checking problem for a modal formula within a pointed Kripke model: its complexity is $O((n + r) \cdot (k + k'))$ [16]. Then, translate back the new atoms to their respective support formulas. Take the partition S_i containing the state of evaluation s. To evaluate $\varphi \rightsquigarrow \psi$, compute the two sets $most(||\varphi||_i)$ and $||\psi||_i$. The latter is straightforward. The former requires at most n^2-steps (assuming the worst case: $||\varphi||_i = S_i = S$). This must be done for all k support formulas, thus, the complexity of the whole procedure is $O(k \cdot n^2)$.

5 Conclusions and Future Works

SBTrust is a logical framework for reasoning about decision trust, built on belief and support. To formalize support, we introduced a novel non-monotonic conditional operator, axiomatizing the flat fragment of the logic $\mathbb{F}$, and based on preference semantics.

Due to the generality of these concepts, SBTrust can integrate elements from the approaches mentioned in Sect. 1.1 within a unified framework. More precisely, SBTrust can naturally and flexibly represent relevant factors for trust through the support conditional. For example, we can express the importance of cognitive conditions for the emergence of trust without having to give an explicit formalization of them but relying on their implicit support of trust. Following up the discussion initiated in Example 1, we illustrate how SBTrust can be used to combine different elements that contribute to establishing trust and match those to the various paradigms described in Sect. 1.1.

Example 13. Assume that to trust $GoodV_i$, customer C seeks to fulfill three conditions: i) a cognitive-based one; ii) a reputation-based one; iii) a policy-based one. The cognitive-based condition could be captured by the notion of occurrence Trust (denoted by formula $OccTV_i$) given in [24], which depends on multiple cognitive features of the agents involved such as the goals of C, the ability and intentions of V_i, and the effects of the actions of V_i on the goals of C. The reputation-based condition could be represented, e.g., by proposition $TopRatingV_{i,j}$, while the policy-based condition by proposition $AuthV_i$. Then, the formula $T_\Gamma(GoodV_i)$ will indicate that customer C trusts V_i as a good vendor for reason Γ, where Γ stands for $(OccTV_i \wedge TopRatingV_{i,j} \wedge AuthV_i)$.

The flexibility of SBTrust to express different conditions comes, however, at the expense of reduced deductive power, e.g., we are unable to use the internal structure of cognitive conditions to derive conclusions.

We also claim that our framework is versatile enough to encompass other notions of trust, particularly those based on supportive information, as in [21]. Notably, it also shares similarities with argumentation-based formalizations of trust [5].

Our future research will proceed in two main directions: exploring potential applications of SBTrust and the support operator, and extending the technical results presented in this paper. Application-wise, we plan to use SBTrust to describe trust dynamics in edge computing scenarios [19], by integrating various existing approaches to trust generation - policies, reputation, and cognitive features - within our unified framework. We also expect to leverage SBTrust to model the combination of the various (explainability, legal, privacy- and security-related) factors influencing the trust perception when using generative AI models [31]. From the technical point of view, we plan to study the derived operator T in isolation and identify its properties independently of support and belief. Moreover, we intend to extend SBTrust by: (i) using beliefs within support statements; (ii) providing a proof calculus, along the lines of that in [14], equipped with a prover; (iii) moving towards a quantitative, dynamic, and multi-agent setting.

Acknowledgements. Work partially supported by the FWF project LoDEx Grant-DOI 10.55776/I6372 and by the Italian Ministry of University and Research programme "PRIN 2022" - grant number 2022598LMZ - AsCoT-SCE - CUP: H53D23003430006.

References

1. Abadi, M.: Logic in access control (tutorial notes). In: Aldini, A., Barthe, G., Gorrieri, R. (eds.) FOSAD 2007-2009. LNCS, vol. 5705, pp. 145–165. Springer, Heidelberg (2009). https://doi.org/10.1007/978-3-642-03829-7_5
2. Aldini, A., Curzi, G., Graziani, P., Tagliaferri, M.: A probabilistic modal logic for context-aware trust based on evidence. Int. J. Approx. Reason. **169**, 109167 (2024)
3. Aldini, A., Tagliaferri, M.: A taxonomy of computational models for trust computing in decision-making procedures. In: Jøsang, A. (ed.) European Conference on Information Warfare and Security, pp. 571–578 (2018)
4. Aldini, A., Tagliaferri, M.: Logics to reason formally about trust computation and manipulation. In: Saracino, A., Mori, P. (eds.) ETAA 2019. LNCS, vol. 11967, pp. 1–15. Springer, Cham (2020). https://doi.org/10.1007/978-3-030-39749-4_1
5. Amgoud, L., Demolombe, R.: An argumentation-based approach for reasoning about trust in information sources. Argum. Comput. **5**(2–3), 191–215 (2014)
6. Amoozadeh, M., et al.: Trust in generative AI among students: an exploratory study. In: 55th ACM Technical Symposium on Computer Science Education, vol. 1, SIGCSE 2024, pp. 67–73. ACM (2024)
7. Åqvist, L.: Deontic logic. In: Gabbay, D., Guenthner, F. (eds.) Handbook of Philosophical Logic, vol. II, pp. 605–714. Springer, Heidelberg (1984)
8. Artz, D., Gil, Y.: A survey of trust in computer science and the semantic web. J. Web Semant. **5**(2), 58–71 (2007)
9. Benlahbib, A., Nfaoui, E.H.: AmazonRep: a reputation system to support Amazon's customers purchase decision making process based on mining product reviews. In: 4th International Conference on Intelligent Computing in Data Sciences (ICDS), pp. 1–8 (2020)
10. Birnbaum, M.H., Navarro-Martinez, D., Ungemach, C., Stewart, N., Quispe-Torreblanca, E.G.: Risky decision making: testing for violations of transitivity predicted by an editing mechanism. Judgm. Decis. Mak. **11**(1), 75–91 (2016)
11. Bochman, A.: A logic for causal reasoning. In: Gottlob, G., Walsh, T. (eds.) Proceedings of IJCAI-03, pp. 141–146. Morgan Kaufmann (2003)
12. Braga, D., Niemann, M., Hellingrath, B., Neto, F.: Survey on computational trust and reputation models. ACM Comput. Surv. **51**(5), 1–40 (2018)
13. Castelfranchi, C., Falcone, R.: Trust Theory: A Socio-Cognitive and Computational Model. John Wiley and Sons, Hoboken (2010)
14. Ciabattoni, A., Rozplokhas,: Streamlining input/output logics with sequent calculi. In: Principles of Knowledge Representation and Reasoning, pp. 146–155 (2023)
15. Crupi, V., Iacona, A.: Three ways of being non-material. Stud. Logica. **110**, 47–93 (2022)
16. Fagin, R., Halpern, J.Y., Moses, Y., Vardi, M.: Reasoning about Knowledge. The MIT Press, Cambridge (2003)
17. Liu, F., Lorini, E.: Reasoning about belief, evidence and trust in a multi-agent setting. In: An, B., Bazzan, A., Leite, J., Villata, S., van der Torre, L. (eds.) PRIMA 2017. LNCS (LNAI), vol. 10621, pp. 71–89. Springer, Cham (2017). https://doi.org/10.1007/978-3-319-69131-2_5
18. Fishburn, P.C.: Nontransitive preferences in decision theory. J. Risk Uncertain. **4**(2), 113–134 (1991)
19. Fotia, L., Delicato, F., Fortino, G.: Trust in edge-based Internet of Things architectures: state of the art and research challenges. ACM Comput. Surv. **55**(9), 1–34 (2023)
20. Frost-Arnold, K.: The cognitive attitude of rational trust. Synthese **191**(9), 1957–1974 (2014)
21. Gambetta, D. (ed.) Trust: Making and Breaking Cooperative Relations. Blackwell (1988)

22. Grossi, D., van der Hoek, W., Kuijer, L.B.: Reasoning about general preference relations. Artif. Intell. **313**, 103793 (2022)
23. Halpern, J.Y., Moses, Y.: A guide to completeness and complexity for modal logics of knowledge and belief. Artif. Intell. **54**(3), 319–379 (1992)
24. Herzig, A., Lorini, E., Hübner, J.F., Vercouter, L.: A logic of trust and reputation. Logic J. IGPL **18**(1), 214–244 (2009)
25. Humberstone, L.: Philosophical Applications of Modal Logic. College Publications (2016)
26. Jacovi, A., Marasović, A., Miller, T., Goldberg, Y.: Formalizing trust in artificial intelligence: prerequisites, causes and goals of human trust in AI. In: Fairness, Accountability, and Transparency, pp. 624-635. ACM (2021)
27. Jensen, C.D.: The importance of trust in computer security. In: Zhou, J., Gal-Oz, N., Zhang, J., Gudes, E. (eds.) IFIPTM 2014. IAICT, vol. 430, pp. 1–12. Springer, Heidelberg (2014). https://doi.org/10.1007/978-3-662-43813-8_1
28. Kamvar, S.D., Schlosser, M.T., Garcia-Molina, H.: The eigentrust algorithm for reputation management in p2p networks. In: 12th International Conference on World Wide Web, pp. 640–651 (2003)
29. Kraus, S., Lehmann, D., Magidor, M.: Nonmonotonic reasoning, preferential models and cumulative logics. Artif. Intell. **44**(1), 167–207 (1990)
30. Kyburg, H.E.: Probability and the logic of rational belief. Wesleyan University Press (1961)
31. Leschanowsky, A., Rech, S., Popp, B., Bäckström, T.: Evaluating privacy, security, and trust perceptions in conversational AI: a systematic review. Comput. Hum. Beh. **159**, 108344 (2024)
32. Makinson, D., van der Torre, L.: Input/output logics. J. Phil. Logic **29**, 383–408 (2000)
33. Marsh, S.: Formalising Trust as a Computational Concept. University of Stirling (1994)
34. Parent, X.: Completeness of Åqvist's systems E and F. Rev. Symb. Logic **8**(1), 164–177 (2015)
35. Poole, D.: What the lottery paradox tells us about default reasoning. In: Principles of Knowledge Representation and Reasoning, pp. 333–340. Morgan Kaufmann (1989)
36. Rozplokhas, D.: LEGO-like small-model constructions for Åqvist's logics. In: Proceedings of AIML 2024 (2024)
37. Schumm, G.F.: Transitivity, preference and indifference. Phil. Stud. Int. J. Phil. Anal. Tradit. **52**(3), 435–437 (1987)
38. Schwartz, T.: Cycles and Social Choice: The True and Unabridged Story of a Most Protean Paradox. Cambridge University Press, Cambridge (2018)
39. Seamons, K.E., et al.: Negotiating trust on the web. IEEE Internet Comput. **6**(06), 30–37 (2002)
40. Shoham, Y.: A semantical approach to non-monotonic logics. In: Symposium on Logic in Computer Science (LICS 1987), pp. 275–279. IEEE CS Press (1987)
41. Smith, A.: What Amazon key teaches us about trust in customer relationships. Forbes (2017)
42. Soleimani, M.: Buyers' trust and mistrust in e-commerce platforms: a synthesizing literature review. Inf. Syst. E-bus. Manag. **20**(1), 57–78 (2022)
43. Tagliaferri, M., Aldini, A.: From belief to trust: a quantitative framework based on modal logic. J. Log. Comput. **32**(6), 1017–1047 (2022)

Asynchronous Byzantine Consensus with Trusted Monotonic Counters

Yackolley Amoussou-Guenou[1], Maurice Herlihy[2], and Maria Potop-Butucaru[3(✉)]

[1] Université Paris-Panthéon-Assas, CRED, Paris, France
[2] Computer Science Department, Brown University, Providence, RI 02912, USA
[3] Sorbonne Université, CNRS, LIP6, Paris, France
maria.potop-butucaru@lip6.fr

Abstract. The paper promotes a new design paradigm for Byzantine tolerant distributed algorithms using trusted abstractions (oracles) specified in a functional manner. The contribution of the paper is conceptual. The objective here is to design distributed fundamental algorithms such as reliable broadcast and asynchronous byzantine consensus using trusted execution environments and to help designers to compare various solutions on a common ground. In this framework we revisit Bracha's seminal work on Asynchronous Byzantine Consensus. Our solution uses trusted monotonic counters abstraction and tolerates t Byzantine processes in a system with n processes, $n \geq 2t + 1$. The keystone of our construction is a novel and elegant Byzantine Reliable Broadcast algorithm resilient to $t < n$ Byzantine processes.

Keywords: Asynchronous · Byzantine agreement · Trusted abstractions

1 Introduction

Byzantine Agreement problem introduced in the seminal paper [28] has been studied for decades in various models ranging from synchronous (e.g. [27]), to asynchronous (e.g. [7,34]).

The blockchains era revived the interest for the problem. In Algorand [12] for example the authors propose solutions for synchronous and partially synchronous systems constructing their solution on top of Verifiable Random Function abstraction [32]. In [1] the authors define and address the complexity issues. More recently, in [17] the authors use recent advances in cryptography in order to reduce the word complexity. In [35] authors analyze the complexity of randomized algorithms for binary consensus under different common coins. However, none of these works focus on breaking the $3f + 1$ bound for the asynchronous settings.

One of the main problems to be solved in order to break the $3f + 1$ bound in Byzantine prone environments is the ability of Byzantine nodes to equivocate (i.e. a faulty node may send different messages to different nodes). *Trusted*

U. Schmid and R. Kuznets (Eds.): SIROCCO 2025, LNCS 15671, pp. 23–38, 2025.
https://doi.org/10.1007/978-3-031-91736-3_2

execution components (e.g. A2M [13], TrInc [30], USIG [37]) are reputed to be powerful tools for avoiding equivocation.

Trusted components have been heavily used to increase the resilience of PBFT like protocols [10] and therefore they found applications in many recent blockchain algorithms. Among the first to attack the execution of PBFT in trusted execution environments are Correia *et al.* [18,19]. They introduced TTCB wormhole, a distributed component with local parts (local TTCBs) in nodes and its own bounded secure communication channel (i.e. a channel that cannot be affected by malicious faults where all operations have a bounded delay). By using this wormhole, the authors proved that PBFT can support a fraction of half Byzantine nodes. In other words, they circumvent FLP [23] impossibility by relying on a synchronous and secure distributed subsystem. Although this method allows to increase fault tolerance, its practical implementation is too difficult to set up.

In the quest of practicality, Chun *et al.* [13] introduced Attested Append-Only Memory (A2M), a trusted system that targets to remove from the faulty nodes the ability to equivocate. An A2M is a set of trusted ordered append-only logs that provide an attestation for each entry. Furthermore, they propose PBFT-EA, a modified PBFT [9] that uses A2M for each message exchanged; the message is appended to a log and the attestation produced is sent along with the message. The use of this abstraction increases the resilience to half. Compared to TTCB that requires a secure and synchronous communication channel, A2M requires no stronger assumptions on network than PBFT. However, A2M needs large secure storage (for the log).

An alternative to this is the use of a monotonic counters implemented in a tamperproof module. Levin *et al.* propose TrInc [30], a trusted monotonic counter that deals with equivocation in large distributed systems by providing a primitive: once-in-a-lifetime attestations. They also prove that TrInc can implement A2M.

Later, Veronese *et al.* in [37] propose a specific monotonic trusted counter, USIG (Unique Sequential Identifier Generator), a local service available in each node that signs a message and assigns it the value of a counter. The service offers two functions: one that returns a certificate, and one that validates certificates. These certificates are based on a secure counter: the counter value is never duplicated, and successive counter values are successive integers. This service has to be implemented in a tamper-proof module. Furthermore, they propose two algorithms (MinBFT and its speculative version MinZyzzyva) that implement following the same pattern as PBFT *state machine replication* which consists of replicating a service in a group of servers with strong consistency guaranties. Each server maintains a set of state variables, which are modified by a set of operations. The operations are deterministic and atomic. The initial state of the servers is the same. The properties that the two proposed algorithms satisfy are: safety- all correct servers execute the same requests in the same order; liveness- all correct clients' requests are eventually executed. MinBFT and its speculative version MinZyzzyva implement state machine replication using USIG in systems with a minority of Byzantines.

Another line of research combines speculative methods and trusted environments (e.g. CheapBFT [26] and ReBFT [22]). In a normal execution case (when there are no Byzantine nodes), $f+1$ nodes are enough to guarantee the agreement. In case of detected or suspected Byzantine nodes the protocol switches to a PBFT inspired protocol with trusted hardware and activates f extra passive replicas.

Interestingly, in the context of blockchains, the use of trusted environments in order to increase the resilience is very recent (e.g. *VABA* [39], the asynchronous version of *HotStuff*, *Damysus* [21], and TenderTee [3]). The first use of it was proposed in [39]. The authors enhance HotStuff blockchain in order to tolerate a minority of corruptions in a PBFT style protocol. Their algorithm builds on top of an underlying expander graphs and use threshold signatures. The small trusted component used has A2M flavour. Another recent paper introduces *Damysus* ([21]), a PBFT protocol that uses trusted environments to improve Hotstuff resilience. In this paper, the authors introduce two trusted services *Checker* and *Accumulator* that respectively increase resilience and reduce latency. The correctness of the proposed protocol has been proven for the partially synchronous environments. In [3] the authors propose a methodology to automatically plug A2M in the Tendermint protocol to increase its resilience. Furthermore, they prove that the same methodology can be applied to the repeated consensus abstraction with the same results in terms of resilience. This work still needs a *partially synchronous* execution environment.

In this paper we are interested in designing trusted distributed algorithms (e.g. reliable broadcast, consensus) having optimal resilience to Byzantine faults in asynchronous settings without the use of threshold signatures or assumptions on the underlying topology as in [39]. We focus here in solving Byzantine Consensus problem [7] in *asynchronous* settings. Probabilistic consensus enhanced with trusted environments in order to avoid equivocation seems to be a good compromise in this direction. One of the key building blocks of our Probabilistic Byzantine Consensus is a deterministic Byzantine Reliable broadcast algorithm tolerant to $t < n$ Byzantine processes. *Byzantine reliable broadcast* is a fundamental problem in fault-tolerant distributed systems. It consists of ensuring that a correct initiator process broadcasts its value to all correct processes, even in the presence of malicious Byzantine processes. For decades, Byzantine Reliable Broadcast has been at the core of various consensus protocols, and more recently, at the core of certain blockchains.

Algorithms solving the Byzantine Reliable Broadcast problem have been proposed in various environments: with static or dynamic Byzantine nodes, or in conjunction with transient faults. Byzantine Reliable Broadcast solutions (e.g. [5,6,24,31,36]) achieve resilience of at least $n \geq 3t+1$ processes, where t is the maximum number of Byzantine processes. In this paper we continue the line of work opened by [20] which proposes a reliable broadcast algorithm tolerant with $n \geq 2f+1$ processes where f is the number of Byzantine faults. Contrarily to us, they use failure detectors. In [38] authors proposed an algorithm similar to ours but which only tolerates $f < n/3$ Byzantine processes. This bound has been ameliorated to $f < n/2$ in [2]. Our work improves further the bound to $f < n$.

Tackling another problem, [29] proposes an algorithm for the atomic broadcast problem which uses a trusted execution environment with a monotonic counter similarly to us. However, differently from us the focus of the authors is not on the minimal trust assumptions is expected from the trusted execution environment. Moreover, their solution builds on the failure detector based reliable broadcast of [20].

Another line of research related to the use of trusted execution environments for improving the resilience of distributed algorithms is the line initiated by Clement *et al.* [16]. Although the trusted execution environments make protocols immune to equivocation (where the initiator sends different messages to different processes), Clement *et al.* [16] show that non-equivocation is not enough to provide $n \geq 2f + 1$ resilience, nor to support the equivalent of digital signatures. The authors prove that it is possible to use non-equivocation to transform any protocol that works under the crash fault model into a protocol that tolerates Byzantine faults by adding the ability to guarantee the transferable authentication of network messages (e.g., using digital signatures). In [4] the authors revisited and extend the work of Clement *et al.* [16] providing a transformer with a polynomial communication overhead instead of exponential and covering also randomized distributed algorithms.

Our approach is more conceptual in the sense that we would like to design Byzantine tolerant distributed algorithms on top of trusted oracles that allow us to abstract low level (trusted hardware) assumptions and provide a separation between the conceptual and the technical part. Notice that, the trusted oracles we define may have different implementations. Specifying the functional property that is needed at the application level we need a generic solution that can be implemented using different technologies or physical architectures. Oracles in distributed systems have been used for decades starting with the seminal work on failure detectors [11] that paved the way of designing protocols with formal proofs of correctness and the investigation of precise lower bounds and impossibility results.

Our Contribution. Our work extends the line of research related to using trusted execution environments in order to increase the resilience of distributed algorithms in Byzantine prone environments. The novelty of our approach is in identifying the minimal assumptions on trusted abstraction needed to solve fundamental building blocks in distributed computing (e.g. consensus and reliable broadcast). In this paper we revisit the original simple and elegant solutions proposed by Bracha [7] in an environment where processes are equipped with a trusted monotonic counter abstraction that provides a non-falsifiable, verifiable, unique, monotonic, and sequential counter. The use of this abstraction in a clever way allows us to first implement a Reliable Broadcast resilient to $t < n$ Byzantine processes in asynchronous communication environments. Moreover, on top of this optimized Reliable Broadcast primitive we construct a Probabilistic Byzantine Consensus resilient to t Byzantine processes in systems of size $n \geq 2t+1$. Differently from the transformer-based approach where the transformations are obtained with an important communication overhead (exponential in the case of [16] and polynomial in the case of [4]), our design uses only a constant overhead[1] with the respect of the Bracha's original solutions while improving both the resilience (from $n/3$ to $n/2$) and the number of communication steps (from 3 steps per round to 2 steps per round), while the expected number of rounds remains the same. Our work opens a new direction of research similar to oracle-based distributed computing [11]. The trusted environment is encapsulated in a trusted abstraction (oracle) providing a set of guarantees. The methodology used in this paper for Byzantine Reliable Broadcast and Probabilistic Byzantine Consensus can be easily extended to other Byzantine tolerant distributed algorithms.

2 System Model and Problems Definition

We consider a set of n asynchronous sequential processes, of which up to t can be *Byzantine*, meaning they can deviate from the given protocol. The rest are *correct* processes. We consider static adversary, in the sense that the Byzantine processes do not corrupt other processes during the protocol execution.

Processes communicate by exchanging messages through an asynchronous network. We make the usual assumptions that there is a public key infrastructure (PKI) where public keys are distributed, each process has a (universally known) public key a matching private key, and each message is signed by its creator. Messages are not lost or spuriously generated. Each process can send messages directly to any other process, and each process can identify the sender of every message it receives.

We assume that the communication is asynchronous, and that processes have access to a communication primitive which ensures that any message m sent by a correct process is received by every correct process in a finite (but unknown) time.

Following Bracha, [7], we define *Byzantine Reliable Broadcast* and *Probabilistic Byzantine Consensus* as follows:

Definition 1 (Byzantine Reliable Broadcast). *We say that an algorithm implements* Byzantine reliable broadcast *if:*

- brb-CorrectInit: *If the initiator is correct, all correct processes deliver the initiator's value.*
- brb-ByzantineInit: *If the initiator is Byzantine, then either no correct process delivers any value, or all correct processes deliver the same value.*

[1] For this comparison, we did not take into account the cost of implementing the trusted monotonic counter abstraction nor any transformer.

In the binary consensus problem every process p has some initial value $v_p \in \{0, 1\}$, and every correct process must decide on the same value v that is one of the proposed values.

Definition 2 (Probabilistic Byzantine Consensus). *A protocol implements* probabilistic Byzantine consensus *if:*

- *Agreement: all correct processes decide on the same value.*
- *Validity: if all processes start with the same value v, then all correct processes decide on v.*
- *Termination (probabilistic). Every correct process eventually decides with probability 1.*

In the context of round-based randomized algorithms, the termination property can be restated as follows: the probability that a correct process is undecided after r rounds approaches zero as r approaches infinity.

3 Trusted Monotonic Counter Object

The Trusted Monotonic Counter Oracle abstraction TMC-Object defined below is the core of our novel Byzantine Reliable Broadcast protocol that supports t Byzantine failures among n processes, where $n > t$, a great improvement on the classical $n \geq 3t + 1$ algorithms.

TMC-Object provides each process with a read-only local variable, called trustedCounter. Whenever the TMC-Object is invoked, it returns a value for trustedCounter strictly greater than any previous value it returned. The difference between two successive counter values is exactly 1.

The TMC-Object supports the operation **get_certificate()**. A process p invokes **get_certificate(m)** with a message m. The object returns a *certificate* and the value of the *counter*. The certificate object certifies that the returned value was created by the tamper-proof TMC-Object object for the message m. That is, it the value of the monotonic counter trustedCounter, which is incremented whenever **get_certificate(m)** is called. The TMC-Object object guarantees the following properties:

- **Uniqueness:** The **get_certificate()** operation of TMC-Object will never assign the same identifier to two different messages.
- **Sequentiality:** The **get_certificate()** operation of TMC-Object will always assign an identifier that is the successor of the previous one.

Note that the sequentiality property implies the *Strict Monotonicity:* The **get_certificate()** operation of TMC-Object will always assign an identifier that is strictly greater than the identifier assigned to the preceding invocation. Note also that the sequentiality property implies uniqueness.

To send a message u certified by TMC-Object, a process p first invokes the **get_certificate()** operation of TMC-Object, which creates a certificate $C_{(p,u)}$

corresponding to the value of the trustedCounter c_p,[2] then the process sends the tuple $(u, \mathcal{C}_{(p,u)}, c_p)$, which can be verified by any other process receiving the message. The verification is done using the **check_certificate()** operation of the TMC-Object, which confirms whether the given certificate was produced for the given message and has given the value of the counter. Each invocation of the **get_certificate()** operation of TMC-Object increments the value of the trustedCounter c_p of process p. We call that sequence of operations TMC-Object-Send u.

When receiving a message $(u, \mathcal{C}_{(p,u)}, c_p)$, a process must check if the certificate $\mathcal{C}_{(p,u)}$ for message u corresponds to the value of the counter c_p. If not, the message is considered invalid and is ignored. If they correspond, the message is said to be valid according to the TMC-Object.

4 Asynchronous Byzantine Reliable Broadcast

We describe Algorithm 1, a Byzantine reliable broadcast algorithm which uses a unique invocation of **get_certificate()** operation of the TMC-Object(with the counter initialized at 0). The initiator of the broadcast uses the TMC-Object-Send operation to send the value to be broadcast. Our Byzantine Reliable Broadcast is resilient to any number, $t < n$, of Byzantine processes.

The protocol works in two sequential asynchronous steps. In the initial step (Step 0) of the protocol, when a process p wants to broadcast a value v, it TMC-Object-Sends an initial message for v ($\langle initial, v \rangle$) to all other processes. The process initiating the broadcast is called the *initiator*. The initiator sends the message as well as the associated certificate and counter, which can be checked by all other processes.

In Step 1, when receiving a valid initial message from the proposer, say with value v, a process sends the initiator message (with the certificate and value of the counter) to all the processes. In such a way, it ensures that all other processes will eventually receive the initiator's certified message. First notice that only messages with counter value of 1 are considered valid, since a message with a higher counter value means that the initiator already sent another value. More generally, the value of the counter should be 1 more than the previously known value of the counter.

Theorem 1. *Let n be the number of processes, and t be an upper bound of the Byzantine processes. If $n > t$, Algorithm 1 implements Byzantine Reliable Broadcast in $O(n^2)$ messages and using a unique invocation of the* ***get_certificate()*** *operation of the TMC-Object.*

Proof. First, notice that when a correct process p delivers a message u, it means that p received a valid initiator message u. Eventually every other correct process q will receive that same initiator message thanks to the network guarantees. Therefore, if there is no equivocation, they must deliver that same value.

[2] When the value of a counter can be verified against a given certificate, we say that the certificate corresponds to that value of the counter, similarly to digital signatures.

Algorithm 1. Byzantine Reliable Broadcast with a unique TMC-Object

```
procedure TMC-Object_Broadcast(v)
    Step 0
     if p is the initiator then
       TMC-Object-Send ⟨initial, v, p⟩ to all
    Step 1
     Upon reception of (⟨initial, u, id_initiator⟩, C, 1) message
       Send (⟨initial, u, id_initiator⟩, C, 1) to all
       Deliver u
```

If the initiator is correct, then all other correct processes eventually receive and deliver the initiator's message.

It remains to show that when two different correct processes deliver a message, they deliver the same message.

Recall that when a correct process receives a (TMC-Object) message, it checks the value of the trusted counter associated to the message. Moreover, it checks whether the message and the value of the counter are coherent with the corresponding certificate, i.e., checking if the **check_certificate()** operation returns true on the message, certificate, and counter value.

By way of contradiction, assume there exist two correct processes p and q which deliver distinct messages $u \neq v$. Without loss of generality, assume that p delivers message u and q delivers message v. To do so, p must have received $(\langle initial, u, id\_initiator\rangle, \mathcal{C}_{\text{initiator}}, 1)$ and q must have received the message $(\langle initial, v, id\_initiator\rangle, \mathcal{C}'_{\text{initiator}}, 1)$, which is impossible because of the *Uniqueness* property of the TMC-Object. Therefore, p and q must have delivered the same message, which concludes the proof.

In Algorithm 1, the number of messages sent is exactly $n+n^2$, which is $O(n^2)$. In more details, the initiator sends n messages (one to each other process), and each process when receiving the initiator message sends it back which is $n \times n$. □

Notice that we consider only the message complexity, therefore, we did not count the cost of implementing the TMC-Object, which has no impact on the message complexity, but rather could have an impact of the bit complexity.

5 Asynchronous Byzantine Consensus

In this section we propose a Byzantine Probabilistic Consensus algorithm (Algorithm 3) resilient to $t < n/2$ Byzantine processes. The keystone of the solution is the Byzantine Reliable Broadcast resilient to any number, $t < n$, of Byzantine processes (Algorithm 1).

Remark 1. Notice that when there are two different initiators, there are two instances of the broadcast.

Our probabilistic Byzantine consensus protocol (Algorithm 3) is inspired from the Bracha's probabilistic consensus [7]. The protocol works in asynchronous sequential *rounds* where each round, $k \geq 0$ is split in two sequential steps, $2k$ and $2k+1$.

Algorithm 2. Byzantine Reliable Broadcast invoked by process p_i

```
procedure TMC-OBJECT_BROADCAST(p_i, v_i, instance)
    TMC-Object-Send ⟨initial, v_i, p⟩ to all

    Upon reception of (⟨initial, u, id_initiator⟩, C, instance + 1) message
      Send (⟨initial, u, id_initiator⟩, C, instance + 1) to all
      Deliver u
```

Algorithm 3. Probabilistic Byzantine Consensus invoked by process p_i with value v_i

```
procedure PROBABILISTIC_CONSENSUS(v_i)
  Step 2k
    TMC-Object_Broadcast(p_i, v_i, 2k)
      wait until n − t messages from Step 2k or ∃k' ≤ k, v :
  (#(d, v) received from Step 2k' + 1) > n/2:
      if ∃v : (#v received from Step 2k > n/2) then
        v_i = (d, v)
      if ∃k' ≤ k, v : (#(d, v) received from Step 2k' + 1) > n/2 then
        Decide v and Terminate
  Step 2k + 1
    TMC-Object_Broadcast(p_i, v_i, 2k + 1)
      wait until n − t messages from Step 2k + 1 or ∃k' ≤ k, v :
  (#(d, v) received from Step 2k' + 1) > n/2:
      if ∃k' ≤ k, v : (#(d, v) received from Step 2k' + 1) > n/2 then
        Decide v and Terminate
      if ∃v : (#(d, v) received from Step 2k + 1) ≥ 1) then
        v_i = v
      otherwise
        v_i = flip()
    k = k + 1
    go to Round k
```

In each even step $2k$, of round $k \geq 0$, after proposing its local value, each process waits to collect $n-t$ messages from this step. If the process delivers the same message, say v, strictly more than $n/2$ times from different processes, then the process locally tags v, by setting its local value to $(\mathtt{d}, v)$. $\mathtt{d}$ is the tag marker. It means the process is ready to decide v and is letting the other processes know about it.

In each odd step $2k+1$ of round $k \geq 0$, after broadcasting its local value, each process waits to collect $n-t$ messages from the current step. If among the delivered messages, the process delivers at least one tagged value from the current step, say $(\mathtt{d}, v)$, the process sets its local value to the value v which will be proposed during the next step. Otherwise, if the process delivers no tagged value from the current step, it randomly selects a value which will be its proposal for the next step.

In each step, if a process delivers the tagged message $(\mathtt{d}, v)$ from strictly more than $n/2$ different processes from the same round, the process decides value v and terminates.

Note that a process executing the Byzantine Probabilistic Consensus algorithm (Algorithm 3) invokes the Byzantine Reliable Broadcast primitive two times in each round. Each invocation will change the value of the trusted counter. For completeness of the presentation, Algorithm 2 presents the parameterized version of Algorithm 1.

To prove the correctness of Algorithm 3, we emphasize the main guarantees we rely on. These guarantees need not be taken as axiomatic, because they could be implemented as described below. Moreover, we also make explicit for which property each guarantee is necessary.

Hypothesis 1. *Byzantine processes cannot lie about the output of their randomness.*

Hypothesis 1 is achievable by the use of verifiable random functions (VRF [12,33]). We rely on this hypothesis to prove the (probabilistic) Termination property.

Hypothesis 2. *All messages sent are causally valid, in the sense that each message is accompanied by the elements that caused it.*

We rely on Hypothesis 2 to provide the Agreement property. Hypothesis 2 can be implemented by requiring each process to accompany each message with the prior messages that caused the process to compute that message. In the case of Algorithm 3, this hypothesis simply means that a message sent at a step $s > 0$ should be accompanied by the messages the process received at the previous step $s-1$. Any message that is not causally valid is ignored.

Lemma 1. *Suppose there are n processes, of which at most t are Byzantine. If $n \geq 2t+1$, for any $k \geq 0$, if all correct processes at step $2k$ proposes value v, then at the end of step $2k$ all correct processes will have as value either $(\mathtt{d}, v)$ or v.*

Proof. Assume there exists a correct process q that has a value u or $(\mathtt{d}, u)$ with $u \neq v$ at the end of step $2k$.

Case 1: q received more than $n/2$ messages with value u, which is impossible since all correct processes started with the same value v and the number of Byzantine processes is limited to $t < n/2$.

Case 2: q changed its value to some value u. This is impossible since q received $n - t$ messages. Among the $n - t$ messages there at most t messages with an erroneous value u forged by fewer than $n/2$ Byzantine processes, and at least one message from correct processes with value v. In this case, q keeps its initial value v. □

Lemma 2. *Let n be the number of processes, and t be an upper bound of the Byzantine processes. If $n \geq 2t + 1$, for any $k \geq 0$, if all processes at step $2k$ propose value v, then at the end of the step $2k$ all correct processes will have as value $(\mathtt{d}, v)$.*

Proof. By Lemma 1 if all correct processes start with value v then at the end of step $2k$ all correct processes will have as value either $(\mathtt{d}, v)$ or v.

Assume that there exists a correct process q which ends step $2k$ with v instead of $(\mathtt{d}, v)$. This means that q did not receive more than $n/2$ messages with the same value v. This is impossible since q waits for $n - t$ messages. Over the $n - t$ messages either all of them come from correct processes (hence same value v) or some of them (up to t) come from Byzantine processes. However, Byzantine processes start with v (by assumption) and cannot change their initial value without proof of modification.

Overall, if all processes start with value v then at the end of step $2k$ all correct processes will have value $(\mathtt{d}, v)$. □

Lemma 3. *For any $k \geq 0$, if all processes propose value v at step $2k$, then at step $2k+1$, if a Byzantine process broadcasts a value then its only causally valid message is $(\mathtt{d}, v)$.*

Proof. By Lemma 2, a correct process at the end of step $2k$ will have as value $(\mathtt{d}, v)$. Hence in the next step it will broadcast $(\mathtt{d}, v)$. If at step $2k+1$ a Byzantine process broadcasts a message, then that message should be $(\mathtt{d}, v)$. □

Lemma 4 (Validity). *For any $k \geq 0$, if all processes propose value v at step $2k$, then all correct processes decide on v.*

Proof. If all processes start with the same value v then at the end of step 0 (which is a $2k$ step with $k = 0$) all correct processes end with the value $(\mathtt{d}, v)$ (see Lemma 2). In step 1 (which is $2k + 1$) each correct process sends $(\mathtt{d}, v)$ and waits for $n - t$ messages from step $2k + 1$. Based on Hypothesis #2 and Lemma 3, Byzantine processes either send $(\mathtt{d}, v)$ or stay silent. Therefore, all correct processes will gather at least $n - t$ messages with v, hence all correct processes decide v. □

Lemma 5 (Agreement). *Let n be the number of processes, and t be an upper bound of the Byzantine processes. If $n \geq 2t + 1$, if two correct processes decide, they decide the same value. More generally, if one correct process decides, all correct processes decide the same value.*

Proof. Let p and q be two correct processes that decided. Let p decide at some round k and q at some round $k' > k$. It follows that p received more than $n/2$ messages $(\mathtt{d}, v)$ in round k. Therefore, all correct processes receive at least one $(\mathtt{d}, v)$ message in round k, in particular, q received at least one $(\mathtt{d}, v)$ in round k. Since all processes send the causal proof of their messages (to satisfy Hypothesis #2), it follows that all processes, including q, will start round $k+1$ with the same value v. Following Lemma 4 all correct processes, including q, decide v in round $k+1$. □

Lemma 6 (Termination). *The probability that a correct process is undecided after r rounds approaches zero as r approaches infinity.*

Proof. As shown in Lemma 4, if all processes propose the same value v, then all correct processes eventually decide value v.

If one correct process decides, then thanks to Lemma 5, all correct processes eventually decide. Therefore, let us assume that no correct process decides yet. There are two cases. Either (i) no process enters line 14 and all processes randomly flip their value, or (ii) at least one correct process received one $(\mathtt{d}, v)$ message from its current step and so does not flip its value.

- First, consider the case where no process ever receives a $(\mathtt{d}, v)$ message corresponding to its current step and round. Therefore, they will always flip their coins at each round. Notice, however, that thanks to Lemma 4, when all processes will have the same proposal, then all correct processes decide that value, which guarantees Termination. The probability of such an event happening for each round k is low, *i.e.*, p^n, where p the probability of having any value. Since the Byzantine processes cannot control (nor lie about, by Hypothesis #1) their random input value, the probability of having a value for any Byzantine process is the same as for any correct process. However, the probability this event never occurs in an infinite execution is the limit of $(1-p^n)^k$ when k goes to infinity, which is equal to 0, since $p \in (0, 1)$, and $n > 0$. Hence, if processes always flip their value, they will terminate with probability 1.
- There remains to discuss what happens if at least one process does not flip their value.
  If a process does not flip its value, it means that such process received a $(\mathtt{d}, v)$ message from its current step (say, $2k+1$). It means no other value $(v' \neq v)$ could be sent as $(\mathtt{d}, v')$, in fact, by Hypothesis 2, since Byzantine processes must causally justify their votes, they cannot send a $(\mathtt{d}, v')$ with $v' \neq v$. Therefore, all the processes that will receive such message and will not flip will have v as value for the next round. Notice that if one correct process decides, then by Lemma 5, all correct processes eventually decide. Suppose there is no decision yet. Either all correct processes do not flip and have the same value v for the next round (and terminate at that round by Lemma 4), or some do flip.
  Therefore, if some processes flip repeatedly, and the others do not flip but have the same value; then as in the case above, with probability 1 over the infinite

execution, there will be a situation where the processes that flipped will end up with the same value as those that do not flip, and hence, by Lemma 4 they will terminate. Notice that the probability of having all processes having the same value must be higher in this case (than in the above case) since some are already proposing the same value.

In both cases, therefore, with probability 1, all correct processes eventually terminate. □

A direct corollary of Lemma 4 is that any correct process eventually decides with probability 1. We can then state the correctness of Algorithm 3.

Theorem 2. *Let n be the number of processes, and t be an upper bound of the Byzantine processes. If $n \geq 2t + 1$, Algorithm 3 implements Probabilistic Byzantine Consensus with $O(n^3)$ messages.*

6 Conclusions and Discussions

In this paper we propose a novel solution for Probabilistic Byzantine Consensus in an environment where processes are equipped with a trusted monotonic counter abstraction that provides a non-falsifiable, verifiable, unique, monotonic, and sequential counter. Our solution tolerates t Byzantine processes with $n \geq 2t + 1$ in asynchronous settings. The keystone of our construction is an elegant deterministic Byzantine Reliable Broadcast algorithm that implements a Reliable Broadcast resilient to $t < n$ Byzantine processes. Our work continues the line of research opened by [18,19] and continued by [25,37] that promotes the benefits of using trusted hardware in improving the resilience of distributed algorithms. Our work has similar flavor to the line of work of oracle-based distributed computing [11] and opens several research directions including the modelisation of distributed algorithms executed in trusted environments or their composition [8,14,15]. We believe that our oracle-based approach may be easily refined in order to address other distributed problems.

References

1. Abraham, I., Malkhi, D., Spiegelman, A.: Asymptotically optimal validated asynchronous byzantine agreement. In: Robinson, P., Ellen, F. (eds.) Proceedings of the 2019 ACM Symposium on Principles of Distributed Computing, PODC 2019, Toronto, ON, Canada, 29 July–2 August 2019, pp. 337–346. ACM (2019). https://doi.org/10.1145/3293611.3331612
2. Amoussou-Guenou, Y., Beltrando, L., Herlihy, M., Potop-Butucaru, M.: Byzantine reliable broadcast with one trusted monotonic counter. In: Masuzawa, T., Katayama, Y., Kakugawa, H., Nakamura, J., Kim, Y. (eds.) Stabilization, Safety, and Security of Distributed Systems - 26th International Symposium, SSS 2024, Nagoya, Japan, 20–22 October 2024, Proceedings. Lecture Notes in Computer Science, vol. 14931, pp. 360–374. Springer, Heidelberg (2024). https://doi.org/10.1007/978-3-031-74498-3_26

3. Beltrando, L., Potop-Butucaru, M., Alfaro, J.: Tendertee: increasing the resilience of tendermint by using trusted environments. In: 24th International Conference on Distributed Computing and Networking, ICDCN 2023, Kharagpur, India, 4–7 January 2023, pp. 90–99. ACM (2023). https://doi.org/10.1145/3571306.3571394
4. Ben-David, N., Chan, B.Y., Shi, E.: Revisiting the power of non-equivocation in distributed protocols. In: Milani, A., Woelfel, P. (eds.) PODC '22: ACM Symposium on Principles of Distributed Computing, Salerno, Italy, 25–29 July 2022, pp. 450–459. ACM (2022)
5. Bonomi, S., Decouchant, J., Farina, G., Rahli, V., Tixeuil, S.: Practical byzantine reliable broadcast on partially connected networks. In: 2021 IEEE 41st International Conference on Distributed Computing Systems (ICDCS), pp. 506–516. IEEE (2021)
6. Bonomi, S., Farina, G., Tixeuil, S.: Reliable broadcast despite mobile byzantine faults. In: Bessani, A., Défago, X., Nakamura, J., Wada, K., Yamauchi, Y. (eds.) 27th International Conference on Principles of Distributed Systems, OPODIS 2023, Tokyo, Japan, 6–8 December 2023. LIPIcs, vol. 286, pp. 18:1–18:23. Schloss Dagstuhl - Leibniz-Zentrum für Informatik (2023). https://doi.org/10.4230/LIPICS.OPODIS.2023.18
7. Bracha, G.: Asynchronous byzantine agreement protocols. Inf. Comput. **75**(2), 130–143 (1987). https://doi.org/10.1016/0890-5401(87)90054-X
8. Canetti, R., Cheung, L., Kaynar, D.K., Lynch, N.A., Pereira, O.: Compositional security for task-pioas. In: 20th IEEE Computer Security Foundations Symposium, CSF 2007, Venice, Italy, 6–8 July 2007, pp. 125–139. IEEE Computer Society (2007)
9. Castro, M., Liskov, B.: Practical byzantine fault tolerance. In: Proceedings of the Third USENIX Symposium on Operating Systems Design and Implementation (OSDI), New Orleans, Louisiana, USA, 22–25 February 1999, pp. 173–186. USENIX Association (1999). https://dl.acm.org/citation.cfm?id=296824
10. Castro, M., Liskov, B.: Practical byzantine fault tolerance and proactive recovery. ACM Trans. Comput. Syst. **20**(4), 398–461 (2002)
11. Chandra, T.D., Toueg, S.: Unreliable failure detectors for reliable distributed systems. J. ACM **43**(2), 225–267 (1996)
12. Chen, J., Micali, S.: Algorand: a secure and efficient distributed ledger. Theor. Comput. Sci. **777**, 155–183 (2019). https://doi.org/10.1016/J.TCS.2019.02.001
13. Chun, B., Maniatis, P., Shenker, S., Kubiatowicz, J.: Attested append-only memory: making adversaries stick to their word. In: Proceedings of the 21st ACM Symposium on Operating Systems Principles 2007, SOSP 2007, Stevenson, Washington, USA, 14–17 October 2007, pp. 189–204. ACM (2007). https://doi.org/10.1145/1294261.1294280
14. Civit, P., Potop-Butucaru, M.: Brief announcement: composable dynamic secure emulation. In: Agrawal, K., Lee, I.A. (eds.) SPAA '22: 34th ACM Symposium on Parallelism in Algorithms and Architectures, Philadelphia, PA, USA, 11–14 July 2022, pp. 103–105. ACM (2022)
15. Civit, P., Potop-Butucaru, M.: Dynamic probabilistic input output automata. In: Scheideler, C. (ed.) 36th International Symposium on Distributed Computing, DISC 2022, Augusta, Georgia, USA, 25–27 October 2022. LIPIcs, vol. 246, pp. 15:1–15:18. Schloss Dagstuhl - Leibniz-Zentrum für Informatik (2022)
16. Clement, A., Junqueira, F., Kate, A., Rodrigues, R.: On the (limited) power of non-equivocation. In: ACM Symposium on Principles of Distributed Computing, PODC '12, Funchal, Madeira, Portugal, 16–18 July 2012, pp. 301–308. ACM (2012). https://doi.org/10.1145/2332432.2332490

17. Cohen, S., Keidar, I.: Brief announcement: subquadratic multivalued asynchronous byzantine agreement WHP. In: Oshman, R. (ed.) 37th International Symposium on Distributed Computing, DISC 2023, L'Aquila, Italy, 10–12 October 2023. LIPIcs, vol. 281, pp. 39:1–39:6. Schloss Dagstuhl - Leibniz-Zentrum für Informatik (2023)
18. Correia, M., Neves, N.F., Veríssimo, P.: How to tolerate half less one byzantine nodes in practical distributed systems. In: 23rd International Symposium on Reliable Distributed Systems (SRDS 2004), Florianpolis, Brazil, 18–20 October 2004, pp. 174–183. IEEE Computer Society (2004). https://doi.org/10.1109/RELDIS.2004.1353018
19. Correia, M., Neves, N.F., Veríssimo, P.: BFT-TO: intrusion tolerance with less replicas. Comput. J. **56**(6), 693–715 (2013). https://doi.org/10.1093/comjnl/bxs148
20. Correia, M., Veronese, G.S., Lung, L.C.: Asynchronous byzantine consensus with 2f+1 processes. In: Shin, S.Y., Ossowski, S., Schumacher, M., Palakal, M.J., Hung, C. (eds.) Proceedings of the 2010 ACM Symposium on Applied Computing (SAC), Sierre, Switzerland, 22–26 March 2010, pp. 475–480. ACM (2010). https://doi.org/10.1145/1774088.1774187
21. Decouchant, J., Kozhaya, D., Rahli, V., Yu, J.: DAMYSUS: streamlined BFT consensus leveraging trusted components. In: EuroSys '22: Seventeenth European Conference on Computer Systems, Rennes, France, 5–8 April 2022, pp. 1–16. ACM (2022). https://doi.org/10.1145/3492321.3519568
22. Distler, T., Cachin, C., Kapitza, R.: Resource-efficient byzantine fault tolerance. IEEE Trans. Comput. **65**(9), 2807–2819 (2016). https://doi.org/10.1109/TC.2015.2495213
23. Fischer, M.J., Lynch, N.A., Paterson, M.: Impossibility of distributed consensus with one faulty process. In: Proceedings of the Second ACM SIGACT-SIGMOD Symposium on Principles of Database Systems, Colony Square Hotel, Atlanta, Georgia, USA, 21–23 March 1983, pp. 1–7. ACM (1983). https://doi.org/10.1145/588058.588060
24. Guerraoui, R., Komatovic, J., Kuznetsov, P., Pignolet, Y.A., Seredinschi, D.A., Tonkikh, A.: Dynamic byzantine reliable broadcast [technical report]. arXiv preprint arXiv:2001.06271 (2020)
25. Gupta, S., Rahnama, S., Pandey, S., Crooks, N., Sadoghi, M.: Dissecting BFT consensus: in trusted components we trust! In: Luna, G.A.D., Querzoni, L., Fedorova, A., Narayanan, D. (eds.) Proceedings of the Eighteenth European Conference on Computer Systems, EuroSys 2023, Rome, Italy, 8–12 May 2023, pp. 521–539. ACM (2023)
26. Kapitza, R., et al.: Cheapbft: resource-efficient byzantine fault tolerance. In: European Conference on Computer Systems, Proceedings of the Seventh EuroSys Conference 2012, EuroSys '12, Bern, Switzerland, 10–13 April 2012, pp. 295–308. ACM (2012). https://doi.org/10.1145/2168836.2168866
27. King, V., Saia, J.: Breaking the $O(n^2)$ bit barrier: scalable byzantine agreement with an adaptive adversary. J. ACM **58**(4), 18:1–18:24 (2011)
28. Lamport, L., Shostak, R.E., Pease, M.C.: The byzantine generals problem. ACM Trans. Program. Lang. Syst. **4**(3), 382–401 (1982). https://doi.org/10.1145/357172.357176
29. Leinweber, M., Hartenstein, H.: Brief announcement: Let it TEE: asynchronous byzantine atomic broadcast with n $\geq$ 2f+1. In: Oshman, R. (ed.) 37th International Symposium on Distributed Computing, DISC 2023, L'Aquila, Italy, 10–12 October 2023. LIPIcs, vol. 281, pp. 43:1–43:7. Schloss Dagstuhl - Leibniz-Zentrum für Informatik (2023). https://doi.org/10.4230/LIPIcs.DISC.2023.43

30. Levin, D., Douceur, J.R., Lorch, J.R., Moscibroda, T.: Trinc: small trusted hardware for large distributed systems. In: Proceedings of the 6th USENIX Symposium on Networked Systems Design and Implementation, NSDI 2009, Boston, MA, USA, 22–24 April 2009, pp. 1–14. USENIX Association (2009). http://www.usenix.org/events/nsdi09/tech/full_papers/levin/levin.pdf
31. Maurer, A., Tixeuil, S.: Self-stabilizing byzantine broadcast. In: 2014 IEEE 33rd International Symposium on Reliable Distributed Systems, pp. 152–160. IEEE (2014)
32. Micali, S., Rabin, M.O., Vadhan, S.P.: Verifiable random functions. In: 40th Annual Symposium on Foundations of Computer Science, FOCS '99, New York, NY, USA, 17–18 October 1999, pp. 120–130. IEEE Computer Society (1999). https://doi.org/10.1109/SFFCS.1999.814584
33. Micali, S., Vadhan, S., Rabin, M.: Verifiable random functions. In: Proceedings of the 40th Annual Symposium on Foundations of Computer Science, FOCS '99, p. 120. IEEE Computer Society (1999). https://doi.org/10.1109/SFFCS.1999.814584
34. Mostéfaoui, A., Moumen, H., Raynal, M.: Signature-free asynchronous binary byzantine consensus with t ¡ n/3, o(n2) messages, and O(1) expected time. J. ACM **62**(4), 31:1–31:21 (2015). https://doi.org/10.1145/2785953
35. Mostefaoui, A., Perrin, M., Weibel, J.: Randomized consensus: Common coins are not the holy grail! Technical report, LS2N-Nantes Université (2024)
36. Raynal, M.: Fault-Tolerant Message-Passing Distributed Systems: An Algorithmic Approach. Springer, Heidelberg (2018)
37. Veronese, G.S., Correia, M., Bessani, A.N., Lung, L.C., Veríssimo, P.: Efficient byzantine fault-tolerance. IEEE Trans. Comput. **62**(1), 16–30 (2013). https://doi.org/10.1109/TC.2011.221
38. Wattenhofer, R.: Distributed systems. Lecture notes (2022). https://disco.ethz.ch/courses/hs21/distsys/lnotes/DistSys_Script.pdf
39. Yandamuri, S., Abraham, I., Nayak, K., Reiter, M.K.: Communication-efficient BFT protocols using small trusted hardware to tolerate minority corruption. IACR Cryptol. ePrint Arch., 184 (2021). https://eprint.iacr.org/2021/184

Sublinear-Time Collision Detection with a Polynomial Number of States in Population Protocols

Takumi Araya and Yuichi Sudo(✉)

Hosei University, Tokyo, Japan
sudo@hosei.ac.jp

Abstract. This paper addresses the collision detection problem in population protocols. The network consists of state machines called agents. At each time step, exactly one pair of agents is chosen uniformly at random to have an interaction, changing the states of the two agents. The collision detection problem involves each agent starting with an input integer between 1 and n, where n is the number of agents, and requires those agents to determine whether there are any duplicate input values among all agents. Specifically, the goal is for all agents to output false if all input values are distinct, and true otherwise.

In this paper, we present an algorithm that requires a polynomial number of states per agent and solves the collision detection problem with probability one in sub-linear parallel time, both with high probability and in expectation. To the best of our knowledge, this algorithm is the first to solve the collision detection problem using a polynomial number of states within sublinear parallel time, affirmatively answering the question raised by Burman, Chen, Chen, Doty, Nowak, Severson, and Xu [PODC 2021] for the first time.

Keywords: population protocols · collision detection · ranking

1 Introduction

In this paper, we explore the *population protocol* model, introduced in 2004 and studied extensively since then [2,3,5,6,10,21,22,24,28,32]. The model consists of a network, or *population*, of n state-machines, referred to as *agents*. At each time step, a pair of agents is selected uniformly at random to engage in an *interaction* (i.e., pairwise communication), during which they update their states. Agents are anonymous, i.e., they lack unique identifiers. In the population protocol model, time complexity is commonly measured in *parallel time*, defined as the number of time steps divided by the population size n (the number of agents). In practice, interactions often occur simultaneously across the population; parallel time approximately captures this behavior. Throughout the remainder of this section (i.e., Sect. 1), we assume parallel time when discussing time complexity.

U. Schmid and R. Kuznets (Eds.): SIROCCO 2025, LNCS 15671, pp. 39–55, 2025.
https://doi.org/10.1007/978-3-031-91736-3_3

In this paper, we study the collision detection problem in the population protocol model. Before presenting our main results, we briefly explain prior studies on problems closely related to collision detection, namely self-stabilizing leader election and self-stabilizing ranking.

Leader election has been extensively studied in the population protocol model. Leader election can be solved by a simple two-state protocol [5], where initially, all agents are leaders. The protocol employs only one transition rule: when two leaders meet, one of them becomes a follower (i.e., a non-leader). This simple protocol elects a unique leader in $O(n)$ parallel time. Furthermore, this protocol is time-optimal: Doty and Soloveichik [19] demonstrated that any constant-space protocol requires linear time to elect a unique leader. Later, in 2015, Alistarh and Gelashvili [3] developed a leader election protocol that converges in $O(\log^3 n)$ parallel time and uses $O(\log^3 n)$ states per agent. Subsequently, numerous papers have focused on fast leader election, including [2,10,21,22,28]. Gąsieniec, Staehowiak, and Uznanski [22] developed an algorithm that converges in $O(\log n \log\log n)$ time and uses a surprisingly small number of states: only $O(\log\log n)$ per agent. This is considered space-optimal because it is established that every leader election protocol requiring $O(n/\mathrm{polylog}(n))$ time also requires $\Omega(\log\log n)$ states [1]. Sudo et al. [28] presented a simple protocol that elects a unique leader within $O(\log n)$ time and utilizes $O(\log n)$ states per agent. This is time-optimal, as any leader election protocol requires $\Omega(\log n)$ time, even if it uses an arbitrarily large number of states and the agents know the exact size of the population [25]. (This lower bound may appear obvious, yet it does not directly follow from a simple coupon collector argument because we can specify an initial configuration where all agents are followers.) Finally, in 2020, Berenbrink et al. [10] provided a time and space-optimal protocol, i.e., an $O(\log n)$-time and $O(\log\log n)$-states leader election protocol.

Self-stabilizing leader election (SS-LE) has garnered significant attention within this model. This variant of leader election stipulates that (i) starting from any configuration, the population must reach a safe configuration where exactly one leader exists; and (ii) once a safe configuration is reached, the unique leader must be maintained indefinitely. These conditions ensure tolerance against finitely many transient faults, which is critical since many protocols (both self-stabilizing and non-self-stabilizing) assume the presence of a unique leader. Consequently, SS-LE is essential for enhancing the fault tolerance of the population protocol model itself. However, no protocol can solve SS-LE unless each agent in the population knows the *exact size* n of the population [13,20][1].

Numerous studies have focused on overcoming this impossibility by employing various strategies, including assuming oracles [8,14,20], assuming that agents precisely know the population size n [12,13], restricting the topology [4,16,17,34,35], or slightly relaxing the requirements of SS-LE [23,24,26,27,29–33]. Among these, all algorithms that adopt the n-knowledge approach [12,13]

[1] Strictly speaking, the cited works proves a slightly weaker impossibility. Nevertheless, this impossibility can be proved using a similar technique: a simple partitioning argument. See [32] for details (page 618, footnote).

elect the unique leader by solving a more general problem, called *self-stabilizing ranking*. This problem stipulates that: (i) each agent v maintains an output variable $v.\texttt{rank} \in \{1, 2, \ldots, n\}$, (ii) starting from any configuration, the population must reach a safe configuration where no two agents share the same rank value; and (iii) once a safe configuration is reached, no agent updates its rank. SS-LE can be straightforwardly reduced to self-stabilizing ranking because, once ranking is achieved, there is exactly one agent with $\texttt{rank} = 1$. This agent can thus be regarded as the unique leader. Cai, Izumi, and Wada [13] present an algorithm that solves the self-stabilizing ranking problem within $O(n^2)$ parallel time, using n states per agent. In contrast, Burman, Chen, Chen, Doty, Nowak, Severson, and Xu [12] introduce significantly faster algorithms that, however, require more states. Specifically, they offer an $O(n)$ parallel time algorithm with $O(n)$ states per agent and an $O(\log n)$ parallel time algorithm with a super-exponential number of states. This leads to a natural question: Can self-stabilizing ranking be solved in sublinear parallel time using only a polynomial number of states per agent? To solve self-stabilizing ranking, agents must detect whether any distinct agents share the same rank. Consequently, Burman et al. raise an open question: Is there a sublinear parallel time algorithm with $poly(n)$ states that can solve the following problem, which we refer to as the *collision detection problem* in this paper (See Sect. 2.3 for the formal definition of this problem):

> Each agent v is given an input $\texttt{rank} \in \{1, 2, \ldots, n\}$. The goal for each agent is to decide whether at least one pair of agents have the same input value in rank.

Our Contribution

In this paper, we affirmatively answer the open question raised by Burman et al. [12]. Specifically, we introduce a collision detection algorithm that stabilizes within $O(\sqrt{n \log n})$ parallel time in expectation and $O(n^{1/2} \cdot \log^{3/2} n)$ parallel time with high probability. This algorithm uses $\tilde{O}(n)$ states per agent with high probability, excluding the input variable rank, which requires $O(n)$ states. The proposed algorithm is always correct; that is, it eventually reaches a stable configuration where all agents output the correct answer with probability 1.

It remains an open question whether self-stabilizing ranking, or weaker variants such as loosely-stabilizing ranking, can be solved within sublinear parallel time using a polynomial number of states per agent. This question persists primarily because our collision detection protocol is not self-stabilizing.

Organization of this Paper

Section 2 introduces the preliminaries, including key terminologies and the definition of the model. Section 3 describes the basic submodules that are instrumental in designing the proposed algorithm CollisionDetection. Section 4 presents CollisionDetection, proves its correctness, and bounds its time and space complexities.

In the remainder of this paper, we will not use parallel time; instead, we will discuss stabilization time in terms of the number of time steps (or interactions).

2 Preliminaries

Throughout this paper, we denote the set of *non-negative* integers by $\mathbb{N}$ and represent the set $\{z \in \mathbb{N} \mid x \leq z \leq y\}$ by $[x, y]$. When the base of a logarithm is omitted, it is assumed to be 2. For any positive integer i, we denote the i-th projection map by π_i, i.e., for any element $\mathbf{x} = (x_1, x_2, \dots, x_k) \in X_1 \times X_2 \times \dots \times X_k$, we define $\pi_i(\mathbf{x}) = x_i$. The $\tilde{O}$-notation hides a poly-logarithmic factor; that is, for any function $f(x)$, $\tilde{O}(f(x))$ denotes $O(f(x) \cdot \log^c x)$ for some constant c.

2.1 Model

A *population* is a network consisting of *agents*. We denote the set of all the agents by V and let $n = |V|$. We assume that a population is a complete graph, thus every pair of agents (u, v) can interact, where u serves as the *initiator* and v serves as the *responder* of the interaction.

A *protocol* $P(Q, s_{\text{init}}, X, Y, T, f_{\text{out}})$ consists of a set Q of states, the initial state s_{init}, a set X of input symbols, a set Y of output symbols, a transition function $T : (Q \times X)^2 \to Q^2$, and an output function $f_{\text{out}} : Q \times X \to Y$. Initially, all agents are in state $s_{\text{init}} \in Q$. When two agents interact, T determines their next states according to their current states and inputs. The *output* of an agent is determined by f_{out}: the output of an agent in state $q \in Q$ and input $x \in X$ is $f_{\text{out}}(q, x)$. A protocol may be given design parameters, such as a lower bound n_L and an upper bound n_U on the population size n. In this case, each component of the protocol, namely Q, s_{init}, X, Y, T, and f_{out}, may depend on these parameters.

A *configuration* is a mapping $C : V \to Q \times X$ that specifies the states and the inputs of all agents in the population. We say that a configuration C *changes* to another configuration C' via an interaction $e = (u, v)$, denoted by $C \stackrel{P,e}{\to} C'$, if $(\pi_1(C'(u)), \pi_1(C'(v))) = T(C(u), C(v))$, $\pi_2(C'(u)) = \pi_2(C(u))$, $\pi_2(C'(v)) = \pi_2(C(v))$, and $C'(w) = C(w)$ for all $w \in V \setminus \{u, v\}$.[2] We say that a configuration C' is reachable from a configuration C if there is a sequence of configurations $C = C_0, C_1, \dots, C_k$ such that C_i changes to C_{i+1} via some interaction for each $i \in [0, k-1]$. A configuration C is *stable* if, for every configuration C' reachable from C, the output of each agent remains the same in both C and C', i.e., $f_{\text{out}}(C(v)) = f_{\text{out}}(C'(v))$.

We assume the *uniformly random scheduler* $\mathbf{\Gamma}$, which selects two agents to interact at each step uniformly at random from all ordered pairs of agents. Specifically, $\mathbf{\Gamma} = \Gamma_0, \Gamma_1, \dots$ where each Γ_t is a random variable such that $\Pr(\Gamma_t = (u, v)) = \frac{1}{n(n-1)}$ for any $t \geq 0$ and any distinct $u, v \in V$. The *execution* of protocol P starting from a configuration C_0 under the uniformly random scheduler $\mathbf{\Gamma}$ is defined as the sequence of configurations $\Xi_P(C_0, \mathbf{\Gamma}) = C_0, C_1, \dots$ such that $C_t \stackrel{P,\Gamma_t}{\to} C_{t+1}$ holds for all $t \geq 0$. Note that each C_i is also a random variable.

[2] In this paper, we assume that the inputs provided to the agents do not change.

In this paper, we use the notation $v.\mathtt{var}$ to denote the value of a variable var maintained by an agent v. By abuse of notation, we use $C(v).\mathtt{var}$ to represent the value of var in a configuration C.

2.2 With High Probability

In this paper, we frequently use the term "with high probability," abbreviated as w.h.p. Various definitions for this term exist, such as with probability $1 - n^{-1}$ or with probability $1 - n^{\Omega(1)}$. Following [9], we adopt the below definition.

Definition 1 (with high probability). *A property of a protocol P holds* with high probability *if, for any constant $\eta = \Theta(1)$, we can ensure the property is satisfied with a probability of $1 - O(n^{-\eta})$ by adjusting the constant parameters of P, if necessary.*

For example, if a protocol P_τ with a design parameter $\tau = \Theta(1)$ elects a leader in $\tau \cdot n \log n = O(n \log n)$ time steps with probability $1 - n^{-\tau}$, we say that P_τ elects a leader in $O(n \log n)$ time steps w.h.p. In this case, we can also explicitly state the constant factor of the running time and say that P_τ elects a leader in $\tau \cdot n \log n$ time steps w.h.p. Note that the requirement for the success probability (i.e., $1 - O(n^{-\eta})$) is defined asymptotically; therefore, we only need to consider sufficiently large population sizes n.

2.3 Collision Detection Problem

The collision detection problem requires a protocol $P = (Q, s_{\mathrm{init}}, X, Y, T, f_{\mathrm{out}})$ to accept the input symbol $[1, n]$, i.e., $[1, n] \subseteq X$, where $n = |V|$ is the population size. Each agent v is assigned an input symbol in $[1, n]$, denoted by $v.\mathtt{rank}$. A configuration $C : V \to Q \times X$ is termed *initialized* if, for every $v \in V$, v is in the initial state s_{init} and its input $v.\mathtt{rank}$ is within $[1, n]$ in C. More formally, C is initialized if $\pi_1(C(v)) = s_{\mathrm{init}}$ and $\pi_2(C(v)) \in [1, n]$ for all $v \in V$. We define $\mathcal{C}_{\mathrm{init}}(P)$ as the set of these initialized configurations. The goal of this problem is to detect whether or not a *collision* exists, i.e., $u.\mathtt{rank} = v.\mathtt{rank}$ for some distinct agents $u, v \in V$. Each agent must output either 1 or 0, representing "there is at least one collision among the ranks of the agents" and "there is no such collision," respectively. For simplicity, we assume every protocol P includes a variable $\mathtt{collision} \in \{0, 1\}$, and an agent v outputs 1 if and only if $v.\mathtt{collision} = 1$. That is, $f_{\mathrm{out}}(C(v)) = v.\mathtt{collision}$ for any configuration C and any agent $v \in V$.

Definition 2 (Collision Detection Problem). *Let $P = (Q, s_{\mathrm{init}}, X, Y, T, f_{\mathrm{out}})$ be a protocol, where $[1, n] \subseteq X$. A protocol P* solves *the collision detection problem if for any initialized configuration $C_0 \in \mathcal{C}_{\mathrm{init}}(P)$, both of the following conditions are satisfied:*

- *For any stable configuration C reachable from C_0, $C(v).\mathtt{collision} = 1$ for all $v \in V$ if $C_0(u).\mathtt{rank} = C_0(w).\mathtt{rank}$ for some distinct agents $u, w \in V$; otherwise, $C(v).\mathtt{collision} = 0$ holds for all $v \in V$.*

- *The execution $\Xi_P(C_0, \mathbf{\Gamma})$ eventually reaches a stable configuration with probability 1.*

For any execution $\Xi = C_0, C_1, \ldots$, we define the *stabilization time* of Ξ as the minimum t such that C_t is a stable configuration. We say that a protocol P *stabilizes* within t time steps with high probability (respectively, in expectation) if, for any initialized configuration $C_0 \in \mathcal{C}_{\mathrm{init}}(P)$, the stabilization time of $\Xi_P(C_0, \mathbf{\Gamma})$ is at most t with high probability (respectively, in expectation).

2.4 Uniform Protocols

A protocol is *uniform* if it does not depend on the population size at all, that is, if it does use any knowledge on the population size. In the field of population protocols, originally, the number of states of a protocol $\mathcal{P} = (Q, \rho, Y, \pi, \delta)$ is simply defined as $|Q|$. Thus, the number of states for uniform protocols must be either $\Theta(1)$ or infinite. Some problems inherently require memory space of a non-constant size.[3] This means that uniform protocols devoted for those problems always require an infinite number of states by definition, while each execution of these protocols may use only small number of states depending on population size. (A uniform protocol must be defined independently from the population size, but of course, its execution may depend on the population size.) Thus, we require another way to evaluate the number of states for uniform protocols as several works do [9,11,15,18]. This paper adopts the following simple definition of *the number of states* of a protocol $P = (Q, s_{\mathrm{init}}, X, Y, T, f_{\mathrm{out}})$.

- An agent maintains a constant number of variables $x_1, x_2, \ldots, x_s$, and the combination of their values constitutes the state of the agent. The first element Q of protocol $P = (Q, s_{\mathrm{init}}, X, Y, T, f_{\mathrm{out}})$ can be regarded as the set of all such states. The domain of each variable may not be bounded, thus Q may be an infinite set.
- For any agent $v \in V$, define v's amount of information at time step t as $f(v,t) = \sum_{i=1}^{s} \iota_i$, where each ι_i is the number of bits required to encode the value of variable x_i. The number of states of an execution Ξ is defined as $\#(\Xi) = \max\{2^{f(v,t)} \mid v \in V, t = 0, 1, \ldots\}$ in the execution. We say that the number of states of a protocol P is at most z with high probability (respectively, in expectation) if for any initialized configuration $C_0 \in \mathcal{C}_{\mathrm{init}}(P)$, $\#(\Xi_P(C_0, \mathbf{\Gamma})) \le z$ holds with high probability (respectively, in expectation).

3 Tools

3.1 One-Way Epidemic

The one-way epidemic protocol was introduced and analyzed by Angluin, Aspnes, and Eisenstat [6] and has been widely used thereafter. The goal of this protocol is to propagate the maximum value of a given variable `var` to all agents.

[3] For example, Doty and Soloveichik [19] proved that $\omega(1)$ states are necessary to solve the leader election problem within $o(n^2)$ expected interactions under the uniformly random scheduler.

Algorithm 1: EPIDEMIC(var) at an interaction where a and b are an initiator and a responder, respectively.

$b.\texttt{var} \leftarrow \max(a.\texttt{var}, b.\texttt{var})$

Algorithm 2: PHASECLOCK(F) at an interaction where a and b are an initiator and a responder, respectively.

Variables: $\texttt{timer} \in [0, m-1], \texttt{epoch} \in [0, F], \texttt{leader} \in \{0,1\}$
Note: m is a sufficiently large constant. (See Lemma 2.)
Notation: $\max_m(x,y) = \begin{cases} x & x \in [0, y - \lceil m/2 \rceil] \cup [y+1, y + \lfloor m/2 \rfloor] \\ y & \textbf{otherwise} \end{cases}$

```
if b.leader = 1 ∧ a.timer = b.timer then
    b.timer ← (b.timer + 1) mod m
    if b.timer = 0 then
        b.epoch = min(b.epoch + 1, F)
else if b.leader = 0 then
    b.timer ← max_m(a.timer, b.timer)
Execute EPIDEMIC(epoch)
```

The strategy is simple: when an initiator a and a responder b meet, and if $a.\texttt{var} > b.\texttt{var}$ holds, $b.\texttt{var}$ is updated to $a.\texttt{var}$. (See Algorithm 1.) Angluin et al. [6] prove the following useful lemma, which guarantees that the maximum value is propagated to the entire population in $O(n \log n)$ steps w.h.p.

Lemma 1. *Suppose that an execution of the protocol* EPIDEMIC(var) *starts from a configuration C where $M = \max_{v \in V} v.\texttt{var}$. Then, for any fixed $\eta > 0$, there exists a constant d such that for sufficiently large population size n, with probability $1 - n^{-\eta}$, every agent $v \in V$ satisfies $v.\texttt{var} \geq M$ within $dn \log n$ time steps.*

3.2 Phase Clock with a Leader

Together with the epidemic protocol, Angluin et al. [6] introduce the phase clock protocol. This protocol requires the presence of a unique leader. Specifically, it assumes that each agent maintains a variable $\texttt{leader} \in \{0,1\}$, and there exists an agent $a_L \in V$ such that $a_L.\texttt{leader} = 1$ and $b.\texttt{leader} = 0$ for all $b \in V \setminus \{a_L\}$ consistently. The goal of this protocol is allowing the leader a_L to determine whether $\Theta(n \log n)$ time steps has passed or not. In this paper, we modify this protocol slightly so that all agents are synchronized with a variable $\texttt{epoch} \in [0, F]$, where F is a given (possibly non-constant) integer.

In addition to epoch, this protocol maintains a variable $\texttt{timer} \in [0, m-1]$, where m is a constant parameter. Initially, $a_L.\texttt{timer} = 0$ and $v.\texttt{timer} = m-1$ for all $v \in V \setminus \{a_L\}$, while the epoch value of all agents are 0. In this protocol, only the responder updates its variables at each interaction. Lines 2–8 in Algorithm 2 specify how the responder updates `timer` and `epoch`. The leader a_L increments its `timer` by one modulo m when it encounters an initiator with the same timer value. The timer of a non-leader b is overwritten when it encounters an initiator with a `timer` value from $b.\mathtt{timer}+1$ to $b.\mathtt{timer}+\lfloor m/2 \rfloor$ modulo m, i.e., a `timer` value in $[0, y - \lceil m/2 \rceil] \cup [y+1, y+\lfloor m/2 \rfloor]$, where $y = b.\mathtt{timer}$. The leader a_L increments its `epoch` by one when it increments its `timer` from $m-1$ to 0 unless $a_L.\mathtt{epoch}$ has already reached the given maximum value F. The increased `epoch` value is propagated to all agents via the epidemic protocol.

The following lemma directly follows from Lemma 1 and the analysis of the phase clock protocol presented by Angluin et al. [6].

Lemma 2. *Let C_0 be a configuration with exactly one leader a_L, where the variables of all agents are initialized as specified above. Let $\Xi_m = \Xi_{\text{PhaseClock}(F)}(C_0, \mathbf{\Gamma})$, where* PhaseClock *is the protocol specified in Algorithm 2 with the parameter m. For any fixed constants $c, d_1, \eta \geq 0$, there exist constants m and d_2 such that, for all sufficiently large population sizes n, the execution Ξ_m satisfies the following properties with probability $1 - n^{-\eta}$:*

- *For any $i \in [0, \min(n^c, F)]$, there are at least $d_1 \log n$ time steps where $v.\mathtt{epoch} = i$ holds for all agents v, and*
- *For any $i \in [0, \min(n^c, F) - 1]$, each agent v maintains $v.\mathtt{epoch} = i$ for at most $d_2 n \log n$ steps.*

By this lemma, as long as $F = poly(n)$, we can ensure that with high probability, all agents simultaneously experience each of the epochs $0, 1, \ldots, F$ during an arbitrarily large $\Omega(n \log n)$ time steps with a controllable constant factor d_1, while each epoch $0, 1, \ldots, F-1$ finishes in $O(n \log n)$ time steps.

4 Collision Detection

In this section, we present a collision detection protocol CollisionDetection, which stabilizes within $O(n^{3/2} \cdot \sqrt{\log n})$ time steps (respectively, $O(n^{3/2} \cdot \log^{3/2} n)$ time steps) in expectation (respectively, with high probability) and uses $\tilde{O}(n)$ states per agent with high probability, excluding the input variable `rank`, which requires $O(n)$ states. This protocol is always correct, i.e., it eventually reaches a stable configuration where all agents output the correct answer with probabilit 1.

In Sect. 4.1, we present a collision detection protocol under the assumption that (i) there is a pre-elected leader, and (ii) all agents are aware of a common lower bound and a common upper bound on n, both of which are asymptotically tight. In Sect. 4.2, we eliminate these assumptions.

4.1 Collision Detection with a Leader and Rough Knowledge of n

In this section, we present a protocol CollisionDetectionWithBounds(n_L, n_U), abbreviated as CDWB(n_L, n_U). This protocol requires that

Algorithm 3: CDWB(n_L, n_U) at an interaction where a and b are an initiator and a responder, respectively.

9 **if** a.rank $= b$.rank **then**
10   b.collision $\leftarrow 1$
11 Let $r = \lceil 3\eta \log n_U \rceil$, $\ell = \lceil \sqrt{n_L \log n_L} \rceil$, and $z = \lceil n_U / \ell \rceil$
12 Execute PhaseClock($r \cdot z + 1$)
13 **if** b.epoch has increased in PhaseClock **then**
14   Let $k = (b.\texttt{epoch} - 1) \bmod z$
15   **if** b.rank $\in [k \cdot \ell + 1, (k+1) \cdot \ell]$ **then**
16   Choose χ uniformly at random from $\{0, 1\}$
17   b.gid $\leftarrow (b.\texttt{rank} - k \cdot \ell, \chi)$
18   b.infectivity $\leftarrow \lfloor \log n_L - \log \ell \rfloor - 2$
19   **else**
20   b.gid $\leftarrow \perp$
21   b.infectivity $\leftarrow 0$
22 **if** a.epoch $= b$.epoch > 0 **then**
23   **if** a.gid $\neq \perp \land a$.infectivity $> 0 \land b$.gid $= \perp$ **then**
24   b.gid $\leftarrow a$.gid
25   a.infectivity $\leftarrow b$.infectivity $\leftarrow a$.infectivity $- 1$
26   **if** $\exists x, y, z : a.\texttt{gid} = (x, y) \land b.\texttt{gid} = (x, z) \land y \neq z$ **then**
27   b.collision $\leftarrow 1$
28 Execute Epidemic(collision)

there is always a unique leader $a_L \in V$ and that the two arguments n_L and n_U are asymptotically tight lower and upper bounds on n. Specifically, $c_1 n \le n_L \le n \le n_U \le c_2 n$ must hold for some constants c_1 and c_2. We assume that the agents are aware of these constants c_1 and c_2, allowing us to incorporate them into the design of this protocol.

The pseudocode of this protocol is presented in Algorithm 3. The algorithm consists of three parts. The first part, referred to as *the backup protocol*, detects collisions through direct interactions (Lines 9–10): when two agents with the same rank meet, the responder b raises a collision flag, i.e., b.collision is set to 1. If only one pair of agents u, v $\in V$ shares a common rank, it can be easily observed that this process requires $\binom{n}{2} = \Theta(n^2)$ time steps in expectation. The second part detects collisions much more quickly, requiring $O(n^{3/2}\sqrt{\log n})$ time steps in expectation and $O(n^{3/2} \log^{3/2} n)$ time steps with high probability, using a square root decomposition technique (Lines 11–27). Both the first and second parts allow *some* agent to raise a collision flag if a collision is detected. The third part propagates the raised flag to the entire population via the epidemic protocol (Lines 28).

Although the backup protocol is much slower, it is still necessary because the second part may fail to detect a collision with non-zero probability. More-

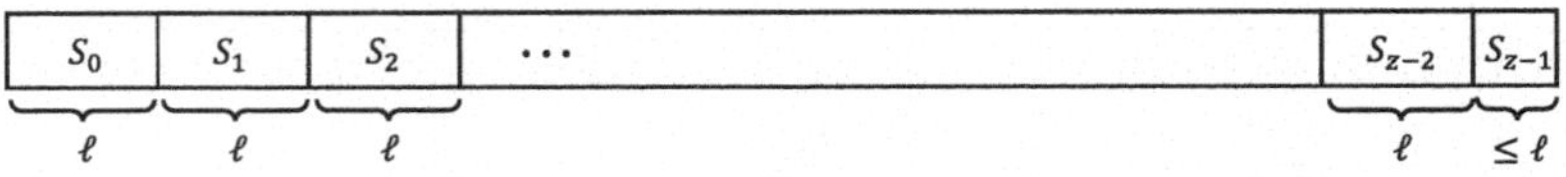

Fig. 1. Segments of the second part of CDWB(n_L, n_U)

over, the backup protocol assists the second part in quickly detecting collisions. Specifically, we present the following lemma:

Lemma 3. *Let C_0 be a configuration where there are $x = \Omega(\sqrt{n \log n})$ pairs of colliding agents, i.e., $|\{(u, v) \in V^2 \mid u \neq v, u.\mathtt{rank} = v.\mathtt{rank}\}| = \Omega(\sqrt{n \log n})$ holds in C_0. Then, in an execution of the backup protocol starting from C_0, at least one agent raises a collision flag within $O(n^{3/2})$ time steps, both in expectation and with high probability.*

Proof. At each time step, one of the x colliding pairs is selected to interact with a probability of $x/\binom{n}{2} = \Omega(\log n/n^{3/2})$, resulting in the responder of the pair raising a collision flag. Therefore, a collision flag is raised within $O(n^{3/2}/\log n)$ steps in expectation, and within $O(n^{3/2})$ steps with high probability. □

This lemma implies that, regardless of the performance of the second part, the proposed protocol solves the collision detection problem within $O(n^{3/2})$ time steps, both in expectation and with high probability, if the number of colliding pairs is $\Omega(\sqrt{n \log n})$. Therefore, for the remainder of this section, we will assume that the number of colliding pairs is $o(\sqrt{n \log n})$.

Table 1. Notations used in the second part of CDWB(n_L, n_U)

Notation	Definition	Explanation
ℓ	$\lceil \sqrt{n_L \log n_L} \rceil$	Length of a segment
z	$\lceil n_U/\ell \rceil$	Number of segments
S_k	$[k \cdot \ell + 1, (k+1) \cdot \ell] \cap [1, n_U]$	k-th segment ($0 \leq k < z$)
r	$\lceil 3\eta \cdot \log n_U \rceil$	Number of epochs dedicated to each segment

We will now explain how the agents behave in the second part, which is the main focus of the proposed protocol. Utilizing the unique leader assumption, the agents selected for each interaction execute the phase clock protocol presented in Sect. 3 (Line 12). Let $\ell = \lceil \sqrt{n_L \log n_L} \rceil$ and $z = \lceil n_U/\ell \rceil$. In addition, let $r = \lceil 3\eta \cdot \log n_U \rceil$, where η is a hidden design parameter ensuring that the targeted time complexity, i.e., $O(n^{3/2} \cdot \log^{3/2} n)$ time steps, are achieved with probability $1 - O(n^{-\eta})$. The argument F of the phase clock protocol is set to $r \cdot z + 1$. By Lemma 2, it is guaranteed with high probability that for each $i = 1, 2, \ldots, r \cdot z$, all agents are simultaneously in epoch i for at least $d_1 n \log n$ steps with any sufficiently large constant d_1, and all agents enter the final epoch $r \cdot \ell + 1$ within

$\Theta(rzn \log n) = \Theta(n^{3/2} \log^{3/2} n)$ steps. We divide the set of integers $[1, n_U]$ into z segments with at most size ℓ, i.e., for any $k = 0, 1, 2, \ldots, z-1$, the k-th segment is $S_k = [k \cdot \ell + 1, (k+1) \cdot \ell] \cap [1, n_U]$ (Fig. 1). We assign r epochs $k-1, k-1+z, k-1+2z, \ldots, k-1+(r-1)z$ to detect collisions of ranks in S_k. Conversely, each epoch $i \in [1, r \cdot z]$ is dedicated to collision detection in $S_{(i-1) \bmod z}$. The notations introduced here are summarized in Table 1.

The goal of epoch $i \in [1, r \cdot z]$ is to detect whether or not there exists $x \in S_{(i-1) \bmod z}$ such that $|\{v \in V \mid v.\mathtt{rank} = x\}| \geq 2$, in $O(n \log n)$ time steps with probability at least 1/3 for a sufficiently large n, while ensuring that no agent mistakenly detect a collision if no such x exists. By the achievement of this goal, all r epochs devoted to each segment results in the correct decision with probability $1 - (1/3)^r \geq 1 - n^{-\eta}$. To achieve this goal, each agent $v \in V$ maintains a variable $v.\mathtt{gid} \in ([1, \ell] \times \{0, 1\}) \cup \{\bot\}$ and $v.\mathtt{infectivity} \in [0, \lfloor \log n_L - \log \ell \rfloor - 2]$. Suppose that an agent v enters epoch i now. If $v.\mathtt{rank} \in S_k$, where $k = (i-1) \bmod z$, it generates a group identifier $(v.\mathtt{rank} - k \cdot \ell, \chi)$, where χ is a random number (or *nonce*) chosen uniformly at random in $[0, \lceil {n_U}^{\eta/r} \rceil - 1]$, and stores it in $v.\mathtt{gid}$ (Lines 15–18). In addition, $v.\mathtt{infectivity}$ is reset to the maximum value $\lfloor \log n_L - \log \ell \rfloor - 2$. Otherwise, $v.\mathtt{gid}$ and $v.\mathtt{infectivity}$ are set to $\bot$ and 0, respectively (Lines 19–21). We say that an agent with $\mathtt{gid} = \bot$ a *null* agent. A non-null agent with $\mathtt{infectivity} = x$ has ability to make 2^x copies of non-null agent: whenever a non-null agent a with $a.\mathtt{infectivity} > 0$ and a null agent b have an interaction as the initiator and the responder, respectively, $a.\mathtt{gid}$ is copied to $b.\mathtt{gid}$ and both $a.\mathtt{infectivity}$ and $b.\mathtt{infectivity}$ are set to $a.\mathtt{infectivity} - 1$ (Lines 23–25). If two agents have an interaction at which their group identifier share a common rank, but have different nonce, they notice the existence of a collision. Then, the responder of the interaction raises a collision flag (Line 27).

The proliferation of group identifiers resembles the epidemic protocol [6]. However, unlike the epidemic protocol, this process limits the maximum number of non-null agents. This constraint enhances the speed of proliferation, ensuring that proliferation completes in every epoch with high probability. This is formalized in the following lemma:

Lemma 4. *Let* $P = \mathrm{CDWB}(n_L, n_U)$, $i \in [1, r \cdot z]$, *and* $k = (i-1) \bmod z$. *Let* $C_0 \in \mathcal{C}_{\mathrm{init}}(P)$ *be any initialized configuration where there are at most* $2 \cdot \ell$ *agents whose rank is in* S_k, *and let* $\Xi_P(C_0, \mathbf{\Gamma}) = C_0, C_1, \ldots$. *Let* t_b *be the minimum integer such that all agents are in epoch* k *in the* t_b*-th configuration* C_{t_b}. *Then, with high probability, there exists a non-negative integer* $t_{\mathrm{done}} = t_b + O(n \log n)$ *such that for all* $v \in V$, $v.\mathtt{epoch} = i$ *and* $v.\mathtt{infectivity} = 0$ *in* $C_{t_{\mathrm{done}}}$.

Proof. Let t_e be the maximum integer such that all agents are in epoch i in configuration C_{t_e}, and let V_k be the set of agents whose rank is in S_k in configuration C_0. Since $|V_k| \leq 2\ell$, during epoch i, i.e., in configurations $C_{t_b}, C_{t_b+1}, \ldots, C_{t_e}$, there are always at most $2\ell \cdot 2^{\lfloor \log n_L - \log \ell \rfloor - 2} \leq n/2$ non-null agents, thus there are at least $n/2$ null agents.

We will prove the lemma using the concept of *virtual agents* introduced by [27] with slight modifications. Let $U(t)$ be the set of null agents at time step t.

From the above discussion, $U(t) \geq n/2$ holds for all $t \in [t_b, t_e]$. Define $\iota(v) = C(v).\mathtt{infectivity}$ for any $v \in V \setminus U(t_b)$. If proliferation completes correctly in epoch i, each $v \in V \setminus U(t_b)$ must have generated $2^{\iota(v)}$ non-null agents, including itself. In this proof, we map those agents to distinct integers from 0 to $2^{\iota(v)} - 1$. For any $v \in V \setminus U(t_b)$, $\alpha \geq 0$, and $j \geq 1$, we denote the j-th bit from the end of α by $b(v, \alpha, j)$, i.e., $b(v, \alpha, j) = \lfloor \alpha / 2^{j-1} \rfloor \bmod 2$. By definition, $b(v, \alpha, j) = 0$ for any $j \geq \iota(v) + 1$. Then, for any $t \geq t_b$, we define the *virtual agent* $\nu(v, \alpha, t)$ and its number of divisions $\kappa(v, \alpha, t)$ as follows: $\nu(v, \alpha, t_b) = v$, $\kappa(v, \alpha, t_b) = 0$, and

$$\nu(v, \alpha, t+1) = \begin{cases} u & \text{if } b(v, \alpha, \kappa(v, \alpha, t) + 1) = 1 \wedge \exists u \in U(t) : \Gamma_t = (\nu(v, \alpha, t), u) \\ \nu(v, \alpha, t) & \text{otherwise,} \end{cases}$$

$$\kappa(v, \alpha, t+1) = \begin{cases} \min(\kappa(v, \alpha, t) + 1, \iota(v)) & \text{if } \exists u \in U(t) : \Gamma_t = (\nu(v, \alpha, t), u) \\ \kappa(v, \alpha, t) & \text{otherwise.} \end{cases}$$

Clearly, proliferation completes before time step t if $\kappa(v, \alpha, t) = \iota(v)$ for all $v \in V \setminus U(t_b)$ and $\alpha \in [0, 2^{\iota(v)} - 1]$. Note that a virtual agent $\nu(v, \alpha, t)$ is determined solely by C_{t_b} and $\Gamma_{t_b}, \Gamma_{t_b+1}, \ldots, \Gamma_{t-1}$. Therefore, at any time step $t \in [t_b, t_e]$, the virtual agent $\nu(v, \alpha, t)$ interacts with a non-null agent as the initiator with a probability of $\frac{1}{n} \cdot \frac{U(t)}{n-1} \geq \frac{1}{2(n-1)}$. Consequently, by the Chernoff Bound, $\kappa(v, \alpha, t) = \iota(v)$ holds with high probability for some $t_{\mathrm{done}} = t_b + O(n \log n)$. By applying the union bound, with high probability, this condition is met for all $v \in V \setminus U(t_b)$ and $\alpha \in [0, 2^{\iota(v)} - 1]$, indicating that the proliferation completes within $O(n \log n)$ steps with high probability. Note that we can disregard the probability that $t_e < t_{\mathrm{done}}$, as we can choose arbitrarily large constant d_1 in Lemma 2. □

Lemma 5. *Let* $P = \mathrm{CDWB}(n_L, n_U)$. *Starting from any initialized configuration* $C_0 \in \mathcal{C}_{\mathrm{init}}(P)$, *execution* $\Xi_P(C_0, \mathbf{\Gamma})$ *solves the collision detection. It stabilizes within* $O(n^{3/2} \cdot \sqrt{\log n})$ *time steps (respectively,* $O(n^{3/2} \cdot \log^{3/2} n)$ *time steps) in expectation (respectively, with high probability). This protocol uses* $\tilde{O}(n)$ *states per agent.*

Proof. We can easily verify that the number of states is $\tilde{O}(n)$ by checking the domain of the variables, excluding the input variable `rank`: `gid` uses $O(\sqrt{n})$ states and `epoch` uses $O(r \cdot z) = \tilde{O}(\sqrt{n})$ states, while all other variables use only a poly-logarithmic number of states.

Proving the correctness is also easy. The first part of the protocol (i.e., the backup protocol) eventually makes an agent raise a flag if and only if there is at least one colliding pair of agents. The raised flag is propagated to the entire population via the epidemic protocol. False-positive errors are not permitted also in the second part of CDWB. In each epoch $i \in [1, r \cdot z]$, the group identifiers are initialized. When no agents share a common rank, all non-null agents v with $\pi_1(v.\mathtt{gid}) = \alpha$ originate from a common agent with a rank $k \cdot \ell + \alpha$, where $k = (i-1) \bmod z$. Therefore, all such agents have the same nonce, thus preventing any of them from raising a collision flag. Thus, starting from any initialized configuration $C_0 \in \mathcal{C}_{\mathrm{init}}(P)$, execution $\Xi_P(C_0, \mathbf{\Gamma})$ eventually reaches a stable configuration with probability 1, where all agents output the correct answer.

Next, we bound the stabilization time of P. From the above discussion, the stabilization time is zero when there is no rank collision in C_0. Thus, assume the presence of a colliding pair $u, v \in V$, i.e., $u.\mathtt{rank} = v.\mathtt{rank}$, in C_0. Let $k \in \{0, 1, \ldots, z-1\}$ be the integer such that $C_0(u).\mathtt{rank} \in S_k$. By Lemma 3, we assume that there are only $o(\sqrt{n \log n})$ colliding pairs of agents without loss of generality. Under this assumption, $|V_k| \le 2\ell$, where V_k is the set of agents whose ranks are in S_k in configuration C_0. Otherwise, there are $\sum_{j=1}^{\ell} \binom{x_j}{2} \ge \sum_{j=1}^{\ell} (x_j - 1) > \ell = \Omega(\sqrt{n \log n})$ colliding pairs, yielding a contradiction, where x_j is the number of agents whose rank is $k \cdot \ell + j$. With probability $1/2$, u and v generate different nonces and thus obtain distinct group identifiers when they enter epoch i. By Lemma 4, with high probability, each of their group identifiers is copied to $2^{\lfloor \log n_L - \log \ell \rfloor - 2} \ge n_L/(8\ell)$ agents, respectively, in the first half period of epoch i, at least $\frac{d_1}{2} n \log n$ time steps, where d_1 is the constant that appears in Lemma 2. Thereafter, during the next $\frac{d_1}{2} \cdot n \log n$ time steps, there is at least one time step at which one agent in u's group and one agent in v's group have an interaction with probability at least

$$1 - \left(1 - \frac{\binom{n_L/8\ell}{2}}{\binom{n}{2}}\right)^{\frac{d_1}{2} \cdot n \log n} = 1 - \left(1 - \Omega\left(\frac{1}{n \log n}\right)\right)^{\frac{d_1}{2} \cdot n \log n} \ge 1 - \epsilon$$

for any constant ϵ because we can choose an arbitrarily large constant for d_1. To conclude, some agent raises a collision flag in epoch i with probability $1/2 - \epsilon \ge 1/3$. There are r such epochs dedicated to S_k, and those epochs appear in every z epochs. Thus, $\Xi(C_0, \mathbf{\Gamma})$ stabilizes within $O(r \cdot z \cdot n \log n) = O(n^{3/2} \log^{3/2} n)$ steps with probability at least $1 - (1 - 1/3)^r \ge 1 - 2^{-r} \ge 1 - n^{-\eta}$, i.e., with high probability. The expected stabilization time is bounded by

$$\sum_{j=1}^{r} \left(1 - \frac{1}{3}\right)^{j-1} \cdot \frac{jz}{3} \cdot O(n \log n) + \left(1 - \frac{1}{3}\right)^{r} \cdot O(n^2) = O(zn \log n) = O\left(n^{\frac{3}{2}} \sqrt{\log n}\right).$$

□

Remark 1 (No false-positive error even without the assumptions). The epoch $v.\mathtt{epoch}$ of each agent $v \in V$ is monotonically non-decreasing. Therefore, no false-positive errors occur in the execution of CDWB(n_L, n_U) even without the previously stated assumption that there is always a unique leader a_L and the parameters satisfy $n_L \le n \le n_U$, $n_L = \Theta(n)$, and $n_U = \Theta(n)$.

Remark 2 (Derandomization). One might consider CDWB a randomized algorithm because it generates a binary nonce at Line 16, seemingly deviating from the model defined in Sect. 2.1. However, we can easily derandomize our protocol by exploiting the randomness from the uniformly random scheduler $\mathbf{\Gamma}$. The derandomization can proceed as follows: (i) When an agent increases its epoch, it enters the *waiting mode* instead of initializing `gid` and `infectivity`; (ii) When an agent in the waiting mode has an interaction, the agent can generate a 1-bit nonce χ, setting $\chi = 0$ if it is the initiator of the interaction, and $\chi = 1$ otherwise., thus it initializes `gid` and `infectivity` with the generated nonce, and returns to the *normal mode*. Let u and v be any two agents with the same rank. Note that the nonce generated by u and v may not be independent, which could potentially complicate the analysis. However, their nonces are dependent only if they have previously met, at which point they would have raised a collision flag via the backup protocol. Therefore, this dependency does not pose a problem.

4.2 Collision Detection Without Assumptions

To run CDWB(n_L, n_U), we require a unique leader $a_L \in V$ and two parameters, n_L and n_U, that satisfy $n_L \le n \le n_U$, $n_L = \Theta(n)$, and $n_U = \Theta(n)$. We address the collision detection problem without this assumption by running this protocol and the counting protocol Approximate, given by Berenbrink, Kaaser, and Radzik [11], in parallel. The protocol Approximate elects a leader using the leader election protocol presented by Gąsieniec and Stachowiak [21], and computes $\lfloor \log n \rfloor$ or $\lceil \log n \rceil$ with high probability. This protocol uses $O(\log n \cdot \log\log n)$ states with high probability.

The protocol Approximate maintains four variables: $\mathtt{logNum} \in \mathbb{N}$, $\mathtt{level} \in \mathbb{N}$, $\mathtt{countFin} \in \{0,1\}$, and $\mathtt{leader} \in \{0,1\}$[4]. The following proposition holds by the definition of Approximate, which enables us to integrate these two protocols. (See [11] for the definition of Approximate.)

Proposition 1. *An execution of the protocol* Approximate *has the following properties:*

- *For all* $v \in V$, $v.\mathtt{level}$ *is monotonically non-decreasing and increases only when* v *joins an interaction as the responder,*
- *For all* $v \in V$, $v.\mathtt{logNum}$ *never changes while* $v.\mathtt{countFin} = 1$,
- *For all* $v \in V$, $v.\mathtt{countFin}$ *reverts from* 1 *to* 0 *only if* $v.\mathtt{level}$ *increases, and*
- *When two agents with the same level meet, the* `countFin` *value is updated according to the one-way epidemic from the initiator to the responder. Only an agent with* $\mathtt{leader} = 1$ *can set* `countFin` *to* 1, *in addition to this rule,*

For any variable `var`, we say that a configuration C is `var`-stable if, for every configuration C' reachable from C and every agent $v \in V$, the value of `var` for v is the same in both C and C', i.e., $C(v).\mathtt{var} = C'(v).\mathtt{var}$. For any set of variables χ, we say that a configuration C is χ-stable if C is `var`-stable for every `var` in χ. Due to the analysis in [11], we have the following lemma:

[4] We have changed the names of the variables for consistency with the notations of this paper.

Lemma 6 ([11]). *Let $C_0 \in \mathcal{C}_{\mathrm{init}}(P)$ be any initialized configuration and let $\Xi_P(C_0, \mathbf{\Gamma}) = C_0, C_1, \dots,$ where $P =$ APPROXIMATE. With high probability, there is an integer $t = O(n \log^2 n)$ such that:*

- *C_t is a $\{\mathtt{logNum}, \mathtt{leader}, \mathtt{level}\}$-stable configuration,*
- *All agents share the same value in* `logNum`*, which is either $\lfloor \log n \rfloor$ or $\lceil \log n \rceil$, in C_t,*
- *There exists exactly one agent a_L with* `leader` *$= 1$ in C_t,*
- *All agents satisfy* `countFin` *$= 0$ in C_{t-1},*
- *Exactly two agents satisfy* `countFin` *$= 1$, one of whom is a_L, in C_t.*

The protocol APPROXIMATE *uses $O(\log n \cdot \log \log n)$ states with high probability.*

If agents could determine whether they have already reached a configuration C_t, integrating CDWB and APPROXIMATE would be straightforward: when two agents a and b meet, they execute APPROXIMATE if either $a.\mathtt{countFin} = 0$ or $b.\mathtt{countFin} = 0$; otherwise, they execute CDWB(n_L, n_U), setting $n_L = 2^{a.\mathtt{logNum}-1}$ and $n_U = 2^{a.\mathtt{logNum}+1}$. However, such straightforward integration is not feasible, as agents cannot detect this condition. In particular, the value of the variable `countFin` in APPROXIMATE may revert from 1 back to 0. Fortunately, this issue can be resolved, roughly speaking, by the following modification: each time an agent's `level` increases, it terminates the current execution of CDWB and resets its variables maintained by CDWB to their initial values. We call the resulting protocol COLLISIONDETECTION.

We have the following theorem. See the arXiv version [7] for the details of COLLISIONDETECTION, including its pseudocode, and the proof of Theorem 1.

Theorem 1. *Protocol* COLLISIONDETECTION *solves the collision detection problem with probability* 1*. The execution of* COLLISIONDETECTION *under the uniformly random scheduler stabilizes within $O(n^{3/2} \log^{3/2} n)$ time steps and uses $\tilde{O}(n)$ states per agent (excluding the space needed to store the input variable* `rank`*) with high probability. It stabilizes within $O(n^{3/2}\sqrt{\log n})$ time steps in expectation.*

Acknowledgments. This work is supported by JST FOREST Program JPMJFR226U and JSPS KAKENHI Grant Numbers JP19K11826 and JP20KK0232.

Disclosure of Interests. The authors have no competing interests to declare that are relevant to the content of this article.

References

1. Alistarh, D., Aspnes, J., Eisenstat, D., Gelashvili, R., Rivest, R.L.: Time-space trade-offs in population protocols. In: Proceedings of the Twenty-Eighth Annual ACM-SIAM Symposium on Discrete Algorithms, pp. 2560–2579. SIAM (2017)
2. Alistarh, D., Aspnes, J., Gelashvili, R.: Space-optimal majority in population protocols. In: Proceedings of the Twenty-Ninth Annual ACM-SIAM Symposium on Discrete Algorithms, pp. 2221–2239. SIAM (2018)

3. Alistarh, D., Gelashvili, R.: Polylogarithmic-time leader election in population protocols. In: Proceedings of the 42nd International Colloquium on Automata, Languages, and Programming, pp. 479–491 (2015)
4. Angluin, D., Aspnes, J., Fischer, M.J., Jiang, H.: Self-stabilizing population protocols. ACM Trans. Auton. Adapt. Syst. **3**(4), 1–28 (2008)
5. Angluin, D., Aspnes, J., Diamadi, Z., Fischer, M.J., Peralta, R.: Computation in networks of passively mobile finite-state sensors. Distrib. Comput. **18**(4), 235–253 (2006)
6. Angluin, D., Aspnes, J., Eisenstat, D.: Fast computation by population protocols with a leader. Distrib. Comput. **21**(3), 183–199 (2008)
7. Araya, T., Sudo, Y.: Sublinear-time collision detection with a polynomial number of states in population protocols (2024). https://arxiv.org/abs/2411.09957
8. Beauquier, J., Blanchard, P., Burman, J.: Self-stabilizing leader election in population protocols over arbitrary communication graphs. In: International Conference on Principles of Distributed Systems, pp. 38–52 (2013)
9. Berenbrink, P., Elsässer, R., Friedetzky, T., Kaaser, D., Kling, P., Radzik, T.: Time-space trade-offs in population protocols for the majority problem. Distrib. Comput. **34**, 91–111 (2021)
10. Berenbrink, P., Giakkoupis, G., Kling, P.: Optimal time and space leader election in population protocols. In: Proceedings of the 52nd Annual ACM SIGACT Symposium on Theory of Computing, pp. 119–129 (2020)
11. Berenbrink, P., Kaaser, D., Radzik, T.: On counting the population size. In: Proceedings of the 2019 ACM Symposium on Principles of Distributed Computing, pp. 43–52 (2019)
12. Burman, J., et al.: Time-optimal self-stabilizing leader election in population protocols. In: Proceedings of the 2021 ACM Symposium on Principles of Distributed Computing, pp. 33–44 (2021)
13. Cai, S., Izumi, T., Wada, K.: How to prove impossibility under global fairness: on space complexity of self-stabilizing leader election on a population protocol model. Theory Comput. Syst. **50**(3), 433–445 (2012)
14. Canepa, D., Potop-Butucaru, M.G.: Stabilizing leader election in population protocols (2007). http://hal.inria.fr/inria-00166632
15. Chatzigiannakis, I., Michail, O., Nikolaou, S., Pavlogiannis, A., Spirakis, P.G.: Passively mobile communicating machines that use restricted space. In: Proceedings of the 7th ACM ACM SIGACT/SIGMOBILE International Workshop on Foundations of Mobile Computing, pp. 6–15 (2011)
16. Chen, H.P., Chen, H.L.: Self-stabilizing leader election. In: Proceedings of the 38th ACM Symposium on Principles of Distributed Computing, pp. 53–59 (2019)
17. Chen, H.P., Chen, H.L.: Self-stabilizing leader election in regular graphs. In: Proceedings of the 39th ACM Symposium on Principles of Distributed Computing, pp. 210–217 (2020)
18. Doty, D., Eftekhari, M.: Efficient size estimation and impossibility of termination in uniform dense population protocols. In: Proceedings of the 2019 ACM Symposium on Principles of Distributed Computing, pp. 34–42 (2019)
19. Doty, D., Soloveichik, D.: Stable leader election in population protocols requires linear time. Distrib. Comput. **31**(4), 257–271 (2018)
20. Fischer, M.J., Jiang, H.: Self-stabilizing leader election in networks of finite-state anonymous agents. In: International Conference on Principles of Distributed Systems, pp. 395–409 (2006)

21. Gąsieniec, L., Stachowiak, G.: Fast space optimal leader election in population protocols. In: Proceedings of the Twenty-Ninth Annual ACM-SIAM Symposium on Discrete Algorithms, pp. 2653–2667. SIAM (2018)
22. Gąsieniec, L., Stachowiak, G., Uznanski, P.: Almost logarithmic-time space optimal leader election in population protocols. In: The 31st ACM on Symposium on Parallelism in Algorithms and Architectures, pp. 93–102. ACM (2019)
23. Izumi, T.: On space and time complexity of loosely-stabilizing leader election. In: International Colloquium on Structural Information and Communication Complexity, pp. 299–312 (2015)
24. Sudo, Y., Eguchi, R., Izumi, T., Masuzawa, T.: Time-optimal loosely-stabilizing leader election in population protocols. In: 35nd International Symposium on Distributed Computing (DISC 2021), pp. 40:1–40:17 (2021)
25. Sudo, Y., Masuzawa, T.: Leader election requires logarithmic time in population protocols. Parallel Process. Lett. **30**(01), 2050005 (2020)
26. Sudo, Y., Masuzawa, T., Datta, A.K., Larmore, L.L.: The same speed timer in population protocols. In: the 36th IEEE International Conference on Distributed Computing Systems, pp. 252–261 (2016)
27. Sudo, Y., Nakamura, J., Yamauchi, Y., Ooshita, F., Kakugawa, H., Masuzawa, T.: Loosely-stabilizing leader election in a population protocol model. Theor. Comput. Sci. **444**, 100–112 (2012)
28. Sudo, Y., Ooshita, F., Izumi, T., Kakugawa, H., Masuzawa, T.: Time-optimal leader election in population protocols. IEEE Trans. Parallel Distrib. Syst. **31**(11), 2620–2632 (2020)
29. Sudo, Y., Ooshita, F., Kakugawa, H., Masuzawa, T.: Loosely-stabilizing leader election on arbitrary graphs in population protocols. In: International Conference on Principles of Distributed Systems, pp. 339–354 (2014)
30. Sudo, Y., Ooshita, F., Kakugawa, H., Masuzawa, T.: Loosely stabilizing leader election on arbitrary graphs in population protocols without identifiers or random numbers. IEICE Trans. Inf. Syst. **103**(3), 489–499 (2020)
31. Sudo, Y., Ooshita, F., Kakugawa, H., Masuzawa, T., Datta, A.K., Larmore, L.L.: Loosely-stabilizing leader election for arbitrary graphs in population protocol model. IEEE Trans. Parallel Distrib. Syst. **30**(6), 1359–1373 (2018)
32. Sudo, Y., Ooshita, F., Kakugawa, H., Masuzawa, T., Datta, A.K., Larmore, L.L.: Loosely-stabilizing leader election with polylogarithmic convergence time. Theor. Comput. Sci. **806**, 617–631 (2020)
33. Sudo, Y., Shibata, M., Nakamura, J., Kim, Y., Masuzawa, T.: Self-stabilizing population protocols with global knowledge. IEEE Trans. Parallel Distrib. Syst. **32**(12), 3011–3023 (2021)
34. Yokota, D., Sudo, Y., Masuzawa, T.: Time-optimal self-stabilizing leader election on rings in population protocols. IEICE Trans. Fundam. Electron. Commun. Comput. Sci. **104**(12), 1675–1684 (2021)
35. Yokota, D., Sudo, Y., Ooshita, F., Masuzawa, T.: A near time-optimal population protocol for self-stabilizing leader election on rings with a poly-logarithmic number of states. In: Proceedings of the 2023 ACM Symposium on Principles of Distributed Computing, pp. 2–12 (2023)

On the Existence of Extension-Based Proofs of Impossibility for Set-Agreement

Hagit Attiya[1], Pierre Fraigniaud[2(✉)], Ami Paz[3], and Sergio Rajsbaum[4]

[1] Department of Computer Science, Technion, Haifa, Israel
[2] IRIF, Université Paris Cité and CNRS, Paris, France
pierre.fraigniaud@irif.fr
[3] LISN, Université Paris Saclay and CNRS, Paris, France
[4] Inst. de Matematicas, Universidad Nacional Autonoma de Mexico, Mexico City, Mexico

Abstract. A recent breakthrough by Alistarh, Aspnes, Ellen, Gelashvili, and Zhu [STOC 2019] established that there are no *extension-based proofs* of impossibility for set-agreement within the class of *non-uniform* iterated immediate snapshot (NIIS) algorithms. An extension-based proof can be modeled as a game between a *prover* and an *algorithm* claiming to solve set-agreement, where this algorithm can be from a class $\mathcal{C}$ of algorithms. Note that the non-existence of extension-based proofs of impossibility for a class $\mathcal{C}$ of algorithms implies the non-existence of extension-based proofs of impossibility for all classes of algorithms containing $\mathcal{C}$. This result has then been revisited by Attiya, Castañeda, and Rajsbaum [OPODIS 2020] who showed that the same holds for the *smaller* class of *uniform* iterated immediate snapshot (IIS) algorithms, and even with a slightly stronger prover.

The main takeaway message of our work is that these previous results, which show the non-existence of extension-based proofs of impossibility for set-agreement within smaller and smaller universal classes of algorithms, do not necessarily preclude the existence of an extension-based proof of impossibility for set-agreement for an even smaller class of algorithms. To illustrate this, we exhibit an extension-based proof of impossibility for set-agreement for the class of *memoryless* IIS algorithms. This latter class may not be universal but is strong enough to solve non-trivial tasks such as approximate agreement and renaming. Moreover, we show that the result by Attiya et al. for IIS algorithms, with the stronger prover, does not extend to *colorless* IIS algorithms. These two results underline the fact that the existence of an extension-based proof of impossibility for a task strongly depends on the considered class of algorithms, and is still open even for set-agreement.

H. Attiya—Supported by the Israel Science Foundation (grant number 22/1425).
P. Fraigniaud—Additional support from the ANR projects DUCAT (ANR-20-CE48-0006), PREDICTIONS (ANR-23-CE48-0010), and ENEDISC (ANR-24-CE48-7768).
S. Rajsbaum—This work was done while S. Rajsbaum was visiting IRIF, partially supported by ANR Project DUCAT (ANR-20-CE48-0006).

U. Schmid and R. Kuznets (Eds.): SIROCCO 2025, LNCS 15671, pp. 56–73, 2025.
https://doi.org/10.1007/978-3-031-91736-3_4

Keywords: Distributed Computing · Set-Agreement · Extension-Based Proof

1 Introduction

1.1 Context and Objective

The celebrated seminal result by Fischer, Lynch, and Paterson [10] (henceforth, FLP) demonstrating the impossibility of solving (binary) consensus in asynchronous systems with one crash-prone process, is extremely significant not only by itself, but also for the proof strategy leading to it. Among many interesting properties, the proof possesses one extremely desirable feature: it provides an explicit way of catching an algorithm pretending to solve consensus. As a consequence, the proof may contribute to help designers of algorithms claiming to solve consensus in finding the missing pieces in their code.

Roughly, the FLP impossibility result is obtained as follows (see, e.g., [5] for more details). Assuming the existence of an algorithm ALG pretending to solve consensus, the proof explicitly constructs an execution in which ALG never terminates. This is done by (1) identifying an initial input configuration for which ALG cannot terminate without taking a step, and (2) iteratively *extending* a partial t-step execution in which ALG cannot terminate, into a partial $(t+1)$-step execution in which ALG cannot terminate either, for every $t \geq 0$. This is achieved using a so-called *valency* argument, which enables to exhibit, for every $t \geq 0$, and for any *bivalent* configuration of the system after t steps, a set of processes such that, if these processes concurrently take one more step then the resulting configuration is also bivalent. A bivalent configuration (w.r.t. an algorithm ALG) is a configuration of the system satisfying that, starting from this configuration, ALG terminates with consensus 0 in some executions (i.e., some interleaving of the communications), but terminates with consensus 1 in other executions. In particular, the algorithm cannot terminate in a bivalent configuration. The valency argument is used to show that, for every $t \geq 0$, if the system is in a bivalent configuration after ALG has executed t steps, then the execution can be extended to $t+1$ steps while keeping the system in a bivalent configuration.

Although this style of *extension-based proofs* can be used for other tasks beyond consensus [2,11], such proofs of impossibility are not known for several central distributed computing tasks, such as set-agreement, weak symmetry breaking, and renaming. In particular, current proofs of impossibility for set-agreement refer to sophisticated arguments from algebraic topology related to, e.g., shellability and the Nerve Lemma. This is perfectly fine as far as establishing impossibility of set-agreement is concerned, yet these proofs are not constructive in the sense that a designer of an algorithm ALG claiming to solve set-agreement learns that ALG must fail, but this designer is not provided with any clue of why ALG fails.

There is at least one good reason explaining why, intuitively, designing a more informative proof of impossibility for set-agreement may be difficult, if at all possible. It can be observed that consensus has strong links with graph connectivity, i.e., the existence or absence of a path connecting any two vertices in a graph capturing the executions of any algorithm. Instead, set-agreement is inherently linked to higher dimensional notions of connectivity, like the ability to contract loops (for 2-set-agreement), spheres (3-set-agreement), and $(k-1)$-dimensional spheres in general (for k-set-agreement), in simplicial complexes capturing the executions of any algorithm. Yet, the question remains: Is there an informative proof of impossibility for set-agreement? Specifically:

Is there an extension-based proof of impossibility for set-agreement?

1.2 Previous Work

A recent breakthrough result by Alistarh, Aspnes, Ellen, Gelashvili, Zhu [1] exhibited strong evidence that there may well be no extension-based proofs of impossibility for set-agreement. More specifically, Alistarh et al. formalize an extension-based proof as a *2-player game* between a *wait-free* algorithm[1] claiming to solve set-agreement, and a *prover* aiming at demonstrating that the algorithm is wrong by gradually constructing an execution in which it produces illegal outputs or never terminates. As per the FLP impossibility proof for consensus, the construction of a non-terminating execution for set-agreement proceeds iteratively, as a sequence of phases. At phase $t \geq 0$, the system is in some configuration σ_t, and the prover selects a finite set of finite executions for which it asks the algorithm to reveal the *valency* (i.e., the set of possible output values) of the system configuration resulting from each of these executions starting from σ_t. Then, based on these valencies, the prover selects one of the configurations, σ_{t+1}, and if the algorithm cannot terminate in σ_{t+1}, another phase begins. The prover wins if the algorithm can never terminate, i.e., the game continues forever, and otherwise the algorithm wins.

Provers differ in their choices of executions, and in their choice of the configuration σ_{t+1} at the t-th phase of the game. If a prover wins against all algorithms, then this prover provides an extension-based proof of impossibility for set-agreement. However, Alistarh et al. [1] have shown that, in fact, each prover is defeated by at least one algorithm. In other words, no prover can win agains all algorithms. One may be tempted to conclude from this result that no extension-based proofs of impossibility exist for set-agreement. This conclusion would however be an overstatement, even if one agrees on the fact that the 2-player game of Alistarh et al. does capture the nature of an extension-based proof.

Indeed, the result by Alistarh et al. [1] holds for the class of so-called *non-uniform iterated immediate snapshot* (NIIS) algorithms (see [14]). This class of algorithms is a sub-class of the general class of read-write algorithms for crash-prone asynchronous shared-memory systems. This is perfectly fine, as the class

[1] An algorithm ALG is wait-free whenever it can tolerate up to $n-1$ failures in n-process systems. That is, independently from the number of (crash) failures, every correct process must terminate in ALG.

of NIIS algorithms is *universal* for asynchronous wait-free shared-memory computing with crash failures, in the sense that if a task is solvable wait-free by a read-write algorithm, then it is solvable wait-free by an NIIS algorithm (see [6]). However, a crucial point is that NIIS algorithms form a *large* class of algorithms. As a consequence, a prover has to win the game against a large number of opponents (i.e., a large number of algorithms).

To make this point clearer, let us consider an analogy. The world champions in chess were, and still are not asked to win games against all chess players in the world. Each one was actually requested to win against a single opponent. This is considered sufficient for winning the world title as the opponent is chosen (thanks to various forms of qualification) to be "universal", in the sense that he or she is at least as good as any other chess player in the world (apart from the world champion). To continue the analogy, the class of read-write algorithms can be viewed as the class of all chess players, and the class of NIIS algorithms can be viewed as a smaller, but still pretty large class of players, say the class of all chess grandmasters. It would be an extraordinary difficult challenge, even for a chess world champion, to defeat all grandmasters simultaneously in a tournament. Analogously, it can be considered as an extraordinary difficult challenge for a prover to defeat all NIIS algorithms.

Attiya, Castañeda, and Rajsbaum [3] have considered the smaller class of *uniform* iterated immediate snapshot (IIS) algorithms (see [15]). A crucial point is that the class of IIS algorithms is indeed still universal [6], but smaller than the class of NIIS algorithms (all IIS algorithms are NIIS algorithms, but not vice versa). Yet, Attiya et al. showed that, as for the case of NIIS algorithms, there are no extension-based proofs of impossibility for set-agreement within the class of IIS algorithms, in the sense that no prover can win agains all IIS algorithms. The proof in [3] actually reveals a striking power of the IIS algorithms as far as defeating a prover is concerned: a T-round IIS algorithm could even be asked by the prover to reveal the valencies of *all* the configurations of the system reachable after $T-1$ rounds of iterated immediate snapshot (i.e., just one round before the algorithm is supposed to terminate), and the prover may be allowed to select a fraction of these configurations, and still the prover cannot win agains all IIS algorithms! Note that the result in [3] can be reformulated in the "standard" framework of extension-based proofs, in which the prover queries the algorithm by asking the valencies of a finite set of executions (but cannot ask the algorithm to reveal its number of rounds T). Indeed, the algorithm can answer the successive queries of the prover by merely providing it with the valencies induced by those its would reveal after round $T-1$.

This discussion raises the following question:

Problem 1: Is there a universal class $\mathcal{X}$ of algorithms for which there exists an extension-based proof of impossibility for set-agreement?

Before investigating this question further, let us underline that recent work have studied the connection between reductions between problems and the ability to establish impossibility results using extension-based proofs [7]. This latter work also considers anonymous protocols, i.e., the state of a process after a scan is the multiset of values seen in the view, rather than the set of values, as in colorless protocols. Other work have considered the synchronous crash-prone message-passing model [16]. Very recently, [17] has given a necessary and sufficient condition for colorless tasks to have no extension-based proofs of impossibility in the NIIS model.

1.3 Our Results

Memoryless Algorithms. The aforementioned results [1,3] provide strong evidence that the answer to Problem 1 may be negative. Nevertheless, we show that for the class of *memoryless* IIS algorithms (see Sect. 4), there is an extension-based proof of impossibility for set-agreement. Roughly speaking, a memoryless algorithm is an algorithm in which the state of a process is an output value (together with the current round number). At each round, a process writes its current output value to the shared memory, snapshots the memory, and based on the result of the snapshot, updates its output value.

The class of memoryless IIS algorithms seem to be an interesting class of algorithms to be studied in itself. Approximate agreement algorithms are frequently of this type (see, e.g., [9]), they are frequently used in control theory-style algorithms (e.g. [8]), and can be used to solve $(2n-1)$-renaming in n-process systems (see, e.g., [5]). Furthermore, we show in Sect. 6 that they are *universal* for 1-dimensional colorless tasks described next.

Since memoryless algorithms may not be universal, our result regarding them does not necessarily answers positively to Problem 1. Nevertheless, it illustrates our point: depending on the class of algorithms that one considers, extension-based proofs for set-agreement may exist or not.

Colorless Algorithms. Next, we focus on the class of *colorless algorithms.* Roughly speaking, colorless algorithms are algorithms that do not use information on which process holds which value or multiplicities of values, but act solely as a function of the *set* of values read in the shared memory. The class of colorless IIS algorithms is universal for *colorless tasks* (see [13]), where colorless tasks are ones defined solely in terms of sets of input and sets of output values, without referring to the number of processes. Consensus and set-agreement are actually archetypical examples of colorless tasks.

In the case where the input and output sets are of size at most two, we have a 1-dimensional colorless task. These are the tasks considered in Sect. 6, such as binary consensus and binary approximate agreement. They have been studied in detail in the past. It is known that if a 1-dimensional colorless task is not wait-free solvable by $n \geq 2$ processes, the impossibility can be proved by extension-based proofs, and by another technique called round-reduction [4].

We prove that the result by Attiya et al. [3], which shows that there are no extension-based proofs of impossibility for set-agreement in IIS algorithms, does not apply literally to the class of colorless IIS algorithms. Specifically, they showed that a T-round (chromatic) IIS algorithm claiming to solve set-agreement could even be asked by the prover to reveal the valencies of all the system configurations reachable after $T-1$ rounds of iterated immediate snapshot (i.e., just one rounds before the algorithm is supposed to terminate), and the prover may be allowed to select a fraction of these configurations, yet the prover still cannot win. We show that this property does not hold for colorless IIS algorithms. Colorless algorithms are therefore (slightly) stronger than general IIS algorithms w.r.t. extension-based proofs, illustrating again that the proof of non-existence of extension-based proofs of impossibility for set-agreement is strongly impacted by the class of algorithms under scrutiny.

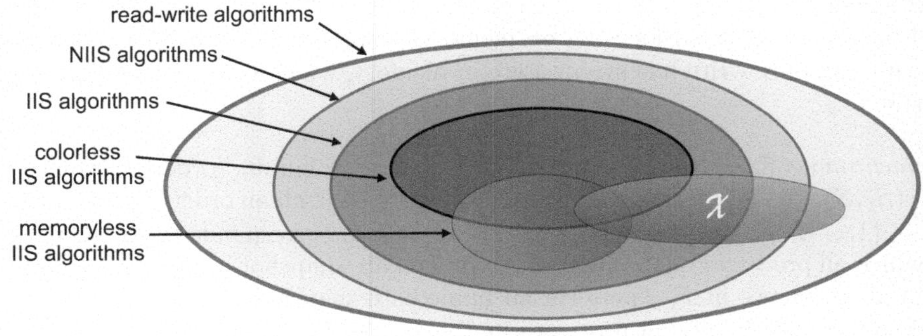

Fig. 1. There are no extension-based proofs of impossibility for set-agreement in the universal class of IIS algorithms, and therefore in all classes containing it (e.g., read-write, and NIIS). On the other hand, we show that there is an extension-based proof of impossibility for set-agreement in the class of memoryless IIS algorithms. However, this (non-trivial) class is not universal for wait-free computing. The main open problem in this line of work is whether or not there exists a universal class $\mathcal{X}$ of algorithms for which there is an extension-based proof of impossibility for set-agreement.

Take-Away Message. The take-away message of this paper is that, while there are more and more evidences that no extension-based proof of impossibility for set-agreement exist, for smaller and smaller universal classes of algorithms, these evidences do not necessarily preclude the existence of an extension-based proof of impossibility for set-agreement for *some* universal class $\mathcal{X}$ of algorithms, as far as wait-free computing is concerned (see Fig. 1). Problem 1 is questioning the existence of such a class $\mathcal{X}$, and this problem is yet to be solved.

2 Model and Notation

2.1 Computational Models

We focus on wait-free computing in the standard shared-memory model [5], where $n \geq 2$ asynchronous processes communicate via a shared memory composed of single-writer/multiple-readers registers. More specifically, we consider the iterated immediate snapshot (IIS) model, that is, the shared memory is organized as an infinite number of memory level $M_r, r \geq 1$, each composed of n registers $M_r[1], \ldots, M_r[n]$, one for each process p_i, $i \in \{1, \ldots, n\}$. All registers are initialized to $\perp$. Computation proceeds in (asynchronous) rounds. At round $r \geq 1$, every process p_i performs an immediate snapshot of the level M_r of the memory, i.e., (1) p_i writes in its register $M_r[i]$, and (2) p_i snapshots all registers $M_r[1], \ldots, M_r[n]$ in M_r, where these two steps being performed as a single atomic operation. Note, however, that several processes may perform these steps simultaneously, in which case they all first write and then read, thus they see each other's updates in their snapshots. Unless specified otherwise, we assume full-information protocols, i.e., when a process writes in a register, it writes its entire history since the beginning of the computation. Of course, this assumption will be discarded when focusing on memoryless protocols as, in this setting, a process can only write its current state in memory, which belongs to a finite set of states.

Asynchronous Executions. An execution is a finite or infinite (ordered) sequence $e = (S_1, S_2, \ldots)$ where, for every $r \geq 1$, the set S_r is itself an ordered partition of $[n] = \{1, \ldots, n\}$. A set $S_r = (P_{r,1}, \ldots, P_{r,k})$, $k \geq 1$, corresponds to the scenario in which all processes in $P_{r,1}$ perform an immediate snapshot in M_r concurrently, then all processes in $P_{r,2}$ perform an immediate snapshot in M_r concurrently, and so on, until finally all processes in $P_{r,k}$ perform an immediate snapshot in M_r concurrently. Therefore, at round r, all processes in $P_{r,1}$ see only themselves, all processes in $P_{r,2}$ see themselves and all the processes in $P_{r,1}$, and all processes in $P_{r,k}$ see all the n processes. For instance, an execution $e = (S_1, S_2, \ldots)$ where $S_r = (\{i\}, P_{r,2}, \ldots, P_{r,k})$ for every $r \geq 1$ is an execution in which process p_i does not see any other processes during the execution; we call it a *solo* execution of p_i (even though more processes my participate). The wait-free condition imposes that process p_i must nevertheless terminate and decide an output value. The correctness condition of the task at hand (e.g., set-agreement) imposes that this output value must be legal for p_i's input value with respect to the input-output specification of the task.

2.2 Tasks, Protocol Complexes, and Decision Maps

We use the standard formulation specifying a task for $n \geq 2$ processes as a triple $\Pi = (\mathcal{I}, \mathcal{O}, \Delta)$ where $\mathcal{I}$ and $\mathcal{O}$ are the $(n-1)$-dimensional chromatic input and output complexes, and $\Delta : \mathcal{I} \to 2^{\mathcal{O}}$ is the input-output specification of the task (see, e.g., [12]). That is, for every input simplex $\sigma \in \mathcal{I}$, which describes a possible input configuration of the system for the processes in σ, $\Delta(\sigma)$ is a set $\tau_1, \ldots, \tau_k$ of simplices in $\mathcal{O}$, each describing an output configuration that is legal with respect to σ. In particular, for each $i \in [k]$, $|\tau_i| = |\sigma|$ and the sets of processes in τ_i and σ are identical. Since the complexes are chromatic, every vertex v of these complexes is a pair (i, x) where $i \in [n]$ is the *color* of v, and x is a value. We interpret (i, x) as indicating process p_i having the value or state x, and specifically, the input value x if $v \in \mathcal{I}$, the output value x if $v \in \mathcal{O}$, and

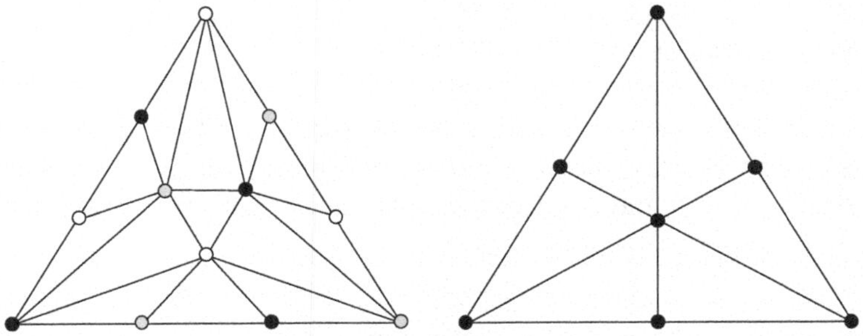

Fig. 2. On the left, the chromatic subdivision of a chromatic triangle. On the right, the barycentric subdivision of a (monochromatic) triangle.

the view x resulting from r immediate snapshot operations if v is a vertex of the protocol complex $\mathcal{P}^{(r)}$ (see below).

Example: k-Set-Agreement. Let $k \geq 1$. Informally, the k-set-agreement task for $n \geq k+1$ processes, each process p_i, $i \in [n]$, starts with input $x_i \in [k+1]$, and must output a value $y_i \in [k+1]$ among the set of values it has seen during its execution, such that the set of all output values has cardinality at most k. That is, at least one input value must be eliminated whenever $k+1$ processes start with $k+1$ different values. More formally, the n-process k-set-agreement task is the task $\Pi = (\mathcal{I}, \mathcal{O}, \Delta)$ where, for every $(x_1, \ldots, x_n) \in [k+1]^n$, the input complex $\mathcal{I}$ has a facet[2] $\phi = \{(i, x_i) \mid i \in [n]]\}$, and the facets of the output complex $\mathcal{O}$ are all sets $\psi = \{(i, y_i) \mid i \in [n]\}$ such that $(y_1, \ldots, y_n) \in [k+1]^n$, and $|\{y_i \mid i \in [n]\}| = k$. That is, no more than k different values must be present in any output simplex $\sigma \in \mathcal{O}$. For every $\sigma = \{(i, x_i) \mid i \in I\}$ with $I \subseteq [n]$, the input-output specification Δ of k-set-agreement stipulates that, for every $\tau \in \mathcal{O}$,

$$\tau \in \Delta(\sigma) \iff \tau = \{(j, y_j) \mid j \in I\} \text{ with } y_j \in \{x_i \mid i \in I\} \text{ for every } j \in I.$$

Protocol Complex. Still using the formalism from [12], for every $r \geq 0$, the protocol complex $\mathcal{P}^{(r)}$ captures all possible states of the system after r iterations of iterated immediate snapshots performed by every process. That is, given a task $\Pi = (\mathcal{I}, \mathcal{O}, \Delta)$, $\mathcal{P}^{(0)} = \mathcal{I}$, and, for every $r \geq 1$, every simplex $\varrho \in \mathcal{P}^{(r)}$ is a possible global state of the system after r rounds. For every $\sigma \in \mathcal{I}$, and every $r \geq 0$, let us denote by $\mathcal{P}^{(r)}(\sigma)$ the sub-complex of $\mathcal{P}^{(r)}$ corresponding of r rounds of iterated immediate snapshots starting from input σ. More generally, for every $t \geq r \geq 0$ and every $\varrho \in \mathcal{P}^{(r)}$, let us denote by $\mathcal{P}^{(t)}(\varrho)$ the sub-complex of $\mathcal{P}^{(t)}$ corresponding of $t - r$ rounds of iterated immediate snapshots starting from configuration ϱ.

A important property used for the analysis of algorithms in the IIS model is that one iteration of immediate snapshot corresponds to a *chromatic subdivision*

[2] Recall that a *facet* of a simplicial complex is a simplex that is maximal for inclusion.

(see Fig. 2), that is, for every $t \geq 0$, $\mathcal{P}^{(t+1)} = \mathsf{ch}(\mathcal{P}^{(t)})$, and thus $\mathcal{P}^{(t)} = \mathsf{ch}^{(t)}(\mathcal{I})$. See [12] for more details.

Decision Map. A T-round algorithm for a task $\Pi = (\mathcal{I}, \mathcal{O}, \Delta)$ is a simplicial map $f : \mathsf{ch}^{(T)}(\mathcal{I}) \to \mathcal{O}$, satisfying that (1) f is chromatic (i.e., it preserves the colors of the vertices), and (2) f agrees with Δ, i.e., for every $\sigma \in \mathcal{I}$, and every $\varrho \in \mathsf{ch}^{(T)}(\sigma)$ with $|\sigma| = |\varrho|$, $f(\varrho) \in \Delta(\sigma)$.

2.3 Colorless Tasks and Colorless Algorithms

A large class of tasks, including set-agreement, do not require using chromatic complexes for being described, i.e., monochromatic complexes suffice. A *colorless* task is defined as a triple $\Pi = (\mathcal{I}, \mathcal{O}, \Delta)$ where $\mathcal{I}$ and $\mathcal{O}$ are the input and output complexes, and $\Delta : \mathcal{I} \to 2^{\mathcal{O}}$ is the input-output specification of the task (see, e.g., [12]). That is, for every input simplex $\sigma \in \mathcal{I}$, which describes a possible set of input values, $\Delta(\sigma)$ is a sub-complex of $\mathcal{O}$ containing all possible sets of output values that are legal for the inputs in σ. Every vertex v of these complexes is single value, i.e., an input value, an output value, or a view of a process after $r \geq 1$ rounds if v is a vertex of the protocol complex $\mathcal{P}^{(r)}$.

Example: k-Set-Agreement. For every $k \geq 1$, the colorless specification of k-set-agreement is the task $\Pi = (\mathcal{I}, \mathcal{O}, \Delta)$ where the input complex $\mathcal{I}$ has a single facet $\sigma = [k+1]$, and the facets of the output complex $\mathcal{O}$ are all non-empty sets $\psi \subset [k+1]$ with $|\psi| = k$. For every $\sigma \subseteq [k+1]$, all simplices $\tau \subseteq \sigma$ of size at most k are in $\Delta(\sigma)$, i.e., $\Delta(\sigma) = \{\tau \subseteq \sigma \mid 1 \leq |\tau| \leq k\}$. Note that $\Delta(\sigma)$ is a sub-complex of the output complex $\mathcal{O}$.

Colorless Algorithms. A core property of colorless tasks is that, if a colorless task is solvable wait-free, then it is solvable by a *colorless algorithm* [13]. A colorless algorithm is an algorithm which ignores the multiplicities of the values that it reads in memory when performing its snapshots. For instance, let us consider n processes, let $(A_1, \ldots, A_s)$ be a partition of $[n]$, and let $\{x_1, \ldots, x_s\}$ be a set of s distinct input values. Let us assume that, for every $i \in [s]$, all processes in A_i start with the same input x_i. Then, if a process p happens to snapshot all these inputs in memory, p ignores multiplicity whenever performing a colorless algorithm, and it considers its view as the *set* $w(p) = \{x_i \mid i \in [s]\}$. This is in contrast to a general (chromatic) algorithm in which the view of p would be $w(p) = \bigcup_{i=1}^{s}\{(j, x_i) \mid j \in A_i\}$ In a colorless algorithm, one iteration of immediate snapshot corresponds to a *barycentric* subdivision (see Fig. 2), that is, for every $t \geq 0$, $\mathcal{P}^{(t+1)} = \mathsf{bary}(\mathcal{P}^{(t)})$. See [12] for more details. A T-round colorless algorithm for a colorless task $\Pi = (\mathcal{I}, \mathcal{O}, \Delta)$ is a simplicial map $f : \mathsf{bary}^{(T)}(\mathcal{I}) \to \mathcal{O}$, that agrees with Δ, that is, for every $\sigma \in \mathcal{I}$, and every $\varrho \in \mathsf{bary}^{(T)}(\sigma)$, $f(\varrho) \in \Delta(\sigma)$.

3 Extension-Based Proofs

3.1 Definition

An *extension-based proof* is a generic methodology for establishing the impossibility of solving a task Π (in a given computing model). It was introduced in [1], inspired by the seminal FLP impossibility result for consensus [10]. It can be viewed as a 2-player game between a *prover*, and an *algorithm*. The algorithm is claiming to solve Π, and the goal of the prover is to corner the algorithm in some contradiction, forcing the latter to admit that its claim does not hold. We formalize such a game in the framework of colorless IIS algorithms.

During the game, the (colorless) algorithm is asked by the prover to provide *valencies* of system configurations, i.e., of simplices in the protocol complex. In the definition below, f is an hypothetic algorithm for Π, i.e., a simplicial map $f : \mathcal{P}^{(T)} \to \mathcal{O}$ that agrees with Δ, for some $T \geq 0$.

Definition 1. *Let $\Pi = (\mathcal{I}, \mathcal{O}, \Delta)$ be a colorless task, and let $f : \mathcal{P}^{(T)} \to \mathcal{O}$ be a T-round colorless algorithm for Π. For every $t \in \{0, \dots, T\}$ and every $\varrho \in \mathcal{P}^{(t)}$, the* valency *of ϱ, denoted by $\mathsf{val}(\varrho)$, is the set of all output values that can be produced by the algorithm f when the processes proceed from state ϱ. That is,*

$$\mathsf{val}(\varrho) = f\big(V(\mathcal{P}^{(T-t)}(\varrho))\big) = \{y \in V(\mathcal{O}) \mid \exists w \in V(\mathcal{P}^{(T-t)}(\varrho)),\ f(w) = y\}.$$

Definition 2. *A* prover *is a pair* (query, choice) *where:*

- *given a global state $\varrho \in \mathcal{P}^{(t)}$ of the system after $t \geq 0$ rounds,* $\mathsf{query}(\varrho)$ *returns a set of finite executions, by all or some of the processes;*
- *given a set $S = \{\varrho_1, \dots, \varrho_s\} \subseteq \mathcal{P}^{(t)}$ of global states of the system after $t \geq 0$ rounds and their valencies,* $\mathsf{choice}(S)$ *returns a state $\varrho \in S$. based solely on the valencies of these states.*

We note that this definition is a bit weaker than in prior work, since the queries are non-adaptive: a query in some round cannot be chosen based on answers to other queries of this round, e.g., ϱ_2 cannot depend on the valency of ϱ_1. This is not a problem, however, since we only present positive results and the prover can choose to ask such non-adaptive queries.

A query strategy is therefore a mechanism allowing the prover to probe different executions of the algorithm, starting from state $\varrho \in \mathcal{P}^{(t)}$. An archetypal example is $\mathsf{query}(\varrho) = \{e_i \mid i \in U\}$ where U is the set of values held by processes in ϱ and $e_i = (\{i\}, U \setminus \{i\})$ for every $i \in U$. That is, e_i is the 1-round execution where processes with value i perform first, while all the other processes concurrently perform right after them. Every execution $e \in \mathsf{query}(\varrho)$ is "probed" in the sense that the algorithm will be asked to reveal the valencies of the state ϱ' resulting from performing e starting from ϱ. A choice strategy is a mechanism allowing the prover to select one state among those resulting from all the executions in $\mathsf{query}(\varrho)$, based solely on the valency of these states. The following definition is directly inspired by [1].

Algorithm 1. An extension-based proof for $\Pi = (\mathcal{I}, \mathcal{O}, \Delta)$ is a game between PROVER and ALG.

ALG reveals $\mathsf{val}(\varrho)$ for every $\varrho \in \mathcal{I}$ ▷ *the valency of each and every input simplex*
PROVER chooses $\sigma = \mathsf{choice}(\{(\varrho, \mathsf{val}(\varrho)) \mid \varrho \in \mathcal{I}\})$ ▷ *σ is fixed for the whole game*
$\varrho \leftarrow \sigma$ ▷ *ϱ is the current simplex in the protocol complex, updated at each iteration*
while *true* **do**
    PROVER computes $\mathsf{query}(\varrho) = \{e_1, \ldots, e_s\}$ ▷ *a set of finite executions*
    for $i = 1$ **to** s **do** $\varrho_i \leftarrow \mathrm{ALG}(\varrho, e_i)$ ▷ ALG *executes e_i starting from ϱ*
    ALG reveals $\mathsf{val}(\varrho_i)$ for every $i \in \{1, \ldots, s\}$ ▷ *valency of each configuration ϱ_i*
    PROVER chooses $\varrho = \mathsf{choice}(\{\mathsf{val}(\varrho_1), \ldots, \mathsf{val}(\varrho_s)\})$ ▷ *one ϱ_i is selected*
    ALG reveals $f(\varrho)$ ▷ *if* ALG *does not terminate in ϱ, its output is $\perp$*
    if $f(\varrho) \in \Delta(\sigma)$ **then** ▷ *true if all proc. terminate, and output correctly*
        if (f and val are consistent and compatible) **then**
            ALG wins, and the game terminates
        else PROVER wins, and the game terminates
        end if
    end if ▷ *if no players won, the game continues for another iteration*
end while ▷ *if the game carries on forever then* PROVER *wins*

Definition 3. *An* extension-based proof *is a game between a prover and an algorithm* ALG. *This game is played according to the rules specified in Algorithm 1. Prover wins if Algorithm 1 does not terminates, or if* ALG *is caught cheating at some point in the execution of Algorithm 1. Otherwise* ALG *wins, i.e., for* ALG *to win, it must behave truthfully, and lead the game specified by Algorithm 1 to terminate.*

Let us detail the features of Algorithm 1. Initially, the algorithm reveals the valencies of all the input simplices, and the prover chooses an input simplex σ according to its choice strategy. The game then proceeds as a sequence of phases, where the current system configuration ϱ is initialized to σ. At each phase, given the current system configuration ϱ, the prover asks the algorithm to perform a set of executions $\mathsf{query}(\varrho) = \{e_1, \ldots, e_s\}$ computed thanks to its query strategy, and to return the valencies of the resulting set of configurations $\{\varrho_1, \ldots, \varrho_s\}$. The prover then chooses one of these configuration thanks to its choice strategy choice, and it updates ϱ accordingly.

ALG wins if, at some iteration t of the game, it can produce an output value of each process involved in the current state ϱ, such that (1) the set of output values satisfies the specification of the task for the initial input state σ chosen by the prover, and (2) this set of output values is consistent with the set of valencies provided to the prover by the algorithm at each of the t iterations of the game, and these sets of valencies are consistent between the different rounds. Instead, if the game carries on forever, i.e., if the algorithm is unable to produce consistent outputs at any iteration of the game, then the prover wins.

Remark. Given an algorithm ALG supposed to solve a task Π, a prover winning against ALG provides an extension-based proof that ALG fails to solve Π. If the prover wins against all algorithms in a certain class $\mathcal{A}$ of algorithms, then it is an extension-based proof of impossibility for Π for algorithms in $\mathcal{A}$. Note that a prover might not succeed against all algorithms in a class $\mathcal{A}$, but may be able to win against all algorithms in a subclass $\mathcal{A}' \subset \mathcal{A}$. Conversely, if no prover can win against all algorithms in $\mathcal{A}'$, then no prover can win against all algorithms in $\mathcal{A}$. Therefore, the absence of extension-based proofs for the class $\mathcal{A}'$ implies the same for any class $\mathcal{A} \supseteq \mathcal{A}'$.

3.2 Examples

Alistarh et al. [1] have proved that there are no extension-based proofs of impossibility for set-agreement, for the (universal) class of *non-uniform* iterated immediate snapshot (NIIS) algorithms. That is, for every prover (query, choice), there exists an NIIS algorithm ALG claiming to solve set-agreement that wins against (query, choice). This result is strong in the sense that the considered provers (query, choice) are even stronger than the one defined by Algorithm 1, as, in particular, it can probe infinite solo executions. Nevertheless, even if the considered class $\mathcal{A}$ of algorithms, namely the NIIS algorithms, is smaller than the class of all read-write algorithms, it is still pretty large, which makes it easier for some ALG to win.

Attiya et al. [3] have strengthened the result of Alistarh et al., by considering the smaller yet universal class of *uniform* iterated immediate snapshot (IIS) algorithms, in which all processes terminate at the same round. That is, for every IIS algorithm ALG, there exists $T \geq 0$ such that all executions of ALG perform in exactly T rounds, independently from the input. It has been proved in [3] that there are no extension-based proofs of impossibility for IIS algorithms claiming to solve set-agreement. This result was achieved by establishing an even stronger result. Specifically, even if the IIS algorithm ALG were asked to reveal its round complexity T to the prover, and even if the prover could then choose an input configuration $\sigma \in \mathcal{I}$, and query all $(T-1)$-round executions starting from σ at once, i.e., all simplices of $\mathcal{P}^{(T-1)}(\sigma)$, still every IIS algorithm ALG wins.

In contrast to these two (negative) examples, in the next section we provide an example where the prover wins the 2-player game specified by Algorithm 1. This is, we show that, for a specific class of algorithms $\mathcal{A}$, there is an extension-based proof of impossibility of set-agreement for $\mathcal{A}$.

4 Memoryless Algorithms

Roughly, a memoryless algorithm for a task $\Pi = (\mathcal{I}, \mathcal{O}, \Delta)$ is an algorithm in which every process p_i starts each round of IIS in a state $y \in V(\mathcal{O})$, and updates its state to $y' \in V(\mathcal{O})$ according to what was snapshot in the shared memory at this round. This goes on until T rounds have been performed, and then p_i outputs its current state y. For simplicity, we focus on memoryless *colorless* algorithms (that can solve, e.g., approximate agreement, but not renaming). Algorithm 2 provides the general form of a memoryless colorless algorithm[3] that is entirely defined by the functions

$$\mathsf{start} : V(\mathcal{I}) \to V(\mathcal{O}), \text{ and } \mathsf{update} : 2^{V(\mathcal{O})} \times \mathbb{N} \to V(\mathcal{O}),$$

where the former depends solely on the input value x, and the latter depends on the current state of the shared memory gathered in the snapshot and on the round number. Given an input x, process p_i computes $y = \mathsf{start}(x)$ that will be its initial output value to be written in memory. Given a snapshot Y of the memory, the behavior of process p_i depends on whether T rounds have been performed or not. If not, then p_i updates its state to $y = \mathsf{update}(Y)$, and proceeds to the next round. Otherwise, it outputs its current state y, and terminates.

Algorithm 2. Memoryless colorless IIS algorithm for process p_i with input x

Require: Each memory level M_r, $r \geq 1$, initialized to $(\bot, \ldots, \bot)$

1: $y \leftarrow \mathsf{start}(x)$ ▷ *p_i computes its initial output value*
2: **for** $r = 1$ **to** T **do**
3: $\quad \mathsf{write}(y)$ in $M_r[i]$ ▷ *at rth round, writes occurs at level r of memory*
4: $\quad Y \leftarrow \{\mathsf{snapshot}(M_r)\}$ ▷ *Y gets the set of values in $(M_r[1], \ldots, M_r[n])$*
5: $\quad y \leftarrow \mathsf{update}(Y, r)$ ▷ *update y based on the current global knowledge Y*
6: **end for**
7: decide y ▷ *p_i decides the final value of y*

The following result shows that there is an extension-based proof of impossibility of set-agreement for the class of memoryless IIS algorithms.

Theorem 1. *There exists a prover that wins against every colorless memoryless IIS algorithm pretending to solve set-agreement.*

Proof. Let $k \geq 1$, and consider a memoryless colorless algorithm ALG that claims to solve k-set-agreement. According to Algorithm 1, ALG is first asked by the prover to reveal $\mathsf{val}(\varrho)$ for every $\varrho \in \mathcal{I}$. The prover chooses the facet $\sigma = \{1, \ldots, k+1\}$. Note that $\mathsf{val}(\sigma) = \{1, \ldots, k+1\}$ since ALG must output $x \in \{1, \ldots, k+1\}$ whenever a process starts with x and runs solo, which forces $\mathsf{val}(\{x\}) = x$ for every input $x \in \{1, \ldots, k+1\}$. Moreover, this condition actually defines the choice mechanism of the prover: At every iteration $t \geq 1$ of the while loop in Algorithm 1, the prover systematically chooses a simplex $\varrho_t \in \mathsf{bary}^{(t)}(\sigma)$ such that $\mathsf{val}(\varrho_t) = \{1, \ldots, k+1\}$. We next show that such a simplex always exists, whatever ALG does.

Initially, let $\varrho_0 = \sigma$. At the t-th iteration of Algorithm 1 for $t \geq 1$, let ϱ_{t-1} be the current simplex. The prover queries all 1-round executions starting from ϱ_{t-1}, and asks ALG to reveal the valencies of all system configurations resulting from these executions. That is, the prover asks ALG to reveal the valencies of all

[3] Instructions 3 and 4 are performed atomically as a single instruction in the IIS model.

simplices in $\mathsf{bary}(\varrho_{t-1})$. A crucial point is that the valencies of the simplices in $\mathsf{bary}(\varrho_{t-1})$ are subject to constraints that force the algorithm to either produce inconsistent outputs, or create a Sperner coloring which allows the prover to extend the execution (see Fig. 3).

For a T-round algorithm, define mappings $\pi_t : [k+1] \to [k+1]$ by $\pi_0 = \mathsf{start}$ and $\pi_t(i) = \mathsf{update}(\{i\}, t)$ for $1 \leq t \leq T$. That is, π_0 is just the initialization of each process separately, and π_t is the update a process makes if it only reads a single value (which it must have also written) at round t. For t', t with $0 \leq t' < t \leq T$, we further denote the composition of the mappings from π'_t to π_t as $\pi_{t'\cdots t} = \pi_t \circ \ldots \circ \pi'_t$, and when $\pi_{t'\cdots t}$ is a permutation (which we next show to always be the case), denote its inverse by $\pi^{-1}_{t'\cdots t}$. We first show that each π_i is a permutation.

Claim. For every $0 \leq t \leq T$, π_t is a permutation on $[k+1]$.

Proof. Assume otherwise, fix the minimal t such that π_t is not a permutation and i, j with $i \neq j$ such that $\pi_t(i) = \pi_t(j)$. Let $x_i = \pi^{-1}_{0\cdots t-1}(i)$ and $x_j = \pi^{-1}_{0\cdots t-1}(j)$, that are both defined since t is minimal. Moreover $x_i \neq x_j$. Consider a solo execution of a process starting with the input x_i. After t steps, it has the value $\pi_{0\cdots t}(x_i) = \pi_t(i)$, and after all the T steps, its output is $\pi_{t+1\cdots T}(\pi_t(i))$. Similarly, a solo execution of a process starting with the input x_j gets to $\pi_t(j) = \pi_t(i)$ after t steps, and terminates with the output $\pi_{t+1\cdots T}(\pi_t(i))$. The validity condition of set-agreement requires the first execution to end with output x_i and the second with x_j, a contradiction. □

Lemma 1. *Fix* $S \subseteq [k+1]$, $0 \leq t \leq T$, *and let* $X = \{\pi^{-1}_{0\cdots t-1}(s) \mid s \in S\}$. *Then* $\mathsf{update}(S, t) \in \{\pi_{0\cdots t}(x) \mid x \in X\}$.

Any simple attempt for a memoryless set-agreement algorithm would have the trivial permutations (that is, a process that sees only one value keeps this value), and more generally, have each process adopt one of the values it sees. If this was the case, the above lemma would be much simpler to state and prove. Unfortunately, we cannot assume ALG is so nicely behaved, as illustrated by the algorithm in Fig. 3. This causes great complication to our proof, but the reader can read it assuming all the permutations are the identity permutation and get good intuition.

Proof. Assume the lemma does not hold, let t be the first round when it fails, and S be the smallest set for which it fails (breaking ties arbitrarily if there are several such sets). Let X as in the statement of the lemma. Let $z = \pi^{-1}_{0\cdots t}(\mathsf{update}(S, t))$, i.e., $z \notin X$ and satisfies $\mathsf{update}(S, t) = \pi_{0\cdots t}(z)$. Consider all $(t-1)$-round executions where at each round, all the processes with values in X take a step before all other processes. This set of executions is represented by a face of the $(t-1)$-round protocol complex $\mathcal{P}^{(t-1)}([\sigma])$, supported by X. The minimality of t and S assures that the lemma holds for this face, and hence, Sperner's lemma (see, e.g., [12, Chapter 9]) guarantees the existence of a face in it with the values S. At round t, the center of the barycentric subdivision of this face represents a state of some process. Concretely, it is the state of a process p, which reads S in its t-th snapshot, after an execution α of the above family of executions. Consider an execution α' where all processes have inputs in X, and behave as in α (i.e., all processes with the same value rune concurrently). Consider the execution $\alpha'\{p\}^{T-t}$ composed of α' and then $T-t$ steps by p only (formally, after every step of p there is a step of all other processes in some order, but this does not affect p's output). In this execution only processe with inputs from X participate, so p must output a value from X. Hence, $\pi_{t+1\cdots T}(\mathsf{update}(S,t)) \in X$. On the other hand, in a solo execution of a process with input z, its output must satisfy $\pi_{t+1\cdots T}(\mathsf{update}(S,t)) = \pi_{0\cdots T}(z) = z$ by the validity condition. But $z \notin X$, a contradiction. □

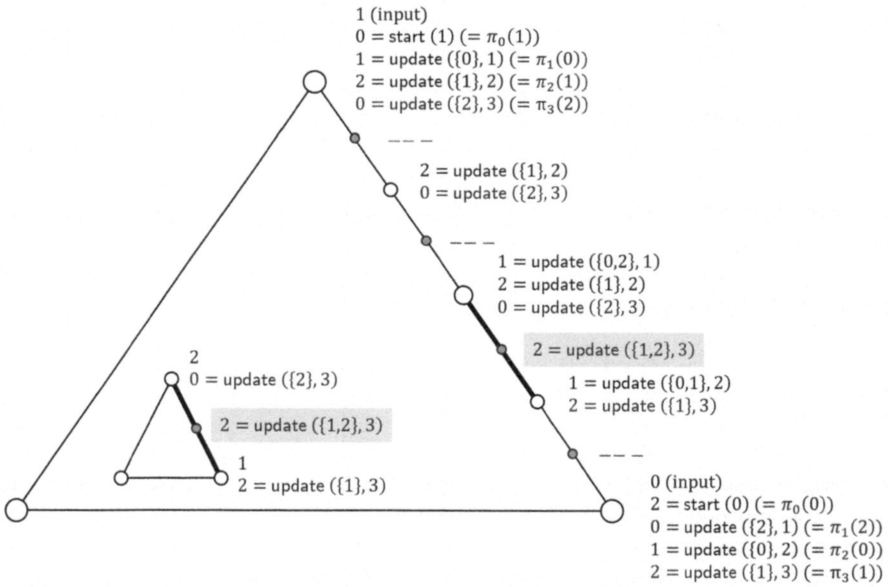

Fig. 3. Illustration of 3 rounds of a 2-set-agreement algorithm with inputs $\{0, 1, 2\}$, and update rules $\mathsf{start}(i) = i - 1 \mod 3$ and $\mathsf{update}(S, t) = \min S + 1 \mod 3$. Note that after 4 rounds the algorithm produces valid outputs (clearly, it cannot also comply with the agreement condition). This also illustrates the proof of Lemma 1 for $t = 2$: to prove that after round 2, the internal edge $S = \{1, 2\}$ (thick line) satisfies $\mathsf{update}(\{1, 2\}, 3) \in \{\pi_3(1), \pi_3(2)\}$, we find a boundary edge with the same values after round 2, which must exist by Sperner's lemma, and prove the claim for that edge.

A direct consequence of Lemma 1 is that the valency $\mathsf{val}(Y)$ of every vertex Y of $\mathsf{bary}(\varrho_{t-1})$ satisfies $\mathsf{val}(Y) \in \{x \mid solo(x, t-1) \in Y\}$, where $solo(x, t-1)$ is the value of a process starting with x after $t-1$ solo steps (see Fig. 3). In other words, the function $c : V(\mathsf{bary}(\varrho_{t-1})) \to [k+1]$ which associates the "color" $c(Y) = \mathsf{val}(Y)$ to every vertex $Y \in \mathsf{bary}(\varrho_{t-1})$ is a Sperner coloring of the triangulation $\mathsf{bary}(\varrho_{t-1})$ of ϱ_{t-1}. By Sperner's Lemma [12, Chapter 9], there exists a simplex $\varrho_t \in \mathsf{bary}(\varrho_{t-1})$ such that $\mathsf{val}(\varrho_t) = [k+1]$. The simplex ϱ_t is the simplex that the prover chooses. This choice strategy allows the prover to construct an infinite sequence $(\varrho_t)_{t\geq 0}$ of successive configurations such that, for every $t \geq 0$, ALG cannot terminates in ϱ_t. Hence the prover wins the game. □

Remark. The sole difference between *chromatic* memoryless IIS algorithms, and *colorless* memoryless IIS algorithms is that chromatic memoryless IIS algorithms view the result Y of a snapshot as a set of values colored by the process IDs, whereas colorless memoryless IIS algorithms view Y as a set of values, ignoring process IDs and multiplicities. This gives slightly more power to the chromatic algorithms. Nevertheless, by the same arguments as in the proof of Theorem 1, one can show that there exists a prover that wins against every (chromatic) memoryless IIS algorithm pretending solving set-agreement.

5 Colorless Algorithms

In the previous section we showed one case in which extension-based proofs exist: when the algorithm is memoryless. We now consider another case where such a proof exists, but in the model suggested by Attiya et al. [3]. There, the prover receives the valencies of all the $(T-1)$-round states of the system, for a T-round algorithm claiming to solve set-agreement. This prover has much more information than the one of [1], as it has more valencies and since the algorithm is uniform. While Attiya et al. [3] show that even in this case no extension-based proof exists, we prove that further restricting the algorithm changes the picture. When considering 2-set-agreement with only *colorless* algorithms, an extension-based proof à la Attiya et al. [3] exists.

Theorem 2. *There exists a prover in the style of [3] that wins against every colorless IIS algorithm pretending to solve 2-set-agreement.*

6 Universality of Memoryless Colorless Algorithms

We conclude the paper by stressing the fact that memoryless colorless algorithms do not form a small class of algorithms. For 1-dimensional colorless tasks, this class is in fact universal.

Theorem 3. *If a 1-dimensional colorless task is wait-free solvable then it is wait-free solvable by a memoryless colorless algorithm.*

7 Conclusion

The main purpose of this paper was to stress the fact that, despite the breakthrough result by Alistarh et al. [1], and the follow up result by Attiya et al. [3], there are still interesting questions regarding the existence of an extension-based proof of impossibility for set-agreement is still open. Establishing that there are no extension-based proofs of impossibility for set-agreement requires to show that, for every *universal* class $\mathcal{C}$ of read-write wait-free algorithms, there are no prover capable to defeat all algorithms taken from $\mathcal{C}$ pretending to solve set-agreement. We believe that, beyond the intellectual curiosity leading to questioning whether this result holds, proving or disproving the existence of a universal class of algorithms for which there exists an extension-based proof of impossibility for set-agreement would require to improve our knowledge about the design of lower bounds and impossibility results for distributed computing, a research direction of the utmost importance in the current technological trend in which the nature of computation becomes inherently distributed, whether it be at the microscopic scale (e.g., multi-core computing), or at the macroscopic scale (e.g., server farms and data centers).

References

1. Alistarh, D., Aspnes, J., Ellen, F., Gelashvili, R., Zhu, L.: Why extension-based proofs fail. SIAM J. Comput. **52**(4), 913–944 (2023)
2. Attiya, H., Castañeda, A., Hendler, D., Perrin, M.: Separating lock-freedom from wait-freedom at every level of the consensus hierarchy. J. Parallel Distrib. Comput. **163**, 181–197 (2022)
3. Attiya, H., Castañeda, A., Rajsbaum, S.: Locally solvable tasks and the limitations of valency arguments. J. Parallel Distrib. Comput. **176**, 28–40 (2023)
4. Attiya, H., Fraigniaud, P., Paz, A., Rajsbaum, S.: One step forward, one step back: FLP-style proofs and the round-reduction technique for colorless tasks. In: DISC. LIPIcs, vol. 281, pp. 4:1–4:23. Schloss Dagstuhl - Leibniz-Zentrum für Informatik (2023)
5. Attiya, H., Welch, J.: Distributed Computing: Fundamentals, Simulations and Advanced Topics. John Wiley & Sons, Hoboken (2004)
6. Borowsky, E., Gafni, E.: A simple algorithmically reasoned characterization of wait-free computations. In: 16th ACM Symposium on Principles of Distributed Computing (PODC), pp. 189–198 (1997)
7. Brusse, K., Ellen, F.: Reductions and extension-based proofs. In: 40th ACM Symposium on Principles of Distributed Computing (PODC), pp. 497–507 (2021)
8. Cavorsi, M., Sabattini, L., Gil, S.: Multirobot adversarial resilience using control barrier functions. IEEE Trans. Rob. **40**, 797–815 (2024)
9. Charron-Bost, B., Függer, M., Nowak, T.: Approximate consensus in highly dynamic networks: the role of averaging algorithms. In: Halldórsson, M.M., Iwama, K., Kobayashi, N., Speckmann, B. (eds.) ICALP 2015. LNCS, vol. 9135, pp. 528–539. Springer, Heidelberg (2015). https://doi.org/10.1007/978-3-662-47666-6_42
10. Fischer, M.J., Lynch, N.A., Paterson, M.: Impossibility of distributed consensus with one faulty process. J. ACM **32**(2), 374–382 (1985)
11. Herlihy, M.: Impossibility results for asynchronous PRAM. In: 3rd ACM Symposium on Parallel Algorithms and Architectures (SPAA), pp. 327–336 (1991)
12. Herlihy, M., Kozlov, D.N., Rajsbaum, S.: Distributed Computing Through Combinatorial Topology. Morgan Kaufmann, Burlington (2013)
13. Herlihy, M., Rajsbaum, S., Raynal, M., Stainer, J.: From wait-free to arbitrary concurrent solo executions in colorless distributed computing. Theor. Comput. Sci. **683**, 1–21 (2017)

14. Hoest, G., Shavit, N.: Toward a topological characterization of asynchronous complexity. SIAM J. Comput. **36**(2), 457–497 (2006)
15. Rajsbaum, S.: Iterated shared memory models. In: López-Ortiz, A. (ed.) LATIN 2010. LNCS, vol. 6034, pp. 407–416. Springer, Heidelberg (2010). https://doi.org/10.1007/978-3-642-12200-2_36
16. Sheng, Y., Ellen, F.: Extension-based proofs for synchronous message passing. In: DISC. LIPIcs, vol. 209, pp. 36:1–36:17. Schloss Dagstuhl - Leibniz-Zentrum für Informatik (2021)
17. Shi, Y., Liu, W.: Brief announcement: colorless tasks and extension-based proofs. In: DISC. LIPIcs, vol. 319, pp. 54:1–54:6. Schloss Dagstuhl - Leibniz-Zentrum für Informatik (2024)

Exploration of Convex Terrains by a Deterministic Automaton with Pebbles

Mohamed Anouar Baaziz and Andrzej Pelc(✉)

Université du Québec en Outaouais, Gatineau, Canada
{baam03,pelc}@uqo.ca

Abstract. A mobile agent, modeled as a deterministic finite automaton, has to explore a convex terrain in the plane. The agent starts at an unknown point of the terrain and makes a series of moves. Before each move it takes a snapshot, which is the part of the terrain included in the disc of radius 1 centered at the current position of the agent. This snapshot is an input that causes the agent to possibly change state and make the next move in a chosen direction at a chosen distance less than 1, inside the terrain. The agent explores the terrain if every point of it can be eventually "seen" by the agent, i.e., is eventually in some snapshot. For many convex terrains, such exploration is impossible without marking the terrain in any way. Hence we allow the agent to use movable pebbles. A pebble can be dropped, subsequently seen by the agent if it returns to it, and possibly picked again.

We consider the problem of what is the minimum number of pebbles sufficient for an agent to explore convex terrains. Our main result is a complete solution of this problem. We classify all convex terrains into five types, depending on whether the terrain contains a (infinite straight) line and whether it is bounded. For each of these types, we determine the minimum number of pebbles sufficient for exploration of all terrains of a given type. The agent is given the type of the terrain but does not know in which terrain of the given type it is operating and does not know its starting point. In all cases we determine the minimum number of pebbles sufficient for exploration. Each result has two parts. In the positive part, we design an algorithm that explores all terrains of a given type by an agent, using a given number of pebbles. In the negative part, we show that no agent using fewer pebbles can explore all terrains of the given type.

Keywords: exploration · convex terrain · deterministic finite automaton · pebble · mobile agent

A. Pelc—Partially supported by NSERC discovery grant RGPIN-2024-03767 and by the Research Chair in Distributed Computing at the Université du Québec en Outaouais.

U. Schmid and R. Kuznets (Eds.): SIROCCO 2025, LNCS 15671, pp. 74–91, 2025.
https://doi.org/10.1007/978-3-031-91736-3_5

1 Introduction

The Background. Exploring terrains in the plane is an important task with many applications. They range from the mundane task of grass mowing to searching for a stranded person in a forest or decontaminating a terrain subject to a chemical attack. In all these applications, an agent that can be a human or a robot, must guarantee that it eventually sees every point of the terrain, i.e., that every point of the terrain is at some time within some vision radius (normalized to 1) from the agent. In many cases, such as chemical contamination, or the presence of mines put by an enemy, the task of exploration is too dangerous to be executed by a human, and hence we need to rely on a robot. Due to cost considerations, the memory and computational power of such a robot is often limited, hence it is natural to model it by a finite deterministic automaton.

The Model and the Problem. A terrain is defined as an open subset of the plane, i.e., such a subset T that if $x \in T$ then there exists a positive real ϵ such that the disc with center x and radius ϵ is included in T. The border of the terrain T consists of all points $b \notin T$ such that any disc with center b contains some points of T. In this paper we consider convex terrains: a terrain is *convex* if for any two points $x, y \in T$, the segment $[x, y]$ is included in T.

A mobile agent exploring the terrain is represented as a moving point whose actions are modeled by a finite deterministic automaton. It has a compass showing the cardinal directions. The agent starts at an unknown point of the terrain, and makes a series of *steps*. At the beginning of each step, it takes a *snapshot*, which is the intersection of the terrain with the disc of radius 1 centered at the current position of the agent. In other words, the agent "sees" the part of the terrain within distance 1 from its current position. This snapshot is a part of the input given to the automaton that causes the agent to transit to the next state and to make the next move in a chosen direction at a chosen distance less than 1, inside the terrain. The agent *explores* the terrain if every point of it can be eventually "seen" by the agent, i.e., is eventually in some snapshot. For many convex terrains, such exploration is impossible without marking the terrain in any way. This will be justified further in the paper. Hence we allow the agent to use *movable pebbles* that can be dropped at its current position, subsequently seen by the agent if it returns to the same point, and picked again. More precisely, the agent starts carrying some number $p \geq 0$ of pebbles. In any step, the agent is in some state, it is located in some point z of the terrain and carries a given number m (possibly 0) of pebbles. In the snapshot Σ taken in the current step, the agent sees the part of the terrain equal to Σ. It also learns if there is a pebble at its current position. If there is a pebble at its current position, we say that the agent *hits a pebble*. This information, together with the number m of pebbles currently carried by the agent, is an input I given to the automaton in the current step. The input I causes the agent to transit from state S to some state S'. Also, an agent in state S with input I produces an output (α, ϕ, d, n), where $\alpha \in \{pick, drop, leave\}$ is a *pebble action*, $d < 1$ is a distance, $0 \leq \phi < 2\pi$ is an angle, and n is a natural number indicating how many pebbles the agent will carry in the next step. This output causes the following behavior of the agent in the given step. First we describe the pebble action α. If the agent sees a pebble at its current position then it may either *pick* it (removing the pebble and increasing the number of carried pebbles by 1, i.e., $n = m + 1$) or perform the action *leave*, not changing the pebble configuration. If the agent does not see a pebble at its current position z and it carries at least one pebble, it may either *drop* a pebble in point z (decreasing the number of carried pebbles by 1, i.e., $n = m - 1$), or perform the action *leave*, not changing the pebble configuration. The agent cannot drop a pebble if there is already a pebble at its current position, or if it does not carry any pebble, and it cannot pick a pebble if there is no pebble at its current position. The other two parameters of the output describe the move of the agent in the current step. The agent goes at distance d in the direction given by angle ϕ with respect to North. We assume that the output is always such that the point z' at distance d from z in the prescribed direction is in the terrain. Hence, due to convexity, the whole move of the agent will be inside the terrain. At the beginning of the next step, the agent is at point z' in state S'. The trajectory of the agent is the infinite sequence of above described steps.

Notice that the output of the agent and the state to which it transits depend both on the current input and the current state. It is important to explain the dependence on the current snapshot (that is a part of the input). Consider the example of an agent that should explore all half-planes. Suppose that it is exploring the half-plane whose border is the line L, and assume that the agent is at distance $1/2$ from L. The current snapshot is the part of the disc of radius 1 centered at the agent, included in the half-plane. In this situation, the agent can navigate indefinitely in the direction *parallel* to the line L. Indeed, it sees a segment of L, hence it knows its angle γ with respect to North, and it can go, say, at distance $1/2$ in this direction (although the angle γ was not known when the automaton was constructed). In the next step, it will see another segment of L with the same angle γ with respect to North, so it can make another step in the same direction, using the same state, etc. The situation is drastically different, if we wanted the agent to navigate indefinitely in the direction *perpendicular* to the line L. As long as the agent sees a segment of L in its current snapshot, it can compute the angle $\gamma + \pi/2$ and go in this direction. More precisely, in the current state S, with this input that includes the information about the angle γ, the output (α, ϕ, d, n) can satisfy $\phi = \gamma + \pi/2$. However, as soon as the agent loses the sight of L, and its snapshot becomes a full disc of radius 1, then the agent cannot continue in direction $\gamma + \pi/2$ because it has no state that would give such output with this input. Of course, in the design of the automaton we can prepare states corresponding to moves in any finite set of directions (such as, e.g., the cardinal directions) but since γ is not known in advance at the time of constructing the automaton, there is no way to construct an automaton that will navigate indefinitely in a direction perpendicular to the border of every half-plane.

We now give a formalization of the above intuitive description. Let *Snapshots* be the family of all possible snapshots, included in the family of all subsets of the terrain. Let $\{e, f\}$ be the set of two possibilities for the current position z of the agent: e indicates that there is no pebble in z (z is empty), and f indicates that there is a pebble in z (z is full). Let $PebbleActions = \{pick, drop, leave\}$. Let $P = \{0, 1, \ldots, p\}$.

The actions of the mobile agent are formalized as a finite deterministic Mealy automaton $\mathcal{A} = (X, Y, Q, \delta, \lambda, S_0)$. $X = Snapshots \times \{e, f\} \times P$ is the input alphabet, $Y = PebbleActions \times [0, 1) \times [0, 2\pi) \times P$ is the output alphabet. Q is a finite set of states with a special state S_0 called initial. $\delta : Q \times X \to Q$ is the state transition function, and $\lambda : Q \times X \to Y$ is the output function.

The meaning of the input and output symbols is the following. At each step of its functioning, the agent is in some point z of the terrain, in some state S, and carries some number $m \in P$ of pebbles. It sees the snapshot $\Sigma \in Snapshots$ at the current point and learns the bit $b \in \{e, f\}$. The input $I = (\Sigma, b, m) \in X$ gives the automaton information about these facts. Given the state S and the input I, the agent outputs the symbol $\lambda(S, I) \in PebbleActions \times [0, 1) \times [0, 2\pi) \times P$ with the following meaning. The first term indicates the pebble action performed by the agent in the current step. The second and third term indicate the distance at which the agent should go and the angle between North and the move direction, respectively. The fourth term indicates with how many pebbles the agent transits to the next step. Since we imposed restrictions on pebble actions, linked them to the number of carried pebbles, and the agent can never leave the terrain, we have the following restrictions on the possible values of the output function λ. Let $I = (\Sigma, b, m) \in X$, and $\lambda(S, I) = (\alpha, d, \phi, n)$. Let z be the current position of the agent.

- if $b = f$ or $m = 0$ (there is a pebble at the current position of the agent or the agent carries no pebbles) then $\alpha \neq drop$;
- if $b = e$ (there is no pebble at the current position of the agent) then $\alpha \neq pick$;
- if $\alpha = drop$ then $n = m - 1$; if $\alpha = pick$ then $n = m + 1$.
- if z' is the point at distance d from z, such that the segment $[z, z']$ forms angle ϕ with North (i.e., z' is the destination of the agent in the next step) then $z' \in T$.

It is important to note that the Mealy automaton formalizing the agent is finite in the sense that is has a finite number of states, i.e., a finite memory, but both its input and output alphabets are infinite. This is obvious in the case of the input alphabet, as there are infinitely many possible snapshots. As for the output alphabet, it must also be infinite, in order to allow the agent to go in an arbitrary direction, based on the current snapshot. For example, if the agent sees in its snapshot a line with angle γ with respect to North, bordering the terrain, then it can go in parallel to this line, although the angle γ could not be predicted at the time of the algorithm construction.

Given the input symbol I and being in a current state S, the agent makes the changes indicated by the output function (it possibly drops or picks a pebble as indicated, possibly changes the number of carried pebbles as indicated, and goes to the indicated point of the terrain), and transits to state $\delta(S, I)$. The agent starts with p pebbles in the initial state S_0, and there are initially no pebbles in the terrain (hence its initial input symbol is $(\cdot, e, p)$).

Terminology. We will use the following geometric notions. A (straight, infinite) line in direction North-South is called *vertical* and a (straight, infinite) line in direction East-West is called *horizontal.* A *half-plane* with border L, where L is a line, is a maximal convex subset of the complement of L. (There are two half-planes with border L). A *belt* with borders L_1 and L_2, where L_1 and L_2 are two parallel lines, is the union of all open segments (u, v), where u is a point of L_1 and v is a point of L_2. Clearly, the plane, all half-planes and all belts are convex terrains.

A *disc* with center O and radius r, where r is any positive real number, is the set of points at distance less than r from O. A convex terrain is *bounded* if it is contained in some disc. A convex terrain that is unbounded but does not contain any line is called a *cup.*

Our Results. We first provide a full classification of all convex terrains into five disjoint types: bounded convex terrains, cups, belts, half-planes and the plane. Our main result establishes the minimum number of pebbles sufficient for a mobile agent modeled by a deterministic finite automaton to explore all convex terrains of a given type. The automaton is designed for a given type of terrains, without knowing which convex terrain of this type it will have to explore and what is its starting point.

We show that all bounded convex terrains can be explored by an automaton without pebbles. The minimum number of pebbles sufficient to explore all cups, or to explore all belts, is 2. Finally, the minimum number of pebbles sufficient to explore all half-planes, or to explore the plane, is 3.

In all cases, the minimum number of pebbles is established as follows. On the positive side, we show that some automaton using this number of pebbles can explore all convex terrains of a given type. We do this by designing an algorithm that accomplishes this task and can be executed by a deterministic finite automaton. On the negative side, we show that no automaton using fewer pebbles can explore all terrains of the given type.

The main methodological difficulty comes from the fact that, while the terrains to be explored are unbounded (apart from the easiest case of type 1 terrains), the agent exploring them has only finite memory and does not know which terrain it is exploring (the agent is designed to explore all terrains of a given type). Thus the exploration algorithms must be designed in a way to prevent the agent from "getting lost" in the terrain by entering a loop that would strand it in one direction, leaving some parts of the terrain unexplored. This has to be done using only few pebbles as markers. Organizing exploration in a way to prevent this danger, not knowing the explored terrain of a given type or the starting point, is our main algorithmic contribution.

Due to lack of space, all proofs and detailed descriptions of algorithms are omitted and will appear in the journal version of this paper.

Related Work. The task of exploration of various environments by mobile agents has been studied for several decades (cf. [27,33]). There are two prevalent models of the explored environment: it is either a geometric terrain, i.e., a part of the plane, possibly with opaque obstacles, or a network modeled as a graph. In the first case, the agents are modeled as points moving in the terrain with a certain radius of vision (possibly infinite), and the aim is that every point in the terrain has to be eventually "seen" by at least one agent. In the second case, mobile agents move along edges of a graph and the aim is that every node has to be eventually visited by at least one agent.

In the geometric scenario, most of the papers were concerned with exploration of bounded terrains, usually polygons with obstacles [2,14,15,17,29]. In this case, exploration is always feasible and the focus was on complexity of exploration [14], or on its competitive ratio [15,17,29]. A problem related to exploration of terrains is that of searching for a point [31] or for a line [4] in the plane. Another task accomplished by mobile agents in the plane is that of pattern formation [16,20,35]. Mobile agents, modeled as points moving in the plane have to form a pattern given to them *a priori* as input. This task has been mostly studied under the assumption that mobile agents are asynchronous robots without persistent memory, having full visibility of other robots' positions.

In the network scenario, the network can be either modeled as a strongly connected directed graph, where mobile agents can move only from tail to head of a directed edge [1,5,6,18,26], or it can be modeled as an undirected graph, where mobile agents can traverse each edge in both directions [3,8,11,19,22,23,32,34]. Another important modeling dichotomy concerns labeling of nodes. In some papers it is assumed that nodes have unique labels which agents can recognize [18,23,32], in others – nodes are anonymous [5,6,11,12,34].

Studies concerning exploration of graphs also differ regarding the adopted efficiency measure. One of them is the cost of exploration, i.e., the number of edge traversals, (cf., e.g., [32]), and the other is the memory size of the agent, measured either in bits or by the number of states of the finite automaton modeling the agent (cf., e.g., [19,25,26]). Some authors were concerned only with the feasibility of exploration of some classes of graphs by teams of automata, or by a single automaton with movable pebbles [7,9,11,19,25,30,34]. The explored graphs were usually finite and the task was to construct automata that explore a given class of these graphs. It was shown in [34] that no finite set of finite automata can cooperatively explore all cubic planar graphs. It was shown in [11] that a single automaton cannot explore all mazes, i.e., finite connected subgraphs of the infinite grid. Hoffmann [28] proved that this cannot be done even by an automaton with a single pebble. By contrast, Blum and Kozen [7] showed that this task can be accomplished by two cooperating automata or by a single automaton with two pebbles. The authors of [21] proved that an agent with $\Theta(\log \log n)$ distinguishable pebbles and bits of memory can explore any bounded-degree graph with at most n nodes, and that these bounds are tight. The size of port-labeled graphs which cannot be explored by a given automaton was studied in [25]. In [9], the authors investigated the minimum number of pebbles sufficient for an automaton to explore a single "wedge" of the infinite anonymous oriented grid. The latter paper differed from the previous ones in that there was a single graph, known in advance, to be explored but the graph was infinite. Many authors studied the problem of exploring the infinite anonymous oriented grid by cooperating agents modeled as either deterministic or probabilistic automata [10,13,24].

2 Classification of Convex Terrains

In this section, we provide a classification of all convex terrains in the plane into five disjoint types. The main criterion of our classification is whether the convex terrain contains a line or not. We have the following result.

Theorem 1. *If a convex terrain contains a line then it is either the plane, or a half-plane, or a belt.*

Theorem 1 implies the following classification of all convex terrains into five disjoint types: the plane, half-planes, belts, cups, and bounded convex terrains.

3 Basic Procedures and Conditions

Following the established custom in the literature on automata navigating in various environments, we will present the behavior of our agent by designing algorithms that can be executed by a finite deterministic automaton, rather than formally describing the construction of an automaton by defining its output and state transition functions.

We define three basic instructions that will be used in our algorithms: *drop a pebble*, *pick a pebble*, and $go(d, \phi)$. The first two do not need any explanation, and the third instructs the agent to go at distance d (where $d < 1$) in the direction forming angle ϕ with direction North (where $0 \leq \phi < 2\pi$). Thus, for example, $go(1/3, \pi)$ means "go South at distance 1/3" and $go(1/2, 3\pi/2)$ means "go East at distance 1/2". Since in many cases the distance traveled during a move will be 1/2, we will use the shorthand $step(\phi)$, instead of $go(1/2, \phi)$. Also, in many cases, the moves will be in the cardinal directions and we will use $step(N)$, $step(W)$, $step(S)$, $step(E)$, instead of $step(0)$, $step(\pi/2)$, $step(\pi)$, $step(3\pi/2)$, respectively.

Using the instruction $go(d, \phi)$, we define the simple procedure `go-to` c, where c is some point in the current snapshot of the agent.

Procedure `go-to` c

$\phi :=$ the angle with respect to North of the segment between the current position and c;
$d :=$ the distance from the current position to c;
$go(d, \phi)$;

When the current snapshot intersects a line L, we will often use the following procedure that moves the agent to a point of this line.

Procedure `Move-to`(L)

(a, b) := the intersection of the snapshot and L; c := the midpoint of (a, b); `go-to` c;

Moreover, in our algorithms, we will often use the following types of conditions regarding the current input of the agent. Condition "no pebble" means that there is no pebble at the current location of the agent. Condition "no border" means that the current snapshot of the agent is a full disc of radius 1 (i.e., the agent does not "see" the border). These conditions can be verified at each step of the algorithm.

For an angle $0 \leq \phi < 2\pi$, we say that the snapshot of an agent is *ϕ-free* if the point at distance 1/2 from the position of the agent, in direction ϕ, is in the snapshot. We will use expressions "N-free", "W-free", "S-free", "E-free" instead of "0-free", "$\pi/2$-free", "π-free", "$3\pi/2$-free", respectively. The third type of condition is "snapshot ϕ-free": it means that the current snapshot is ϕ-free. Thus we will use conditions "snapshot N-free", "snapshot W-free", "snapshot S-free", "snapshot E-free" with the above meaning.

Consider the system of Cartesian coordinates with origin at the current position of the agent, the x-axis pointing East and the y-axis pointing North. For two points $a = (x, y)$ and $a' = (x', y')$ in a snapshot, we say that:

a is above a', if $y > y'$; a is below a', if $y < y'$; a is left of a', if $x < x'$; a is right of a', if $x > x'$.

We will use these relations as conditions regarding the current input of the agent.

4 Bounded Convex Terrains

In this section, we show that all bounded convex terrains can be explored by a single automaton without pebbles. We start with a high-level description of the Algorithm `Exploration of Bounded Terrains`. It is divided into two parts: the agent executes first the procedure `Trip South` and then the procedure `Trip North`.

Procedure `Trip South` starts at the starting point of the agent. It proceeds in phases. In each phase, the agent goes West by steps of length 1/2, as long as its snapshot is W-free, and then goes East by steps of length 1/2, as long as its snapshot is E-free. Then the agent starts going West, again by steps of length 1/2. Let L be the horizontal line at distance 1/2 South of the current position of the agent. At each step, the agent verifies if its snapshot intersects the line L. If it does, this intersection is a segment (a, b) (a "window"), because of the convexity of the terrain. At the first step when this happens, the agent goes to the midpoint c of the segment (a, b) and a new phase starts. If no snapshot taken in the current phase intersects the line L, it means that the agent has explored all points of the terrain South of the current horizontal line. This concludes procedure `Trip South` and the agent starts procedure `Trip North`. Since the terrain is bounded, procedure `Trip South` must end in some phase.

Procedure `Trip North` results from replacing all occurrences of the word "South" by the word "North" in procedure `Trip South`. Upon completion of procedure `Trip North`, the agent stops. As before, procedure `Trip North` must end in some phase because the terrain is bounded. At this time, the entire terrain is explored.

We now give a pseudo-code of procedure `Trip South`

Procedure `Trip South`

continue := *true*;
while *continue* **do**
  while snapshot W-free **do**
    step(W);
  while snapshot E-free **do**
    step(E);
  $no - window$:= *true*; (** this indicates that there is no window in the snapshot **)
  while snapshot W-free **and** $no - window$ **do** (** try vertical **)
    L := horizontal line South of the current position, at distance 1/2 from it;
    if snapshot intersects L **then**
      `Move-to`(L); $no - window$:= *false*;
    else
      step(W);
  if $no - window$ **then**
    continue := *false*;

The pseudo-code of procedure `Trip North` differs from the pseudo-code of procedure `Trip South` by replacing the single occurrence of the word "South" by the word "North".
Now the exploration algorithm can be succinctly formulated as follows.

Algorithm `Exploration of Bounded Terrains`

`Trip South`; `Trip North`

The following theorem is the main result of this section.

Theorem 2. *Algorithm* `Exploration of Bounded Terrains` *can be executed by an automaton without pebbles and explores any bounded convex terrain.*

5 Cups

In this section, we show that the minimum number of pebbles sufficient to explore all cups is 2. We start with the high-level description of Algorithm `Exploration of Cups` that can be executed by an automaton using two pebbles. This algorithm is by far the most complicated of all our exploration algorithms. This is due to the fact that cups are on the one hand unbounded terrains, and on the other hand, they may have a border that does not consist of straight segments.

In view of the first fact, it is impossible to use the simple idea from the previous section that aims at finding the southern and northern "limits" of the terrain. On the contrary, the border of the terrain must somehow be followed to find the "bottom" of the cup. In view of the second fact, this is not easy to do, as the agent cannot travel "parallel to the border", as it will be able to do in the case of belts. Thus more complicated ideas must be used. The algorithm consists of three parts.

In the first part, which is the procedure `Find Border`, the agent travels along the vertical line containing its starting point p. First it puts the two pebbles at distance 1/2 from p, one North of p and the other South of p. Then it goes up and down by steps of length 1/2, from pebble to pebble, each time pushing the northern pebble North at distance 1/2 and pushing the southern pebble South at distance 1/2. Since the terrain is a cup, it does not contain any line and hence, at some point q, the snapshot of the agent must stop being N-free when the agent goes North, or stop being S-free, when the agent goes South. The agent drops the pebble at q, brings the other pebble and picks the pebble at q. The agent is at distance less than 1/2 from the border and carries both pebbles. This concludes procedure `Find Border`.

The second part of the algorithm is either the procedure `Orthogonal Expand Down`, or the procedure `Orthogonal Expand Up`, depending on how the procedure `Find Border` ended. Procedure `Orthogonal Expand Down` is called in the case when the snapshot of the agent stopped being N-free when the agent went North, and procedure `Orthogonal Expand Up` is called in the case when the snapshot of the agent stopped being S-free when the agent went South. Both procedures are executed in phases. In each phase, the aim of each of the procedures is to push a single pebble away from the border of the terrain (thus moving it down in the case of procedure `Orthogonal Expand Down` and moving it up in the case of procedure `Orthogonal Expand Up`) and to travel along perpendicular segments between the pebble and the border.

As both procedures are very similar and symmetric with respect to each other, we only describe procedure `Orthogonal Expand Down` (cf. Fig. 1). It starts at the above defined point q, where the agent drops a pebble. The agent carries the other pebble during the entire execution of the procedure: this part requires only one pebble. The procedure has three variants, depending on the characteristics of the snapshot at the point q. The variants are coded by the angle γ whose values can be $3\pi/4$, π or $5\pi/4$. For a given value of γ, we define $\gamma_1 = \gamma - 3\pi/4$ and $\gamma_2 = \gamma + 3\pi/4$. Procedure `Orthogonal Expand Down` ($\gamma$) works in phases. At the beginning of each phase, the agent is at the pebble. It moves in direction γ_1 by steps of length 1/2 until its snapshot stops being γ_1-free. Then it moves back to the pebble by steps of length 1/2. Then it moves in direction γ_2 by steps of length 1/2 until its snapshot stops being γ_2-free. Then it moves back to the pebble by steps of length 1/2. Finally the agent pushes the pebble in direction γ at distance 1/2 and drops the pebble. At this time a new, identical phase begins.

We now explain how the angle γ is decided on the basis of the snapshot at point q. At this point, the snapshot is not N-free. Consider the disc D of radius

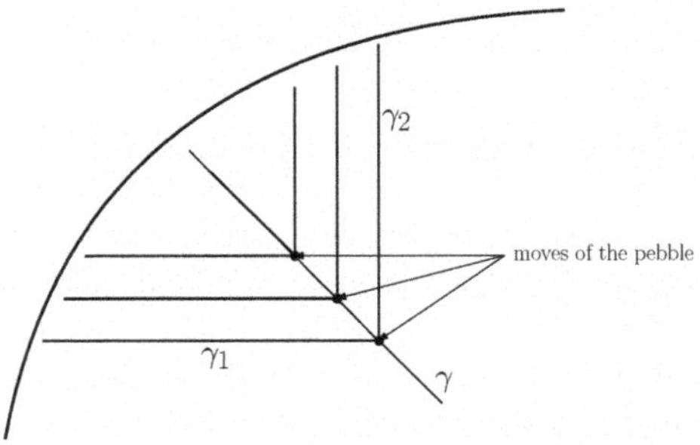

Fig. 1. Procedure `Orthogonal Expand Down`(γ)

1, centered at q and the circle C limiting this disc. Let B be the connected component of the intersection of the border with D that intersects the half-line H originating at q and going North. Consider the points a and b, where B intersects C, such that a is West of H and b is East of H. There are 3 cases: a can be above b, the segment (a, b) can be horizontal, or a can be below b. In the first case we take $\gamma = 3\pi/4$, in the second case we take $\gamma = \pi$, and in the third case we take $\gamma = 5\pi/4$. Note that, due to convexity of the terrain, in each of the three cases, the directions γ_1 and γ_2 must be obstructed by the border of the terrain, and thus the conditions "until its snapshot stops being γ_1-free" and "until its snapshot stops being γ_2-free" must be satisfied at some point.

There are two cases concerning the execution of procedure `Orthogonal Expand Down`. Either the phases are repeated indefinitely, if the pebble can always be pushed in direction γ at distance 1/2, or there is the last phase after which the snapshot is not γ-free, and hence this move is impossible. In the first case, due to the fact that consecutive routes of the agents in directions γ_1 and γ_2 are close, and they are expanded indefinitely, with their intersection moving farther and farther from the border of the terrain, the entire cup is eventually explored. In the second case, the procedure is stopped because of the obstruction by the southern part of the border. In this case, the agent starts the third part of the algorithm, which is the procedure `Swipe Vertical Down`. At this point, the agent carries one pebble and has just dropped the other pebble which will be called the eastern pebble.

The idea of procedure `Swipe Vertical Down` is similar to that of procedure `Trip South` from Sect. 4. As before, the aim is to get to the "limit" of the terrain, this time either to the "West limit" or to the "East limit". However, there is a major difference: while in bounded terrains we were sure to get to the "southern limit" (and then to the "northern limit", using the symmetric procedure `Trip North`), now we only know that we must reach either the "western limit" or the "eastern limit" but we do not know which. Hence, choosing only one direction and proceeding in it would be dangerous, as (with bad luck) this may never end and some points in the other direction might be left unexplored. To avoid this danger, we use two pebbles that are pushed one to the West and one to the East, and the agent expands its exploration in both directions. This way the agent eventually reaches the limit that actually exists, and then keeps exploring the terrain in the other direction indefinitely.

The first part of Procedure `Swipe Vertical Down` is procedure `Initiation`. It starts with the agent going North by steps of length 1/2. Let L be the vertical line at distance 1/2 West of the current position of the agent. At each step, the agent verifies if its snapshot intersects the line L. If it does, this intersection is a segment (a, b) (a "western window"), because of the convexity of the terrain. If this does not happen before the snapshot stops being N-free, the agent has explored the entire part of the terrain West of the current vertical line (it reached the "western limit"). Then it starts the procedure `Final Swipe Vertical East`, see below. Otherwise, at the first step when this happens, the agent goes to the midpoint c of the segment (a, b) (called the center of the window), and then South by steps of length 1/2, until its snapshot stops being S-free. At this point, the agent drops a pebble which will be called the western pebble. Then it goes North by steps of length 1/2, at each step verifying if its snapshot intersects the vertical line at distance 1/2 East of the current position of the agent. At some point this must happen. This intersection is a segment (a, b) (an "eastern window"), because of the convexity of the terrain. At the first step when this happens, the agent goes to the midpoint c of the segment (a, b) (called the center of the window), and then South by steps of length 1/2, until it hits the eastern pebble. Then the agent goes North looking for an eastern window. If this does not happen before the snapshot stops being N-free, the agent has explored the entire part of the terrain East of the current vertical line. Then it starts the procedure `Final Swipe Vertical West`, described below. If an eastern window is found, the agent goes South by steps of length 1/2, until it hits the eastern pebble again. The western pebble is also close to the southern part of the border, on a vertical line at distance 1/2 West from the current one. This concludes procedure `Initiation`.

The rest of procedure `Swipe Vertical Down` is executed in phases, each of them formalized as procedure `Phase of Expansion` (cf. Fig. 2). At the beginning of each phase, there are two pebbles close to the southern part of the border. The pebbles are on two vertical lines L_1 and L_2, where L_2 is at distance x, East from L_1, and the agent is at the eastern pebble. The aim of the phase is to relocate the eastern pebble and situate it close to the southern part of the border at a vertical line L'_2 at distance 1/2 East of L_2, and to relocate the western pebble and situate it close to the southern part of the border at a vertical line L'_1 at distance 1/2 West of L_1, as long as such actions are possible. (The lines L'_1 and L'_2 are at distance $x + 1$). This is done as follows.

The agent picks the eastern pebble and goes North looking for an eastern window. If this does not happen before the snapshot stops being N-free, the agent has explored the entire part of the terrain East of the current vertical line. Then it starts the procedure `Final Swipe Vertical West`, described below. If an eastern window is found, the agent goes to the center of the window (reaching line L'_2), and then South by steps of length 1/2, until its snapshot stops being S-free. At this point, the agent drops a pebble. Then the agent repeats the following actions, until it hits the western pebble: it goes North until it finds a western window. Then it goes to the center of this window, and then South by steps of length 1/2, until its snapshot stops being S-free. At some repetition it must hit the western pebble. The agent picks it and goes North until it finds a western window. If this does not happen before the snapshot stops being N-free, the agent has explored the entire part of the terrain West of the current vertical line. Then it starts the procedure `Final Swipe Vertical East`, see below. If a western window is found, the agent goes to its center (reaching line L_1'), and then South by steps of length 1/2, until its snapshot stops being S-free. At this point, the agent drops a pebble. Then the agent repeats the following actions, until it hits the eastern pebble: it goes North until it finds an eastern window. Then it goes to the center of this window, and then South by steps of length 1/2, until its snapshot stops being S-free. At some repetition it must hit the eastern pebble. This finishes the phase, and a new identical phase starts, if the current phase was not interrupted by switching either to procedure `Final Swipe Vertical West` or to procedure `Final Swipe Vertical East`.

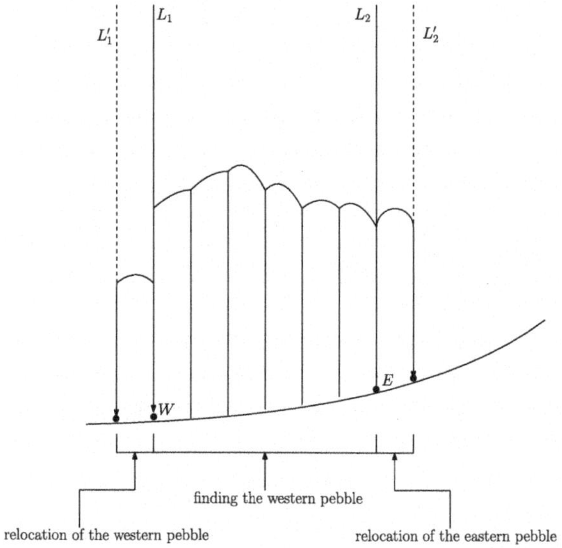

Fig. 2. Procedure `Phase of Expansion`

Since the terrain is a cup, at some phase the agent must switch either to procedure `Final Swipe Vertical West` (at which point it has explored the entire part of the terrain East of the current vertical line), or to procedure `Final Swipe Vertical East` (at which point it has explored the entire part of the terrain West of the current vertical line). The aim of these procedures is to complete the exploration of the terrain. We only describe procedure `Final Swipe Vertical West`, as procedure `Final Swipe Vertical East` is analogous.

Procedure `Final Swipe Vertical West` is executed in phases. No pebbles are used anymore. In each phase, the agent goes North by steps of length 1/2, as long as its snapshot is N-free. Then it goes South by steps of length 1/2, as long as its snapshot is S-free. Then it goes North by steps of length 1/2 until it finds a western window. This must happen because otherwise the terrain would be bounded. Now the agent goes to the center of this window. This finishes the current phase and a new identical phase begins. Phases are repeated indefinitely, and each point of the terrain is eventually in some snapshot of the agent.

This concludes the high-level description of the exploration algorithm.

Theorem 3. *Algorithm* `Exploration of Cups` *can be executed by an automaton with two pebbles, and explores every cup.*

In order to prove that the number of pebbles used by Algorithm `Exploration of Cups` cannot be improved, we use Lemma 1. We need the following notion. A *U-shape* is a subset U of the plane obtained as follows. Let I be any closed segment in the plane with extremities a and b. Let H_a and H_b be half-lines perpendicular to I with the same slope (i.e., perpendicular to I and "on the same side" of I), with origins a and b, respectively. U is the convex hull of $I \cup H_a \cup H_b$.

Lemma 1. *Consider any automaton using at most one pebble and navigating in the plane. Then the trajectory of the automaton is contained in some U-shape.*

Using Lemma 1 we can now prove the negative result of this section.

Lemma 2. *No automaton using at most one pebble can explore all cups.*

Theorem 3 and Lemma 2 imply the main result of this section.

Corollary 1. *The minimum number of pebbles sufficient for an automaton to explore all cups is 2.*

6 Belts

In this section, we show that the minimum number of pebbles sufficient to explore all belts is 2. We start with the high-level description of Algorithm `Exploration of Belts` that can be executed by an automaton using two pebbles. The algorithm consists of two parts.

In the first part, which is the procedure `Explore quadrant`, the agent explores a quarter of the plane until it sees the border of the belt. This is done in phases. In each phase, the agent pushes one pebble North at a distance 1/2, the other pebble East at a distance 1/2, and navigates between pebbles along a "staircase" consisting of "steps" of length 1/2 going West and North. Once the agent sees the border, i.e., when its snapshot is not a full disc, it learns the slope of it, which is the angle $0 \leq \gamma < \pi$ between the North and the border line, picks both pebbles, and starts the second part which depends on this slope.

If the border is vertical ($\gamma = 0$), the agent executes procedure `Horizontal swipes`. First, the agent approaches the western border at distance less than 1/2 and places the pebbles at distance 1 from each other, at the same distance from the western border. The rest of the exploration proceeds in phases that are repeated indefinitely. At the beginning of each phase, the agent is at the northern pebble. It goes East by steps of length 1/2 until the snapshot stops being E-free, and returns to the pebble by steps of length 1/2. Then it pushes the pebble North at distance 1/2 and goes South by steps of length 1/2 until it hits the southern pebble. Next, the agent goes East by steps of length 1/2 until the snapshot stops being E-free, and returns to the pebble by steps of length 1/2. Then it pushes the pebble South at distance 1/2 and goes North by steps of length 1/2 until it hits the northern pebble. This ends the current phase and the next identical phase starts.

If the border is not vertical, ($\gamma \neq 0$), the agent executes procedure `Vertical swipes` by traversing consecutive vertical segments from border to border. Let γ be the slope of the border, that the agent learned at the end of part one. First the agent approaches the southern border at distance less than 1/2. It places both pebbles at the same distance from this border and at distance 1 from each other. The rest of the exploration proceeds in phases that are repeated indefinitely. At the beginning of each phase, the agent is at the western pebble. It goes North by steps of length 1/2 until its snapshot stops being N-free, and then returns to the pebble by steps of length 1/2. Then it pushes the pebble in direction γ at distance 1/2. Next the agent goes in direction $\gamma + \pi$ by steps of length 1/2, until it hits the eastern pebble. It goes North by steps of length 1/2 until its snapshot stops being N-free, and then returns to the pebble by steps of length 1/2. Then it pushes the pebble in direction $\gamma + \pi$ at distance 1/2 and goes in direction γ by steps of length 1/2, until it hits the western pebble. This ends the current phase, and the next phase starts.

It should be stressed that when travelling (in both directions) along a line with slope γ, the agent is close to the border, i.e., it sees the slope of the border in its snapshot at each step. Hence (without remembering the parameter γ) it can execute a move *step*(γ) or *step*($\gamma + \pi$).

This concludes the high-level description of Algorithm `Exploration of Belts`.

Theorem 4. *Algorithm* `Exploration of Belts` *can be executed by an automaton with two pebbles, and explores every belt.*

In order to prove that the number of pebbles used by Algorithm `Exploration of Belts` cannot be improved, we use Lemma 3 whose proof is based on Lemma 1.

Lemma 3. *No automaton using at most one pebble can explore all belts.*

Theorem 4 and Lemma 3 imply the main result of this section.

Corollary 2. *The minimum number of pebbles sufficient for an automaton to explore all belts is 2.*

7 The Plane and the Half-Planes

In this section, we consider the two remaining types of convex terrains: the (single-element) type consisting of the plane and the type consisting of all half-planes. In both cases, we show that the minimum number of pebbles sufficient to explore all terrains of each of these types is 3. The positive parts of these results are a consequence of [9], and the negative parts are a consequence of [10]. While both above papers deal with exploration of the oriented grid, rather than the plane, the transition between these scenarios is fairly straightforward. The next result follows from [9].

Theorem 5. *1. There exists an automaton that explores the plane using 3 pebbles.*
2. There exists an automaton that explores all half-planes using 3 pebbles.

It remains to show matching lower bounds on the number of pebbles sufficient to explore the plane and on the number of pebbles sufficient to explore all half-planes. Here we use the result from [10] that implies the following theorem.

Theorem 6. *1. No automaton using at most 2 pebbles can explore the plane.*
2. No automaton using at most 2 pebbles can explore all the half-planes.

Theorems 5 and 6 imply the main result of this section.

Corollary 3. *1. The minimum number of pebbles sufficient for an automaton to explore the plane is 3.*
2. The minimum number of pebbles sufficient for an automaton to explore all half-planes is 3.

8 Conclusion

We classified all convex terrains into five types and, for each type, we established the minimum number of pebbles that are sufficient for a deterministic finite automaton to explore all terrains of this type. These minimum numbers are 0, 2, and 3, depending on the type. A natural open problem is to generalize these results to arbitrary, not necessarily convex, terrains. More precisely, propose a natural classification of all terrains into types, and, for each type, establish the minimum number of pebbles that are sufficient for an automaton to explore all terrains of this type. A more restricted but apparently still challenging question is the following: does there exist a deterministic finite automaton using a finite number of pebbles that can explore all terrains?

References

1. Albers, S., Henzinger, M.R.: Exploring unknown environments. SIAM J. Comput. **29**, 1164–1188 (2000)

2. Albers, S., Kursawe, K., Schuierer, S.: Exploring unknown environments with obstacles. Algorithmica **32**, 123–143 (2002)
3. Awerbuch, B., Betke, M., Rivest, R., Singh, M.: Piecemeal graph learning by a mobile robot. Inf. Comput. **152**(2), 155–172 (1999)
4. Baeza-Yates, R., Culberson, J., Rawlins, J.: Searching the plane. Inf. Comput. **106**, 234–252 (1993)
5. Bender, M., Fernandez, A., Ron, D., Sahai, A., Vadhan, S.: The power of a pebble: exploring and mapping directed graphs. Inf. Comput. **176**(1), 1–21 (2002)
6. Bender, M., Slonim, D.: The power of team exploration: two robots can learn unlabeled directed graphs. In: 35th Annual Symposium on Foundations of Computer Science (FOCS), pp. 75–85 (1994)
7. Blum, M., Kozen, D.: On the power of the compass (or, why mazes are easier to search than graphs). In: 19th Symposium on Foundations of Computer Science (FOCS), pp. 132–142 (1978)
8. Betke, M., Rivest, R., Singh, M.: Piecemeal learning of an unknown environment. Mach. Learn. **18**, 231–254 (1995)
9. Bhagat, S., Pelc, A.: Exploring wedges of an oriented grid by an automaton with pebbles. CoRR arxiv:2302.00052 (2023)
10. Brandt, S., Uitto, J., Wattenhofer, R.: A tight lower bound for semi-synchronous collaborative grid exploration. In: 32nd International Symposium on Distributed Computing (DISC), pp. 13:1–13:17 (2018)
11. Budach, L.: Automata and labyrinths. Math. Nachrichten **86**, 195–282 (1978)
12. Chalopin, J., Das, S., Kosowski, A.: Constructing a map of an anonymous graph: applications of universal sequences. In: 14th International Conference on Principles of Distributed Systems (OPODIS), pp. 119–134 (2010)
13. Cohen, L., Emek, Y., Louidor, O., Uitto, J.: Exploring an infinite space with finite memory scouts. In: 28th Annual ACM-SIAM Symposium on Discrete Algorithms (SODA), pp. 207–224 (2017)
14. Czyzowicz, J., Ilcinkas, D., Labourel, A., Pelc, A.: Worst-case optimal exploration of terrains with obstacles. Inf. Comput. **225**, 16–28 (2013)
15. Czyzowicz, J., Labourel, A., Pelc, A.: Optimality and competitiveness of exploring polygons by mobile robots. Inf. Comput. **209**, 74–88 (2011)
16. Das, S., Flocchini, P., Santoro, N., Yamashita, M.: On the computational power of oblivious robots: forming a series of geometric patterns. In: 29th Annual ACM Symposium on Principles of Distributed Computing (PODC), pp. 67–76 (2010)
17. Deng, X., Kameda, T., Papadimitriou, C.H.: How to learn an unknown environment I: the rectilinear case. J. ACM **45**(2), 215–245 (1998)
18. Deng, X., Papadimitriou, C.H.: Exploring an unknown graph. J. Graph Theory **32**(3), 265–297 (1999)
19. Diks, K., Fraigniaud, P., Kranakis, E., Pelc, A.: Tree exploration with little memory. J. Algor. **51**(1), 38–63 (2004)
20. Dieudonné, Y., Petit F., Villain, V.: Leader election problem versus pattern formation problem. In: 24th International Symposium on Distributed Computing (DISC), pp. 267–281 (2010)
21. Disser, Y., Hackfeld, J., Klimm, M.: Undirected graph exploration with $Theta(\log\log n)$ pebbles. In: 27th Annual ACM-SIAM Symposium on Discrete Algorithms (SODA), pp. 25–39 (2016)
22. Dudek, G., Jenkins, M., Milios, E., Wilkes, D.: Robotic exploration as graph construction. IEEE Trans. Rob. Autom. **7**(6), 859–865 (1991)

23. Duncan, C., Kobourov, S., Kumar, V.: Optimal constrained graph exploration. In: 12th Annual ACM-SIAM Symposium on Discrete Algorithms (SODA), pp. 807–814 (2001)
24. Emek, Y., Langner, T., Stolz, D., Uitto, J., Wattenhofer, R.: How many ants does it take to find the food? Theor. Comput. Sci. **608**, 255–267 (2015)
25. Fraigniaud, P., Ilcinkas, D., Peer, G., Pelc, A., Peleg, D.: Graph exploration by a finite automaton. In: Fiala, J., Koubek, V., Kratochvíl, J. (eds.) MFCS 2004. LNCS, vol. 3153, pp. 451–462. Springer, Heidelberg (2004). https://doi.org/10.1007/978-3-540-28629-5_34
26. Fraigniaud, P., Ilcinkas, D.: Directed graphs exploration with little memory. In: 21st Symposium on Theoretical Aspects of Computer Science (STACS). LNCS 1996, pp. 246–257 (2004)
27. Hemmerling, A.: Labyrinth Problems: Labyrinth-Searching Abilities of Automata. Teubner-Texte zur Mathematik, vol. 114. B. G. Teubner Verlagsgesellschaft, Leipzig (1989)
28. Hoffmann, F.: One pebble does not suffice to search plane labyrinths. In: Gécseg, F. (ed.) FCT 1981. LNCS, vol. 117, pp. 433–444. Springer, Heidelberg (1981). https://doi.org/10.1007/3-540-10854-8_47
29. Hoffmann, F., Icking, C., Klein, R., Kriegel, K.: The polygon exploration problem. SIAM J. Comput. **31**, 577–600 (2001)
30. Kozen, D.: Automata and planar graphs. Fund. Computat. Theory (FCT), 243–254 (1979)
31. Langetepe, E.: On the optimality of spiral search. In: Proceedings of 21st Annual ACM-SIAM Symposium Discrete Algorithm (SODA 2010), pp. 1–12 (2010)
32. Panaite, P., Pelc, A.: Exploring unknown undirected graphs. J. Algor. **33**(2), 81–295 (1999)
33. Rao, N., Kareti, S., Shi, W., Iyengar, S.: Robot navigation in unknown terrains: introductory survey of non-heuristic algorithms. Technical Report ORNL/TM-12410, Oak Ridge National Lab. (1993)
34. Rollik, H.A.: Automaten in planaren Graphen. Acta Informatica **13**, 287–298 (1980)
35. Suzuki, I., Yamashita, M.: Distributed anonymous mobile robots: formation of geometric patterns. SIAM J. Comput. **28**(4), 1347–1363 (1999)

Explicit Token-Based Communication for Mobile Entities

Balasingham Balamohan[1], Stefan Dobrev[2], Paola Flocchini[3], and Nicola Santoro[4](✉)

[1] Toronto, Canada
balamohan@alumni.utoronto.ca
[2] Slovak Academy of Science, Bratislava, Slovakia
Stefan.Dobrev@savba.sk
[3] School of Electrical Engineering and Computer Science, University of Ottawa, Ottawa, Canada
pflocchi@uottawa.ca
[4] School of Computer Science, Carleton University, Ottawa, Canada
santoro@scs.carleton.ca

Abstract. Consider a team of mobile computational entities, called *agents*, operating in a networked environment. To cooperatively solve problems and perform assigned tasks, the agents need to coordinate and communicate. The interaction mechanisms available to the agents have a strong impact on both computability and complexity.

In the literature, a wide range of communication mechanisms have been explored, ranging from powerful options that require significant system resources (such as *whiteboards*) to very basic and cost-effective methods, like identical passive objects known as *tokens*.

While algorithms that use whiteboards can effectively and explicitly communicate complex and resource-intensive information (like network maps), protocols that rely on token manipulation avoid to perform complex communication. Instead, they employ coordination and limited communication in an ad-hoc and implicit manner.

The contribution of this paper consists in the design of simple, efficient communication protocols employing only tokens (and space). These protocols can be used as subroutines for explicitly communicating complex information in systems providing only token-based communication.

This allows designers of token-based algorithms to explicitly employ complex communication between agents, enhancing their design freedom and flexibility. Additionally, our protocols facilitate the translation of results from models that depend solely on face-to-face communication to those that operate without such interactions but use tokens instead.

B. Balamohan—Independent Researcher.

U. Schmid and R. Kuznets (Eds.): SIROCCO 2025, LNCS 15671, pp. 92–108, 2025.
https://doi.org/10.1007/978-3-031-91736-3_6

1 Introduction

1.1 Framework

Teams of autonomous mobile computational entities, usually called *agents*, operating in a discrete networked space (e.g., a graph) have been investigated in a variety of fields.

In distributed computing, the agents operate without any external control, can move from node to neighbouring node, and execute the same protocol to achieve a common goal (e.g., performing a given task, solving a given problem).

The research interest is on the computability and complexity aspects of cooperative problem-solving by such teams. The investigations have focused on a variety of problems, including *Exploration*, where the agents must collaboratively visit all the nodes of the graph (e.g., [6,10,17,18,21]), *Gathering*, where the agents have to move to the same node not determined in advance (e.g., [2,7,8,13]), *Fault Localization*, where they have to determine the location of a faulty node which is harmful to the agents (e.g., [4,5,10–12,14,20]), etc.

To cooperatively solve a problem, the entities need to interact and communicate; how communication is performed has clearly a strong impact on the algorithmic techniques to be employed, and on the cost of the solution protocols.

In the literature, sometimes the communication issues are entirely circumvented by abstracting them away, thus avoiding to consider their design details and their costs. This is for example the case of the `Face-to-Face` communication model, where agents can explicitly exchange any arbitrary amount of information when they are at the same node (e.g., [3,8,9]).

In contrast, the main communication models considered in the literature do not avoid the issue; rather, they explicitly define the communication mechanism provided by the system to support the interaction and coordination among the entities.

For example, in the common `Whiteboard` model (e.g., see [2,6,11,15,19]), every node is equipped with a shared memory area, the whiteboard, accessible in fair mutual exclusion to any entity currently at that node; communication is achieved by reading and writing on the whiteboard. It is important to note that the `Whiteboard` model offers a direct and explicit communication mechanism, which is however very expensive in terms of network resources.

The theoretical quest has been for interaction and communication mechanisms that are less demanding in terms of system resources. One such mechanism is that used in the `Token` model (e.g., [5,7,10,12,17]): the system provides a (small) number of identical objects, called pebbles or tokens, which can be placed on nodes, picked up, and carried by the agents. Clearly, the `Token` model uses fewer resources (just a few bits per node) than the `Whiteboard` model. However, since the tokens are identical and do not contain explicit information, in the `Token` model the semantics of the communication (i.e., the information being communicated) is embedded in the manipulation of the tokens (i.e. pick up, drop) performed by the agents. It is thus natural that protocols that use tokens do so in a totally ad-hoc and specific way, and the information is always given in an implicit way (e.g., see [1,10,14]).

Not surprisingly, to date there is no efficient specific mechanism which offers explicit means of communication like those of a whiteboard, but instead uses only tokens. To fill this gap is precisely the main motivation of this paper.

1.2 Contributions

The central problem we address is: *How can an agent possessing information transfer this information efficiently to another (receiver) agent, using tokens only?* We call this 2- *party unidirection communication* (2-UC in short).

We investigate 2-UC thoroughly: We explore the boundaries of its solvability, present lower bounds on the communication cost, as well as present protocols matching the lower bounds in almost all cases. The crucial system parameters influencing the results are the number p of tokens, and the size n of the subset of nodes, called *rooms*, used for the communication. We take special care to ensure that our protocols can be seamlessly used as subroutines in higher-level agent computations, while minimizing the impact of this requirement on the cost of communication. In order to achieve that, we define a *strict* property of a protocol: *The terminal configuration must not occur during the execution.* This allows any agents not participating in the subroutine call to the protocol to recognize whether the protocol has completed its work, and prevent any undesired interaction messing-up the execution. A *normal* protocol drops the strictness requirement, as in some cases this safety is not needed or can be achieved by other means.

While the protocols we show are simple, deliberate design and careful attention to detail was needed to achieve matching the upper and lower bounds, as well as to minimize/eliminate the penalty for satisfying strictness.

Let $q = \binom{p+n-1}{n}$ and $q' = \binom{p+n-2}{n}$. The following table captures our results for the two communication problems we studied in detail. All non-bold results are tight. The values in the table represent the *communication base*: i means that in one round of interaction (both sender and receiver observing and modifying tokens) a number between 0 and $i-1$ can be communicated.

Table 1. Summary of our results. The values in the brackets are for normal termination. All non-bold results are tight.

problem	2-UC			Broadcasting		
tokens\\ rooms	$n=1$	$n=2$	$n>2$	$n=1$	$n=2$	$n>2$
$p=1$	*impossible*					
$p=2$	*imposs.*	2(3)	q	*imposs.*	2	q
$p=3$	1 (2)	**3** (6)	q	imposs. (1)	**3**	q
$p>3$	$p-2$ $(p-1)$	$\mathbf{q'}$ (q)	q	$p-3$ $(p-2)$	$\mathbf{q'}$	q

We then turn our attention to *multi-agents* communication. The protocols developed for that case of two agents can be easily employed as subroutines with which to design multi-agents protocols for efficient communication using tokens. In particular, we show how to use them to efficiently solve the most basic general multi-agents problem, *broadcasting*. The solutions for other general multi-agents communication problems (e.g., *convergecast*, *gossip*) can be built using techniques analogous to the ones we show here.

2 Model

2.1 The System

The system is composed of a set $\mathcal{A} = \{A_0, A_1, \ldots, A_{k-1}\}$ of k identical mobile computational entities, called *agents*, operating in a network viewed as an undirected connected graph $G = (V, E)$. The *communication* sub-system is composed of a set $\mathcal{R} = \{R_0, R_1, \ldots, R_{n-1}\} \subseteq V$ of $n \geq 1$ distinguished nodes, called *rooms*, and of $p \geq 1$ identical *tokens*, initially placed in room R_0, which can be picked up, held, carried and placed into rooms by the agents. More than one agent can stay in the same room at the same time, but they do not see other agents. Since G is connected, each room is obviously reachable from any other room. For ease of description, in the following we assume that agents can move directly from any room to any other.

When in a room, an agent sees how many tokens are placed in this room; it can pick up any subset of the tokens placed there, and it can drop there any subset of the tokens it holds. The operation of an agent accessing the tokens in a room and possibly altering their number is *atomic*, and granted in fair mutual exclusion (weak fairness is sufficient for our purposes).

The agents do not have any means of direct communication other than the tokens.

The timing and activations of the agents is controlled by a *semi-synchronous* adversary, called *scheduler*. Time is divided in discrete intervals, called steps; in each time step, the scheduler selects a non-empty subset of agents to activate. If several co-located agents are activated, only one of them succeeds accessing the room and remains active. The still active agents then apply (concurrently) their transition function $[r, a, s] \mapsto [r', a', s', l]$ where r is the number of tokens the agent sees in the current node, a is the number of tokens it holds, s is the agent's current state, r' is the new number of tokens in the current node, a' is the new number of tokens the agent holds, s' is the agent's new state, and l is the room (location) where the agent should move (ϵ means staying in the current room). Naturally, $r + a = r' + a'$ must be satisfied. Observe that the amount of time an agent spends performing any action (e.g., moving from a room to another, staying in a room before requesting access to the tokens placed there, etc.) is finite but otherwise unpredictable, and controlled by the scheduler. The only restriction on the adversarial power of the scheduler is that each agent is activated infinitely often; however, there is no finite bound on the number of consecutive steps an agent is activated.

We assume that the agents have unbounded memory, sufficient to store and reasonably manipulate the messages to be transmitted.

2.2 The Basic Problem: 2-Party Unidirectional Communication

The building block of our communication mechanism is an efficient solution of the *2-Party Unidirectional Communication* problem (2-UC) described here.

Consider the case of two distinct agents, a *sender* A_0 and a *receiver* A_1, each with its own protocol, operating in n rooms and using p tokens.

A *token configuration* $C = [r_0, r_1, ..., r_{n-1}; a_0, a_1]$ is a $n+2$-tuple where $0 \le r_i \le p$ indicates the number of tokens placed in R_i, and $0 \le a_j \le p$ indicates the number of tokens held by A_j; by definition, $\sum r_i + \sum a_j = p$. A *system configuration* includes also the locations of both agents. Finally, *a full configuration* includes also the states of both agents and the receiver's output tape. Conversely, a *room configuration* strips from the token configuration the information about the tokens each agent holds. Note that an agent observing the room configuration (and knowing how many tokens it holds in its hand) can calculate how many tokens are held by the other agent; i.e., it can deduce the token configuration. Hence, in the remainder of the paper, we will interchangeably use token-configuration/room-configuration/configuration.

Let $\mathcal{M}$ be a (possibly infinite but) countable set of *messages*. Consider an arbitrary ordering of elements of $\mathcal{M}$ and let us represent each message by its rank. Hence, w.l.o.g. $\mathcal{M} = \mathbb{Z}^+$.

The sender A_0 has a message $M \in \mathcal{M}$ it wants to transmit to the receiver, stored as part of its initial state. The receiver A_1 has a write-only output tape.

We say that the sender has *successfully transmitted* the input message M iff the receiver has written M to its output tape, both agents have entered a terminal state, and the system is in a terminal configuration[1]. By definition, in a starting configuration no agent holds any tokens.

A protocol *implements* (or *solves*) 2-UC if it allows the sender to successfully transmit any $M \in \mathcal{M}$ to the receiver within finite time.

Any solution protocol of 2-UC can be straightforwardly used to solve the more general problem of *2-Party Bidirectional Conversation* (2-BC) where the two agents alternate sending messages to each other. A solution for 2-BC can be implemented as a sequence of unidirectional transmissions, with the agents exchanging their roles as sender/receiver, for example as follows: $[p, 0, \dots, 0; 0, 0]$ is used as both starting and terminal configuration for 2-UC (allowing the agents to easily switch their roles on each termination of 2-UC and restart it again), a new message *pass* (indicating an empty message) is added to $\mathcal{M}$, and the termination is by both agents upon detecting two *pass* messages in a row.

Indeed, a protocol solving 2-UC can be used as the building block of a variety of token-based explicit communication mechanisms, each solving a more complex form of communication, possibly involving multiple agents.

While using 2-UC to solve 2-BC is straightforward, as in both protocols there are only two agents involved and the connection between invocations of 2-UC is trivial, in the general case (as e.g. in one-to-all *Broadcasting*) there are more agents involved that should not interfere while the subroutine call to 2-UC is ongoing.

[1] A terminal configuration need not to be different from a starting configuration.

In order to facilitate that, we define the following properties of a protocol:

- *strict*: the terminal configuration does not appear during the execution of the protocol, but only as its very last configuration
- *normal*: the terminal configuration could appear during the execution
- *relaxed*: the agents can hold tokens in the terminal configuration

Strict protocols are the easiest ones to use as a subroutine in more complex scenarios: in the case of *multi-agent* communication problems, the agents can reliably detect when the protocol terminates by observing the terminal configuration. Hence, they can avoid interference while the protocol is running, and seamlessly continue once the result of the subroutine is ready. The *normal* variant allows the two involved agents to continue working together, as they both know from their terminating states when the subroutine is completed; however other agents in the system might have problems cooperating with them as termination can be confused with an intermediate state of execution. The *relaxed* variant allows for one-off execution, without the need for follow-up work. For our purposes it is relevant only with regards to boundaries of when is the problem solvable at all. By default, we assume strict variant; when normal/relaxed variants are used/handled, we discuss this explicitly.

2.3 Basic Properties and Costs

The following notions capture crucial properties and behavioral aspects of token-based communication protocols.

We say that an agent A is *waiting* at time step t if, should it be exclusively activated in all successive steps, the system configuration remains unchanged; if it is not waiting the agent will be said to be *busy* at that time step.

We say that a protocol P is *bounded* if, in every sequence of activations, including unfair ones, the number of transformative operations (i.e., token manipulations and/or agent movements) performed by a busy agent before interacting with the other one is bounded.

Clearly, we are only interested in bounded protocols; an unbounded protocol does not really solve the problem, as it incurs unbounded cost. Hence, in the remainder of the paper, a solution protocol is always a bounded one.

In a bounded protocol, the agent behaviour can be viewed as cycling through the following phases:

- **waiting phase:** wait in some room, until a specific change in the number of tokens is observed
- **reading phase:** traverse the rooms and learn the token configuration
- **writing phase:** modify the token configuration
- **moving phase:** (optionally) move to a different room

We say that a solution protocol P is *alternating* if, in every execution E, in every time step: (1) at most one agent is in the combined reading/writing phase and (2) not all agents are in waiting phase. Note that this is a property ensured by the protocol, not by the scheduler.

Observe that it is possible to design a non-alternating protocol that is bounded, in which the agents concurrently manipulate the tokens. However, it is easier to design and prove correctness of alternating protocols. Furthermore, as we will show later, non-alternation cannot really help in reducing the communication cost. Therefore, we focus on designing alternating protocols.

Given a solution protocol P, we are interested in the amount of coordination and synchronization between the agents required by P for the communication of a message M. The *cost* is measured in terms of the number of *interactions* occurring between the two agents to perform the communication. In other words, the cost is the number of times the agent performing transformative activities (token manipulations and movements) changes, creating what we call a *context switch* or *alternation*.

Consider an execution E of protocol P using n rooms and p tokens, transmitting message M under activation schedule S. The cost of E, denoted by $\text{cost}_{n,p}(P, M, S)$ is the number of context switches in E. Since n and p are fixed and clear from context, we will omit them for the rest of the paper and simply use $\text{cost}(P, M, S)$.

Let $\text{cost}(P, M) = \max_S(\text{cost}(P, M, S))$, i.e. the worst case cost over all possible activation schedules.

As the protocol can decide to efficiently handle some specific large messages, we have to define the cost more robustly as follows: Let $m \in \mathbb{Z}^+$. Then

$$\text{cost}(P, m) = \max_{M \leq m} \text{cost}(P, M)$$

It turns out that our protocols (but also lower-bound proof arguments) are *uniform*, in the sense that the amount of information transmitted in each round (pair of context switches) is the same for all rounds. Hence, we can view the protocols as communicating in base q: One q-ary digit per round, where q is the number of possible values transmitted in one round (refer to Table 1). Let $\text{base}(P)$ denote the communication base of P. The translation between these two measures is straightforward:

$$\text{cost}(P, m) = 2 \left\lceil \log_{\text{base}(P)} m \right\rceil$$

3 Lower Bound and Impossibility Results for 2-UC

3.1 Impossibility Results

We say that a 2-agent scheduler is *wait-alternating* if it keeps activating the same agent until it starts waiting, and then switches to activating the other agent.

Let $\mathcal{M}$ be a set of m messages. Consider an execution of protocol P transmitting message $M \in \mathcal{M}$ under a wait-alternating scheduler. Let $\pi(P, M) = \{C_i\}$ be the sequence of token configurations at the moments when the scheduler switches activation between the agents. Note that as P is a valid protocol, $\pi(P, M)$ is finite.

Lemma 1. *If $\pi(P, M_0)$ is a prefix of $\pi(P, M_1)$ from some $M_0, M_1 \in \mathcal{M}$ then P does not solve 2-UC, even in the relaxed variant.*

Proof. By contradiction. Consider the states of the receiver agent A_1 during the two executions: first, A_0 transmitting M_0 and second, A_0 transmitting M_1, both under the same activation schedule. In the first case, A_1 must terminate, with M_0 on its output tape. As A_1 observes the same room configurations in both of these executions, until and including the very last token configuration of $\pi(P, M_0)$, it must terminate with M_0 as an output, even when the input message has been M_1.

Let us call the two configurations $D_0 = [0, \ldots, 0; p, 0]$ and $D_1 = [0, \ldots, 0; 0, p]$ *dead configurations*. Observe that a dead configuration can appear in $\pi(P, M)$ only as the last one: D_0 can only be reached by A_0 picking-up all remaining tokens, but then A_1 is unable to modify the token configuration (and vice versa for D_1). Hence, having a dead configuration as a terminal one prevents the protocol from being used as a subroutine with follow-up computation.

Let T_P be the rooted tree, rooted at the initial configuration, induced by the union of $\pi(P, M)$ for all $M \in \mathcal{M}$.

Theorem 1. *There is no solution protocol for the (non-relaxed) 2-UC problem when $n = 1$ and $p = 2$, even if the scheduler is fixed to be wait-alternating one.*

Proof. Observe that the initial token configuration $[2; 0, 0]$, as well as each of the remaining non-dead configurations ($[1; 1, 0]$, $[1; 0, 1]$ and $[0; 1, 1]$), have a single non-dead child in T_P. As solving 2-UC problem means reaching a terminal configuration with all tokens in rooms (i.e. unreachable from a dead configuration), all $\pi(P, M)$ are prefixes of the same path

$$[2; 0, 0] \mapsto ([1; 1, 0] \mapsto [0; 1, 1] \mapsto [1; 0, 1] \mapsto [2; 0, 0])^+$$

However, by Lemma 1 that means that P does not solve 2-UC.

As 2-party bidirectional conversation involves alternating at least two 2-party unidirectional transmissions, with only the last one being possibly relaxed (and only if the conversation is relaxed), this immediately yields:

Corollary 1. *No 2-party bidirectional conversation (even relaxed-one) is possible with 1 room and 2 agents.*

While having single room for communication is a problem, having a. single token for communication is serious trouble.

We say that $v \in T_P$ is *wait-friendly* if there exists a room r such that for each child u of v, the number of tokens in r differs between v and u. We say that the tree T_P is *wait-friendly* if all its nodes are wait-friendly.

Intuitively, if T_P is not wait-friendly, there is a scenario under which an agent cannot wait in a single room to learn that the other agent has executed its action, it has to cyclically check several rooms, ultimately incurring unbounded cost.

Lemma 2. *If T_P is not* wait-friendly *then P is not bounded.*

Theorem 2. *If $p = 1$, there is no protocol P solving the relaxed 2-UC, regardless of the number of rooms, even if the scheduler is wait-alternating one.*

Proof. Assume P solves the relaxed version of 2-UC. Consider its T_P. If T_P does not branch, by Lemma 1 P does not solve 2-UC. If T_P does branch, then the node v in which T_P branches is not *wait-friendly*: having single token means the children's configurations contain the token in different rooms. The theorem then follows from Lemma 2.

3.2 Lower Bound on Cost

Let $\mathcal{M}$ be a set of m messages and let l be the first level (the root is of level 0) of T_P containing at least m branches. Then, by Lemma 1, any protocol communicating all messages of $\mathcal{M}$ must perform at least l alternations, in the worst case.

In order to compute l as a function of m, we need to estimate the number of children of any even-level node $v \in T_P$. Note that, as the protocol is deterministic and the scheduler fixed, the nodes of even level branch are only due to different M's, not due to different activation schedules. Meanwhile, the nodes of odd level do not branch at all, for the same reason.

Let $v = [r_0, r_1, \ldots, r_{n-1}; a_0, a_1]$ be a node of T_P. Then, by definition of T_P, the number of v's children is at most $\binom{p-a_1+n}{n} - 1$: all $p - a_1$ tokens can be freely distributed among n rooms and A_0's hand, with the previous configuration being forbidden. This already leads to a substantial lower bound; however, we can tighten it more by careful consideration of how can agents efficiently coordinate/alternate token manipulation and by introducing a meaner scheduler, as shown in the following.

We say that a scheduler is *early-alternating* if it activates the agents according to the following algorithm:

Algorithm 1. Scheduler AlternateEarly

```
while true do
    activate A_1 until it starts waiting
    activate A_0 until A_1 wakes-up
end while
```

Observe that the early-alternating scheduler is asymmetric w.r.t. agents A_0 and A_1; similarly as wait-alternating scheduler, this scheduler is also not valid as it can get stuck in a busy agent.

Let us define $\pi'(P, M)$ and T'_P analogously to $\pi(P, M)$ and T_P, but using early-alternating scheduler. Note that Lemma 1 still applies, as its proof does not depend on the scheduler.

Lemma 3. *Consider an even-level node* $v = [r_0, r_1, \ldots, r_{p-1}; a_0, a_1] \in T'_P$. *Let* R *be the room in which* A_1 *waits (when the execution has reached* v*) until* A_0 *wakes it up, and let* $\beta = [b_0, b_1, \ldots, b_{p-1}]$ *be a tuple fully describing the wake-up logic of* A_1 *at this moment:* $\forall i : b_i \in \{0, 1\}$, *with* $b_i = 1$ *iff* A_1 *wakes-up upon observing* i *tokens in* R. *Let* j *be the smallest* i *for which* $b_i = 0$. *Let* $p' = p - a_1$ *be the number of tokens* A_0 *can manipulate and let* $\chi_v = 1$ *iff the protocol is strict and the terminal configuration would wake-up* A_1, 0 *otherwise.*

Then the number of children v *can have in* T'_P *is at most*

$$c(p', j, \beta) = \binom{p' - j + n - 1}{n} - \sum_{i \neq j : b_i = 0} \binom{p' - r_i + n - 1}{n - 1} + j\binom{p' - j + n - 1}{n - 1} - \chi_v$$

Let $\chi = 1$ for strict protocols, 0 otherwise. Using Lemma 3 allows us to prove the lower bound on the communication cost:

Theorem 3. *Let* P *be a solution protocol for* 2*-UC. Then* $base(P) \geq \binom{p+n-1}{n}$ *for* $n \geq 2$ *and* $base(P) \geq q = p - 1 - \chi$ *for* $n = 1$.

Proof. By Lemma 3 the number of children is maximized as $q = \binom{p+n-1}{n}$, obtained for $a_1 = j = 0$ and $\beta = [1, 0, \ldots, 0]$. If $n > 1$, it is possible to ensure $\chi_v = 0$: choose a terminal configuration containing all tokens in one room different from R. If $n = 1$, for strict protocols we have to subtract $\chi_v = 1$.

4 Alternating Protocols Implementing 2-UC

In this section we show alternating protocols solving 2-UC for the cases not covered by Theorems 1 and 2. We present the solution for strict termination, and point out when relaxing this might lead to higher efficiency.

For strict protocols, one might be tempted to think that the terminal state can be used to signal the end of message. However, that signalling would be done by the sender at the moment when the receiver has not yet terminated, violating the strictness requirement.

As the best way to indicate the end of message depends on the particular domain (n, p as well as the message size distribution), we will employ self-delimited messages to identify the end of transmission; this is abstracted in functions SELFDELIMITED(), its inverse GETDATA() and ISCOMPLETE(), with self-explanatory meanings.

We are trying to satisfy these design goals:

- ensuring seamless alternation, i.e. the next agent should start examining the room configuration only after the currently active agent has finished modifying it; this ensures that while reading and writing configurations is not atomic w.r.t. to the scheduler, they are always performed in full, without interruption
- maximizing the number of child configurations in each node of T_P, in order to minimize the communication cost, and
- satisfying strictness

As the agents know each other's protocol, after completing its reading phase each agent knows in which room the other agent is waiting, and what change will wake it up. Hence, the first goal is achieved by making sure that the last write of the agent's writing phase occurs in the other agent's waiting room.

The second goal is ensured by minimizing a_1 and j, while the third goal is satisfied by careful design of the receiver's confirmation that it has read the configuration left by the sender.

Let us first consider the general case of $n \geq 3$, $p \geq 2$. The *rank* of a configuration in a set X is defined as its rank (starting from 0) in an arbitrary, but fixed, total order of X (e.g. lexicographical). The high-level description of the resulting algorithm 2P-multiRoom is shown in the following pseudocode:

Algorithm 2. 2P-multiRoom (Receiver)

```
initialize M ← 0, Z ← 1, v ← [p, 0, ..., 0]
while true do
    wait in the R_1 until the number of tokens changes to a non-zero value
    let X be the set of all possible child configurations of v
    traverse all rooms, learn the new configuration and calculate its rank y in X
    M ← yZ + M
    Z ← |X|Z
    pick all tokens from R_1 and drop them in R_2        ▷ signalling receiverReady
    if isComplete(M) then
        output getData(M)
        terminate
    end if
    v ← the room configuration right after signalling receiverReady
end while
```

The algorithm 2P-multiRoom, as presented, works only for $n \geq 3$. If $n = 2$, the algorithm can be modified by replacing all references to R_2 by R_0. This works for normal mode, but does not for strict: If all tokens are in rooms, A_1 would, by signalling receiverReady, create the terminal configuration. The solution is for A_0 to always keep one token in hand, preventing A_1 to ever create a configuration with p tokens in a room. Note that this reduces the number of children to $\binom{p+n-2}{n} = \binom{p}{2}$, i.e. for $p = 2$ the protocol degrades to unary communication. However, by carefully designing a specific algorithm 2P-2Rooms2Tokens, we can achieve binary communication for this case as well.

If $n = 1$, yet another approach to signalling receiverReady needs to be taken: In even iterations of the main while loop (the very first iteration is even) A_1 signals by picking-up a single token, while in odd iteration A_1 signals by dropping its token. The rest of the Sender/Receiver algorithms is accordingly adjusted and actually mostly simplified, as there are no other rooms to deal with. Let us call the resulting algorithm 2P-singleRoom.

Lemma 4. *Algorithm 2P-multiRoom is alternating.*

Algorithm 3. 2P-MULTIROOM (Sender)

initialize $M \leftarrow$ SELFDELIMITED(M), $Z \leftarrow 1$, $v \leftarrow [p, 0, \ldots, 0]$
while true **do**
  let X be the set of all possible child configurations of v
  $y \leftarrow M \bmod |X|$
  $M \leftarrow \lfloor M/|X| \rfloor$
  $Z \leftarrow |X|Z$
  let v' be a room configuration with rank y within X
  create configuration v', while ensuring that the last token manipulation occurs in R_1
  wait in R_2 until a token drop by A_1 is observed ▷ i.e. VERIFYRECEIVERREADY
  if $M = 0$ **then**
    collect all tokens and drop them in R_0 ▷ create terminal configuration
    terminate
  end if
  $v \leftarrow$ the room configuration right after VERIFYRECEIVERREADY
end while

Proof. It follows from the construction, by induction on the number of rounds:

- A_1 always waits in R_1 for a non-zero number of tokens, while A_0 drops a token into R_1 as the last token manipulation of its writing phase.
- A_0 always waits in R_2 for a change in the number of tokens, while A_1 drops tokens only in R_2

As in 2P-SINGLEROOM, each agent performs just one action before starting to wait for a change in token configuration, and we have:

Lemma 5. *Algorithm 2P-SINGLEROOM is alternating.*

Let us give superscript i to the values used by these algorithms at the beginning of round i. Let us give subscripts s and r, to the values used by A_0 and A_1, respectively. Hence, M_s^i denotes the value of M in the sender's algorithm, at the beginning of round i.

The following statements can be easily proven inductively by construction:

- $v_s^i = v_r^i$ (initialized the same, updated in the same way)
- $X_s^i = X_r^i$ (both agents look at the same configuration v to compute its children X)
- $y_s^i = y_r^i$ (from alternation, A_1 reads the configuration written by A_0, both are computing the rank within the same set X)
- $Z_s^i = Z_r^i$ (from both using the same formula $Z \leftarrow |X|Z$ and same X)

From that, by induction and construction we get the loop invariant:

$$M_s^i Z_s^i + M_r^i = \text{SELFDELIMITED}(M)$$

Applying the loop invariant to the very last iteration, and combining with the (assumed) correctness of SELFDELIMITED(), GETDATA() and ISCOMPLETE() yields the main correctness theorem:

Theorem 4. *Algorithms 2P-*MULTIROOM *and 2P-*SINGLEROOM *correctly implement 2-UC*

Let us now focus on the complexity. For that purpose, we will calculate the cost for transmitting SELFDELIMITED(M) and not M directly (as that depends on the encoding).

Theorem 5. *Let* $q = \binom{p+n-1}{n}$ *and* $q' = \binom{p}{2}$. *Then*

- *for* $n > 2$, $base(2P - multiRoom) = q$, *even if strict*
- *for* $n = 2$, $p > 2$, $base(2P - multiRoom) = q$, *if normal*
- *for* $n = 2$, $p > 2$, $base(2P - multiRoom) = q'$, *if strict*
- *for* $n = 2$, $p = 2$, $base(2P - 2Rooms2Tokens) = 2$, *even if super strict*[2]
- *for* $n = 1$, $base(2P - singleRoom) = p - 1$, *if normal*
- *for* $n = 1$, $base(2P - singleRoom) = p - 2$, *if strict*

Observe that except for $n = 2$ and strict, the above upper bounds on the communication cost match the lower bounds precisely.

5 Multi-agents Communication

In the previous section, we have considered explicit communications using tokens when there are only two agents involved and developed optimal solution protocols. The protocol developed for that case forms a building block from which efficient protocols using tokens can be easily constructed for communications among multiple agents. The most common (and investigated) forms of general communication among multiple agents are: *one-to-all*, a problem known as *broadcast*; *all-to-one*, known as *convergecast*; and *all-to-all*, known as *gossip*; note that the more generic problems of *one-to-many*, *many-to-one* and *many-to-many* communication are special instances where the number of participants in the communication is restricted.

In this section we shall show how to use the building block to solve the most basic general multi-agents problem, *broadcasting*. The solutions for the other general problems can be built using techniques analogous to the ones we show here.

5.1 Broadcasting: Basics and Protocol

Consider the *Broadcasting* problem among k agents, using p tokens and n rooms. One agent (the sender, A_0) has a message M. The goal is for all other agents to learn the message M. Initially, all p tokens are placed in R_0. At the end, all tokens must end in R_0 and all agents are in termination state.

Because of the semi-synchronous activation scheduler, we get the following basic limitation:

[2] with both $[2, 0]$ and $[0, 2]$ forbidden.

Theorem 6. *If the number of agents in not known, then the broadcasting problem cannot be solved.*

Proof. Assume protocol P solves the broadcasting of some message M without knowing the number of agents in a scenario with k agents, terminating after t activations. Consider now a scenario where one additional receiver is added, but never activated in the first t time steps. All agents except the new one will terminate (not knowing k, nothing changed for them), and the new one has no chance to learn the message.

While knowing the number of agents involved might seem to be a trivial requirement to satisfy, if one wants to use the broadcasting as a subroutine involving a subset of agents satisfying some predicate, it might turn out to be not so trivial after all.

The basic idea is to use a strict version of 2-party communication as a building block. The receivers, one-by-one, communicate with the sender and learn the message. The key to our algorithm is the coordination mechanism between the sender and the receivers, enabling the receivers to communicate with the sender without messing with each other, as well as for the sender to learn when all receivers have learned the message.

We dedicate one configuration C_m to denote *the sender is ready* to transmit the message to the next receiver. The uninformed receivers are waiting until they observe C_m, then try to pick-up tokens. A receiver that is able to gain access and pick-up tokens is the one that will communicate with the sender and learn M, using a strict version of algorithm 2P-MULTIROOM with starting/terminal state set to C_m.

Hence, C_m reappears only after the communication from sender to receiver is completed, allowing the sender to proceed to sending to the next receiver. As we present the strict version of the broadcasting protocol, we employ a separate configuration C_s as its starting/terminal configuration. This means that also C_s must be excluded by 2P-MULTIROOM as a possible child configuration, i.e. if the broadcasting needs to be strict, the subroutine protocol for 2-UC needs to be *super strict.*

For $n \geq 4$, we choose $C_m = [0, p, 0, \ldots, 0]$ and $C_s = [p, 0, \ldots, 0]$. Algorithm 2P-MULTIROOM is modified by replacing R_i by $R_{i \oplus 1}$ where $\oplus$ is $+$ modulo n. Observe that, with such a choice, the number of child configurations is not reduced at all: Neither C_m nor C_s are valid child configurations, because they have 0 tokens in R_2, where A_1 waits for non-zero tokens to appear.

Algorithm 4. MP-MULTIROOM (Receiver)

1: wait until the number of tokens in R_1 is p and you gain access to R_1
2: pick-up one token from R_1 and drop it in R_0 ▷ receiver ready
3: execute (super) strict 2P-MULTIROOM as receiver to learn M
4: return to R_0, if required
5: **terminate**

Algorithm 5. MP-MULTIROOM (Sender)

```
initialize x ← k − 1
pick-up the tokens from R0 and drop them in R1                ▷ sender ready
while x > 0 do
    wait until a token appears in R0                          ▷ receiver ready
    pick-up the tokens from R0 and R1
    execute strict 2P-MULTIROOM as sender
    x ← x − 1                  ▷ the number of tokens in R1 is now restored to p
end while
collect all tokens and drop them in R0       ▷ create the terminal configuration
terminate
```

Note that the same technique of using extra rooms for terminal configurations with all tokens in a single room can be used for deeper nesting of subroutines, while paying nothing for super strictness – as long as there are rooms available.

If the number of rooms is $n = 3$, there are two problems to deal with:

- if $r_2 + r_0 = p$, RECEIVERREADY would create C_s (p tokens in r_0); the solution in such case is to distribute some tokens from R_2 also to R_1
- however, if $r_0 = p - 1$, A_0 waiting in R_0 would not observe any change; the solution in such case is for A_1 to move the token from R_2 to R_1 and for A_0 to wait in R_1 instead of R_0

Note that no reduction in efficiency (number of children) was necessary.

If the number of rooms is $n = 2$, we will use the strict 2-room version of 2P-MULTIROOM. Note that, as both C_s and C_m use p tokens in rooms, and the strict 2-room version of 2P-MULTIROOM uses only children with $p-1$ tokens, we do not have to worry that C_s or C_m will be accidentally created. For $p = 2$, protocol 2P-2ROOMS2TOKENS (already super strict) will be used as a subroutine, yielding 2 children per round.

Finally, if the number of rooms is $n = 1$, the same approach can be applied, with $C_s = [p]$ and $C_m = [p - 1]$ and using 2P-SINGLEROOM as a subroutine. However, that will make the solution for $p = 3$ infeasible: having only two usable room configurations, that must alternate, means that one agent will be collecting tokens while the other will be losing them – that can't continue for long. This can be partially saved if we do not insist on strict termination of the broadcasting algorithm. In such case, we can set $C_s = C_m$, and 2P-SINGLEROOM will work, working with 2 children each round.

5.2 Broadcasting: Correctness and Complexity

The key to the correctness of the algorithm is the fact that the communication is serialised:

Lemma 6. *At any moment of time, at most one receiver is in steps 2 or 3 of Algorithm 4.*

Proof. A receiver proceeds to step only after observing p tokens in R_1 and succesfully gaining access to R_1. Since gaining access and token manipulation is atomic, no other agent is able to proceed to step 2 before the number of tokens in R_1 is reduced. As a strict version of 2P-MULTIROOM is used in step 3, the configuration with p tokens in R_1 is restored only at the very end of step 3. Combining with construction (step 1 of receiver) this completes the proof.

The combination of step 2 of receiver and step 4 of sender algorithm, together with the correctness of 2P-MULTIROOM ensures that once a receiver succeeds in step 1, it will learn the message.

The fact that the sender's main loop is executed $k - 1$ times and that the starting and terminal configuration of 2P-MULTIROOM are the same ensures that all receivers learn the message. For the same reason, the sender terminates as well; its step 4 then ensures the correct terminal configuration.

Theorem 7. *Algorithm MP-MULTIROOM correctly solves the broadcasting problem.*

Regarding the communication cost: The only overhead is the initial move of tokens from R_0 to R_1 and the receiver's acknowledgment. The token manipulation in step 5 can be subsumed into 2P-MULTIROOM (more precisely, line 7 of Algorithm 3).

Theorem 8.
$cost(MP - multiRoom, M) = 2 + (k - 1) \times cost(2P - multiRoom, M)$

Acknowledgments.. This work has been partially supported by the Natural Sciences and Engineering Council of Canada, under Discovery Grants A2415 and 203254, and by VEGA 1/0117/25 of Slovak Republic.

References

1. Balamohan, B., Dobrev, S., Flocchini, P., Santoro, N.: Exploring an unknown dangerous graph with a constant number of tokens. Theor. Comput. Sci. **610**, 169–181 (2016)
2. Barrière, L., Flocchini, P., Fraigniaud, P., Santoro, N.: Rendezvous and election of mobile agents: Impact of sense of direction. Theory Comput. Syst. **40**(2), 143–162 (2007)
3. Barrière, L., et al.: Connected graph searching. Inf. Comput. **219**, 1–16 (2012)
4. Cai, J., Flocchini, P., Santoro, N.: Decontaminating a network from a black virus. Int. J. Network. Comput. **4**(1), 151–173 (2014)
5. Chalopin, J., Das, S., Labourel, A., Markou, E.: Tight bounds for black hole search with scattered agents in a synchronous ring. Theor. Comput. Sci. **509**, 70–85 (2013)
6. Das, S., Flocchini, P., Kutten, S., Nayak, A., Santoro, N.: Map construction of unknown graphs by multiple agents. Theor. Comput. Sci. **385**(1–3), 34–48 (2007)
7. Di Luna, G.A., Flocchini, P., Pagli, L., Prencipe, G., Santoro, N., Viglietta, G.: Gathering in dynamic rings. Theor. Comput. Sci. **811**, 79–98 (2020)

8. Dieudonne, Y., Pelc, A., Peleg, D.: Gathering despite mischief. ACM Trans. Algor. **11**(1), 1–28 (2014)
9. Dissaux, T., Fioravantes, F., Gahlawat, H., Nisse, N.: Recontamination helps a lot to hunt a rabbit. In: 48th International Symposium on Mathematical Foundations of Computer Science (MFCS), pp. 1–14 (2023)
10. Dobrev, S., Flocchini, P., Královič, R., Santoro, N.: Exploring an unknown dangerous graph using tokens. Theor. Comput. Sci. **472**, 28–45 (2013)
11. Dobrev, S., Flocchini, P., Prencipe, G., Santoro, N.: Searching for a black hole in arbitrary networks: optimal mobile agents protocolss. Distrib. Comput. **19**(1), 1–19 (2006)
12. Dobrev, S., Santoro, N., Shi, W.: Using scattered mobile agents to locate a black hole in an un-oriented ring with tokens. Int. J. Found. Comput. Sci. **19**(6), 1355–1372 (2008)
13. Eguchi, R., Kitamura, N., Izumi, T.: Fast neighborhood rendezvous. IEICE Trans. Inf. Syst. **E105–D**(3), 597–610 (2022)
14. Flocchini, P., Ilcinkas, D., Santoro, N.: Ping pong in dangerous graphs: optimal black hole search with pebbles. Algorithmica **62**(3–4), 1006–1033 (2012)
15. Flocchini, P., Kellett, M., Mason, P., Santoro, N.: Map construction and exploration by mobile agents scattered in a dangerous network. In: 23rd IEEE International Symposium on Parallel and Distributed Processing (IPDPS), pp. 1–10 (2009)
16. Flocchini, P., Prencipe, G., Santoro, N. (eds.): Distributed Computing by Mobile Entities. Springer, Heidelberg (2019)
17. Gotoh, T., Flocchini, P., Masuzawa, T., Santoro, N.: Exploration of dynamic networks: tight bounds on the number of agents. J. Comput. Syst. Sci. **122**, 1–18 (2021)
18. Gotoh, T., Sudo, Y., Ooshita, F., Kakugawa, H., Masuzawa, T.: Exploration of dynamic tori by multiple agents. Theor. Comput. Sci. **850**, 202–220 (2021)
19. Klasing, R., Markou, E., Radzik, T., Sarracco, F.: Approximation bounds for black hole search problems. Networks **52**(4), 216–226 (2008)
20. Markou, E., Shi, W.: Dangerous Graphs, Chapter 16 of [16]. Springer, Heidelberg (2019)
21. Nagahama, S., Ooshita, F., Inoue, M.: Ring exploration of myopic luminous robots with visibility more than one. Inf. Comput. **292**, 105036 (2023)

Red-Blue Pebbling with Multiple Processors: Time, Communication and Memory Trade-Offs

Toni Böhnlein(✉), Pál András Papp(✉), and Albert-Jan N. Yzelman

Computing Systems Lab, Huawei Zurich Research Center, Zurich, Switzerland
{toni.boehnlein,pal.andras.papp,albertjan.yzelman}@huawei.com

Abstract. The well-studied red-blue pebble game models the execution of an arbitrary computational DAG by a single processor over a two-level memory hierarchy. We present a natural generalization to a multiprocessor setting where each processor has its own limited fast memory, and all processors share unlimited slow memory. To our knowledge, this is the first thorough study that combines pebbling and DAG scheduling problems, capturing the computation of general workloads on multiple processors with memory constraints and communication costs. Our pebbling model enables us to analyze trade-offs between workload balancing, communication and memory limitations, and it captures real-world factors such as superlinear speedups due to parallelization.

Our results include upper and lower bounds on the pebbling cost, an analysis of a greedy pebbling strategy, and an extension of NP-hardness results for specific DAG classes from simpler models. For our main technical contribution, we show two inapproximability results that already hold for the long-standing problem of standard red-blue pebbling: (i) the optimal I/O cost cannot be approximated to any finite factor, and (ii) the optimal total cost (I/O+computation) can only be approximated to a limited constant factor, i.e., it does not allow for a polynomial-time approximation scheme. These results also carry over naturally to our multiprocessor pebbling model.

Keywords: Red-blue pebble game · Scheduling · Parallel computing · Limited memory · Approximation

1 Introduction

The computational requirements of modern applications, ranging from scientific simulations to artificial intelligence, necessitate parallel data processing. However, developing parallel algorithms that efficiently divide (sub)tasks, manage memory, and organize communication between processors presents significant difficulties to computer scientists. One particular bottleneck for the performance of parallel computations is due to data locality, i.e., the memory of modern hardware features a hierarchical structure, and to process data, it has

U. Schmid and R. Kuznets (Eds.): SIROCCO 2025, LNCS 15671, pp. 109–126, 2025.
https://doi.org/10.1007/978-3-031-91736-3_7

to be stored in the lowest layer. Since the layers' memory is limited in size, data movements between them (called I/O operations) are necessary and impact the execution time of computations significantly. The phenomenon becomes especially noticeable when dealing with tasks involving basic operations on large amounts of input data, e.g., the training of neural networks.

In this paper, we introduce a model offering improved insights into the challenges regarding I/O costs and limited memory in the context of parallel computing. Our focus is directed toward the efficient execution of a specific computation by several processors (rather than the design of a parallel algorithm tailored for a particular problem). We consider directed acyclic graphs (DAGs) as the model for a computation. The nodes correspond to single operations, while the directed edges express that the output data of a node is required as input for another node, enforcing precedence constraints on their order of execution. We are concerned with devising effective and efficient schedules for a given DAG and number of processors with limited working memory.

Hong and Kung [21] introduced the red-blue pebble game to study the I/O complexity of computations when executed by a *single* processor with a two-level memory hierarchy. The processor has *fast memory* (working memory) of limited capacity. Computing a node (and storing its output data in fast memory) requires that its predecessors' output data is stored in fast memory. To free up space in fast memory, I/O operations can transfer data to (and back from) *slow memory*, which has unlimited capacity.

In this model, available data is indicated by placing a *pebble* on the corresponding node. Data stored in fast memory is represented by a red pebble, and data in slow memory by a blue pebble. The game starts with an empty DAG. A red pebble can be placed on a node (i.e., computing the node) if all its predecessors have red pebbles on them. A red pebble can be placed on a node if it already has a blue pebble, and vice versa (I/O operations). Any pebble can be removed for free. The goal is to place pebbles on the sinks (outputs of the computation) while minimizing the number of I/O operations. The number of red pebbles used may not exceed a given fast memory capacity at any time.

Red-blue pebbling studies trade-offs between I/O costs and memory size. In our paper, we extend this model to a setting where several processors compute in parallel, and communication is organized via shared memory; this allows us to study trade-offs between time, communication and memory size.

To familiarize ourselves with pebbling, consider the example DAG depicted in Fig. 1. First, assume that we pebble it using only 3 red pebbles on a single processor. We may start by placing red pebbles on nodes v_1 and v_2, which then permits us to place a red pebble on node v_3. The red pebbles on v_1 and v_2 are no longer needed and can be removed. Observe that to pebble node v_4, we again need all 3 red pebbles. Since v_3 has an outgoing edge to node v_5, which we did not pebble yet, we place a blue pebble on v_3 (casting our first I/O operation), and then remove the red pebble from v_3. Next, v_4 (and its predecessors) can be pebbled analogously to v_3, leading to a configuration where we have a red pebble on v_4 and a blue pebble on v_3. We use another I/O operation to put a red

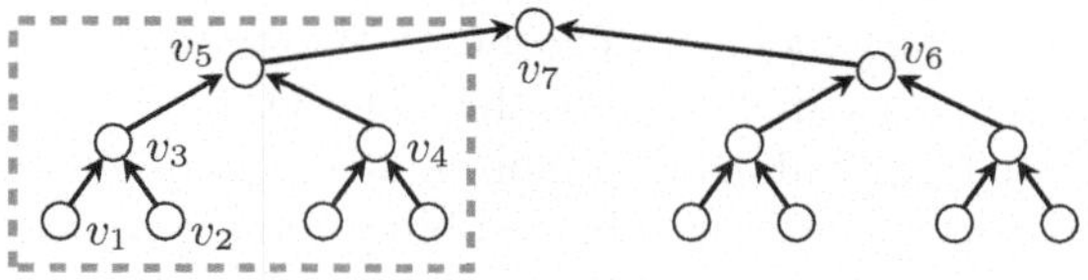

Fig. 1. A simple example DAG for pebbling.

pebble back on v_3, since it has a blue pebble. Now, we can place a red pebble on node v_5, and then remove the red pebbles from v_3 and v_4. Note that we pebbled the sub-graph in the dashed box with 2 I/O operations. Another I/O operation places a blue pebble on v_5. We pebble the subtree rooted v_6 in an identical way to the one rooted at v_5. To finish, we bring a red pebble back to v_5 with an I/O step, and then place a red pebble on v_7.

In contrast, consider pebbling the DAG with multiple processors for an intuitive description of our model. We use two processors p_1 and p_2 each having their own set of 3 red pebbles. To place a red pebble on a node, its predecessors require red pebbles of the *same* processor. We pebble the left and the right subtree of node v_7 with red pebbles of p_1 and p_2, respectively, using the strategy described above for a single processor. All steps are executed in parallel, reducing both the compute and I/O operations by a factor 2. The result is a configuration with a red pebble of p_1 and p_2 on nodes v_5 and v_6, respectively.

Recall that with one processor, we needed two I/O operations to convert the red pebble on v_5 to blue and then back to red. In contrast, with two processors we need I/O operations to ensure that we have red pebbles of the same processor on v_5 and v_6. The required communication between the processors is done via the shared memory (blue pebbles): We replace the red pebble of p_1 on v_5 by a blue pebble, and then the blue pebble by a red pebble of p_2. To finish the computation, we place a red pebble of p_2 on v_7.

Our Contribution. We present the multiprocessor red-blue pebbling game (MPP), which is a natural generalization of red-blue pebbling. MPP essentially combines the areas of DAG pebbling problems (single-processor computation with limited memory) and DAG scheduling problems (multiprocessor computation, but with unlimited memory). The result is a simple yet expressive model that captures computational costs, I/O costs incurred by memory limitations, and inter-processor communication costs. Unlike earlier attempts to generalize red-blue pebbling, MPP naturally combines these aspects into a single cost function, allowing for a convenient study of the trade-offs between these factors. These trade-offs have been analyzed in detail for many concrete computations before (e.g., matrix multiplication [24,25,40]), but not for arbitrary DAGs.

We first discuss the basic properties of MPP, including simple bounds on the pebbling cost, NP-hardness for specific DAG classes, and the analysis of a simple greedy approach. We also analyze how adding more processors (with the same, or reduced amount of fast memory) can affect the pebbling problem, i.e., how the optimal pebbling cost can change in the same DAG.

Then as our main technical result, we present two hardness results for approximating the optimal strategy in red-blue pebbling problems. Both of these proofs already apply to single-processor red-blue pebbling, thus presenting novel results for a problem that has been studied for several decades. Firstly, we show that the minimal number of I/O steps in a pebbling cannot be approximated in polynomial time to any finite multiplicative factor, or any additive $n^{1-\varepsilon}$ term (for any $\varepsilon > 0$), unless P = NP. Secondly, we show that the minimal total cost (including both I/O and computation steps) can only be approximated to a limited factor, i.e., there is a $\delta > 0$ such that the optimum cannot be approximated to a $(1+\delta)$ factor in polynomial time, unless P = NP; in other words, the problem is APX-hard, allowing no PTAS. Both of these results carry over naturally to our multiprocessor pebbling model.

2 Related Work

Pebble games on graphs are used to capture various aspects of computation. For instance, the standard (black) pebble game [12,30] models general time-memory trade-offs, with the results including bounds on achievable trade-offs [22,26,32] and the complexity of computing pebbling strategies [2,13,18,36]. Other pebble games aim to capture different properties of computing, such as non-determinism or reversibility [3,6,11,15].

Hong and Kung [21] introduce the red-blue pebble game to study I/O complexity, and derive lower bounds based on a DAG partition technique. It is applied to specific computations in [17,31]. Further bounds on I/O costs were derived with different methods in [19,23,35]. Other works [16,33,34] extend the red-blue pebble game to single and multiple processors over memory hierarchies. The work of [24,25] refines the technique of [21] and derive improved I/O lower bounds for special DAGs. The bounds are extended to a multiprocessor setting where workload is perfectly balanced. In contrast to MPP, these works do not consider trade-offs between computation and I/O.

As for the complexity of computing optimal pebbling strategies, Demaine and Liu [14] show that standard red-blue pebbling is PSPACE-complete, and propose variations that are shown NP-complete. NP-hardness for red-blue pebbling with computation costs is also shown in [7]. The work of [28] shows an inapproximability to a $(2-\varepsilon)$ factor, assuming the unique games conjecture. The work of [9] presents a bi-criteria approximation for the optimal I/O cost to a $\text{poly}\log(n)$ factor, provided that it is allowed to violate the memory bound by a $\text{poly}\log(n)$ factor.

Naturally, there are also several more realistic models that capture both parallelization and memory limitations, but these exhibit key differences from MPP. The work of [5,10] studies the cost components (computation, different kinds of I/O) separately, mostly assuming that I/O steps cannot be parallelized. Other works assume simpler models regarding computation costs: they are not considered at all in [4], and are handled with artificial balance constraints in [19, 37]. Moreover, these models are primarily used for studying specific computations (e.g., Matrix multiplication, FFT) rather than arbitrary DAGs.

Valiant [39] introduces the bulk-synchronous parallel (BSP) computing model with explicit synchronisation, which was extended later [27,38] to include data locality and communication via shared memory. The model was further generalized to capture multi-processor/core architectures with multi-level memory [40]. These models describe practical compute systems more closely, and also match our MPP model. There are also theoretical studies on DAG scheduling in BSP [29], but without memory restrictions.

3 Model Definition

Our model for a computation is a directed acyclic graph (DAG) $G = (V, E)$. We use the notation $n = |V|$ for the number of nodes. The in/out-degree of a node is the number of its incoming/outgoing edges. We call nodes with in/out-degree 0 the source/sink nodes. Let Δ_{in} denote the highest in-degree in our DAG, i.e., the highest number of inputs to an operation in our computation; several previous works on pebbling assume that Δ_{in} is small, e.g., a constant [14,28]. Moreover, we denote the set of positive integers by $\mathbb{Z}^+$, and use the notation $[a] = \{1, ..., a\}$, for $a \in \mathbb{Z}^+$. Our goal is to execute a computational DAG on k processors, each having limited fast memory of size r, for parameters $k, r \in \mathbb{Z}^+$.

3.1 Single Processor Red-Blue Pebbling

Before introducing our parallel model, we briefly recap the single-processor red-blue pebble game (SPP) [21], where we have a single processor with fast memory of size r and unlimited slow memory. Nodes with red pebbles on them (denoted $R \subseteq V$) and nodes with blue pebbles on them (denoted $B \subseteq V$) correspond to the output data that is currently saved in fast and slow memory, respectively. In order to pebble a DAG, the following rules can be applied:

(R1-S) Place a red pebble on a node that has a blue pebble,
(R2-S) Place a blue pebble on a node that has a red pebble,
(R3-S) Place a red pebble on a node if all its predecessors have red pebbles,
(R4-S) Remove a (red or blue) pebble.

The game starts with an empty DAG; rule (R3-S) allows us to place red pebbles on source nodes. A pebbling strategy is a sequence of the rules which places pebbles on the sink nodes in the end, while the number of red pebbles does not exceed r at any step. The I/O costs are the total number I/O operations, i.e., applications of rules (R1-S) or (R2-S). The goal is to pebble a DAG with minimum I/O costs. Most works on SPP analyze lower bounds on I/O cost for a given DAG and memory limit r.

Additional variations of SPP have also been proposed to simplify SPP, make it more realistic, or resolve the fact that this base SPP variant is not even in NP, but rather PSPACE-complete. This is because the optimal pebbling sequence can be super-polynomially long in extreme cases, since repeatedly deleting and recomputing the same node incurs no cost. Notable SPP variants include:

- *One-shot SPP:* (R3-S) can be applied only once for each node [9],
- *No-deletion SPP:* (R4-S) is not allowed [14],
- *SPP with computation costs:* (R3-S) also incurs a small cost of ε [7,28].

The first two variants feature somewhat artificial restrictions to prohibit the deletion and then recomputation of a node entirely. On the other hand, SPP with computation costs is more realistic, discouraging this in a natural way by ensuring that computation steps also incur some cost, as in practice. This last SPP variant is also the closest one our multiprocessor pebbling model.

We note that besides this, there are also smaller aspects of the definition that vary over different previous works. For instance, some papers assume instead that source nodes already have a blue pebble initially, or that sink nodes specifically need to have a blue pebble in the end; this is also the case in the original work of [21]. For most proof constructions, these model variants can be reduced to each other with some simple tricks; we refer to [28] for a summary. Similarly, some works assume in rule (R1-S) that the new red pebble replaces the blue pebble, and in rule (R2-S) vice versa [19,28]; our SPP-related claims also carry over easily to this variant.

3.2 Multiple Processor Red-Blue Pebbling

We introduce the multiprocessor red-blue pebble game (MPP) which extends red-blue pebbling to a setting where k processors compute a DAG in parallel. Data stored in a processor's fast memory is represented by a red pebble of its *shade*, i.e., there are red pebbles of k different shades. The number of red pebbles of each shade is limited by r. The processors share unlimited slow memory that is used to (i) store data that cannot be kept in fast memory, and (ii) to communicate data between processors. Data available in the slow memory is represented by blue pebbles. Note that we assume that several pebbles of different shade/color can be placed on a node at the same time.

In our parallel version of the game, the rules allow multiple processors to simultaneously either transfer data between fast and slow memory, or compute nodes. More formally, define set $R^j \subseteq V$ as the set of *red pebbles* of *shade* j, for $j \in [k]$, and set $B \subseteq V$ as the set of *blue pebbles*. For $m \in \mathbb{Z}^+$ such that $m \leq k$, we call an injective function $f_m : [m] \to [k]$ a *shaded selection*, and extend the transition rules as follows:

(R1-M) For a shaded selection f_m and vertices $v_1, v_2, \ldots, v_m$ such that $v_i \in R^{f_m(i)}$, add v_i to B, for $i \in [m]$,

(R2-M) For a shaded selection f_m and vertices $\{v_1, v_2, \ldots, v_m\} \subseteq B$, add v_i to $R^{f_m(i)}$, for $i \in [m]$,

(R3-M) For a shaded selection f_m and vertices $v_1, v_2, \ldots, v_m$ such that for all predecessors u of v_i we have $u \in R^{f_m(i)}$, for $i \in [m]$, add v_i to $R^{f_m(i)}$, for $i \in [m]$,

(R4-M) Remove a (red or blue) pebble.

A parallel version of (R4-M) could also be defined; we use this simpler version since this rule will incur no cost. Note that any single rule places at most k pebbles, and that processors can be idle.

A *configuration* is a $(k+1)$-tuple $C_i = (R_i^1, R_i^2, \ldots, R_i^k, B_i) \subseteq V^{k+1}$ defining red and blue pebbles placed on the DAG. A configuration C_i is *valid* if $|R_i^j| \leq r$, for $j \in [k]$, i.e., it respects the fast memory size. An initial configuration C_0 sets $R_0^j = \emptyset$, for $j \in [k]$, and $B_0 = \emptyset$. Let $S \subseteq V$ be the sink nodes. We say configuration C_i is terminal, if $S \subseteq B_i \bigcup_{j=1}^{k} R_i^j$ holds. A *pebbling strategy* $(C_0, C_1, \ldots, C_T)$, for $T \in \mathbb{Z}^+$, is a sequence of valid configurations such that (i) C_i is obtained by applying a transition rule to C_{i-1}, for $i \in [T]$, and (ii) C_0 and C_T are initial and terminal, respectively. Equivalently, the pebbling strategy can be represented by a sequence of transition rules $(t_1, t_2, \ldots, t_T)$, where $t_i \in \{$(R1-M), (R2-M), (R3-M), (R4-M)$\}$ such that C_i is the result of applying t_i to C_{i-1}, for $i \in [T]$. To assign costs to a pebbling strategy, we assign costs to each rule. Let $g \in \mathbb{Z}^+$ be a parameter specifying the cost of an I/O step. We define the cost function for the transition rules as follows:

- $c(t_i) = g$ if $t_i =$ (R1-M) or (R2-M),
- $c(t_i) = 1$ if $t_i =$ (R3-M),
- $c(t_i) = 0$ if $t_i =$ (R4-M).

Then, the cost of a strategy $C(t_1, t_2, \ldots, t_T) = \sum_{i=1}^{T} c(t_i)$ is the total costs of its moves. Given a DAG G and parameters $k, r, g \in \mathbb{Z}^+$, the goal of MPP is to find a minimum cost pebbling strategy. We denote the cost of the optimum pebbling strategy by OPT.

3.3 Model Discussion

We first note that as in SPP, the rules of MPP allow for *recomputation*. Indeed, when I/O is expensive, it can sometimes be beneficial to compute the same node more than once; see Sect. 4 for an example.

Furthermore, the transition rules assume that the processors compute and access memory *synchronously*. This simplifies the analysis greatly, enabling us to formulate the cost function as a linear term of I/O and compute costs. However, in some practical settings, one processor may be computing while another one is accessing memory. This could be modelled by allowing each processor in a step to execute one of the SPP rules independently; however, assigning costs to a pebbling strategy then becomes an intricate matter. We expect that most of the general reasoning about pebbling strategies also carries over to such an asynchronous setting; it has been shown that the improvements from a non-synchronous schedule are limited to a factor 2 [29]. Synchronization of communication is also natural in some hardware architectures, and several parallel programming models, like BSP [39], also feature synchronization steps.

While parallel computing models typically study trade-offs between computation time and communication, SPP studies trade-offs between I/O costs and memory size. Combining these in MPP allows us to study the three-fold trade-off between computation time, communication, and memory size. The I/O steps in MPP can happen either due to (i) communicating data between processors, or (ii) saving data to slow memory to free up space in fast memory.

We note that several models in previous works are closely related to MPP; we discuss these in detail in the full version of the paper [8]. For instance, [19] also outlines a generalization of SPP to multiple processors; however, in contrast to MPP, their work captures computation costs via an artificial balance constraint, which imposes heavy limitations on the model. We also note that with $r = \infty$ and minor adjustments, MPP also becomes equivalent to DAG scheduling in the BSP model [29]. This shows that with small variations, MPP is indeed a generalization of both SPP and DAG scheduling problems.

4 Fundamental Properties of MPP

Straightforward Bounds. Similarly to SPP, if $r \leq \Delta_{in}$, then there can be no valid pebbling strategy, since a node of in-degree Δ_{in} requires $(\Delta_{in} + 1)$ red pebbles of the same shade to be computed: Δ_{in} on its in-neighbors, and one more on the node itself. Thus we always implicitly assume $r \geq \Delta_{in} + 1$.

However, if $r \geq \Delta_{in} + 1$, there is indeed always a valid pebbling. Consider the nodes in any topological order. We can always select any processor p to compute the next node v, load all the (already computed) in-neighbors of v from slow memory to p (at a cost of at most $\Delta_{in} \cdot g$), compute the node on p (at a cost of 1), save the value of v to slow memory (at a cost of g), and then remove the red pebbles from v and its in-neighbors (for free). This strategy incurs a cost of at most $(\Delta_{in} + 1) \cdot g + 1$ for each node. On the other hand, since each (R3-M) step can compute at most k nodes, the number of compute steps is at least $\frac{n}{k}$. This shows the following simple bounds for the optimal cost.

Lemma 1. *For any instance of MPP, we have* $\frac{n}{k} \leq \mathit{OPT} \leq (g \cdot (\Delta_{in} + 1) + 1) \cdot n$.

A Simple Example Gadget. We briefly discuss a simple example, the *zipper gadget*, which highlights many vital aspects of pebbling problems, and different variants of it are used throughout our proofs. We only describe the gadget and these properties here on a high level, leaving the details to the full version [8]. The gadget, shown in Fig. 2, consists of two input groups S_1, S_2 of d nodes each, and a *main chain* of n_0 nodes, with the input groups having edges to the main chain nodes in an alternating fashion. We typically have $n_0 = n - O(1)$ to make S_1 and S_2 asymptotically irrelevant.

- The gadget was used in [28] to analyze trade-offs in SPP. E.g. for $r = 2d+2$, we can always keep red pebbles on both S_1 and S_2, and use the last 2 red pebbles to compute the main chain without any I/O steps. However, for $r = d+2$, we still need 2 red pebbles to compute along the main chain, so we repeatedly need to move the other d red pebbles between S_1 and S_2, yielding a much higher cost.

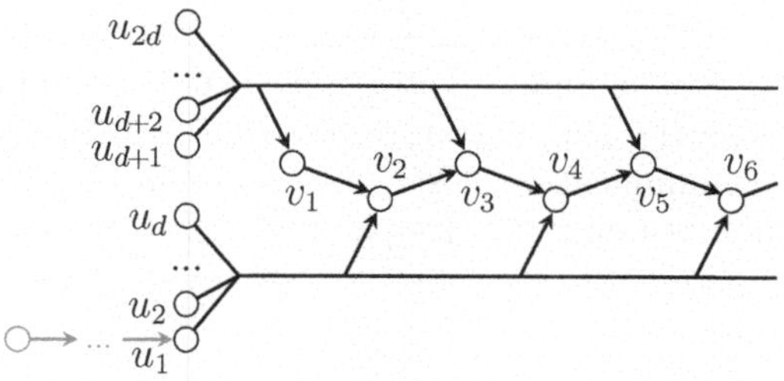

Fig. 2. Zipper gadget consisting of 2 input groups $S_1 = \{u_1, u_2, \ldots, u_d\}$ and $S_2 = \{u_{d+1}, \ldots, u_{2d}\}$, and a main chain $v_1, \ldots, v_{n_0}$. The edges going from the input groups are combined into a single arrow for simplicity. The extension to discourage recomputation is only illustrated for u_1 in gray.

- This example with $r = d+2$ also highlights the role of recomputation: we repeatedly need to move d red pebbles between S_1 and S_2, and doing this with I/O steps (repeatedly loading from slow memory) incurs a cost of $d \cdot g$ for each main chain node. However, since the nodes in S_1 and S_2 are sources, we can also simply compute them again with (R3-M) steps anytime, at a much lower cost of d per main chain node. To ensure that such a recomputation is suboptimal, we can also attach a chain of length $2g$ in front of each u_i: then recomputing u_i from a source node requires $2g + 1$ compute steps, whereas saving u_i to slow memory and loading it back later costs at most $2g$.
- Considering the gadget in MPP, if $k = 1$ and $r = d+2$, we again need to keep moving the d red pebbles between S_1 and S_2, which incurs a cost of $(d \cdot g + 1)$ per main chain node. However, with $k = 2$ processors and $r = d+2$, we can keep S_1 and S_2 in the fast memory of different processors, compute the main chain alternatingly, and only communicate these main chain nodes. This incurs a cost of $(2 \cdot g + 1)$ per chain node, and thus results in a superlinear speedup for larger d values.

NP-Hardness. Regarding the complexity of MPP, it is not surprising that the problem is NP-hard, since it generalizes SPP. However, MPP is already NP-hard on rather simple DAGs.

Lemma 2. *MPP is already NP-hard on the following subclasses of DAGs:*

- *2-layer DAGs (where the longest path has length* 1*),*
- *in-trees (where every out-degree is at most* 1*).*

The lemma follows from [29], where the same results are established for BSP scheduling; however, it requires some further work to adapt these proof constructions to MPP.

A Greedy Algorithm. It is also natural to wonder if we can obtain good solutions e.g. by following a greedy pebbling strategy. We can derive a naive upper bound on this approach, using the long known results that in DAG scheduling without memory limits or communication costs, any non-idle greedy strategy is a 2-approximation of the optimum [20]. Employing such a greedy approach for the computations, and assuming the worst-case strategy discussed for Lemma 1, we get the following bound.

Lemma 3. *Any pebbling where the compute steps follow a non-idle greedy schedule gives a* $2 \cdot (g \cdot (\Delta_{in} + 1) + 1)$*-factor approximation of the optimum.*

Unsurprisingly, greedy strategies can also return rather bad solutions. In general, it is even non-trivial to precisely define a greedy strategy for MPP, since pebbling consist of various aspects. Here we make observations for a general class of greedy strategies: we only assume that processor p always picks the next node to compute as the yet uncomputed node with the largest number (or largest fraction) of in-neighbors having a red pebble of p. We show a lower bound for any such greedy strategy, regardless of how compute steps are parallelized, how ties are broken, or how I/O steps are applied to compute the chosen node.

Lemma 4. *There exist DAGs where any such greedy pebbling algorithm is worse than the optimum*

- *by a* $\frac{1}{5} \cdot \Delta_{in} - 1$ *factor asymptotically (for any* $\Delta_{in} = O(1)$*),*
- *by a* $\frac{2}{3} \cdot g + 1$ *factor asymptotically (for any* $g \geq 2$*).*

Lower Bounds on Pebbling Costs. We now discuss how we can apply lower bounds on the number of required I/O steps from SPP to bound the optimum cost in MPP. The key insight is that a pebbling strategy for MPP can be implemented using a single processor (SPP) with fast memory of size $r \cdot k$. Specifically, each parallel rule can be simulated using k sequential rules.

Lemma 5. *Let G be a DAG such that an SPP pebbling strategy with fast memory of size $k \cdot r$ requires at least L steps of I/O, for some $L \in \mathbb{Z}^+$. Then, an MPP pebbling strategy with k processors, each having fast memory of size r, requires at least L/k steps of I/O.*

Corollary 1. *Let G be a DAG such that an SPP pebbling strategy with fast memory of size $k \cdot r$ requires at least L steps of I/O, for some $L \in \mathbb{Z}^+$. Then, an MPP strategy with k processors and fast memory of size r each has cost of at least $g \cdot L/k + n/k$.*

It follows that we can translate lower bounds for a single processor to k processors. Indeed, many previous works [21,25] derive I/O lower bounds for specific computations in SPP, which are obtained utilizing a special DAG partition. For example, Hong and Kung [21] derive a lower bound of $\frac{n \log n}{\log(rk)}$ on the number of required I/Os in SPP (with fast memory size $r \cdot k$) for the n-point FFT DAG.

This translates to a lower bound of $\frac{n}{k} \cdot (g \cdot \frac{\log n}{\log(rk)} + 1)$ on the cost in MPP for the same DAG. Another well-studied computation is matrix-matrix multiplication. Kwasniewski et al. [25] improve the technique of [21] and derive a lower bound of $\frac{2n^3}{\sqrt{rk}} + n^2$ in the single processor case, resulting in a lower bound of $\frac{n}{k} \cdot (g \cdot (\frac{2n^2}{\sqrt{rk}} + n) + 1)$ on the costs of matrix-matrix multiplication in MPP.

Finally, we show that there are instances where this bound is essentially tight.

Lemma 6. *For any n, there is a DAG construction with $\mathtt{OPT} \leq g \cdot L/k + n/k + O(1)$ in MPP.*

5 The Impact of More Processors: Trade-Offs Between k, r and OPT

We next analyze how more available processors can affect the optimal pebbling strategy for a DAG, which captures fundamental trade-offs for parallel computing. For convenience, we compare the simplest case of 1 processor to k processors, but our proofs are easy to carry over to any k-fold increase in the number of processors. We use $\mathtt{OPT}^{(k)}$ as a short notation for the optimal pebbling cost with k processors.

There are two natural ways to do this comparison. Let r_0 denote the amount of fast memory in the 1-processor case. One option is to compare this to a setting with k processors and $r := \frac{r_0}{k}$ fast memory on each; we call this the *fair* comparison, as the total fast memory over all processors remains unchanged. Another option is to compare to a setting with k processors and simply $r := r_0$ for each, i.e., all processors having the same fast memory r_0 as before. This *practical* comparison is more relevant for applications, where computations are often parallelized by simply employing more processors of the same kind.

"Fair" Comparison. We first compare the case of 1 processor with $r = r_0$, to the case of k processors with $r = \frac{r_0}{k}$. This is an interesting setting: on the one hand, the k processors allow for parallelization of computations and I/O, but on the other hand, processors have less fast memory, possibly resulting in further I/O steps to save and reload data. We first show that the optimum cost can decrease by a factor k at most; intuitively, this is because in the fair case, any pebbling strategy with k processors can be transformed into a 1-processor schedule with at most k times the cost. The bound is tight, as can be seen in e.g. a DAG with k independent chains of length $\frac{n}{k}$.

Lemma 7. *In the fair case, we have $\frac{\mathtt{OPT}^{(k)}}{\mathtt{OPT}^{(1)}} \geq \frac{1}{k}$, and there are DAGs such that $\frac{\mathtt{OPT}^{(k)}}{\mathtt{OPT}^{(1)}} = \frac{1}{k}$.*

With the same fast memory scattered over k processors, the optimum can also increase notably.

Lemma 8. *In the fair case, there is a construction with*

$$\frac{OPT^{(k)}}{OPT^{(1)}} \geq \frac{k-1}{k} \cdot g \cdot (\Delta_{in} - 1) + 1 - o(1)\,.$$

Recall that $\mathtt{OPT}^{(1)} \geq n$ and $\mathtt{OPT}^{(k)} \leq (g \cdot (\Delta_{in} + 1) + 1) \cdot n$, so this bound is essentially tight for large k and Δ_{in}. Another construction shows that the optimum can also be non-monotonic in k.

Lemma 9. *In the fair case, we can have* $OPT^{(2)} \leq OPT^{(1)}$ *and* $OPT^{(2)} \leq OPT^{(4)}$.

"Practical" Comparison. We now compare MPP with 1 and with k processors, with the processor(s) having the same $r = r_0$ in both cases. Here the larger k value only comes with advantages, so the optimum can never increase; on the other hand, it can easily decrease as before, e.g., for k independent chains. What makes this setting more interesting is that the optimum can decrease by a factor larger than k; such a *superlinear speedup* is a well-known (and highly desired) phenomenon in real-world systems. To our knowledge, MPP is the first DAG scheduling or pebbling model that naturally captures this behavior.

Lemma 10. *In the practical case, for any* $\varepsilon > 0$, *we can have* $\frac{OPT^{(1)}}{OPT^{(2)}} \geq \frac{\Delta_{in}-1}{2} - \varepsilon$.

The proof idea has already been outlined in Sect. 4 with the zipper gadget. With the appropriate Δ_{in}, the lemma can achieve a speedup of any constant factor, already for $k = 2$.

The Number of I/O Steps. We also briefly study the number of I/O steps separately: let $\mathtt{OPT}_{I/O}{}^{(k)}$ denote the number of steps (R1-M) and (R2-M) in the optimal pebbling with k processors. Our observations here hold for both the fair and the practical case.

Since $k = 1$ requires no communication, whereas $k = 2$ might, it is easy to construct a DAG where $\mathtt{OPT}_{I/O}{}^{(1)} = 0$, but $\mathtt{OPT}_{I/O}{}^{(2)} = \Theta(n)$. More surprisingly, we can also have a similar decrease in I/O, i.e., a DAG with $\mathtt{OPT}_{I/O}{}^{(1)} = \Theta(n)$, but $\mathtt{OPT}_{I/O}{}^{(2)} = 0$. Intuitively, this can happen when the computations can only be distributed in a very imbalanced way, so with 2 processors, it becomes more beneficial to do a lot of recomputations with one of the processors, instead of using I/O steps.

6 Inapproximability

Given the NP-hardness of MPP, a natural follow-up question is whether the optimum can be approximated to some factor in polynomial time. We show that MPP is also hard from this perspective.

Theorem 1. *MPP is APX-hard: there is a constant* $\delta > 0$ *such that no polynomial-time algorithm can approximate the optimum to a* $(1 + \delta)$ *factor, unless* $P = NP$.

We prove this property already for the simplest case of $k = 1$, i.e., SPP with computation costs. The proof can then easily be extended to any k value.

Lemma 11. *SPP with computation costs is APX-hard.*

Proof (Proof sketch). Our proof is motivated by a construction in [28], which reduces one-shot SPP to vertex cover on a graph of N nodes to show another property. However, this proof considers SPP, where (R3-S) compute steps are free, and hence the proof is rather careless with the number of nodes, using very large gadgets to ensure asymptotic behavior. As a result, the I/O cost is only $\Theta(N) = o(n)$. With computation costs adding up to at least n in SPP, this makes the I/O costs asymptotically irrelevant. Hence the reduction does not work directly for pebbling models with computation cost.

To adapt this idea, we substantially decrease the size of gadgets to ensure that $n = O(N)$, and hence the I/O cost of any solution is in $\Theta(N) = \Theta(n)$. By design, in any reasonable solution, a specific part of this I/O cost is proportional to the size of a vertex cover in the underlying graph; as such, an approximation for the best pebbling could be transformed into an approximation of vertex cover. The main ingredients of the proof are as follows:

- we show how to modify the node gadgets in the construction of [28] to reduce their size to constant, thus ensuring that our construction has $n = \Theta(N)$ altogether,
- we execute some further changes in the gadgets to avoid some other undesired properties, which could be ignored in the original proof of [28] due to their asymptotic analysis,
- we then show that in any reasonable pebbling, the I/O cost has a linear term that is proportional to the size of a vertex cover in the underlying graph.

Altogether, this modified construction allows for an L-reduction to the vertex cover problem in 3-regular graphs, which is known to be APX-hard [1]. This implies that it is already NP-hard to find a pebbling of cost at most $(1+\delta) \cdot \texttt{OPT}$ for an appropriate $\delta > 0$.

In general, the approximability of MPP is an intricate question, as the total cost is often dominated by computation costs, which are at least $\frac{n}{k}$. Thus, even if we have two solutions where I/O costs differ by a large factor, this can translate to only a negligible difference in total cost. To better capture these differences, we introduce an alternative cost metric.

Definition 1. *Given an MPP pebbling of cost C, its* surplus cost *is $C - \frac{n}{k}$.*

Intuitively speaking, surplus cost ignores this unavoidable cost of $\frac{n}{k}$, and instead only measures "imperfections" in the pebbling, such as I/O steps, work imbalance between the processors, and recomputations. In terms of this new metric, finding a good solution is much more challenging: it is not possible to approximate the surplus cost in MPP to any finite factor. We again show this via proving an inapproximability result in standard one-shot SPP (i.e. minimizing I/O costs, with computations being free): we show that it is already NP-hard here to distinguish the cases when $\texttt{OPT} = 0$ and when $\texttt{OPT} \geq n^{1-\varepsilon}$, for any $\varepsilon > 0$.

Theorem 2. *In one-shot SPP, it is NP-hard to approximate the optimum to any finite multiplicative factor, and to any additive $n^{1-\varepsilon}$ term (for any $\varepsilon > 0$).*

Proof Sketch. The main part of our proof is to develop a DAG construction where it is already NP-hard to decide whether one-shot SPP has $\mathtt{OPT} = 0$ or $\mathtt{OPT} \geq 1$. After this, some technical steps allow for extending the same result to an optimum of either 0 or $n^{1-\varepsilon}$.

When looking for a solution of cost 0, pebbling becomes notably simpler: blue pebbles cannot be used at all, since (R1-S) and (R2-S) incur cost. The use of (R4-S) is also simple: without recomputation, a red pebble should be deleted exactly when all out-neighbors have been pebbled. As such, a pebbling is characterized by the order of the n computation steps.

We point out that pebbling with only (R3-S) and (R4-S) is essentially equivalent to one-shot *black pebbling*, which is long known to be NP-hard [36]; as such, the novelty of this part of our theorem is somewhat limited. However, [36] considers a slightly different variant of pebbling, where compute steps can also "slide" a pebble from an in-neighbor. It may also be possible to adapt the proof in [36] to our case with further work; instead, we present a novel, somewhat simpler reduction based on a different problem, and we also devise new gadgets that might be of independent interest for future works on pebbling.

Our construction is organized into consecutive chains of *level gadgets* that form *towers*. Intuitively, a pebbling strategy always needs to keep one level of each tower in fast memory, and it can 'proceed' to the next level, computing the nodes of the next level and deleting pebbles from the current level. There is no benefit to having partially pebbled levels in our DAG, so intuitively, each level can be considered a single entity in our analysis. To pebble the DAG correctly, we need to go through the towers of levels in a carefully designed order, to ensure that the current set of levels never requires more than r pebbles altogether. The level gadgets are shown in Fig. 3, and discussed in the full version [8].

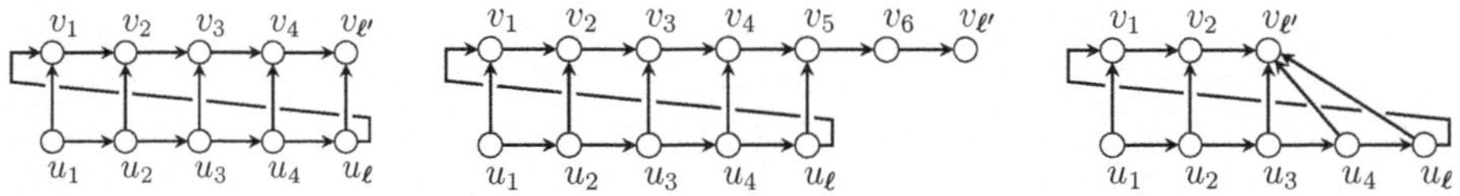

Fig. 3. Examples of consecutive levels. The 1st level (bottom row) always has size $\ell = 5$. The 2nd level (top row) has $\ell' = 5$ on the left side, $\ell' = 7$ in the middle, and $\ell' = 3$ on the right.

Using these gadgets, we present a reduction from finding a clique of size q in a graph G'. Our DAG has a main tower, which is used to control the number of available pebbles at all times. We also add smaller tower gadgets for each node and edge of G', as illustrated in Fig. 4. Furthermore, we draw edges between specific levels of the node and edge gadgets whenever a node is incident to an edge, and also between the main tower and all node and edge gadgets.

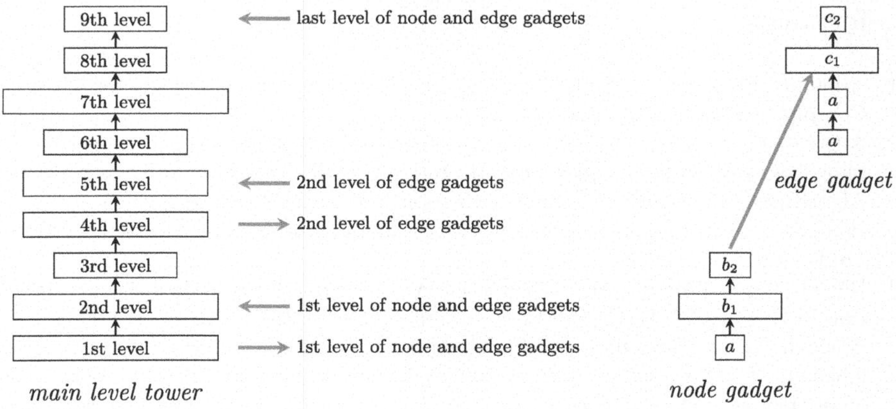

Fig. 4. High-level sketch of our construction: the main tower on the left, and node/edge gadgets on the right, with the level sizes and the dependency between incident node-edge pairs also shown.

In the beginning, we can only pebble the 1st level of the main tower, then the first level of each node/edge gadget, then the 2nd level of the main tower. Proceeding to the 3rd level of the main tower then frees up many pebbles; we can use these to proceed to the 2nd (and 3rd) level of node gadgets. However, due to $b_2 > a$, we can only move to the 3rd level of at most q node gadgets; otherwise, we would not have enough remaining pebbles to reach the 4th level of the main tower afterwards. Thus intuitively, any pebbling strategy corresponds to selecting at most q node gadgets.

The next levels of the main tower allow/require us to move to the 2nd level in edge gadgets. The 6th level again provides more free pebbles, so we can move to the 3rd (and then 4th) level in any edge gadget, provided that both of its incident nodes were chosen earlier. The next (7th) level then allows the fewest free pebbles: we can only proceed to this level if we reached the 4th level in at least $\binom{q}{2}$ edge gadgets. This is only possible if we found a set of q nodes in G' that span $\binom{q}{2}$ edges. Finally, after this point, the pebbling is easy to finish. Hence a pebbling of cost 0 exists if and only if G' has a clique of size q.

The theorem directly carries over to surplus cost in MPP: given the same DAG in MPP, a pebbling of cost 0 will translate to a surplus cost of 0, while a pebbling with $n^{1-\varepsilon}$ I/O steps (or the same amount of recomputation) will translate to a surplus cost of at least $n^{1-\varepsilon}$.

Corollary 2. *In MPP, it is NP-hard to approximate the optimum surplus cost to any finite multiplicative factor, and to any additive $n^{1-\varepsilon}$ term (for any $\varepsilon > 0$).*

References

1. Alimonti, P., Kann, V.: Some APX-completeness results for cubic graphs. Theoret. Comput. Sci. **237**(1–2), 123–134 (2000)
2. Austrin, P., Pitassi, T., Wu, Y.: Inapproximability of treewidth, one-shot pebbling, and related layout problems. In: International Workshop on Approximation Algorithms for Combinatorial Optimization, pp. 13–24. Springer (2012)
3. Bennett, C.H.: Time/space trade-offs for reversible computation. SIAM J. Comput. **18**(4), 766–776 (1989)
4. Bilardi, G., Scquizzato, M., Silvestri, F.: A lower bound technique for communication in BSP. ACM Trans. Parallel Comput. (TOPC) **4**(3), 1–27 (2018)
5. Blelloch, G.E., Chowdhury, R.A., Gibbons, P.B., Ramachandran, V., Chen, S., Kozuch, M.: Provably good multicore cache performance for divide-and-conquer algorithms. In: Proceedings of the 19th Annual ACM-SIAM Symposium on Discrete Algorithms, pp. 501–510 (2008)
6. Blocki, J., Holman, B., Lee, S.: The parallel reversible pebbling game: analyzing the post-quantum security of iMHFs. In: Theory of Cryptography Conference, pp. 52–79 (2022)
7. Blocki, J., Ren, L., Zhou, S.: Bandwidth-hard functions: reductions and lower bounds. In: Proceedings of the 2018 ACM SIGSAC Conference on Computer and Communications Security, pp. 1820–1836 (2018)
8. Böhnlein, T., Papp, P.A., Yzelman, A.N.: Red-blue pebbling with multiple processors: time, communication and memory trade-offs (full version). arXiv preprint arXiv:2409.03898 (2024)
9. Carpenter, T., Rastello, F., Sadayappan, P., Sidiropoulos, A.: Brief announcement: approximating the I/O complexity of one-shot red-blue pebbling. In: Proceedings of the 28th ACM Symposium on Parallelism in Algorithms and Architectures, pp. 161–163 (2016)
10. Chowdhury, R.A., Ramachandran, V.: Cache-efficient dynamic programming algorithms for multicores. In: Proceedings of the 20th ACM Symposium on Parallelism in Algorithms and Architectures, pp. 207–216 (2008)
11. Cook, S., Sethi, R.: Storage requirements for deterministic/polynomial time recognizable languages. In: Proceedings of the Sixth Annual ACM Symposium on Theory of Computing, pp. 33–39 (1974)
12. Cook, S.A.: An observation on time-storage trade off. In: Proceedings of the 5th Annual ACM Symposium on Theory of Computing, pp. 29–33 (1973)
13. Demaine, E.D., Liu, Q.C.: Inapproximability of the standard pebble game and hard to pebble graphs. In: Workshop on Algorithms and Data Structures, pp. 313–324. Springer (2017)
14. Demaine, E.D., Liu, Q.C.: Red-blue pebble game: complexity of computing the trade-off between cache size and memory transfers. In: Proceedings of the 30th ACM Symposium on Parallelism in Algorithms and Architectures, pp. 195–204 (2018)
15. Dymond, P.W., Tompa, M.: Speedups of deterministic machines by synchronous parallel machines. J. Comput. Syst. Sci. **30**(2), 149–161 (1985)
16. Elango, V., Rastello, F., Pouchet, L.N., Ramanujam, J., Sadayappan, P.: On characterizing the data movement complexity of computational DAGs for parallel execution. In: Proceedings of the 26th ACM Symposium on Parallelism in Algorithms and Architectures, pp. 296–306 (2014)

17. Elango, V., Rastello, F., Pouchet, L.N., Ramanujam, J., Sadayappan, P.: On characterizing the data access complexity of programs. In: Proceedings of the 42nd ACM SIGPLAN-SIGACT Symposium on Principles of Programming Languages, pp. 567–580 (2015)
18. Gilbert, J.R., Lengauer, T., Tarjan, R.E.: The pebbling problem is complete in polynomial space. In: Proceedings of the 11th Annual ACM Symposium on Theory of Computing, pp. 237–248 (1979)
19. Gleinig, N., Hoefler, T.: The red-blue pebble game on trees and DAGs with large input. In: International Colloquium on Structural Information and Communication Complexity, pp. 135–153. Springer (2022)
20. Graham, R.L.: Bounds on multiprocessing timing anomalies. SIAM J. Appl. Math. **17**(2), 416–429 (1969)
21. Hong, J.W., Kung, H.T.: I/O complexity: the red-blue pebble game. In: Proceedings of the 13th ACM Symposium on Theory of Computing, pp. 326–333 (1981)
22. Hopcroft, J., Paul, W., Valiant, L.: On time versus space. J. ACM (JACM) **24**(2), 332–337 (1977)
23. Jain, S., Zaharia, M.: Spectral lower bounds on the I/O complexity of computation graphs. In: Proceedings of the 32nd ACM Symposium on Parallelism in Algorithms and Architectures, pp. 329–338 (2020)
24. Kwasniewski, G., et al.: Pebbles, graphs, and a pinch of combinatorics: towards tight I/O lower bounds for statically analyzable programs. In: Proceedings of the 33rd ACM Symposium on Parallelism in Algorithms and Architectures, pp. 328–339 (2021)
25. Kwasniewski, G., Kabić, M., Besta, M., VandeVondele, J., Solcà, R., Hoefler, T.: Red-blue pebbling revisited: near optimal parallel matrix-matrix multiplication. In: Proceedings of the International Conference for High Performance Computing, Networking, Storage and Analysis, pp. 1–22 (2019)
26. Lengauer, T., Tarjan, R.E.: Asymptotically tight bounds on time-space trade-offs in a pebble game. J. ACM (JACM) **29**(4), 1087–1130 (1982)
27. McColl, W., Tiskin, A.: Memory-efficient matrix computations in the BSP model. Algorithmica **24**(3–4), 287–297 (1999)
28. Papp, P.A., Wattenhofer, R.: On the hardness of red-blue pebble games. In: Proceedings of the 32nd ACM Symposium on Parallelism in Algorithms and Architectures, pp. 419–429 (2020)
29. Papp, P.A., Anegg, G., Yzelman, A.N.: DAG scheduling in the BSP model (2023). arXiv preprint arXiv:2303.05989
30. Paterson, M.S., Hewitt, C.E.: Comparative schematology. In: Record of the Project MAC Conference on Concurrent Systems and Parallel Computation, pp. 119–127 (1970)
31. Ranjan, D., Savage, J., Zubair, M.: Upper and lower I/O bounds for pebbling R-pyramids. J. Discrete Algorithms **14**, 2–12 (2012)
32. Reischuk, R.: Improved bounds on the problem of time-space trade-off in the pebble game. J. ACM (JACM) **27**(4), 839–849 (1980)
33. Savage, J.E.: Extending the Hong-Kung model to memory hierarchies. In: International Computing and Combinatorics Conference, pp. 270–281. Springer (1995)
34. Savage, J.E., Zubair, M.: A unified model for multicore architectures. In: Proceedings of the 1st International Forum on Next-Generation Multicore/manycore Technologies, pp. 1–12 (2008)
35. Scott, J., Holtz, O., Schwartz, O.: Matrix multiplication I/O-complexity by path routing. In: Proceedings of the 27th ACM Symposium on Parallelism in Algorithms and Architectures, pp. 35–45 (2015)

36. Sethi, R.: Complete register allocation problems. In: Proceedings of the 5th ACM Symposium on Theory of Computing, pp. 182–195 (1973)
37. Solomonik, E., Carson, E., Knight, N., Demmel, J.: Trade-offs between synchronization, communication, and computation in parallel linear algebra computations. ACM Trans. Parallel Comput. (TOPC) **3**(1), 1–47 (2017)
38. Tiskin, A.: The bulk-synchronous parallel random access machine. Theoret. Comput. Sci. **196**(1–2), 109–130 (1998)
39. Valiant, L.G.: A bridging model for parallel computation. Commun. ACM **33**(8), 103–111 (1990)
40. Valiant, L.G.: A bridging model for multi-core computing. J. Comput. Syst. Sci. **77**(1), 154–166 (2011)

Deterministic Color-Optimal Self-stabilizing Semi-synchronous Gathering: A Certified Algorithm

François Bonnet[1], Quentin Bramas[2], Pierre Courtieu[3], Xavier Défago[1], Lionel Rieg[4], Sébastien Tixeuil[5,6], and Xavier Urbain[7](✉)

[1] Institute of Science Tokyo, Tokyo, Japan
[2] University of Strasbourg, ICUBE, Strasbourg, France
[3] CNAM, Cédric, Paris, France
[4] Grenoble INP - UGA, CNRS (UMR 5104), Verimag, Saint-Martin-d'Hères, France
[5] Sorbonne University, LIP6, Paris, France
[6] Institut Universitaire de France, Paris, France
[7] Universite Claude Bernard Lyon 1, CNRS, INSA Lyon, LIRIS, UMR5205, Villeurbanne, France

xavier.urbain@univ-lyon1.fr

Abstract. We consider the problem of gathering in finite time and at the same location, not known beforehand, a set of deterministic semi-synchronous robots, starting from an arbitrary initial configuration that may even be bivalent (that is, a configuration where the robots are evenly split on two different locations).

This problem is known to be unsolvable when the robots are oblivious, that is, when they cannot remember their past actions. We present a deterministic gathering algorithm where robots may remember and communicate one bit of memory. This bit may be arbitrarily (and adversarially) set in the initial configuration. Our solution is thus memory optimal and self-stabilizing. Its proof of correctness is formally certified by the Coq proof assistant using the Pactole framework.

Keywords: Mobile robots · Proof assistant · Self-stabilization

1 Introduction

Mobile robot networks have garnered significant interest within the distributed computing community due to their wide range of potential applications, including search and rescue operations, environmental monitoring, and surveillance in hazardous environments. In the classical model used to mathematically study such robotic swarms [23], robots operate in Look-Compute-Move (LCM) cycles. During each cycle, a robot observes its surroundings to capture a snapshot of the positions of other robots, computes a destination based on this snapshot, and then moves to the calculated location.

This work was partially supported by ANR project SAPPORO 2019-CE25-0005.

U. Schmid and R. Kuznets (Eds.): SIROCCO 2025, LNCS 15671, pp. 127–143, 2025.
https://doi.org/10.1007/978-3-031-91736-3_8

Oblivious robots execute these cycles without retaining any memory of their past actions. In contrast, luminous robots [12] can retain and exchange a constant amount of information, represented by an individual light visible to other robots, the color of which can be altered by the hosting robot. The execution of these tasks is influenced by the synchronization level among the robots, which can be either synchronous (FSYNC), semi-synchronous (SSYNC), or asynchronous (ASYNC). In this study, we concentrate on the SSYNC model, where robots perform LCM cycles in discrete rounds, with a fair scheduler ensuring that all robots are activated within a finite number of steps.

The robots in this setting have minimal capabilities: they are anonymous, run the same algorithm, and may be disoriented, meaning that their observations are interpreted through local coordinate systems that may not align with those of other robots. In distributed systems like these, an appealing property for resilience against transient failures is self-stabilization [16]. Within the context of mobile LCM robots, self-stabilization entails [1,14,15,19] that the robots can begin from arbitrary initial locations, including configurations with multiplicity points, and for luminous robots, that their memory (including the initial color of their light) can be arbitrarily set.

In this paper, we tackle the fundamental problem of gathering these robots starting from an arbitrary configuration in finite time to a unique location, not known beforehand, and maintain them there forever. Note that the fact that the robots must remain gathered forever implies that any algorithm solving the gathering problem for arbitrary initial configuration is by definition self-stabilizing. Despite its apparent simplicity, the gathering problem has attracted considerable attention [18]. A key result [9,23] establishes that in the SSYNC or ASYNC models, no oblivious deterministic algorithm can solve the gathering problem for two robots without additional assumptions. The core reasoning behind this impossibility is that robots may possess symmetric coordinate systems and can perpetuate this symmetry indefinitely. This finding can be extended [9] to the bivalent scenario, where an even number of robots are initially evenly distributed between two locations. Prencipe [21] further generalized this result, demonstrating that in SSYNC, when $n > 2$ robots start from distinct initial locations, no oblivious deterministic algorithm can solve the gathering problem without additional assumptions, such as multiplicity detection. On the other hand, in the FSYNC model, it is possible to solve the gathering problem without such additional assumptions [3]. In the ASYNC model, gathering becomes feasible if $n > 2$ robots start from distinct initial locations and are equipped with weak multiplicity detection [8].

The fundamental challenge in deterministically solving the gathering problem stems from the possibility of robots starting in a bivalent configuration. Any other starting configuration is solvable [7,10], as maintaining a non-bivalent configuration is an invariant in these algorithms. To overcome this impossibility, various assumptions have been explored in previous research. For instance, Dieudonné *et al.* [15] addressed the problem by assuming an odd number of robots, as this prevents the formation of any bivalent configuration. Défago *et al.* [14] introduced probabilistic elements into the algorithm to resolve symmetrical situations, allowing for the handling of potentially bivalent initial configurations. However, this approach sacrifices deterministic correctness, achieving gathering only in expected finite time. More recently, Asahiro *et al.* [1] explored the use of deterministic robots that do not all execute the same algorithm, proving that two distinct algorithms are both necessary and sufficient to solve the problem. In summary, within the oblivious setting, successful gathering requires compromising on either flexibility [15], deterministic guarantees [14], or uniformity [1].

In the context of luminous robots, a partial affirmative answer to the central question of whether a deterministic self-stabilizing gathering algorithm exists in the SSYNC model has been provided for the case of two robots (i.e., the rendezvous problem [13,25]). With two robots, any non-gathered configuration is bivalent, yet the problem can still be solved deterministically in both the SSYNC [25] and ASYNC [13] scheduling models, starting from any arbitrary state, including light color. It has been established that two colors are both necessary and sufficient to solve the rendezvous problem, making these solutions optimal in that regard. Terai et al. [24] provided a partial answer to the general case of self-stabilizing gathering $n \geq 2$ luminous robots. First, in the SSYNC execution model, two colors are sufficient to gather from an arbitrary placement of the robots, but their solution requires that all robots start with the same color (hence, their solution is not self-stabilizing). Second, assuming a centralized scheduler (that may activate only one robot at a time to execute a Look-Compute-Move cycle), a fully-self-stabilizing solution is feasible. Overall, to date, no self-stabilizing SSYNC solution has been found for scenarios involving more than two robots, leading to clear open questions: In the luminous model, does a self-stabilizing deterministic gathering algorithm exist in the SSYNC execution model for more than two robots? If so, what is the minimum number of colors required?

Our Contribution. We provide a positive answer to the computability question and an optimal answer to the complexity question. In more details, we design and prove correct a self-stabilizing deterministic gathering algorithm that relies on strong multiplicity detection and makes use of two colors per light (a single color is tantamount to requiring the robots to be oblivious, and thus renders the problem impossible to solve). A noteworthy property of our approach lies in the use of formal methods to assess the correctness of our algorithm, and we make use of the PACTOLE library and the COQ proof assistant.

Our use of formal methods is commanded by observing that results in the mobile robots area of Distributed Computing are traditionally described and proved in natural language. However, algorithms, even when presented as pseudo-code, cannot be fully understood without the *precise setting* in which they are executed. *Implicit assumptions* are often considered folklore, but because properties of distributed algorithms depend on very subtle hypotheses, even a small, barely noticeable shift in these assumptions can drastically change an algorithm's behavior. Such shifts still appear in the literature, leading to erroneous results being discovered sometimes years after publication (see recent error discoveries and references therein [5,17]). Formal methods address this issue by using tools that provide a non-ambiguous mathematical language. Given the difficulty of establishing correct results in mobile robotics, using a common, non-ambiguous format for definitions is a significant advantage to ensure trust in the proved result (and improve its re-usability in subsequent work). Among closely related works, most algorithms [1,7,14,15,25] only have pen-and-paper correctness proofs. One notable exception is the luminous rendez-vous algorithm by Defago *et al.* [13] that makes use of model checking to prove its correctness (and also prove the correctness of Viglietta's algorithm [25] as a byproduct). Unfortunately, the approach of Defago *et al.* [13] is tightly coupled to a two-robots system, and is unlikely to scale to arbitrary size, as a key result of Sangnier *et al.* [22] in the context of parameterized model-checking shows that non-trivial properties (namely, liveness properties) are undecidable.

As a result, following the path of Courtieu *et al.* [10], we use a proof assistant, that is not bound by the limitations of the model-checking approach, to assess the correctness of our algorithm. Interestingly, since the algorithm of Courtieu *et al.* [10] solves gathering when the initial configuration is not bivalent, we were not only able to reuse its algorithm as a building block, but also its formal proof for (partially) proving our own solution, saving development time. A general comparison of our solution with previous works is presented in Table 1.

This article describes the algorithm, and gives an intuition of the key steps in the proof of its correctness. The actual complete proof is provided as a Coq development. As explained in [4], assisted proofs are themselves correct by construction and thus, do not need a manual review by the community. However, a remaining crucial step is to ensure that the formal statements match the claimed results. In particular, the formal Coq *definitions* and *final theorem statements* must be carefully checked to make sure that the formal development provides indeed the proofs of the relevant properties.

The source code of our formal development for Coq 8.19 and 8.20 is available at the following URL: https://pactole.liris.cnrs.fr.

2 Model

We consider a distributed system where a set of $n \in \mathbb{N}^*$ entities, called robots, are located on an Euclidean plane. We identify each robot by a number in $[0, n-1] = \mathcal{N}$ for the analysis only as robots do not have access to their identity. Each robot r has a *location* l, represented by coordinates in a global coordinate system, and a color c, represented by a boolean, *true* for the color White and *false* for the color Black. A *configuration* of the system at time t, denoted $C(t)$ is a function that maps each robot identity to its state (l, c). The set of all configurations is denoted $\mathcal{C}$.

The robots are said to be *uniform* as they all execute the same algorithm. They are *disoriented* in the sense that this algorithm takes as input, not the

Table 1. A comparison of our algorithm with existing SSYNC solutions (most desirable properties are in boldface).

Reference	Size	Deterministic	Uniform	Proof	Oblivious	Self-stabilizing
[15]	n is odd	**yes**	**yes**	pen-and-paper	**yes**	**yes**
[25]	$n = 2$	**yes**	**yes**	pen-and-paper	no	**yes**
[7]	**any**	**yes**	**yes**	pen-and-paper	**yes**	no
[10]	**any**	**yes**	**yes**	**proof assistant**	**yes**	no
[14]	**any**	no	**yes**	pen-and-paper	**yes**	**yes**
[1]	**any**	**yes**	no	pen-and-paper	**yes**	**yes**
[13]	$n = 2$	**yes**	**yes**	**model-checker**	no	**yes**
[24]	**any**	**yes**	**yes**	pen-and-paper	no	no
This paper	**any**	**yes**	**yes**	**proof assistant**	no	**yes**

configuration $C(t)$, but a local perception of it called the *observation*. In this paper, the observation is obtained from the configuration $C(t)$ by removing the identity of the robots and by changing the coordinate system of the locations. We consider that each robot has its own coordinate system and that we can transform a location in the global coordinate system to the local coordinate system of r using a similarity φ_r that keeps the origin invariant (that is a combination of rotation, reflection, and scaling). To ensure that the robot knows where it is located, we also translate the locations so that a robot is always located at the origin of its local coordinate system (that is, it is ego-centered). We denote by $\mathbf{t}_r$ the translation of the plane mapping $C(t)(r)$ to $(0,0)$. Formally, the observation of a robot r in configuration $C(t)$ is the multiset

$$obs(C(t), r) = \bigcup_{r' \in \mathcal{N}} \{(\varphi_r \circ \mathbf{t}_r(l),\ c) \mid (l, c) = C(t)(r')\}$$

In practice, the observation can be given to the algorithm in a more convenient way, such as a list of pairs (in arbitrary order), or other structures as described in our formal proof in Sect. 4.

The system evolves when the robots are activated at each round. The update of the configuration is a construction formally defined by the Pactole framework we are using. The intuition behind this construction is as follows. At each round t, a subset $S(t) \subset \mathcal{N}$ of the agents are activated. Each active agent retrieves its observation $obs(C(t), r)$, then computes its destination by executing the same algorithm A that takes as input an observation and returns a pair (l_r, c_r) representing a new location l_r and a new color c_r. From there, the new configuration $C(t+1)$ is simply the function that maps each robot r to its new state (its new location in the global coordinate system and its new color) if it was activated, or its unchanged state otherwise:

$$C(t+1)(r) = \begin{cases} C(t)(r) & \text{if } r \notin S(t) \\ (\mathbf{t}^{-1} \circ \varphi_r^{-1}(l_r), c_r) & \text{if } r \in S(t) \end{cases}$$

An execution is an infinite sequence of configurations. A gathered configuration is a configuration where all the robots share the same location. The set of gathered configurations is denoted *Gathered*. The gathering problem consists in reaching a gathered configuration in a finite number of rounds from any initial configuration, and staying idle afterwards. Formally, an algorithm A solves the gathering problem if is satisfies:

$$\forall C(0) \in \mathcal{C},\ \exists t,\ \forall t' \geq t,\ \ \mathrm{C(t)} = C(t') \in \mathit{Gathered}$$

3 Algorithm Description

In this section, we present our algorithm solving the gathering problem without any restriction on the initial configuration. We first give a textual description of it, then a more formal definition closer to the one written in Coq, presented in the next section.

Informal Description. The algorithm consists of three main parts, depending on whether the configuration is color-bivalent, bivalent but not color-bivalent, or non-bivalent. A color-bivalent configuration consists of two towers of equal size at distinct locations, with as many Black and White robots in each tower. A robot that is activated performs the following:

- If the configuration is color-bivalent, the goal is to break the symmetry.
  - If the activated robot is White, go to the middle of the two towers, change color to Black.
  - If the activated robot is Black, change color to White.
- If the configuration is bivalent but not color-bivalent, the number of Black robots is not the same in the two towers, and the activated robot goes to the tower with the higher number of Black robots.
- If the configuration is not bivalent, use the SSYNC oblivious algorithm of [10] certified in Coq, which never creates a bivalent configuration thereafter.

Algorithm 1: Executed by a robot with observation `obs`

```
if bivalent_obs(obs) then
    c ← observer_lght(obs)
    if color_bivalent_obs(obs) then
        if c == true i.e. is White then
            return (middle(obs), false)
        else
            return ((0,0), true)
    else
        return (find_max_black(obs), c)
else
    execute gatherR2_old_pgm(obs)
```

Pseudo-code. In pseudo-code, the algorithm is defined in Algorithm 1. It makes use of functions to check whether the observation `obs` is bivalent or color-bivalent, with `bivalent_obs` and `color_bivalent_obs` respectively. The functions `middle` and `find_max_black` are used to compute particular points of the observation as destinations, respectively the middle of both towers and the tower with the highest number of BLACK robots.

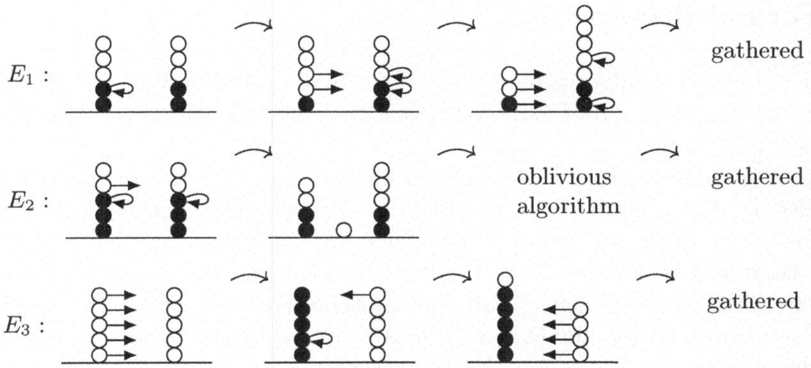

Fig. 1. Execution examples starting from color-bivalent configurations

Example of Executions. Figure 1 illustrates three execution examples starting from color-bivalent configurations. For simplicity, we present the multiplicity points as a tower of robots, to better see the number of BLACK and WHITE robots at each location. In Execution E_1, a single BLACK robot is first activated, so that after one round the number of BLACK robots is not the same at the two locations, and the configuration is no longer color-bivalent (but remains bivalent). A similar result is obtained if more BLACK robots are activated, as soon as a different number of robots is activated on each location. If the same number of BLACK robots is activated, and the configuration remains color-bivalent then eventually all the robots become WHITE, as in execution E_3. The second configuration of execution E_1 is not color-bivalent, as one of the towers has more robots than the other, and every activated robot is ordered to move toward it. Eventually, all the robots gather in the same location.

In Execution E_2, a WHITE robot is activated in one of the towers. Here, regardless of the activation of the BLACK robots, the obtained configuration contains three points so is not bivalent. From there, executing the oblivious algorithm for non-bivalent configurations ensures that the robots eventually gather at a single point.

Consider now Execution E_3, which starts from a color-bivalent configuration where all robots are WHITE. If all are executed, they all move to the middle and join a single multiplicity point. If only a strict subset of one of the towers is activated, then the obtained configuration contains three points and the configuration is not bivalent anymore. Otherwise, as in the illustration, if one tower is activated and not the other, the obtained configuration is bivalent (distance is halved) but one of the towers contains only BLACK robots, and the other only WHITE robots, so the configuration is not color bivalent and the robots eventually gather at a single location.

4 Formal Proof

To ensure that our algorithm is correct we make use of PACTOLE[1], a formal library for the COQ proof assistant[2] that is dedicated to autonomous mobile robots.

Proof assistants are pieces of software that certify the soundness of a proof provided by the user. In the case of COQ, this is achieved using type theory and the Curry-Howard correspondence. In sharp contrast to automated provers, proof assistants do not state that a property holds with full automation. Instead, they check that a proof, usually developed interactively, is correct. With a trade-off between automation and expressiveness, statements can be very close to intuition but proofs can be tedious. Hence, various libraries exist to assist the user in various domains, notably by providing definitions and theorems that are void of implicit assumptions, and specialized proof tactics to help with some (limited form of) automation. PACTOLE is the de facto standard library currently being developed for robotic swarms [11]. It has proven invaluable in establishing formal correctness of several algorithms, among which the one of Courtieu *et al.* [10] we reuse parts of. For an introduction to the COQ proof assistant, the reader may refer to standard expositions [6,20]. A comprehensive presentation of PACTOLE and its uses may be found in a recent survey [11].

The use of a proof assistant like COQ enforces that the context where we reuse theorems and proofs from Courtieu *et al.* [10] is compatible with their soundness (i.e. the current context and the original context of Courtieu *et al.* [10] permit to derive the same results). In other words: all those theorems still hold in our context, and we can build upon them. Hence, the first step is to define the context (evolving space, synchronicity assumptions, etc.) similarly to that of Courtieu *et al.* [10]. Then, we can specify our luminous robots (remember that the robots of Courtieu *et al.* [10] were oblivious).

4.1 Description of Our Algorithm in PACTOLE

The self-stabilizing algorithm `gatherR2` we propose may be seen as consisting of three parts: respectively describing the behavior when confronted to a color-bivalent observation, a bivalent but not color-bivalent observation, and finally, an observation that is not bivalent. In the last case, our algorithm refers to the one

[1] https://pactole.liris.cnrs.fr.

[2] https://coq.inria.fr/.

from Courtieu *et al.* [10], hereafter named `gather_R2_old_pgm`, which has been proven to be sound and not leading to any bivalent configuration.

As a set of only two distinct colors suffices in our algorithm, we choose to instantiate this set `L` with Boolean values, `true` being WHITE, and `false` being BLACK. We can now define the state of a robot to be an ordered pair of type `location * L` consisting of its current location and light color.

As we want to reuse the algorithm of Courtieu *et al.* [10] with as little adaptation work as possible, we want to keep their original observation type (a multiset of location, well-suited to represent a configuration of anonymous robots with full global multiplicity). Yet, in our model, robots have lights and can observe them, *including their own.* Hence we define the new observation as a pair, with the observation from Courtieu *et al.* [10] as the first member, and a record consisting of the light color of the observer and a multiset of locations and light colors as second member.

Now, coherence between the observation with lights and the one without is proven once and for all. Crucially, there is *no need to modify* the algorithm of Courtieu *et al.* [10] (for example, no need to include a projection in their code to prune lights out), and any adaptation of previous results is rather light (see Sect. 4.3).

An algorithm is a function that takes an observation and returns a new robot state; and the code of ours is simply:

```
Definition gatherR2 (obs : observation) : location * L :=
 if bivalent_obs obs (* if bivalent then 2 distinct locations *)
 then let other_loc := find_other_loc (fst obs) origin in
      if color_bivalent_obs obs (* is obs color bivalent? *)
      (* color bivalent: check own light *)
      then if observer_lght (snd obs) (* is own light white? *)
           (* yes: to middle + switch to black *)
           then (middle origin other_loc, false)
           (* no: stay + switch to white *)
           else (origin, true)
      (* bivalent & not color bivalent: go to max black lght *)
      else
       let maj_black := find_max_black obs origin other_loc in
       (maj_black, observer_lght (snd obs))
 (* not bivalent: Courtieu et al. will do + keep light status *)
 else (gatherR2_old_pgm (fst obs), observer_lght (snd obs)).
```

4.2 Structure of the Formal Proof

The formal development itself consists of two main stages.

Firstly, we define an observation taking lights into account; this new context is provided in file `WithMultiplicityLight.v` in the companion code repository. The file contains properties about bivalent and color-bivalent configurations.

Note that, so as to stay coherent with all previous developments on that topic (most notably with Courtieu *et al.* [10]), we use the formal definitions of gathering properties from the PACTOLE library. Let us recall some of them:

Relation `Gather pt e` means that at all rounds of an (infinite) execution `e`, robots are gathered at the same (unchanging) position `pt`.

```
Definition Gather (pt: location) (e : execution) : Prop :=
  Stream.forever (Stream.instant (gathered_at pt)) e.
```

Relation `WillGather e` means that the (infinite) execution `e` is *eventually* in `Gather` for a certain `pt`.

```
Definition WillGather (e : execution) : Prop :=
  Stream.eventually (fun e ⇒ ∃ pt, Gather pt e) e.
```

The definition of a solution for gathering is thus:

```
Definition FullSolGathering (rbg : robogram) (d : demon) :=
  ∀ config : configuration, WillGather (execute rbg d config).
```

An algorithm `rbg` is a solution to Gathering for an activation demon `d` (that is, for some choice of the environment) if for any configuration `config`, the execution eventually leads to a gathered configuration. The main difference from the solution of Courtieu *et al.* [10] is that our current proposal, using `FullSolGathering` has *no condition* on the configuration, it is universally quantified. Further note that it is the statement used in the proof of impossibility of gathering without lights. We are thus dealing with the exact same property.

Secondly, we propose our new algorithm `gatherR2` and, in particular, we establish formally that for any fair semi-synchronous demon and from any configuration, every execution eventually leads to a configuration where all robots are (and remain) gathered. In the code archive this development takes place in file `Algorithm_withLight.v`, together with the adaptation of the results regarding the original algorithm by Courtieu *et al.* [10].

The formal proof itself is organized as follows.

The Overall Structure Follows the Algorithm. Firstly, as the algorithm `gatherR2` differentiates three classes of configuration, namely the color-bivalent, bivalent but not color-bivalent, and the not-bivalent cases, so does the final proof. Its development is thus split into three Coq sections, namely `ColorBivalent`, `BivalentButNotColorBivalent`, and `NonBivalent`. Each of these sections contains, in particular, the general assumptions that define it, and specialized versions of some lemmas, such as the `round_simplify` lemma linking the behavior of the algorithm observed from the local and from the global frames of reference. For example: an obvious assumption in section `NonBivalent` is that the considered configuration is indeed not bivalent, and we can state (and prove) that in this case, robot lights do not change along the execution, that is: for any robot in that configuration, the value of its light after one round for `gatherR2` is the same as before the round. Here, `da` represents the choice of the environment for this round and is introduced earlier in the code.

```
Section NonBivalent.

Variable cfg : configuration.
```

```
Hypothesis Hbivalent : ¬ bivalent cfg.

Lemma round_simplify_Lights : ∀ id,
 get_light (round gatherR2 da cfg id) ≡ get_light (cfg id).
Proof...
```

Notice that we need to repeat neither the configuration `config` nor the hypothesis `¬ bivalent config` in the statement of lemma `round_simplify_Lights`.

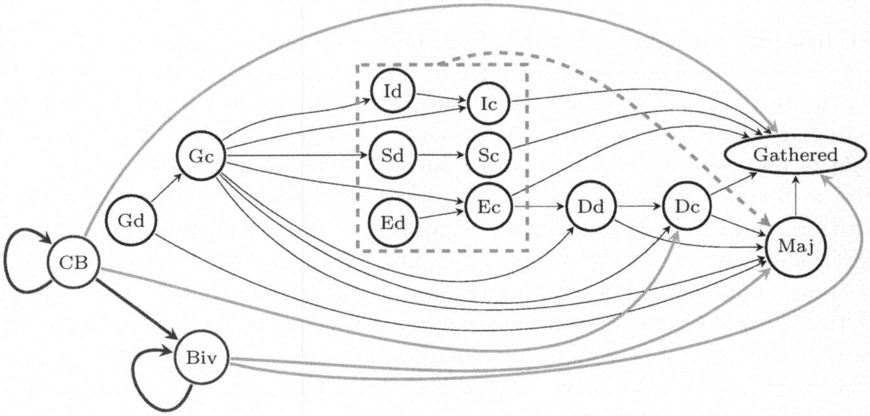

Fig. 2. Reachability graph for the categories of observations. In blue, CB and Biv refer to the Color-Bivalent and Bivalent configurations respectively. In black, the different non-bivalent cases identified in the algorithm of Courtieu *et al.* [10], with self-loops omitted in the non-bivalent cases. The boxed area contains the triangle cases, all linked to Maj. (Color figure online)

Exhaustive Case Analyses. Secondly, the proof explains how a configuration may evolve from one class to another by a careful and exhaustive case analysis. This proof of exhaustivity is a significant effort; although it can be done in only 50 lines of code for the bivalent case as it always evolves either to itself or to a non-bivalent case, the proof for the color-bivalent case is considerably harder (more than 600 lines of code). We identify five sub-cases: one leading to a color bivalent configuration (all active robots are Black and the same number is activated on both towers), two leading to a bivalent but not color bivalent configuration (all active robots are Black but a different number is activated on each tower; all robots are White and a full tower is activated), and two leading to a non-bivalent configuration (all robots are White and more than a full tower moves; some robot does not move on each tower and some robot moves). Getting this exact case distinction correct is tricky and the proof assistant helps a lot in making sure one does not forget a corner case.

The various configuration classes used in the proof of correctness are presented in Fig. 2, where a directed arrow from one class of configuration to another denotes the possibility to change the current configuration class by executing the algorithm. Since the selection of robots that execute their algorithm in a given configuration is decided by an adversary, there may be more than one outgoing arrow in some configuration classes. The classes in black color correspond to those identified by Courtieu *et al.* [10], while the two new vertices: CB (for color-bivalent) and Biv (for bivalent) correspond to new possibilities implied by the algorithm being self-stabilizing.

Termination Through Measure Decrease. Thirdly, it remains to show that along any execution of our algorithm, the successive resulting configurations follow a finite path in this graph that ends up in state Gathered. This result is obtained with an extension of the measure used by Courtieu *et al.* [10]. The *weight* of an observation s is a pair of natural numbers, compared lexicographically, such that the first one is the weight of the corresponding phase p of the algorithm, and the second one is the result of a phase-weight function w_p on the observation. We give to CB and Biv weights greater than for the other phases, with a maximum weight to phase CB. In our case, it is enough to give a weight of 8 to CB and 7 to Biv. Function w_{Biv} is unnecessary as fairness forces the eventual activation of a robot that changes its state, meaning in this particular case that we end up with a tower strictly higher than the other. We may as well choose the constant function 0. Function w_{CB} is simply the number of robots with their light set to BLACK in the observation. Remember that only robots with their light set to WHITE can change location. Obtaining a bivalent configuration after one round is thus either

- activating robots that move to create another tower, and exhaust one of the original towers. This means that all robots at a location were moving, hence all with their light set to WHITE. A resulting *bivalent* configuration consists thus of a tower with only lights set to BLACK, and a tower untouched hence with lights set to WHITE, otherwise the original configuration would not have been color bivalent. This resulting configuration is not color-bivalent.
- activating robots that don't move (i.e. just switch their light color) but in that case, they have to be activated in the same number on both locations otherwise we lose color bivalence.

The only way to stay in CB is indeed to activate the same number of robots with light set to BLACK in each tower.

Let us call old_measure the weighting function of Courtieu *et al.* [10], the global measure is then defined as:

```
Function measure (obs : observation) : nat * nat :=
  if bivalent_obs obs then
    if color_bivalent_obs obs then (8, count_black obs)
    else (7, 0)
      (* By fairness eventually jumps to a maj tower case. *)
```

```
  else old_measure (fst obs).
      (* Courtieu et al.'s measure on lightless obs *)
```

The ordering defined by the lexicographic comparison of the left then right members of these weights is proven to be well-founded.

The arcs in the graphs of Fig. 2 are proven to be all the arcs that can be.

Finally, we can show that the measure of the observation of a resulting configuration is always less than the measure of the observation of the initial configuration provided at least one robot changes state. Then, the proof of the main theorem consists in two nested well-founded inductions: the first one is about fairness ensuring that eventually some robot changes state, and the second one is on the decrease of the chosen measure.

4.3 Modification to the Formal Proof of Courtieu *et al.* [10]

It is interesting to notice that the changes required to the proof of the non-bivalent case borrowed from Courtieu *et al.* [10] are minimal, demonstrating the effectiveness of code reusability with the PACTOLE framework. In particular, by preserving the original observation used as a component inside ours, and using the same notation `!! config` to refer to this component, the change of observation is completely transparent for the old proof.

The main change to the proof of Courtieu *et al.* [10] is the fact that a robot changing its state is not necessarily moving, it may change only its light color. In particular, this is the case for BLACK robots in a color-bivalent configuration. Because of that, we need to replace moving robots with changing ones in several theorem statements. In particular, the fairness of the demon is now used to ensure that eventually, some robot changes its state:

```
Theorem OneMustChange : ∀ config id,
  (* Assuming the configuration is not gathered, ... *)
  ¬gathered_at (get_location (config id)) config →
  (* then, there is a robot gmove that, ... *)
  ∃ gmove, ∀ da, SSYNC_da da →
  (* if it is activated by any SSYNC demon, ... *)
    List.In gmove (active da) →
  (* then its state changes. *)
    List.In gmove (changing gatherR2 da config).
```

The difference in the definitions of `moving` and `changing` only lies in the test: in `changing` we consider the full state of the robots (including their color) whereas in `moving` we only consider their location.

```
Definition moving rbg da config :=
  List.filter
    (fun id ⇒ negb (get_location (round rbg da config id)
                      ≡ b get_location (config id)))
    names.
```

```
Definition changing r da config :=
  List.filter
    (fun id ⇒ negb (round r da config id ≡ b config id))
    names.
```

Nevertheless, we can reuse most of the previous proofs as is because lights are only modified in the color-bivalent case. We can actually prove that moving robots and changing robots are the same whenever the configuration is not color-bivalent:

```
Lemma changing_eq_moving : ∀ config, ¬color_bivalent config →
  changing gatherR2 da config = moving gatherR2 da config.
```

4.4 Final Statement and Assumptions of the Proof

Denoting as a `similarity_demon` a demon where the local frame of reference it provides is centered on the observing robot, the full statement of the final theorem we prove is:

```
Theorem Gathering_in_R2 : ∀ d : similarity_demon,
  (* For any SSYNC fair demon *)
  SSYNC (d : demon) → Fair d →
  (* the gatherR2 algorithm solves the gathering problem. *)
  FullSolGathering gatherR2 d.
```

That is: for any demon `d`, if `d` is in the `SSYNC` relation (i.e., is semi-synchronous) then if `d` is in the `Fair` relation (i.e., is fair) then `FullSolGathering` holds for our algorithm and `d`. Put simply: any execution based on a fair SSYNC demon of our algorithm will eventually lead, from any configuration, to a situation where all robots are, and stay, gathered at a certain location. As previously noted, the initial configuration is truly arbitrary: no assumption is made about the initial placement of the robots, or the initial light color of the robots. So, our algorithm is self-stabilizing.

Using `Print Assumptions Gathering_in_R2`, we can list the 15 axioms that our formal proof relies on. Interestingly, they are exactly the same as the ones of the proof of Courtieu *et al.* [10], meaning that our extension does not add any additional axiom. These axioms can be described as follows:

- The axiomatic description of $\mathbb{R}$ from Coq's standard library (3 axioms);
- An axiomatic description of the smallest enclosing circle together with its properties (8 axioms);
- Geometric properties:
  - Properties of the barycenter of 3 points (3 axioms);
  - Through three points, at most a single circle can be drawn.

Proof Effort. The proof effort in this development remains manageable. Defining observations with lights and the relevant contexts amounts to about 300 lines of specifications and 1000 lines of actual proof in `WithMultiplicityLight.v` in the companion code repository.

Regarding the definition of our new algorithm with the complete proof of its correctness and universality, the file `Algorithm_withLight.v` stretches to about 1100 lines of specifications and 5240 lines of proof. Remember though that this includes for a significant part the proofs of Courtieu *et al.* [10] reformulated to handle observations with lights.

Other changes to improve the Pactole framework are minor: renaming the `moving` property into `changing`, introducing `moving` and `stationary` into the file `Core/Formalism.v` and propagating the renaming to the original algorithm [10]; introducing two predicates `on_loc` and `occupied` to better distinguish robots and their locations in the file `Core/Configurations.v`.

5 Concluding Remarks

We presented the first fully-self-stabilizing (no constraints are put on the number of robots or on the initial configuration) deterministic uniform gathering algorithm for robots operating in the semi-synchronous Look-Compute-Move model. Our execution model, algorithm description, and correctness proof all use a common formal language, ensuring confidence in the claimed results. A byproduct of our approach lies in the adapt and reuse techniques we employed to build upon previously published formal developments in the PACTOLE framework. We believe that future work will also be able to further develop the code we provide to the community.

The most obvious open question raised by our work is the extension of the positive result to weaker scheduling constraints, namely the ASYNC execution model. On the one hand, the possibility for two robots to gather in this setting [13] hints at a positive outcome. On the other hand, the inability to directly scale the model-checking solution of Defago *et al.* [13] implies that many additional formal developments to the PACTOLE framework are necessary, as its support for the ASYNC model is still in early stage of development [2] (in particular, a number of basic lemmas are missing in this model).

References

1. Asahiro, Y., Yamashita, M.: Minimum algorithm sizes for self-stabilizing gathering and related problems of autonomous mobile robots (extended abstract). In: Dolev, S., Schieber, B. (eds.) Stabilization, Safety, and Security of Distributed Systems - SSS 2023. Lecture Notes in Computer Science, vol. 14310, pp. 312–327. Springer, Cham (2023). https://doi.org/10.1007/978-3-031-44274-2_23
2. Balabonski, T., Courtieu, P., Pelle, R., Rieg, L., Tixeuil, S., Urbain, X.: Continuous *vs.* discrete asynchronous moves: a certified approach for mobile robots. In: Atig, M.F., Schwarzmann, A.A. (eds.) NETYS 2019. LNCS, vol. 11704, pp. 93–109. Springer, Cham (2019). https://doi.org/10.1007/978-3-030-31277-0_7
3. Balabonski, T., Delga, A., Rieg, L., Tixeuil, S., Urbain, X.: Synchronous gathering without multiplicity detection: a certified algorithm. Theory Comput. Syst. **63**(2), 200–218 (2019). https://doi.org/10.1007/S00224-017-9828-Z

4. Bauer, A.: How to review formalized mathematics. Blog post (2013). http://math.andrej.com/2013/08/19/how-to-review-formalized-mathematics/
5. Bérard, B., Lafourcade, P., Millet, L., Potop-Butucaru, M., Thierry-Mieg, Y., Tixeuil, S.: Formal verification of mobile robot protocols. Distributed Comput. **29**(6), 459–487 (2016). https://doi.org/10.1007/S00446-016-0271-1
6. Bertot, Y., Castéran, P.: Interactive Theorem Proving and Program Development. Coq'Art: The Calculus of Inductive Constructions. Texts in Theoretical Computer Science. Springer (2004). http://www.labri.fr/publications/l3a/2004/BC04
7. Bramas, Q., Tixeuil, S.: Wait-free gathering without chirality. In: Scheideler, C. (ed.) SIROCCO 2014. LNCS, vol. 9439, pp. 313–327. Springer, Cham (2015). https://doi.org/10.1007/978-3-319-25258-2_22
8. Cieliebak, M., Flocchini, P., Prencipe, G., Santoro, N.: Distributed computing by mobile robots: gathering. SIAM J. Comput. **41**(4), 829–879 (2012). https://doi.org/10.1137/100796534
9. Courtieu, P., Rieg, L., Tixeuil, S., Urbain, X.: Impossibility of gathering, a certification. Inf. Process. Lett. **115**(3), 447–452 (2015). https://doi.org/10.1016/J.IPL.2014.11.001
10. Courtieu, P., Rieg, L., Tixeuil, S., Urbain, X.: Certified universal gathering in $\mathbb{R}^2$ for oblivious mobile robots. In: Gavoille, C., Ilcinkas, D. (eds.) DISC 2016. LNCS, vol. 9888, pp. 187–200. Springer, Heidelberg (2016). https://doi.org/10.1007/978-3-662-53426-7_14
11. Courtieu, P., Rieg, L., Tixeuil, S., Urbain, X.: Swarms of mobile robots: towards versatility with safety. Leibniz Trans. Embed. Syst. **8**(2), 02:1–02:36 (2022). https://drops.dagstuhl.de/entities/document/10.4230/LITES.8.2.2. https://doi.org/10.4230/LITES.8.2.2
12. Das, S., Flocchini, P., Prencipe, G., Santoro, N., Yamashita, M.: Autonomous mobile robots with lights. Theor. Comput. Sci. **609**, 171–184 (2016). https://doi.org/10.1016/J.TCS.2015.09.018
13. Défago, X., Heriban, A., Tixeuil, S., Wada, K.: Using model checking to formally verify rendezvous algorithms for robots with lights in euclidean space. Robotics Auton. Syst. **163**, 104378 (2023). https://doi.org/10.1016/J.ROBOT.2023.104378
14. Défago, X., Potop-Butucaru, M., Parvédy, P.R.: Self-stabilizing gathering of mobile robots under crash or byzantine faults. Distrib. Comput. **33**(5), 393–421 (2020). https://doi.org/10.1007/S00446-019-00359-X
15. Dieudonné, Y., Petit, F.: Self-stabilizing gathering with strong multiplicity detection. Theor. Comput. Sci. **428**, 47–57 (2012). https://doi.org/10.1016/J.TCS.2011.12.010
16. Dijkstra, E.W.: Self-stabilizing systems in spite of distributed control. Commun. ACM **17**(11), 643–644 (1974). https://doi.org/10.1145/361179.361202
17. Doan, H.T.T., Bonnet, F., Ogata, K.: Model checking of robot gathering. In: Aspnes, J., Bessani, A., Felber, P., Leitão, J. (eds.) 21st International Conference on Principles of Distributed Systems, OPODIS 2017, Lisbon, Portugal, 18–20 December 2017. LIPIcs, vol. 95, pp. 12:1–12:16. Schloss Dagstuhl - Leibniz-Zentrum für Informatik (2017). https://doi.org/10.4230/LIPICS.OPODIS.2017.12
18. Flocchini, P.: Gathering. In: Flocchini, P., Prencipe, G., Santoro, N. (eds.) Distributed Computing by Mobile Entities, Current Research in Moving and Computing. Lecture Notes in Computer Science, vol. 11340, pp. 63–82. Springer (2019). https://doi.org/10.1007/978-3-030-11072-7_4
19. Nakajima, K., Takase, K., Wada, K.: Efficient self-stabilizing simulations of energy-restricted mobile robots by asynchronous luminous mobile robots. In: Emek, Y.

(ed.) Structural Information and Communication Complexity - SIROCCO 2024. Lecture Notes in Computer Science, vol. 14662, pp. 494–500. Springer, Cham (2024). https://doi.org/10.1007/978-3-031-60603-8_28
20. Pierce, B.C., et al.: Logical Foundations. Software Foundations series, volume 1. Electronic textbook (2025). https://softwarefoundations.cis.upenn.edu/lf-current/index.html
21. Prencipe, G.: Impossibility of gathering by a set of autonomous mobile robots. Theor. Comput. Sci. **384**(2–3), 222–231 (2007). https://doi.org/10.1016/J.TCS.2007.04.023
22. Sangnier, A., Sznajder, N., Potop-Butucaru, M., Tixeuil, S.: Parameterized verification of algorithms for oblivious robots on a ring. Formal Methods Syst. Des. **56**(1), 55–89 (2020). https://doi.org/10.1007/S10703-019-00335-Y
23. Suzuki, I., Yamashita, M.: Distributed anonymous mobile robots: formation of geometric patterns. SIAM J. Comput. **28**(4), 1347–1363 (1999). https://doi.org/10.1137/S009753979628292X
24. Terai, S., Wada, K., Katayama, Y.: Gathering problems for autonomous mobile robots with lights. Theor. Comput. Sci. **941**, 241–261 (2023). https://doi.org/10.1016/j.tcs.2022.11.018
25. Viglietta, G.: Rendezvous of two robots with visible bits. In: Flocchini, P., Gao, J., Kranakis, E., Meyer auf der Heide, F. (eds.) ALGOSENSORS 2013. LNCS, vol. 8243, pp. 291–306. Springer, Heidelberg (2014). https://doi.org/10.1007/978-3-642-45346-5_21

Approximating Independent Sets in Constant Distributed Rounds

Ravi B. Boppana[1] and Magnús M. Halldórsson[2(✉)]

[1] Department of Mathematics, MIT, Cambridge, USA
rboppana@mit.edu

[2] Department of Computer Science, Reykjavik University, Reykjavik, Iceland
mmh@ru.is

Abstract. A simple randomized one-round distributed algorithm for approximating independent sets in graphs was studied by Boppana, Halldorsson, and Rawitz (SIROCCO 2018). It was shown to attain a $(\Delta + 1)/2$-approximation on unweighted graphs of maximum degree Δ, but only in expectation.

We show here that the same bound holds with high probability, with minimal loss. This means that the algorithm is optimal in the class of 1-round algorithms. It is also better by a factor of $2-o(1)$ than all known $o(\log n)$-round algorithms.

For graphs of constant degree, the algorithm can be derandomized in $O(\log^* n)$ rounds.

Keywords: approximation algorithms · distributed algorithms · randomized algorithms

1 Introduction

We consider the Maximum Independent Set (MaxIS) problem, where the input is a simple graph $G = (V, E)$ and the goal is to find a maximum cardinality *independent set*, namely a subset I of the vertices such that the induced graph $G[I]$ does not contain edges.

A classic lower bound on the independence number $\alpha(G)$ (i.e., the cardinality of the largest independent set) is due to Turán [23]:

$$\alpha(G) \geq \textsc{Turán}(G) \triangleq \frac{n^2}{2m+n} = \frac{n}{\bar{d}(G)+1}, \tag{1}$$

where $n = |V|$ is the number of vertices, $m = |E|$ is the number of edges, and $\bar{d}(G)$ is the average degree in G. It is known to be the best possible bound in terms of the sparsity of G.

A stronger lower bound was given independently by Caro [7] and Wei [24]:

$$\alpha(G) \geq \textsc{CaroWei}(G) \triangleq \sum_{v\in V} \frac{1}{d(v)+1}, \tag{2}$$

U. Schmid and R. Kuznets (Eds.): SIROCCO 2025, LNCS 15671, pp. 144–158, 2025.
https://doi.org/10.1007/978-3-031-91736-3_9

where $d(v)$ is the degree of vertex $v \in V$, It is known to be best possible in terms of the degree sequence of the graph. It is always at least the Turán bound, but can be much stronger. For instance, on the graph H consisting of a clique of size $n/2$ and $n/2$ independent vertices, $\textsc{CaroWei}(H) \geq n/2$, while $\textsc{Turán}(H) \leq 4$.

There are numerous proofs of the Caro-Wei bound, some involving simple greedy algorithms. Arguably the simplest argument known is a probabilistic one:

> *Uniformly randomly permute the vertices, and output the set of vertices that precede all their neighbors in the permutation.* (3)

Each node v precedes its neighbors with probability $1/(d(v)+1)$, so by linearity of expectation the expected size of the output matches exactly $\textsc{CaroWei}(G)$. This argument, which first appeared in the book of Alon and Spencer [2], is due to Boppana [5]. It clearly leads to a very simple local decision rule once the permutation is selected.

An alternative formulation of the algorithm is practical in certain contexts.

> *Each vertex v picks a random real number x_v from $[0,1]$. The vertex joins the independent set if and only if its random number is larger than that of all its neighbors.* (4)

It suffices to select the numbers with precision $1/n^3$, for which collisions are very unlikely. Observe that the computed independent set may not be (inclusion-wise) maximal.

This leads to a simple 1-round distributed algorithm for MaxIS in which each node decides whether to join the independent set after a single round of communication with its neighbors. The same $O(\log n)$ bits transmitted go to all of the node's neighbors, which complies with the Broadcast-CONGEST model of distributed algorithms. Furthermore, the algorithm is asynchronous. This is just about the simplest distributed algorithm one could hope for. The algorithm is alternatively known as the classical ranking algorithm, the first round of Luby [21], or Boppana's algorithm (see [6]). We shall refer to it as Ra, and let $\mathcal{R}(G)$ denote the random variable representing the size of the solution found on graph G.

The performance guarantees of this algorithm were analyzed in [6]. It was shown to attain a $(\Delta+1)/2$-approximation and $\bar{d}/1.657$-approximation on unweighted graphs, but only *in expectation*. Expectation bounds differ from high probability bounds in the distributed settings. Other currently known MaxIS approximation algorithms either give only Δ-approximation or run in $\text{poly}(\Delta) \cdot \log n$ rounds.

1.1 Our Contributions

We show that RA actually attains an essentially equivalent high-probability bound. All but the last result hold with high probability.

The first result follows directly from the application of an appropriate tail bound.

Proposition 1. RA *achieves the Caro-Wei bound within* $1 - o(1)$*-factor, whenever* $\Delta = o(n^{1/3}/\log n)$.

For larger degrees, RA achieves comparable performance guarantees:

Proposition 2. RA *attains a performance guarantee of* $(\Delta+1)/2\cdot(1+o(1))$ *for* MIS, *for* $\Delta \ll n/\log^4 n$. *It guarantees an independent set of size* $(1-o(1))n/(\Delta+1)$.

The key idea behind this result is to round the random values in two steps: only the vertices whose values exceed a certain threshold continue (in the analysis only) to the second step. This reduces the degree of the graph in consideration and thereby the dependencies between the random variables.

We can handle still larger values of Δ by modifying RA slightly.

Proposition 3. *A 2-round variation of* RA *attains a performance guarantee of* $(\Delta + 1)/2 \cdot (1 + o(1))$, *for* $\Delta \leq n/\Omega(\log n)$. *It also guarantees a set of size* $(1 - o(1))n/(\Delta + 1)$.

In this case, we explicitly eliminate from consideration the nodes with too many neighbors that survive the first winnowing.

For $\Delta \geq n/\log n$, Kawarabayashi et al. [19] showed that no constant-round algorithm can guarantee a solution of size $\Omega(n/\Delta)$. Thus, the range we can prove is nearly optimal.

Finally, we can handle all values of Δ with a 4-round algorithm.

Proposition 4. *There is a 4-round* LOCAL *algorithm that attains a performance guarantee of* $(\Delta + 1)/2 \cdot (1 + o(1))$, *for every value of* Δ.

For the weighted case, we obtain partial results.

Proposition 5. *There is 1-round algorithm for* MAXWIS *that attains a* $0.529\Delta(1 + o(1))$*-approximation, on instances where the maximum weight of a node is at most a* $\omega(\Delta^4 \log n))$*-fraction of the total weight.*

On the other side of the spectrum, we show that the Caro-Wei bound (and the resulting performance guarantees) can be obtained *deterministically* very fast when Δ is constant.

Proposition 6. *There is a deterministic algorithm that achieves the (full) Caro-Wei bound in* $O(\Delta^2 + \log^* n)$ *rounds.*

1.2 Related Work

Bar-Yehuda et al. [3] presented distributed implementations for a centralized local ratio Δ-approximation algorithm for Maximum Weight Independent Set (MaxWIS). The first algorithm runs in $O(T_{\mathrm{MIS}(G)} \cdot \log W)$ rounds in the CONGEST model, where $T_{\mathrm{MIS}(G)}$ is the number of rounds needed to compute a maximal independent set in G and W is the ratio between the largest and smallest weights. They also gave a deterministic implementation with the same approximation ratio that runs in $O(\Delta + \log^* n)$ rounds. Kawarabayashi et al. [19] improved the running time while increasing the approximation ratio by a factor of $1 + \varepsilon$. They presented an $O(T_{\mathrm{MIS}(n,\Delta)}/\varepsilon)$-round $(1 + \varepsilon)\Delta$-approximation algorithm, where $T_{\mathrm{MIS}(n,\Delta)}$ is the number of rounds needed to compute a maximal independent set in a graph with n vertices and maximum degree Δ. and a randomized $O(\mathrm{poly}(\log\log n)/\varepsilon)$-round $(1+\varepsilon)\Delta$-approximation algorithm. Faour et al. [11] obtain $(1+\varepsilon)\Delta$-approximation in time $O(\log^2(\Delta W)(\log 1/\epsilon) + \log^* n)$. Their approach uses a fractional solution (for a slightly different LP), but not as part of the local ratio reductions. All the algorithms mentioned work in the CONGEST model (Table 1).

Table 1. Results for MaxWIS.

Source	Approximation	Rounds	Type
[3]	Δ	$O(\log n \log W)$	w.h.p.
	Δ	$O(\Delta + \log^* n)$	Deter.
[6]	$0.53(\Delta + 1)$	1	Expect.
[19]	$(1 + \varepsilon)\Delta$	$O(\mathrm{poly}(\log\log n)/\varepsilon)$	w.h.p.
[11]	$(1 + \varepsilon)\Delta$	$O(\mathrm{poly}(\log(\Delta W), 1/\epsilon) + \log^* n)$	Deter.
[17]	$(1 + \varepsilon)(\Delta + 1)/2$	$O(\mathrm{poly}(\log(\Delta W), 1/\varepsilon) + \log W \log n)$	w.h.p.
	$(1 + \varepsilon)(\Delta + 1)/2$	$O(\mathrm{poly}(\log(\Delta W), 1/\varepsilon) + \Delta + \log^* n)$	Deter.
	$(1 + \varepsilon)(\frac{\Delta}{2} + \frac{1}{4} - \frac{1}{8\Delta-12})$	$O(\mathrm{poly}(\Delta, 1/\varepsilon, \log n))$	Deter.
	$\frac{8}{5} + \varepsilon$, for $\Delta = 3$	$O(\mathrm{poly}(1/\varepsilon, \log n))$	Deter.
This work[a]	$0.53(\Delta + 1)$	1	w.h.p.

[a]Holds only when node weights are not too large compared with the total weight of the graph

Boppana, Halldórsson, and Rawitz [6] analyzed the performance guarantees of Ra and showed it to achieve a tight $(\Delta + 1)/2$ expected approximation ratio (and $(\bar{d} + 2)/1.657$-approximation). A slightly modified algorithm yields an expected approximation which asymptotically approaches $0.529(\Delta + 1)$.

Halldórsson and Rawitz [17] gave a deterministic distributed algorithms for MaxWIS that attains an approximation ratio of $(1 + \varepsilon)(\frac{\Delta}{2} + \frac{1}{4} - \frac{1}{8\Delta-12})$ in $O(\mathrm{poly}(\Delta, 1/\varepsilon, \log n))$ rounds.

In the LOCAL model, $n^{\Theta(1/k)}$-approximation is achievable (in expectation) and best possible for LOCAL algorithms running in k rounds [4] (where the upper bound assumes both unlimited bandwidth and computation), and a $(1+\varepsilon)$-approximation can be achieved in $O(\log n/\varepsilon)$ rounds [14].

Alon [1] gave nearly tight bounds for testing independence properties; his lower bound carries over to distributed algorithms, as argued in [6]. Kuhn et al. [20] showed that achieving any $O(1)$-approximation requires $\Omega(\max(\log\Delta/\log\log\Delta, \sqrt{\log n/\log\log n}))$ rounds of LOCAL. Censor-Hillel, Khoury, and Paz [8] gave a nearly quadratic lower bound on the number of rounds for solving Maximum Independent Set exactly of CONGEST.

The only other work on constant-round algorithms is that of Kawarabayashi et al. [19], who gave a $O(1/\epsilon)$-round algorithm for that computes an independent set of size $(1+\epsilon)n/(\Delta+1)$ for every $\Delta \leq n/\log n$, and showed that this was impossible to achieve for larger values of Δ. Our analysis shows that two rounds suffice, for nearly all values of Δ (Table 2).

Table 2. Results for MaxIS.

Source	Approximation	Rounds	Type
[3]	Δ	$O(\log n)$	w.h.p.
[6]	$(\Delta+1)/2$	1	Expect.
[19]	$(1+\varepsilon)(\Delta+1)$	$O(1/\varepsilon)$	w.h.p.
[11]	$(1+\varepsilon)\Delta$	$O(\text{poly}(\log(\Delta), 1/\epsilon) + \log^* n)$	Deter.
[17]	$(1+\varepsilon)(\Delta+1)/2$	$O(\text{poly}(\log\Delta, 1/\varepsilon) + \log n)$	w.h.p.
	$(1+\varepsilon)(\Delta+1)/2$	$O(\text{poly}(\log\Delta, 1/\varepsilon) + \Delta + \log^* n)$	deter.
This work	$\mathbf{(1+\varepsilon)(\Delta+1)/2}$	1 to 4	w.h.p.
	$\mathbf{(\Delta+1)/2}$	$O(\Delta^2 + \log^* n)$	Deter.

2 Preliminaries

2.1 Definitions and Notation

For a simple graph $G = (V, E)$, let $n = |V|$ be the number of vertices and let $m = |E|$ be the number of edges. The neighborhood of a vertex v is denoted $N(v)$, i.e., $N(v) = \{u : (v, u) \in E\}$. The closed neighborhood of v is define as $N[v] = N(v) \cup \{v\}$. Denote $\deg(v) = |N(v)|$, and let $\Delta = \max_v \deg(v)$ be the maximum degree in G. The average degree is $\bar{d} = \frac{1}{n}\sum_v \deg(v)$. A subset $I \subset V$ is *independent* if the induced graph $G[I]$ does not contain edges of E.

We say that a probabilistic event holds *w.h.p.* (with high probability) if it holds with probability at most n^{-c}, where n is the number of nodes in G and c is an arbitrary constant (set in advance). From here on, c refers to this constant.

We write $a = (1 \pm \delta)b$ to mean that $a \in [b - \delta, b + \delta]$.

2.2 Computation Model

In distributed graph models, each node of the input graph is a processor that has communication ports to its neighboring nodes. In synchronous rounds, the nodes can transmit messages to its neighbors and perform an arbitrary amount of computation. At the end of the protocol, each node announces whether it is in the independent set solution or not. The time complexity of the protocol is the maximum number of rounds used by a node.

In the LOCAL model, nodes can transmit messages of arbitrary length in each round, while in the CONGEST model, the messages must be $O(\log n)$ bits. In the Broadcast-CONGEST model, each node transmits a single message to all its neighbors in a round (but can receive different messages), while in CONGEST and LOCAL, the messages to the neighbors can be all different.

RA runs in a single round, with each node sending only a single $O(\log n)$-bit message. Thus, it is a 1-round Broadcast-CONGEST algorithm.

2.3 Concentration Bounds

Chernoff. We use the following standard tail bound to sparsify the instances (in the analysis only).

Theorem 1 (Chernoff). *Let $\{X_i\}_i$ be independent random variables with $X_i \in [0,1]$. Let $X = \sum_i X_i$. Then, for any $0 < \delta < 1$,*

$$\mathbb{P}[X < (1-\delta)\mathbb{E}[X]], \mathbb{P}[X > (1+\delta)\mathbb{E}[X]] \leq \exp(-\delta^2\mathbb{E}[X]/3) .$$

Read-k. We use the following tail bound to get concentration results for RA for as large Δ as possible. It is somewhat non-standard, but has been used for analyzing distributed algorithms for maximal independent sets [22] and graph coloring [16].

A set $\{X_i\}_{i=1}^r$ of binary random variables is *read-k* if there is a set $\{Y_j\}_{j=1}^m$ of m independent binary random variables and subsets $\{P_i\}_{i=1}^r$ of indices, $P_i \subseteq [m]$, such that X_i is a function of only $\{Y_j\}_{j \in P_i}$, $i \in [r]$, while for each $j \in [m]$, $|\{i : j \in P_i\}| \leq k$. In words, each Y_j influences at most k variables X_i.

Theorem 2 (read-k bound [12]). *Let $\{X_i\}_{i=1}^r$ be a family of read-k binary random variables and let $X = \sum_{i=1}^r X_i$. For any $\delta > 0$,*

$$\mathbb{P}[|X - \mathbb{E}[X]| \geq \delta r] \leq 2\exp(-2\delta^2 r/k) .$$

3 High Probability Bounds

We show how claims on the expected performance of RA translate also to claims that hold with high probability. The issue is non-trivial due to dependencies, and this becomes more pronounced as the maximum degree gets higher.

We first reason about the impact of the loss of imprecision when each node transmits only s bits of the random real x_v.

Lemma 1. *Suppose each node picks a random value $x_v \in [0,1]$, transmits the s most-significant bits of x_v, denoted y_v, to each neighbor. Then, the probability of the event that y_v is strictly larger than all y_u of its neighbors is at least*

$$\mathbb{P}[y_v > \max_{u \in N(v)} y_u] \geq \frac{1}{d(v)+1} - \frac{d(v)}{2^s} .$$

Proof. The probability that x_v is larger than those of all the neighbors of v is exactly $1/(d(v)+1)$. The probability that s_v is equal to that of a specific neighbor is 2^{-s}. The claim follows by the union bound.

Modulo a $(1 - 1/n)$-approximation, it suffices to transmit $3 \log n$ bits.

Recall that $\mathcal{R}(G)$ denotes the random variable representing the size of the solution obtained by RA algorithm on the graph G.

Theorem 3. *For any $\varepsilon > 0$,*

$$\mathcal{R}(G) \geq (1-\varepsilon)\textsc{CaroWei}(G), \quad \textit{with probability } 1 - 2\exp(-2\varepsilon^2 n/(\Delta+1)^3) .$$

Furthermore, RA *achieves a performance ratio of $(1+\epsilon)\Delta/2$, with the same probability.*

Proof. Let Y_v be the sequence of random bits representing the random value x_v by the algorithm. We shall analyze the algorithm as if it only considered the $3 \log n$ most-significant bits, in light of Lemma 1. Let X_v be the indicator random variable if node v is in the independent set produced by our algorithm CW, and let $\mathcal{R}(G) = X = \sum_{v \in V} X_v$ be the size of that solution. Recall that with infinite precision, the expected size of the algorithm solution is exactly $\textsc{CaroWei}(G) \geq n/(\Delta+1)$. Taking the imprecision into account,

$$\mathbb{E}[X] = \sum_{v \in V} \left(\frac{1}{d(v)+1} - \frac{d(v)}{2^{3\log n}} \right) = \textsc{CaroWei}(G) - 2\,m/n^3 \geq \frac{n}{\Delta+1} - 1/n .$$

Observe that each X_v is a function of the Y_u for the nodes $u \in N[v]$ and thus $\{X_v\}_v$ is a read-k family, for $k \leq \Delta$ (as each bit in the representation of a value x_u influences at most Δ variables X_v). By Theorem 2 (with $r = n$, $k = \Delta$, and $\delta = \varepsilon/(\Delta+1)$),

$$\begin{aligned}\mathbb{P}[\mathcal{R}(G) < (1-\varepsilon)\textsc{CaroWei}(G)] &\leq \mathbb{P}[X < \mathbb{E}[X] - \varepsilon n/(\Delta+1)] \\ &\leq 2\exp(-2\varepsilon^2 n/(\Delta+1)^3) .\end{aligned}$$

This implies the first claim, and the performance ratio follows from [6].

Corollary 1. RA *achieves the Caro-Wei bound for graphs with $\Delta + 1 \leq n^{1/3}/\log n$, up to a lower-order term, w.h.p.*

Proof. Apply Theorem 3 with $\varepsilon = 1/(2c\sqrt{\log n})$.

For higher degrees, we modify the analysis.

Theorem 4. *Let G be a graph with n vertices and maximum degree $\Delta = \Delta(G) \leq n/(6\log^4 n)$. Then, $\mathcal{R}(G) \geq (1-\varepsilon)n/\Delta$, for any $\varepsilon \geq 108c\sqrt{\log n}(\Delta/n)^{1/6}$, w.h.p. Furthermore,* RA *achieves a $(1+\varepsilon)(\Delta+1)/2$-approximation, w.h.p.*

Proof. When $\Delta + 1 \leq n^{1/4}$, the claim follows from Theorem 3. Namely, given the bounds on ϵ and Δ it then holds that $\epsilon^2 n/(\Delta+1)^3 \geq c^2 \log n$, so the claim holds w.h.p. by Theorem 3. So assume from now that $\Delta \in [n^{1/4}, n/\log^4 n]$.

Let $\delta = \varepsilon/6$. Let

$$p = \frac{3\delta^{-2} c \log n}{\Delta} .$$

Note that $p < 1$ by the assumption on Δ.

View the random selection x_v of each node v as a two-step process: first, v flips a biased coin y_v with probability p. This determines whether x_v becomes at least $1-p$. Let $V_h = \{v \in V : y_v = 1\}$ and let $H = G[V_h]$ be the graph induced by the nodes selected. In the second step, the node determines the rest of the value x_v. Notice that conditioning on the set V_h, the resulting ordering of V_h (as given by the x_v) is still uniformly random. Since nodes in V_h have precedence over the nodes in $V \setminus V_h$ in the ranking algorithm, it follows that

$$\mathcal{R}(G) \geq \mathcal{R}(H) . \tag{5}$$

By linearity of expectation, the expected size of V_h is pn and the expected degree of each node in H is at most $p\Delta$. Since the node choices are independent, we apply Chernoff (Theorem 1) to obtain that for each vertex v,

$$\mathbb{P}[d_H(v) > (1+\delta)p\Delta] \leq \exp(-\delta^2 p\Delta/3) = \exp(-c\log n) = 1/n^c .$$

For the vertex v_{max} of maximum degree, it similarly holds w.h.p. that $d_H(v_{max}) \geq (1-\delta)p\Delta$. Thus, by a union bound over all n vertices, it holds w.h.p. that

$$\Delta(H) = (1 \pm \delta)p\Delta . \tag{6}$$

By a similar argument we get w.h.p. that

$$|V_h| = (1 \pm \delta)pn . \tag{7}$$

From now on, we condition on the events of Eqs. (6) and (7).

We now apply Theorem 3 (with $G = H$ and $\epsilon = \delta(n)$) to obtain that

$$\mathcal{R}(H) \geq \textsc{CaroWei}(H)(1-\delta) \quad \text{w.p. } 1 - \exp(-\delta^2|V_h|/(\Delta(H)+1)^3) , \tag{8}$$

Observe that

$$\frac{|V_h|}{(\Delta(H)+1)^3} \geq \frac{(1-\delta)pn}{((1+\delta)p\Delta+1)^3} \geq \frac{(1-5\delta)n}{(\Delta+1)(3c\delta^{-2}\log n)^2} = \Theta\left(\frac{\delta^4 n}{\Delta \log^2 n}\right) .$$

Thus, Eq. (8) holds w.h.p., given the lower bound on δ and upper bound on Δ. Similarly, applying Eqs. (7) and (6), it holds w.h.p. that

$$\textsc{CaroWei}(H) \geq \frac{|V_h|}{\Delta(H)+1} \geq \frac{(1-\delta)n}{(1+\delta)\Delta+1} \geq (1-2\delta)\frac{n}{\Delta+1} . \tag{9}$$

Combining Eqs. (8), (5) and (9) yields the first claim.

If $\alpha(G) \leq n/2$, Eq. (9) also implies the claimed approximation ratio. Otherwise, we can obtain by an argument equivalent to Eq. (7) that w.h.p.

$$\alpha(H) \geq (1-\delta)\alpha(G) .$$

Then, combined with Eq. (6), it holds w.h.p. that

$$\textsc{CaroWei}(H) \geq \frac{2}{\Delta(H)} \cdot \alpha(H) \geq \frac{2(1-\delta)}{1+\delta} \cdot \frac{\alpha(G)}{\Delta} \geq (1-2\delta)\frac{\alpha(G)}{(\Delta+1)/2}.$$

3.1 2-Round Algorithm

With additional effort, we can reduce the upper bound on Δ further. This is instructive in that we argue the performance ratio in terms of the Turán bound, rather than the Caro-Wei bound. The reason is that in certain cases, one can reason with high probability about the preservation of edges when similar claims about degrees fail.

We first show that removing high-degree nodes from a sparse graph actually improves the Turán bound.

Lemma 2. *Let G be a graph of average degree d. Let G' be the subgraph of G induced by nodes of degree at most $6d$. Then,* $\textsc{Turán}(G') \geq \textsc{Turán}(G)$.

Proof. Let $t = |V(G)| - |V(G')|$ be the number of nodes removed from G. At least $3dt$ incident edges are removed. Hence,

$$\begin{aligned}\textsc{Turán}(G') &\geq \frac{(n-t)^2}{(dn-3dt)+n-t} = \frac{n^2-2tn+t^2}{d(n-2t)+n-(d+1)t} > \frac{(n-2t)n}{(n-2t)(d+1)} \\ &= \textsc{Turán}(G) .\end{aligned}$$

We modify Ra to exclude some nodes from consideration, as given in Algorithm 1. The idea is limit the maximum degree in the subgraph on which we analyze Ra, resulting in less dependencies between the random variables.

Algorithm 1. Modified Ra (G, Δ, n, ϵ)

Sample each node of $V(G)$ into set V_3 independently with probability $p \cdot q = \delta^{-1}/\Delta$.
Let $V_4 = \{u \in V_3 : |N(v) \cap V_3| \leq 6/\delta\}$ be nodes with few neighbors in V_3
Run Ra on $H_4 = G[V_4]$

Let $\mathcal{MR}(G)$ denote the random variable representing the size of the solution obtained by the Modified Ra.

Theorem 5. *Suppose $\Delta \le n/(C \log n)$, for a sufficiently large constant $C > 0$. Then, $\mathcal{MR}(G) \ge (1-\epsilon)n/\Delta$, for any $\epsilon > 8/((\Delta \log n)/n)^{1/6}$, w.h.p. Furthermore, Modified* Ra *achieves a $(1-\epsilon)\Delta/2$-approximation, w.h.p.*

Proof. We assume $\Delta \ge n/\log^4 n$, in light of the previous theorem. Let $\delta = \epsilon/8$.

We continue with the set V_h sparsified earlier. We sparsify this set further, sampling each node w.p.

$$q = \frac{\delta^{-1}}{3c \log n} ,$$

producing a set V_3 and induced graph H_3. We can no longer obtain a good bound on the maximum degree $\Delta(H_3)$. Instead, we bound the number of edges, $m(H_3)$. Note that $pq = \delta/\Delta$.

Let X_e be the indicator random variable of the event that edge e in H is sampled into H_3. Let $X = m(H_3) = \sum_{e \in H} X_e$ be their sum. For each edge e, the probability that the edge is sampled is q^2. By linearity of expectation, $\mathbb{E}[X] = q^2 m(H) \le q^2 \Delta(H) n(H)$. We apply the read-$k$ bound, noting that $k = \Delta(H) + 1$ (the sampling of each vertex affects at most $\Delta(H)$ edges). Thus, by Theorem 2 with $r = n(H)$, $k = \Delta(H) + 1$,

$$\begin{aligned}
\mathbb{P}[X > (1+\delta)\mathbb{E}[X] &\le \mathbb{P}[X - \mathbb{E}[X] > \delta q^2 \Delta(H) n(H)] \\
&\le 2\exp(-2(\delta q^2)^2 \Delta(H) n(H)) \\
&\le 2\exp(-2\delta^2(1-2\delta)^2 p^2 q^2 \Delta n) \\
&\le 2\exp(-n/\Delta) \\
&\le 2\exp(-C \log n) .
\end{aligned}$$

Setting $C > c$, it holds w.h.p. that

$$m(H_3) = X \le \frac{1+\delta}{2} q^2 \Delta(H) n(H) . \tag{10}$$

We similarly argue as before by Chernoff that w.h.p.,

$$n(H_3) \ge (1-\delta) q n(H) \ge (1-2\delta) pqn . \tag{11}$$

Finally, let H_4 be the subgraph of H_3 induced by nodes with degree at most $12m(H_3)/n(H_3)$. Note that $n(H_4) \ge n(H_3)/2$ (fewer than half the vertices have more than four times the average degree). This is the small subgraph that is actually considered by the algorithm and, perhaps surprisingly, giving almost as large a solution. We now apply Lemma 2 and the definition of the Turán bound to obtain

$$\textsc{Turán}(H_4) \ge \textsc{Turán}(H_3) = \frac{n(H_3)}{2m(H_3)/n(H_3) + 1}$$

Observe that by Eqs. (10) and (11)

$$2m(H_3)/n(H_3) + 1 \le \frac{(1+\delta)q^2 \Delta(H) n(H)}{(1-\delta) q n(H)} + 1 \le \frac{1+\delta}{1-\delta} q \Delta(H) + 1 \le \frac{1+3\delta}{1-\delta} q \Delta(H) ,$$

where we use in the last inequality that $q\Delta(H) \geq 2\delta^{-1}$. Thus, using Eqs. (7) and (6),

$$\text{TURÁN}(H_4) \geq \frac{(1-\delta)^2 n(H)}{(1+3\delta)q\Delta(H)} \geq \frac{(1-7\delta)n}{\Delta}.$$

We finally verify that H_4 satisfies the conditions for Theorem 3 (with $G = H_4$, $n = n(H_4)$, $\Delta = \Delta(H_4)$). We can verify that

$$\frac{\delta^2 n(H_4)}{\Delta(H_4)^3} \sim \frac{\delta^2 n}{12^3 q^2 p^2 \Delta^3} \sim \frac{\delta^4 n}{12^3 \Delta} \geq \frac{\log n}{12^3} \cdot (n/\Delta \log n)^{1/3} \geq \frac{C^{1/3} \log n}{12^3}.$$

By ensuring that $C \geq 12^9 c^3$, it follows by Theorem 3 w.h.p. that

$$\mathcal{MR}(G) \geq \mathcal{R}(H_4) \geq (1-\delta)\text{CAROWEI}(H_4) \geq (1-\delta)\text{TURÁN}(H_4) \geq (1-8\delta)n/\Delta.$$

3.2 Approximation Bound for All Δ

We can also provide a $(1+\epsilon)\Delta/2$-approximation algorithm that works for every Δ. It runs in a constant number of rounds with a combined LOCAL algorithm. Let Lo (Hi) denote the set $Lo = \{u \in V(G) : d(u) \leq n/\log^5 n\}$ ($Hi =$ V(G) $\setminus Lo$) of vertices of degree at most (greater than) $n/\log^5 n$. In light of Theorem 4, we may assume $\Delta \geq n/\log^4 n$. We will write $A \sim B$ to denote $A = (1 \pm \delta)B$.

We first run RA on $G[Lo]$, obtaining an independent set A. We then run the following LOCAL algorithm on each node, obtaining an independent set B. Each node v gathers the 1-ball around it (i.e., all edges of its neighbors, along with its IDs and random values). It then computes an optimal independent set B_v in $G[N(v)]$. It communicates the size of its solution to its distance-3 neighbors and if it is larger than any of those solutions, it commits to that set. If it is larger than the part $A \cap N^2[v]$ contained within its distance-2 neighborhood, it broadcasts the set to $N^2(v)$, effectively replacing the nodes of $A \cap N^2[v]$ with the set B_v. The above ensures that the algorithm always outputs a non-empty set.

We now argue that this combined algorithm achieves the desired performance. Let I be a maximum independent set. If $|I| \leq \Delta/2$, any non-empty set gives a $\Delta/2$-approximation and we are done. So assume $|I| > \Delta/2$. Observe that there are at least $|I \cap B| \cdot n/\log^5 n$ edges incident on the nodes of $I \cap B$, so some vertex sees at least $|I \cap B|/\log^5 n$ of those vertices in its neighborhood. Hence, if $|I \cap B| \geq \log^{10} n$, we obtain a $n/\log^5 n$-approximation, or a $\Delta/\log n$-approximation. Otherwise, $|I \cap A| \geq (1 - 1/\sqrt{n})|I|$, and from Theorem 5, RA obtains a $(1 - o(1))\Delta/2$-approximation.

Finally, observe that the algorithm runs in four rounds. The downside of the algorithm is that it requires super-polynomial local computation and exceeds the communication abilities of CONGEST. Note that by aiming for weaker approximation, once can reduce the computation to a quasipolynomial time.

Corollary 2. *There is a 4-round* LOCAL *algorithm to compute* $(\Delta+1)/2$-*approximate independent sets, for every* Δ.

3.3 Weighted Case

The following modification of RA, MAX, was proposed in [6] (previously considered in an online setting in [10]).

> Each node v picks a random real number x_v uniformly from $[0, 1]$. It broadcasts the values x_v and w_v to its neighbors, who compute from it $r_v = x_v^{1/w_v}$. As before, each node u joins the solution if its value r_u is the highest among its neighbors.

The only difference is the computation of r_v, which now depends on the weight w_v. Again the algorithm runs in a single round of Broadcast-CONGEST, with correctness following as before.

As shown in [6], MAX achieves the following absolute values:

1. $\mathbb{E}[\text{MAX}] \geq \sum_{v \in V} \frac{w(v)^2}{w(N[v])}$
2. $\mathbb{E}[\text{MAX}] \geq w(V)/(1 + wdeg_{avg})$, where $wdeg_{avg} = (\sum_{v \in V} w(N[v]))/(w(V))$,

generalizing the Caro-Wei bound to node-weighted graphs. They showed that its *expected* performance ratio is at most $0.529(\Delta + 1)$.

We use the following tail inequality that is adapted from [9, Thm. 3.2], originally due to Janson [18].

Theorem 6 (Bounded-differences). *Let $\{X_i\}_i$ be random variables with $0 \leq X_i \leq b_i$, where each X_i depends on at most k other random variables X_j. Let $X = \sum_i X_i$. Then, for $t > 0$,*

$$\mathbb{P}[X < \mathbb{E}[X] - t] \leq \exp\left(\frac{-t^2}{k \sum_i b_i}\right) .$$

In our setting, we have the two-valued random variable X_v, whose value is $w(v)$ if v is selected and 0 otherwise. Then, $\sum_v b_v^2 \leq W_{max} w(V)$, where $w(V) = \sum_v w(v)$ and W_{max} is the largest node weight. We may assume that the weights are scaled so that $W_{max} = \Theta(W)$, the ratio between the largest and smallest weight.

When applying Theorem 6, we can only bound $k \leq \Delta^2$, since the events of two nodes u and v joining the algorithm solution are not independent if they share a common neighbor. Note that $\mathbb{E}[X] \in [w(V)/\Delta, w(V)]$.

Corollary 3. *Let $X = \text{MAX}(G)$ on a vertex-weighted graph G. Then,*

$$\mathbb{P}[X < (1 + \delta)\mathbb{E}[X]] \leq \exp\left(\frac{-\delta^2 \mathbb{E}[X]}{\Delta^4 W_{max}}\right) .$$

Corollary 4. *Suppose $W_{max} \leq o(w(V)/(\Delta^4 \log n))$.* MAX *attains a* $0.529\Delta(1 + o(1))$*-approximation for Weighted Independent Sets.*

4 Derandomization

We finish by showing that RA can be derandomized using the method of conditional expectation. This increases the time complexity to $O(\Delta^2 + \Delta \log^* n)$, which is moderate for small Δ.

A general approach for derandomizing distributed algorithms was proposed by Ghaffari, Harris, and Kuhn [13]. Their method is given for the LOCAL model and other details are slightly different, but the idea carries over.

The approach is to first compute a distance-2 coloring of the graph, which is an assignment where nodes with the same color are of distance greater than 2. Halldórsson, Kuhn, and Maus [15] give a CONGEST algorithm that runs in $O(\Delta^2 + \log^* n)$ rounds. We then process the colors in sequence, with the nodes of a given color rounding in parallel.

In more detail, define the random variables $\{X_v\}_{v \in V}$, such that $X_v = 1$ if and only if v is in the solution. The variable X_v satisfies $\mathbb{E}[X_v] = \mathbb{P}[X_v = 1] = 1/(d(v) + 1)$. We want to round these variables to 0-1 values Y_v, where the 1-valued nodes form an independent set, such that $\sum_{v \in V} Y_v \geq \mathbb{E}[X] = \textsc{CaroWei}(G)$, where $X = \sum_{v \in V} X_v$. Suppose a subset R of vertices has been rounded. For a vertex v in $V \setminus R$, let $\mathbb{P}[X_v = 1 \mid R]$ denote the probability of the event $X_v = 1$ conditioned on the values that have been assigned to the neighbors of v in R. The expected value for the whole graph is given by

$$f(R) \doteq \sum_{v \in V \setminus R} \mathbb{P}[X_v = 1 \mid R] + \sum_{v \in R} Y_v \ .$$

We can choose a value Y_v for a vertex $v \in V \setminus R$ that maintains the expectation for the whole set, i.e., such that $f(R \cup \{v\}) \geq f(R)$. By induction, we then have that

$$Y = f(V) \geq f(\emptyset) = \mathbb{E}[X] = \textsc{CaroWei}(G) \ .$$

A similar approach can be used in the weighted setting. In this case $\mathbb{P}[X_v = 1] = w(v)/w(N[v])$ and the expected value is given by

$$f(R) \doteq \sum_{v \in V \setminus R} w(v) \cdot \mathbb{P}[X_v = 1 \mid R] + \sum_{v \in R} w(v) \cdot Y_v \ .$$

Observe that $\mathbb{P}[X_v = 1 \mid R] = \mathbb{P}[X_v = 1 \mid R \cap N(v)]$, for $v \in V \setminus R$. Hence the value Y_v for a node v depends only on its neighbors, and only affects its neighbors. It follows that the decisions for nodes at distance 3 do not affect each other. Hence, a distance-2 color class can be processed in parallel. Observe that when a node chooses a value, it must inform all of its distance-2 neighborhood. Each node has at most one neighbor in a given distance-2 color class, so needs to forward only a single message. Thus, the full method of conditional expectation can be completed in $O(\Delta^2)$ rounds, given a distance-2 coloring using $O(\Delta^2)$ colors.

Theorem 7. *There exists a* Broadcast-CONGEST *algorithm for independent set running in $O(\Delta^2 + \log^* n)$ rounds that: (i) matches the Caro-Wei bound, (ii)* *achieves* $(\Delta + 1)/2$*-approximation, and (iii) achieves asymptotic* $0.529(\Delta + 1)$*-approximation in the weighted setting.*

5 Discussion

It is open whether there is a simple constant-round algorithm that obtains a $c\Delta$-approximation to *weighted* independent sets, for some $c < 1$. It is plausible that one may be able to combine some form of local rules, as in [17], with the ranking approach.

Acknowledgments. Magnús M. Halldórsson was supported by grant no. 217965 from the Icelandic Research Fund.

References

1. Alon, N.: On constant time approximation of parameters of bounded degree graphs. In: Property Testing - Current Research and Surveys, pp. 234–239 (2010)
2. Alon, N., Spencer, J.H.: The Probabilistic Method, 4th edn. Wiley (2016)
3. Bar-Yehuda, R., Censor-Hillel, K., Ghaffari, M., Schwartzman, G.: Distributed approximation of maximum independent set and maximum matching. In: 36th PODC, pp. 165–174 (2017)
4. Bodlaender, M.H., Halldórsson, M.M., Konrad, C., Kuhn, F.: Brief announcement: local independent set approximation. In: PODC, pp. 377–378. ACM (2016)
5. Boppana, R.B.: Personal communication to Joel Spencer (1987)
6. Boppana, R.B., Halldórsson, M.M., Rawitz, D.: Simple and local independent set approximation. In: Lotker, Z., Patt-Shamir, B. (eds.) SIROCCO 2018. LNCS, vol. 11085, pp. 88–101. Springer, Cham (2018). https://doi.org/10.1007/978-3-030-01325-7_12
7. Caro, Y.: New results on the independence number. Technical report, Tel Aviv University (1979)
8. Censor-Hillel, K., Khoury, S., Paz, A.: Quadratic and near-quadratic lower bounds for the CONGEST model. In: 31st DISC, pp. 10:1–10:16 (2017)
9. Dubhashi, D.P., Panconesi, A.: Concentration of measure for the analysis of randomized algorithms. Cambridge University Press (2009)
10. Emek, Y., Halldórsson, M.M., Mansour, Y., Patt-Shamir, B., Radhakrishnan, J., Rawitz, D.: Online set packing. SIAM J. Comput. **41**(4), 728–746 (2012)
11. Faour, S., Ghaffari, M., Grunau, C., Kuhn, F., Rozhon, V.: Local distributed rounding: generalized to MIS, matching, set cover, and beyond. In: 34th SODA (2023)
12. Gavinsky, D., Lovett, S., Saks, M., Srinivasan, S.: A tail bound for read-k families of functions. Random Struct. Algorithms **47**(1), 99–108 (2015)
13. Ghaffari, M., Harris, D.G., Kuhn, F.: On derandomizing local distributed algorithms. In: Proceedings of the 59th Annual IEEE Symposium on Foundations of Computer Science (FOCS 2018) (2018)
14. Ghaffari, M., Kuhn, F., Maus, Y.: On the complexity of local distributed graph problems. In: 49th STOC, pp. 784–797 (2017)
15. Halldórsson, M.M., Kuhn, F., Maus, Y.: Distance-2 coloring in the CONGEST model. In: 39th PODC, pp. 233–242 (2020)

16. Halldórsson, M.M., Kuhn, F., Nolin, A., Tonoyan, T.: Near-optimal distributed degree+1 coloring. In: STOC, pp. 450–463. ACM (2022)
17. Halldórsson, M.M., Rawitz, D.: Distributed fractional local ratio and independent set approximation. In: 31st SIROCCO. LNCS, vol. 14662, pp. 281–299 (2024)
18. Janson, S.: Large deviations for sums of partly dependent random variables. Random Struct. Algorithms **24**(3), 234–248 (2004)
19. Kawarabayashi, K., Khoury, S., Schild, A., Schwartzman, G.: Improved distributed approximations for maximum independent set. In: 34th DISC. LIPIcs, vol. 179, pp. 35:1–35:16 (2020)
20. Kuhn, F., Moscibroda, T., Wattenhofer, R.: Local computation: lower and upper bounds. JACM **63**(2), 17:1–17:44 (2016)
21. Luby, M.: A simple parallel algorithm for the maximal independent set problem. SIAM J. Comput. **15**(4), 1036–1053 (1986)
22. Pemmaraju, S., Riaz, T.: Using read-k inequalities to analyze a distributed MIS algorithm. In: 20th International Conference on Principles of Distributed Systems (OPODIS) (2017)
23. Turán, P.: On an extremal problem in graph theory. Mat. Fiz. Lapok **48**, 436–452 (1941). (in Hungarian)
24. Wei, V.: A lower bound on the stability number of a simple graph. Technical report, Bell Laboratories (1981)

A Visibility vs. Memory Trade-Off for Stand-Up Indulgent Gathering on Lines

Quentin Bramas[1,6], Hirotsugu Kakugawa[2], Sayaka Kamei[3], Anissa Lamani[1(✉)], Fukuhito Ooshita[4], Masahiro Shibata[5], and Sébastien Tixeuil[6]

[1] University of Strasbourg, ICube, CNRS, Strasbourg, France
alamani@unistra.fr
[2] Ryukoku University, Kyoto, Japan
[3] Hiroshima University, Higashihiroshima, Japan
[4] Fukui University of Technology, Fukui, Japan
[5] Kyushu Institute of Technology, Kitakyushu, Japan
[6] Sorbonne University, CNRS, LIP6, IUF, Paris, France

Abstract. We consider the stronger version of the gathering problem for n autonomous mobile robots that evolve in a line graph: if there is no crash, the robots must gather as usual, and if there is a single crash location, the remaining correct robots have to gather at this location. The robots have very weak capabilities: their vision is limited and depends on the initial maximum distance between the robots, they are unaware of n, and they either retain no memory of the past, or retain a fixed number of states.

In this context, we clarify the problem solvability according to visibility range and memory. If the initial number of occupied nodes is odd, we show that *(i)* an oblivious (that retains no past memory) algorithm can solve the problem if the visibility radius is one more hop than the trivial lower bound, and that bound is tight, and *(ii)* robots with one bit of persistent memory can solve the problem with optimal visibility radius, which is also tight with respect to the persistent memory. In the more relaxed setting where the initial number of occupied nodes may be even (but in that case, the distance between the border robots must be even), our one-bit memory algorithm remains valid (and optimal), while we present an oblivious algorithm for the same setting that uses two more hops than our lower bound.

Keywords: Crash failure · Fault-tolerance · LCM robot model · Limited visibility · Lights

1 Introduction

Since the seminal paper of Suzuki and Yamashita [18], the Distributed Computing research community actively studies mobile robot swarms, aiming to characterize what

---

This paper was supported by JSPS KAKENHI No. JP20KK0232, JP23K11059, JP23K28037, JP25K14995, and JP25K03101, and project SAPPORO (Ref. 2019-CE25-0005-1).

U. Schmid and R. Kuznets (Eds.): SIROCCO 2025, LNCS 15671, pp. 159–175, 2025.
https://doi.org/10.1007/978-3-031-91736-3_10

conditions make it possible for robots that are confused (each robot uses a private ego-centered coordinate system) and forgetful (robots may not remember all their past actions) to autonomously move and solve simple problems. One of these conditions is about how the robots coordinate their actions [11]: robots can either act all together (FSYNC), act in synchronized subsets (SSYNC), or arbitrarily (ASYNC). There is a strict hierarchy between FSYNC, SSYNC, and ASYNC: a distributed algorithm for the ASYNC case remains correct for the SSYNC and the FSYNC cases, while an impossibility result for the FSYNC case extends to the SSYNC and the ASYNC cases. Another important condition is the nature of the environment the robots evolve in [11]: robots can either evolve in a continuous environment, or a discrete one. In a discrete space, robots can only be in certain locations and move to adjacent ones. This can be modeled by a graph where the nodes are locations, hence the term "robots on graphs". A discrete space can be considered more realistic for modeling physical constraints or discrete sensors [2]. However, continuous and discrete spaces cannot be compared with respect to computability results: a discrete space has fewer possible robot positions, but a continuous space gives more options to resolve difficult situations (e.g., by moving slightly to break a symmetry). So, possibility and impossibility results do not extend from one model to the other.

One of the benchmark problems is the gathering problem. It is easy to state (robots must meet at the same place in a finite amount of time, without knowing where it is beforehand), but hard to solve (two robots that move according to SSYNC or ASYNC scheduling cannot meet in finite time, unless there are more assumptions). Without faults, gathering admits solutions in a variety of situations [11], *e.g.* continuous space [9], discrete space [16], discrete space with limited visibility [12–15], etc.

Robot failures become more likely as the number of robots increases, or as robots are deployed in dangerous environments. A crash fault occurs when a robot stops following its protocol unexpectedly. For the gathering problem, the desired outcome in case of crash faults must be specified, and two options were considered in the literature: weak gathering requires all non-faulty robots to meet, ignoring the faulty ones, while strong gathering (or stand-up indulgent gathering - SUIG) requires all non-faulty robots to meet at the unique crash location, *e.g.* to collectively repair the crashed robot(s). In continuous Euclidean space, weak gathering is solvable in the SSYNC model [1,3,8,10], while SUIG is only solvable in the FSYNC model [7].

SUIG in a discrete setting was considered either assuming robots have infinite range visibility [5,6] (that is, their sensors are able to obtain the position of all other robots), or assuming the visibility range is limited but robots are endowed with lights [4] (that is, they can remember and communicate a constant amount of information to all robots within their visibility range). All aforementioned works consider the FSYNC model, as even with infinite visibility, Bramas et al. [5] showed that the SUIG problem is solvable only in the FSYNC model. So, the main open question raised by recent results [4–6] is: is memory (*colored lights*) necessary to solve SUIG with limited visibility?

In this paper, we consider the discrete setting, and aim to characterize the solvability of the SUIG problem on lines when robots have limited visibility with respect to *(i)* the visibility range, and *(ii)* the number of colors (that is the memory) used by the robots. Similarly to previous works [4–7], we assume that all crashes, if any, occur at the same location, and that there may be multiple robots at each of some nodes in the initial configuration. The former assumption is necessary for *all* robots to gather at the same location (crashed robots cannot move), while the later assumption may allow a set of robots moving from a same node to split into two sets following a crash without robots recognizing it.

Our Contribution. We clarify the trade-off between the visibility range and the memory (expressed as the number of colors available for the robot lights) to solve SUIG in line graphs.

First, not all initial situations are solvable, and three parameters are used to classify solvable situations: H_{init} denotes the maximum distance between neighboring occupied nodes, M_{init} denotes the number of nodes between two (included) borders (that is, occupied nodes that have one side with no occupied node), and O_{init} denotes the number of occupied nodes. So, O_{init} is smaller than M_{init}, and the visibility range must be at least H_{init} (otherwise, the visibility graph is disconnected, and robots cannot gather).

Our results are summarized in Table 1. As it is known that when O_{init} and M_{init} are both even, SUIG on lines is impossible (regardless of visibility range and number of colors), we first concentrate on the case where O_{init} is odd. We first show (Theorem 1) that any oblivious (that is, using only one color) protocol for SUIG must have a visibility range at least $H_{init} + 1$ when $H_{init} = 3$ (so, one above the trivial lower bound), and we present a matching algorithm for any $H_{init} \geq 3$ (Algorithm 2). As the case of odd M_{init} admits a simple solution [4], we then concentrate on the general case: designing solution for the case where O_{init} *or* M_{init} is odd. Of course, the algorithm is unaware of the starting situation (O_{init} is odd, M_{init}, or both O_{init} and M_{init} are odd). In this case, we provide an optimal visibility range algorithm (H_{init}) with two colors (Algorithm 3), and by Theorem 1, two colors at least are necessary in this case. Finally, still in the general case, we present an oblivious algorithm (Algorithm 1) with visibility range $H_{init} + 3$.

Table 1. Summary of our results.

Initial condition	Visibility range ϕ	Number of colors	Result
O_{init} and M_{init} even	infinite	infinite	Impossible [5]
O_{init} odd	$\phi = H_{init} = 3$	1	Impossible (Thm. 1)
O_{init} odd or M_{init} odd	$\phi \geq H_{init} + 3$	1	Algorithm 1
O_{init} odd	$\phi \geq \max(4, H_{init} + 1)$	1	Algorithm 2
M_{init} odd	$\phi \geq H_{init}$	1	[4]
O_{init} odd or M_{init} odd	$\phi \geq H_{init}$	2	Algorithm 3

Our results show that, contrary to what was hinted by previous work [4–6], persistent memory is *not* necessary to solve SUIG with a finite visibility radius, but it is necessary to solve it with optimal visibility radius (H_{init}). Also, oblivious solutions for the general case where O_{init} *or* M_{init} is odd do exist, and further refining their required visibility range remains open.

Interestingly, the algorithms we present are also the best (with respect to visibility range and number of colors) known algorithms for the fault-free gathering. Yet, our impossibility results only hold for the SUIG problem, even if we conjecture that they also hold for the fault-free gathering.

2 Model

The environment is a line that consists of an infinite number of nodes $\ldots, u_{-2}, u_{-1}, u_0, u_1, u_2, \ldots$, such that a node u_i is connected to both u_{i-1} and u_{i+1}.

Let $\mathcal{R} = \{r_1, r_2, \ldots, r_n\}$ be the set of $n \geq 2$ autonomous robots. Robots are assumed to be anonymous and identical (*i.e.*, they have neither internal nor external identities), uniform (*i.e.*, they all execute the same program, and use no localized parameter such as a particular orientation), and disoriented (i.e., they cannot distinguish left and right). Then, we assume that robots do *not* know n, the number of robots.

A node of the line is *occupied* if it contains at least one robot; otherwise, it is *empty*. If a node contains more than one robot, it is said to have a *multiplicity*. The *distance* between two *nodes* u_i and u_j is the number of edges between them. The *distance* between two *robots* r_1 and r_2 is the distance between two nodes occupied by r_1 and r_2, respectively. Two robots or two nodes are *adjacent* if the distance between them is one. Two robots are *neighboring* if there is no robot between them. A sequence of adjacent occupied nodes is a *block*, and a sequence of adjacent empty nodes is a *hole*. We call a block of size *k* a *k*.block.

Each robot r_i maintains a variable L_i, called *light*, which spans a finite set of states called *colors* of size L. Let $L_i(t)$ be the light color of r_i at time t.

Robots have the ability to sense their environment, including the relative position and color of the other robots. Based on the snapshot resulting from the sensing, they decide whether to move (to an adjacent node) or to stay idle. More precisely, each robot r_i executes the following computational cycle infinitely many times: *(i)* first, r_i takes a snapshot of its environment (LOOK phase), *(ii)* according to the snapshot, r_i decides whether to move or not (COMPUTE phase), and updates (if relevant) its light color, and *(iii)* if r_i decides to move, it moves to one of its adjacent nodes depending on the choice made during the COMPUTE phase (MOVE phase). We call such cycles *LCM cycles*, and assume that light colors persist from one LCM cycle to the next. We consider the *FSYNC* model where at each time instant t, called *round*, each robot executes an LCM cycle synchronously with all other robots, and the *SSYNC* model, where a non-empty subset of robots chosen by an adversarial scheduler executes an LCM cycle synchronously, at each time instant t.

We assume the *full light* model: each robot r_i can see the light color of other robots, and is aware of its own light color. Furthermore, robots do *not* have multiplicity detection capability, even on their current node: if there are multiple robots $r_1, r_2, \ldots r_k$ in a node u, an observing robot r can detect only colors $\{L_i(t) | 1 \leq i \leq k\}$. So, r can detect there are multiple robots at u if and only if at least two robots among $r_1, r_2, \ldots r_k$ have different colors. However, r cannot know how many robots are located at u even if it observes a single color or multiple color at u.

We assume that robots are *myopic*. That is, they have limited visibility: an observing robot r at node u can only sense the robots that occupy nodes within a certain distance, denoted by ϕ, from u. As robots are identical, they share the same ϕ. Let $\mathcal{X}_i(t)$ be the set of colors of robots located at node u_i at time t. If a robot r_j located at u_i takes a snapshot at t, the sensor of r_j outputs a sequence, $\mathcal{V}_j$, of $2\phi + 1$ sets of colors:

$$\mathcal{V}_j \equiv \mathcal{X}_{i-\phi}(t), \ldots, \mathcal{X}_{i-1}(t), [\mathcal{X}_i(t)], \mathcal{X}_{i+1}(t), \ldots, \mathcal{X}_{i+\phi}(t).$$

This sequence $\mathcal{V}_j$ is the *view* of r_j at u_i. To distinguish the sequence center, we use square brackets. If the sequence $\mathcal{X}_{i+1}, \ldots, \mathcal{X}_{i+\phi}$ is equal to the sequence $\mathcal{X}_{i-1}, \ldots, \mathcal{X}_{i-\phi}$, then the view $\mathcal{V}_j$ of r_j is *symmetric*. Otherwise, it is *asymmetric*. In $\mathcal{V}_j$, a node u_k is *occupied* at instant t whenever $|\mathcal{X}_k(t)| > 0$. Conversely, if u_k is *empty* at t, then $\mathcal{X}_k(t) = \emptyset$ holds.

A node u_i is said *singly-colored* at time t, if $|\mathcal{X}_i(t)| = 1$ holds. Note that $|\mathcal{X}_i(t)|$ denotes the number of colors at node u_i, thus even if u_i is singly-colored, it may be occupied by multiple robots (sharing the same color). Now, if u_i is such that $|\mathcal{X}_i(t)| > 1$ holds, u_i is *multiply-colored*. As each robot has a single color, a multiply-colored node always hosts more than one robot.

In the case of a robot r_j located at a singly-colored node u_i, $[\mathcal{X}_i(t)]$ in r_j's view $\mathcal{V}_j$ can be written as $[L_j]$. Then, without loss of generality, if u_i hosts only r_j, the left node of u_i contains one or more robots with color L_k, and the right node of u_i contains one or more robots with color L_l, then $\mathcal{V}_j$ can be written as $L_k[L_j]L_l$.

Our algorithms are driven by observations made on the current view of a robot, so we use *node predicates*: a Boolean function based on the current state of a node. The predicate L_j for a robot r_j matches any set of colors that includes the color of r_j. Some of our algorithm rules expect that a node is singly-colored, *e.g.* with color L_k, in that case, the corresponding predicate is denoted by $L_k!$. To express predicates in a less explicit way, we use character '$*$' to represent any set, including the empty set. The $\neg$ operator is used to negate a particular predicate P (so, $\neg P$ returns *false* whenever P returns *true* and vice versa). Also, the superscript notation P^y represents a sequence of y consecutive sets of colors, each satisfying predicate P. Observe that $y \leq \phi$.

At each time instant t, robots occupy nodes, and their positions and colors form a *configuration* $C(t)$ of the system at time t.

In a given configuration, if the view of a robot r_j at node u_i satisfies predicate $\emptyset^\phi[*]$ or predicate $[*]\emptyset^\phi$, then r_j is a *border robot* and u_i a *border node*.

At instant $t = 0$, let H_{init} be the maximum distance between neighboring occupied nodes, M_{init} be the number of nodes between two borders including border nodes, and $O_{init}(\leq M_{init})$ be the number of occupied nodes. We assume that $\phi \geq H_{init}$, that is, the visibility graph is connected.

As previously stated, no robot is aware of H_{init}, M_{init}, or O_{init}. Let D denote the distance between the two border nodes. Note that, at $t = 0$, $D = M_{init} - 1$ holds. We say $D = 0$ when all robots stay at the same node.

In this paper, each rule in the proposed algorithms is presented using a notation similar to Ooshita and Tixeuil's [17]: $< Label >:< Guard >::< Statement >$. The guard is a predicate on the view $\mathcal{V}_j = \mathcal{X}_{i-\phi}, \ldots, \mathcal{X}_{i-1}, [\mathcal{X}_i], \mathcal{X}_{i+1}, \ldots, \mathcal{X}_{i+\phi}$ obtained by robot r_j at node u_i during the Look phase. If the predicate evaluates to *true*, r_j is *enabled*, otherwise, r_j is *disabled*. In the first case, the corresponding rule $< Label >$ is also said to be *enabled*. If a robot r_j is enabled, r_j may change its color and then move based on the corresponding statement during its subsequent Compute and Move phases. The statement is a pair of (*New color*, *Movement*). *Movement* can be $(i) \rightarrow$, meaning that r_j moves towards node u_{i+1}, $(ii) \leftarrow$, meaning that r_j moves towards node u_{i-1}, and $(iii) \perp$, meaning that r_j does not move. For simplicity, when r_j does not move (resp. r_j does not change its color), we omit *Movement* (resp. *New color*) in the statement.

In the algorithmic descriptions, Rules are presented with decreasing priorities.

To facilitate the understanding of our algorithms, we also present them in a visual manner, where ■ is the observing robot in the view, • is an occupied node, and ∘ is an empty node. The notations are summarized in Table 2.

Table 2. Notations

Formal description	Visual manner	Meaning
$[X]$	■ or [X]	The execution robot with state (color) X
$\emptyset$	○	Empty node
X	●	Occupied node (when $L = 1$)
$X!$	Ⓧ	There are only robots with color X (when $L > 1$)
Z	Ⓩ	A node hosting a multiplicity with color X and Y
$\emptyset^k$	○-○-○- - - -○	Sequence of k empty nodes
X^k	●-●-●- - - -●	Sequence of k occupied nodes
$*$		A node in any state
$\neg(\emptyset^\phi)$		There exists at least one occupied node in the sequence of ϕ empty nodes.

Problem Definition. A robot is said to be *crashed* at time t if it is not activated at any time $t' \geq t$. That is, a crashed robot stops executing the algorithm, and remains with the same color at the same position indefinitely. We assume that robots cannot identify a crashed robot in their snapshots (*i.e.*, they are able to see the crashed robots and their color, but remain unaware of their crashed status). A crash, if any, can occur at any round of the execution.

Furthermore, if more than one crash occurs, all crashes occurs at the same location. In our model, since robots do not have multiplicity detection capability, a location with a single crashed robot and with multiple crashed robots with the same color are indistinguishable. Similarly, multiple non-crashed robots with the same color and the same location have the same behavior (as they share the same view).

We consider the *Stand Up Indulgent Gathering* (SUIG) problem defined by Bramas et al. [7]. An algorithm solves the SUIG problem if, for any initial configuration C_0 (that may contain multiplicities), and for any execution $\mathcal{E} = (C_0, C_1, \ldots)$, there exists a round t such that all robots (including the crashed robots, if any) gather at a single node, not known beforehand, for all $t' \geq t$.

Observe that non-crashed robots located at the crash location may still move, yet a SUIG algorithm guarantees that after finite time, all non-crashed robots are collocated (possibly with crashed robots) and never move thereafter.

3 Impossibility Results

From previous work, some initial configurations where M_{init} and O_{init} are both even prevent the existence of a SUIG solution, even in FSYNC.

Lemma 1 ([5]). *The SUIG problem is unsolvable in FSYNC on line networks starting from an edge-symmetric configuration even with infinite visibility and global strong multiplicity detection.*

Corollary 1 ([15]). *Starting from a configuration where M_{init} is even and O_{init} is even, and all robots have the same color, there exist initial configurations that a deterministic algorithm cannot gather in FSYNC, even for infinite colors luminous robots with infinite visibility and global strong multiplicity detection.*

So, we suppose in the sequel that either M_{init} or O_{init} is odd, thus the initial configuration cannot be edge-symmetric. Previous results show that in this case, SSYNC solutions to SUIG do not exist.

Lemma 2 ([4]). *Even starting from a configuration that is not edge-symmetric, the SUIG problem is unsolvable in SSYNC for infinite colors luminous robots with infinite visibility and global strong multiplicity detection.*

Hence, we assume the FSYNC model in the sequel. We now show a lower bound for the visibility range when robots are oblivious (that is, $L = 1$).

Theorem 1. *The SUIG problem is unsolvable in FSYNC with oblivious robots having visibility range smaller than $H_{init} + 1$, when $H_{init} = 3$.*

Proof. Assume for the sake of contradiction that there exists an algorithm A that solves the SUIG problem in FSYNC with oblivious robots having visibility range $\phi = H_{init} = 3$ (the proof is trivial when ϕ is smaller than H_{init} as the configuration can be initially disconnected).

First we observe that A must contain the following rules:

Rule 1: 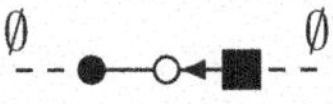

. Indeed, from configuration ●-○-●-○-● where the middle robot is crashed, if the border stays idle, the gathering is never achieved, a contradiction.

Rule 2: 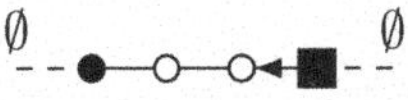

. Same argument as Rule 1.

Rule 3: 

. Indeed, from configuration ●-○-●-○-●-○-●-○-● as the view of the robots is symmetric, if they move, they move to one of their adjacent node, the choice is left to the adversary. The adversary then can choose the destination of the non-border robots to obtain the following configuration: ○-●-○-○-○-●-○-●-○. Then, the first robot being isolated stays idle and the last two robots move toward each other, so the configuration never gather, a contradiction.

Rule 4:  . Indeed, if the rule orders the robot to move left from configuration: ●-○-○-●-○-●-○-○-●-○-●,, the following configuration: ○-●-●-○-○-○-●-●-○-●-○. The first two robots exchange their positions for infinity, and the last three robots eventually gather, so that the whole configuration never gather, a contradiction. The same argument holds when the rule orders the robots to move right.

Rule 5:  . Same argument as Rule 4.

Rule 6: 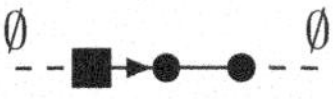 . Indeed, starting from configuration ●-●-● assuming the middle robot is crashed, the border robots must move toward the center. Otherwise, if they stay idle, the configuration remains unchanged for infinity, and if they move away from the center, the configuration remains node-symmetric and they must eventually move towards the center to gather, reaching again the same configuration ●-●-●.

Rule 7:  . Indeed, starting from configuration ●-●-●, since the borders move towards the center, if the center moves, we obtain the edge-symmetric configuration ●-●, a contradiction.

We now consider the three possible rules associated with the following configuration: ●-●-■-○-○-●-○-●

Case left: 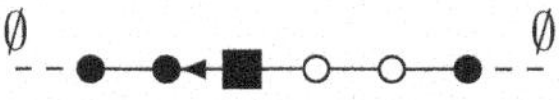 . Then, for the sake of contradiction, consider the following initial configuration: ●-●-●-○-○-●-○-○-● We know that the first robot moves to the right and the second robot stays idle (Rule 6 and 7 described above). The fourth robot stays idle (Rule 5 described above) and the last robot moves to the left (Rule 2 described above), so that after one round we obtain the following configuration ○-●-○-○-○-●-○-●-○. The first robot being isolated stays idle and the last two robots move toward each other, so the configuration never gather, a contradiction.

Case idle:  . From configuration ●-●-●-○-○-●-○-● the first robot moves to the right (Rule 6 described above), fourth robot stays idle (Rule 4 described above) and the last robot moves to the left (Rule 1 described above), so we obtain configuration ●-●-○-○-●-● which is edge-symmetric, a contradiction.

Case right: 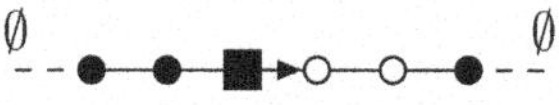 . From configuration: ●-●-●-○-○-●-○-○-●-○-●-○-○-● the first robot moves to the right (Rule 6 described above), the second, fourth, fifth, and sixth robots stay idle (Rule 7, 5, and 4 described above) and the last robot moves to the left (Rule 2 described above), so we obtain configuration ●-○-●-○-●-○-○-●-○-●-○-● which is edge-symmetric, a contradiction.

Thus, we can conclude that the theorem holds. □

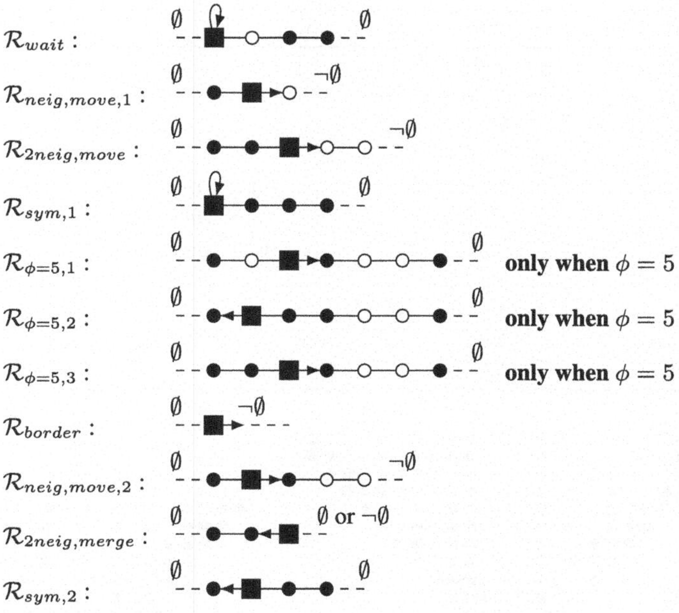

Fig. 1. Visual representation of the rules of Algorithm 1

4 Algorithm with 1 Color

4.1 Algorithm with $\phi \geq H_{init} + 3$

We present in this section Algorithm 1 that solves the SUIG by oblivious robots endowed with visibility range $\phi \geq H_{init} + 3$ from the initial configuration such that either O_{init} or M_{init} is odd.

The overview of the algorithm is as follows. The goal is that the border robots (*i.e.*, robots that do not have neighbors on one side) always move towards the center of the configuration (except in two particular cases). The neighbors at distance 1 and 2 of the border robots also move to avoid creating a multiplicity point until they can together move to the same point so that 3 occupied nodes become 1, hence preserving the parity of the number of occupied nodes. The only two cases a border robot stays idle is when it sees that the configuration is either ●–○–●–● while being on the left side (i.e., its adjacent node is empty, the node two hops away is occupied, and the other border is three hops away) or ●–●–●–● and not sensing any other robots. The first case avoids creating edge symmetric configurations while second case ensures the gathering in a special configuration that can be reached only if there is a crash.

Algorithm 1 presents the formal description of our solution while Fig. 1 presents a visual representation of the rules. In all the rules, the leftmost occupied node is a border and since $\phi \geq H_{init} + 3$, the current robot can see that it is a border. We can see that rule $\mathcal{R}_{border}$ orders the border to move but the rules $\mathcal{R}_{wait}$ and $\mathcal{R}_{sym,1}$ prevent the border from moving in particular configurations. Due to rules $\mathcal{R}_{neig,move,1}$ and $\mathcal{R}_{neig,move,2}$, a robot adjacent to a border moves in the same direction as the border,

Algorithm 1: Algorithm with $\phi \geq H_{init} + 3$

$\mathcal{R}_{wait}$:	$\emptyset^{\phi}\ [X]\emptyset X^2\emptyset^{\phi-3}$	$:: \bot$
$\mathcal{R}_{neig,move,1}$:	$\emptyset^{\phi-1}X\ [X]\emptyset(\neg(\emptyset^{\phi-1}))$	$:: \rightarrow$
$\mathcal{R}_{2neig,move}$:	$\emptyset^{\phi-2}X^2\ [X]\emptyset^2(\neg(\emptyset^{\phi-2}))$	$:: \rightarrow$
$\mathcal{R}_{sym,1}$:	$\emptyset^{\phi}\ [X]X^3\emptyset^{\phi-3}$	$:: \bot$
$\mathcal{R}_{\phi=5,1}$:	$\emptyset^{\phi-2}X\emptyset\ [X]X\emptyset^2X\emptyset$	$:: \rightarrow$ **(only when** $\phi = 5$**)**
$\mathcal{R}_{\phi=5,2}$:	$\emptyset^{\phi-1}X\ [X]X^2\emptyset^2X$	$:: \leftarrow$ **(only when** $\phi = 5$**)**
$\mathcal{R}_{\phi=5,3}$:	$\emptyset^{\phi-2}X^2\ [X]X\emptyset^2X\emptyset$	$:: \rightarrow$ **(only when** $\phi = 5$**)**
$\mathcal{R}_{border}$:	$\emptyset^{\phi}\ [X](\neg(\emptyset^{\phi}))$	$:: \rightarrow$
$\mathcal{R}_{neig,move,2}$:	$\emptyset^{\phi-1}X\ [X]X\emptyset^2(\neg(\emptyset^{\phi-3}))$	$:: \rightarrow$
$\mathcal{R}_{2neig,merge}$:	$\emptyset^{\phi-2}X^2\ [X]*^{\phi}$	$:: \leftarrow$
$\mathcal{R}_{sym,2}$:	$\emptyset^{\phi-1}X\ [X]X^2\emptyset^{\phi-2}$	$:: \leftarrow$

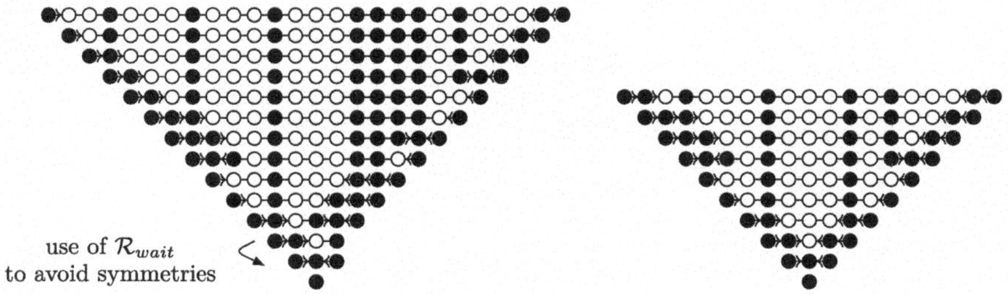

Fig. 2. Examples of executions of Algorithm 1 without crashes ($H_{init} = 4$). On the left, the initial configuration contains an odd number of occupied nodes ($O_{init} = 11$), and on the right, the number of nodes between the two borders is odd ($M_{init} = 19$, and $O_{init} = 8$).

except if the prefix of the configuration is as follows ●-●-●-○-●- - -. In this case, no rule applies, and it stays idle. In this case, the robot at distance 2 from the border moves towards the border to create a multiplicity point by $\mathcal{R}_{2neig,merge}$, while preserving the parity of the number of occupied nodes. Rule $\mathcal{R}_{sym,1}$ and $\mathcal{R}_{sym,2}$ are used only when the configuration is ●-●-●-●, which is an edge-symmetric configuration that occurs only when there is a crash, and is used to avoid reaching configuration ●-● with one of the occupied nodes containing both correct and crashed robots (from which the gathering becomes impossible).

Figure 2 (resp. Figure 3) represents two execution examples of Algorithm 1 without (resp. with) crashes. For each figure, the top is the initial configuration, and just below is the configuration after executing exactly one round.

In the remaining, we consider that $\phi \geq 5$, because if $\phi = 4$, then $H_{init} = 1$ and all robots know that the initial configuration consists of a single block, and in this case a trivial algorithm exists: only the border moves towards their occupied neighbors.

Theorem 2. *Algorithm 1 solves the SUIG problem with oblivious robots assuming* $\phi \geq H_{init} + 3$ *and* O_{init} *or* M_{init} *is odd.*

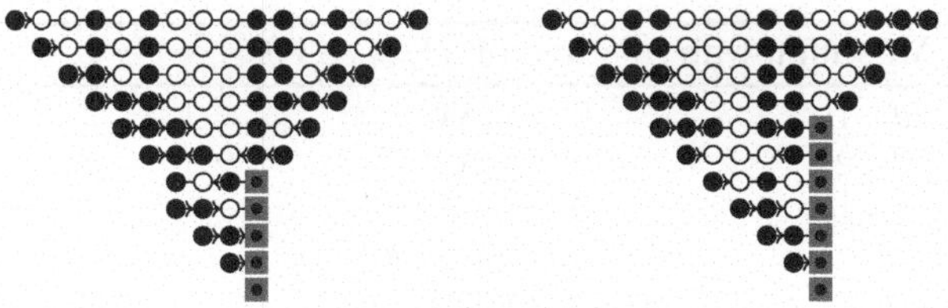

Fig. 3. Examples of executions of Algorithm 1 with a crash occurring at the red node ($H_{init} = 4$). (Color figure online)

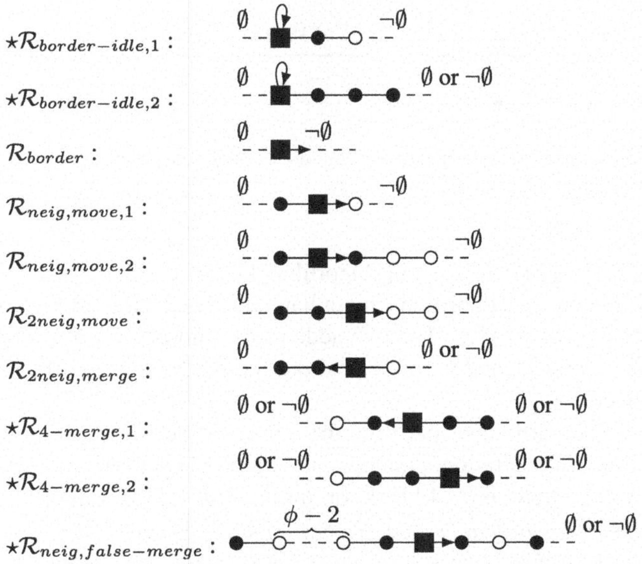

Fig. 4. Visual representation of the rules of Algorithm 2

4.2 Algorithm with $\phi \geq \max(4, H_{init} + 1)$

We present in this section Algorithm 2 that solves the SUIG on line shaped networks by oblivious robots endowed with visibility range $\phi \geq \max(4, H_{init} + 1)$. The number of occupied nodes O_{init} in the initial configuration is assumed to be odd. Algorithm 2 presents the formal description of our solution while Fig. 4 presents a visual representation of the rules.

As robots have limited visibility range, they do not have much information about the overall configuration except that whether they are at the border of the visibility graph. A naive strategy that consists of making only the border robots move within the visibility range does not work since if no precautions are taken, an edge symmetric configuration could be reached, making the gathering unsolvable as shown in Fig. 5.

The idea of Algorithm 2 is to preserve the parity of the number of occupied nodes, as in Algorithm 1. However, we cannot always make the border move as previously, otherwise we are not able to break some unrecoverable symmetries in the end (han-

Algorithm 2: Algorithm with $\phi \geq \max(4, H_{init} + 1)$

$\star\mathcal{R}_border - idle, 1$:	$\emptyset^{\phi}\ [X]X\emptyset(\neg(\emptyset^{\phi-2}))$	$:: \perp$
$\star\mathcal{R}_border - idle, 2$:	$\emptyset^{\phi}\ [X]X^{3}(*^{\phi-3})$	$:: \perp$
$\mathcal{R}_border$:	$\emptyset^{\phi}\ [X](\neg(\emptyset^{\phi}))$	$:: \rightarrow$
$\mathcal{R}_neig, move, 1$:	$\emptyset^{\phi-1}X\ [X]\emptyset(\neg(\emptyset^{\phi-1}))$	$:: \rightarrow$
$\mathcal{R}_neig, move, 2$:	$\emptyset^{\phi-1}X\ [X]X\emptyset^{2}(\neg(\emptyset^{\phi-3}))$	$:: \rightarrow$
$\mathcal{R}_2neig, move$:	$\emptyset^{\phi-2}X^{2}\ [X]\emptyset^{2}(\neg(\emptyset^{\phi-2}))$	$:: \rightarrow$
$\mathcal{R}_2neig, merge$:	$\emptyset^{\phi-2}X^{2}\ [X]\emptyset(*^{\phi-1})$	$:: \leftarrow$
$\star\mathcal{R}_4 - merge, 1$:	$(*^{\phi-2})\emptyset X\ [X]X^{2}(*^{\phi-2})$	$:: \leftarrow$
$\star\mathcal{R}_4 - merge, 2$:	$(*^{\phi-3})\emptyset X^{2}\ [X]X(*^{\phi-1})$	$:: \rightarrow$
$\star\mathcal{R}_neig, false - merge$:	$X\emptyset^{\phi-2}X\ [X]X\emptyset X(*^{\phi-3})$	$:: \rightarrow$

Fig. 5. An example of an edge-symmetric configuration reached by making only the border robots move towards the center.

dled by Rules $\mathcal{R}_{\phi=5,2}$ and $\mathcal{R}_{\phi=5,3}$ in Algorithm 1). Hence the border sometimes has to wait and this is why our algorithm cannot handle initial configuration where O_{init} is even (even if $H_{init}$is also even). The two added rules (noted with a star in the figures) $\mathcal{R}_border - idle, 1$ and $\mathcal{R}_border - idle, 2$ order the border to stay idle, waiting for its neighbor to move.

To preserve the parity, we perform the same merge as in Algorithm 1, when the prefix of the configuration contains a 3.block. However, we perform a different merge when there is a 4.block or more. The added rules to perform this merge are Rules $\mathcal{R}_4 - merge, 1$ and $\mathcal{R}_4 - merge, 2$. When a block of robots contains four or more robots, two robots move away from each other. For example a 4.block - - ○-●-●-●-●-○- - - becomes - - ○-●-○-○-●-○- - -, preserving the parity. Similarly a 5.block - - ○-●-●-●-●-●-○- - - becomes - - ○-●-○-○-●-●-○- - - (the middle robots moves towards an arbitrary direction); a 6.block becomes - - ○-●-○-●-●-○-●-○- - - (the two central robots have exchanged their position); a 7.block becomes - - ○-●-○-○-●-○-○-●-○- - -; and so on. This can occur anywhere in the configuration, while 3.blocks containing a border are merge as in the previous algorithm, but the rule $\mathcal{R}_2neig, merge$ could be executed even if the 3.block does not contain a border, due to the reduced visibility range. This is why rule $\mathcal{R}_neig, false - merge$ is added because the robot in the middle of the 3.block can detect that its neighbor will falsely perform a merge. By making both robots move, no merge is performed and the number of occupied locations remains odd (refer to Fig. 6 for an example).

When a crash occurs, we show that the distance between the border continues to decrease until reaching 3 or 4, and then, regardless of the configuration obtained, the gathering is achieved eventually.

Theorem 3. *Algorithm 2 solves the SUIG problem with oblivious robots assuming $\phi \geq \max(4, H_{init} + 1)$ and O_{init} is odd.*

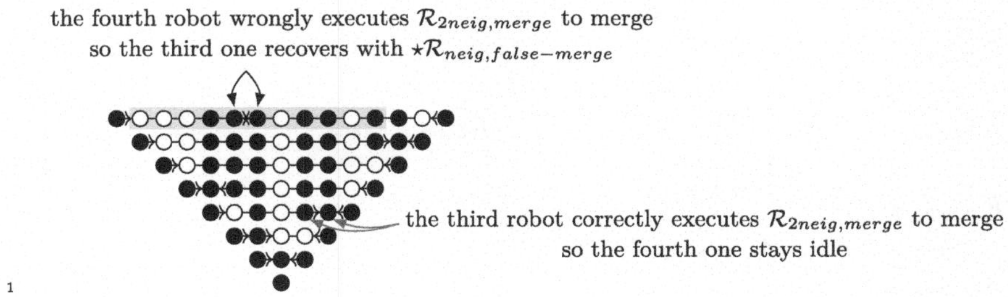

Fig. 6. Example of an execution of Algorithm 2, with $H_{init} = 4$ (hence $\phi \geq 5$). The gray rectangle shows the view of the fourth robot (starting from the left) in the initial configuration. Since it does not see the first robot, it applies rule $\mathcal{R}_2neig, merge$ as it believes that the second robot is a border robot.

5 Algorithm with 2 Colors and $\phi \geq H_{init}$

We present in this section an algorithm that solves the SUIG assuming a visibility range $\phi \geq H_{init}$ and luminous robots endowed with two colors X and Y. The algorithm works assuming that either O_{init} or $M_{init}$is odd. All robots are assumed to have the same color X initially. Let Z denote a multiplicity that consists of robots with both colors X and Y. The formal description of our proposed solution is given in Algorithm 3. To ease the reading of the rules, their visual representations are given in Fig. 7. Figure 8 shows two execution examples of Algorithm 3 without crashes.

To start, let us focus on the case in which there are no crashes. To make the robots gather, the border robots move towards the center of the configuration (refer to Rules $\mathcal{R}_{X-in}$ and $\mathcal{R}_{Z-in}$) and change their respective color whenever they move to an occupied node (refer to Rules $\mathcal{R}_{X-merge}$ and $\mathcal{R}_{Z-merge}$). By alternating between color X and state Z, the borders somehow store the parity of the number of occupied nodes. If O_{init} is odd, a configuration with only two occupied nodes is eventually reached. One node hosts only robots with color X and the other one hosts robots with both colors (state Z). If the occupied nodes are adjacent, the gathering is then achieved by rule $\mathcal{R}_{final}$ and $\mathcal{R}_{Z-merge}$, otherwise it is achieved by rule $\mathcal{R}_{X-in}$ and $\mathcal{R}_{Z-in}$. Next, when O_{init} is even, $M_{init}$is odd. The idea of the algorithm is to make the robots gather on the middle node between the two border robots in the initial configuration. Note that the rules already defined for the case in which O_{init} is odd are sufficient to guarantee the gathering in this case. Indeed, as O_{init} is even, when a configuration with only two occupied nodes is reached, the border robots have not only the same color but they are also at an even distance from each other. Hence, by only moving towards each other (by executing Rules $\mathcal{R}_{X-in}$ or $\mathcal{R}_{Z-in}$), the gathering is eventually achieved.

Let us now take into accounts crashes that might occur during the execution of our algorithm. In order to ensure the gathering, one additional rule is required ($\star\mathcal{R}_{border-move}$) and used when the number of occupied nodes is either two or three. This ensure that a configuration with only two adjacent occupied nodes is reached. When this occurs, one can see that the non-crashed robot are all located on the same node, so that after one or two rounds, they all move towards the crashed location.

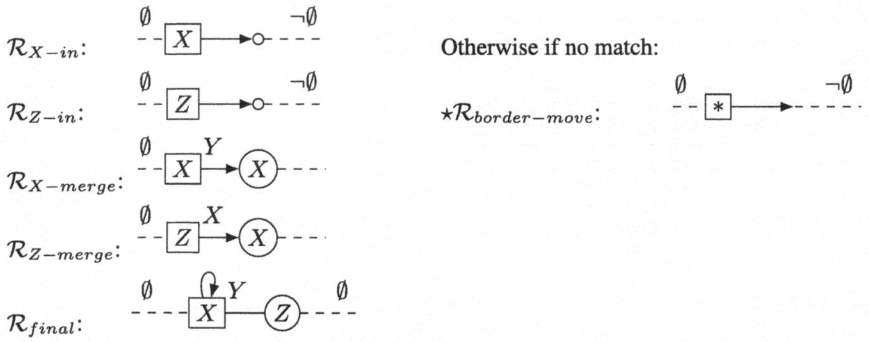

Fig. 7. Visual representation of the rules of Algorithm 3

Algorithm 3: Algorithm with 2 colors and $\phi \geq H_{init}$

$\mathcal{R}_{X-in}$:	$\emptyset^{\phi}\ [X!]\emptyset(\neg(\emptyset^{\phi-1}))$	$::\ \rightarrow$
$\mathcal{R}_{Z-in}$:	$\emptyset^{\phi}\ [Z]\emptyset(\neg(\emptyset^{\phi-1}))$	$::\ \rightarrow$
$\mathcal{R}_{X-merge}$:	$\emptyset^{\phi}\ [X!](X!)(*^{\phi-1})$	$::\ \rightarrow, Y$
$\mathcal{R}_{Z-merge}$:	$\emptyset^{\phi}\ [Z](X!)(*^{\phi-1})$	$::\ \rightarrow, X$
$\mathcal{R}_{final}$:	$\emptyset^{\phi}\ [X!]Z\emptyset^{\phi-1}$	$::\ \bot, Y$
$\star\mathcal{R}_{border-move}$:	$\emptyset^{\phi}\ [*](\neg\emptyset^{\phi})$	$::\ \rightarrow$

To prove the correctness of our solution, we first address the case in which no crashes occur and then discuss the case with crashes. Due to the lack of spaces, some formal proofs have been omitted and moved to the appendix.

We show that if O_{init} is odd in the initial configuration then, in any configuration reached during the execution, if the number of occupied nodes is odd then either both borders have color X, or both color Z. By contrast, if the number of occupied nodes become even (only one border reaches an occupied node) then one border has color X and the other one color Z. This property ensures the gathering as the borders keeps getting closer to either join the same occupied node or two adjacent nodes but not with the same color. By contrast, if O_{init} is even then M_{init} is odd initially. In this case, a border robot always move towards the center of the configuration except when Rule $\mathcal{R}_{final}$ is enabled. However, note that a border robot can only see another robot with state Z only if the robot with state Z is also located at a border node. Since M_{init} is odd and there are no crashes, Rule $\mathcal{R}_{final}$ is never enabled and all robots meet in the node at the center of the configuration. We can hence deduce:

Lemma 3. *If either O_{init} or M_{init} is odd and no crash occurs then, Algorithm 3 ensures the gathering within a finite number of rounds.*

Since in Algorithm 3, only border robots are ordered to move, we can assume in the following that a crash occurs at a border node. We first prove that the algorithm can solve any configuration that consists only in a 2.block or a 3.block:

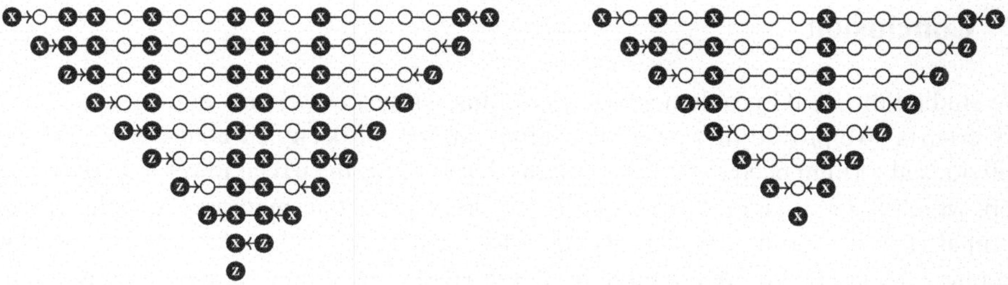

Fig. 8. Examples of executions of Algorithm 3. On the left with $M_{init} = 16$ and $O_{init} = 9$. On the right with $M_{init} = 13$ and $O_{init} = 6$.

Lemma 4. *Algorithm 3 ensures gathering is achieved when starting from any configuration that consists only in a 2.block or a 3.block (regardless of their states), where all the robot at a border are crashed.*

Proof. If the configuration is a 2.block, all the non-crashed robots move towards the crashed location by executing rule $\mathcal{R}_{border-move}$, except if the crashed location has state Z and the correct border has state X. In this case, the correct robots change their color to Y according to rule $\mathcal{R}_{final}$, and then move towards the crashed location by $\mathcal{R}_{border-move}$. So in at most two rounds, the gathering is achieved.

If the configuration is a 3.block, the robots at the correct border move towards the middle, and we reach a 2.block configuration. □

Finally, to address the general case, we first show that the gathering is achieved when the crash occurs at a border with state X and then at the border with state Z. We obtain the following two lemmas:

Lemma 5. *If either O_{init} or M_{init} is odd and a crash occurs at a border node having color X then, Algorithm 3 ensures the gathering within a finite number of rounds.*

Lemma 6. *If either O_{init} or M_{init} is odd and a crash occurs at a border node with state Z then, Algorithm 3 ensures the gathering within a finite number of rounds.*

From Lemmas 5 and 6, we can conclude the following corollary:

Corollary 2. *If either O_{init} or M_{init} is odd and a crash occurs then, Algorithm 3 ensures the gathering within a finite number of rounds.*

Finally, from Lemma 3 and Corollary 2, we can deduce:

Theorem 4. *If either O_{init} or M_{init} is odd, Algorithm 3 solves the stand-up indulgent gathering using two colors and assuming a visibility range $\phi \geq H_{init}$.*

6 Conclusion

We studied the SUIG problem solvability in line graphs based on visibility and memory criteria. We proved that, when the initial number of occupied nodes is odd, *(i)* an oblivious algorithm needs a visibility radius that exceeds the trivial lower bound by one hop, which is tight, and *(ii)* robots with one bit of persistent memory can achieve the optimal visibility radius, which is also tight for the memory. When the initial number of occupied nodes can be even (but the border robots are at an even distance in that case), our one-bit memory algorithm is still optimal, and we present an oblivious algorithm that requires two extra hops than the lower bound.

Some interesting questions remain open:

- In the case of an initial situation where O_{init} is odd or M_{init} is odd, is it possible to restrict the visibility range of an oblivious solution below $H_{init}+3$, the lower bound being $H_{init}+1$? We conjecture that Algorithm 2 (that uses $H_{init}+3$ visibility range) is optimal.
- Are there any algorithms for the case where crashed robots are located at different nodes initially? At least two variants are worth considering: *(i)* requiring all correct robot to gather at the same crashed location, and *(ii)* requiring all correct robots to gather at *any* crashed location.

References

1. Agmon, N., Peleg, D.: Fault-tolerant gathering algorithms for autonomous mobile robots. SIAM J. Comput. **36**(1), 56–82 (2006)
2. Balabonski, T., Courtieu, P., Pelle, R., Rieg, L., Tixeuil, S., Urbain, X.: Continuous *vs.* discrete asynchronous moves: a certified approach for mobile robots. In: Atig, M.F., Schwarzmann, A.A. (eds.) NETYS 2019. LNCS, vol. 11704, pp. 93–109. Springer, Cham (2019). https://doi.org/10.1007/978-3-030-31277-0_7
3. Bouzid, Z., Das, S., Tixeuil, S.: Gathering of mobile robots tolerating multiple crash faults. In: IEEE 33rd International Conference on Distributed Computing Systems, ICDCS 2013, pp. 337–346 (2013)
4. Bramas, Q., et al.: Stand-up indulgent gathering on lines for myopic luminous robots. In: Proceedings of 38th International Conference on Advanced Information Networking and Applications (AINA-2024), pp. 110–121 (2024)
5. Bramas, Q., Kamei, S., Lamani, A., Tixeuil, S.: Stand-up indulgent gathering on lines. In: Dolev, S., Schieber, B. (eds.) Stabilization, Safety, and Security of Distributed Systems - 25th International Symposium, SSS 2023, pp. 451–465 (2023)
6. Bramas, Q., Kamei, S., Lamani, A., Tixeuil, S.: Stand-up indulgent gathering on rings. In: Structural Information and Communication Complexity - 31st International Colloquium, SIROCCO 2024, pp. 119–137 (2024)
7. Bramas, Q., Lamani, A., Tixeuil, S.: Stand up indulgent gathering. Theor. Comput. Sci. **939**, 63–77 (2023)
8. Bramas, Q., Tixeuil, S.: Wait-free gathering without chirality. In: Scheideler, C. (ed.) SIROCCO 2014. LNCS, vol. 9439, pp. 313–327. Springer, Cham (2015). https://doi.org/10.1007/978-3-319-25258-2_22

9. Cieliebak, M., Flocchini, P., Prencipe, G., Santoro, N.: Distributed computing by mobile robots: gathering. SIAM J. Comput. **41**(4), 829–879 (2012)
10. Défago, X., Potop-Butucaru, M., Parvédy, P.R.: Self-stabilizing gathering of mobile robots under crash or byzantine faults. Distrib. Comput. **33**(5), 393–421 (2020)
11. Flocchini, P., Prencipe, G., Santoro, N. (eds.): Distributed Computing by Mobile Entities, Current Researchin Moving and Computing. Springer, Heidelberg (2019)
12. Guilbault, S., Pelc, A.: Gathering asynchronous oblivious agents with local vision in regular bipartite graphs. Theor. Comput. Sci. **509**, 86–96 (2013)
13. Guilbault, S., Pelc, A.: Gathering asynchronous oblivious agents with restricted vision in an infinite line. In: SSS, pp. 296–310 (2013)
14. Kamei, S., Lamani, A., Ooshita, F.: Asynchronous ring gathering by oblivious robots with limited vision. In: 2014 IEEE 33rd International Symposium on Reliable Distributed Systems Workshops, pp. 46–49 (2014)
15. Kamei, S., Lamani, A., Ooshita, F., Tixeuil, S., Wada, K.: Gathering on rings for myopic asynchronous robots with lights. In: Proceedings of 23rd International Conference on Principles of Distributed Systems (OPODIS 2019), pp. 27:1–27:17 (2019)
16. Klasing, R., Markou, E., Pelc, A.: Gathering asynchronous oblivious mobile robots in a ring. Theor. Comput. Sci. **390**(1), 27–39 (2008)
17. Ooshita, F., Tixeuil, S.: Ring exploration with myopic luminous robots. In: Stabilization, Safety, and Security of Distributed Systems, pp. 301–316 (2018)
18. Suzuki, I., Yamashita, M.: Distributed anonymous mobile robots: formation of geometric patterns. SIAM J. Comput. **28**(4), 1347–1363 (1999)

Simplicial Belief

Christian Cachin, David Lehnherr(✉), and Thomas Studer

University of Bern, Bern, Switzerland
{christan.cachin,david.lehnherr,thomas.studer}@unibe.ch

Abstract. Recently, much work has been carried out to study simplicial interpretations of modal logic. While notions of (distributed) knowledge have been well investigated in this context, it has been open how to model belief in simplicial models. We introduce polychromatic simplicial complexes, which naturally impose a plausibility relation on states. From this, we can define various notions of belief.

Keywords: Simplicial complex · Epistemic logic · Plausibility model · Belief modality

1 Introduction

Due to the success of applying methods from combinatorial topology to distributed computing [19], the study of simplicial interpretations for modal logic is currently thriving [11,16,17,21]. At its core lies the epistemic interpretation of simplicial complexes of various kinds. Let $\mathcal{V}$ be a set of vertices. Each vertex corresponds to a local state of an agent, and we say that this vertex is of that agent's color. A simplicial complex $(S, \mathcal{V})$ is a pair where S is a set of non-empty subsets of $\mathcal{V}$ that is closed under set inclusion. Vertices that belong to the same set must be of different colors, and, in the simplest case, maximal elements of S represent global states. An agent a cannot distinguish two global states if its local state is included in both. Hence, simplicial complexes offer sufficient structure for an epistemic interpretation.

The simplicial complex $\mathcal{C}_1$ depicted in Fig. 1 consists of two worlds X and Y (represented by the two solid triangles), each containing the local states of all three agents. Agents a and b have the same local state in X and Y, whereas c's local state differs in the two global states. Therefore, a and b cannot distinguish between X and Y, but c can. That is, a and b always deem both states as possible, while c only considers the actual state. This concept can be extended to groups of agents. In $\mathcal{C}_1$, a and b together cannot distinguish between X and Y because they cannot do so on their own, but a and c can due to c being able to tell X and Y apart. Lastly, we say that the agents a, b, and c are *alive* in X and Y because they both contain vertices with their colors.

An agent a knows a formula ϕ in state X if and only if ϕ is true in all states that a considers as possible when in X. Consequently, if an agent is alive and

U. Schmid and R. Kuznets (Eds.): SIROCCO 2025, LNCS 15671, pp. 176–193, 2025.
https://doi.org/10.1007/978-3-031-91736-3_11

knows ϕ in state X, then ϕ must be true in X. Classical distributed knowledge [18] can be interpreted similarly. A group G knows ϕ in a state X if and only if ϕ is true in all states that G cannot distinguish from X.

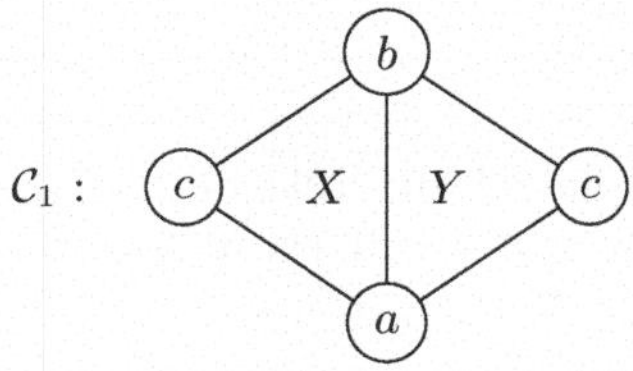

Fig. 1. A simplicial complex C_1 in which the agents a and b cannot distinguish between the worlds X and Y.

While (distributed) knowledge has been studied extensively in this context, it has been open, see [10, Section 4.3], how to model belief on simplicial structures. Unlike knowledge, an agent believing ϕ does not imply that ϕ is true in the actual global state. At first glance, as done in other models, this behavior can be achieved by dropping the requirement that agents must consider the actual state as possible. However, in the simplicial case, an alive agent will always consider the actual global state as possible because it contains its local state. Hence, this property cannot be dropped.

One way of overcoming this issue is to introduce *belief functions* [11]. An agent's belief function f_a maps a state X, in which a is alive, to another state Y, in which a is also present. An agent a believes a formula ϕ in X if and only if it knows ϕ in $f_a(X)$. Hence, belief becomes knowledge if $f_a(X) = X$. In the model C_2 shown in Fig. 2, X is mapped to Y and a believes that c is not alive in the current global state, i.e., it falsely believes that c no longer exists. Disadvantages of this approach are that *i*) belief does not take the topology of the model into account and *ii*) the principle of knowledge-yields-belief, i.e., knowing ϕ implies believing ϕ, does not have a meaningful interpretation.

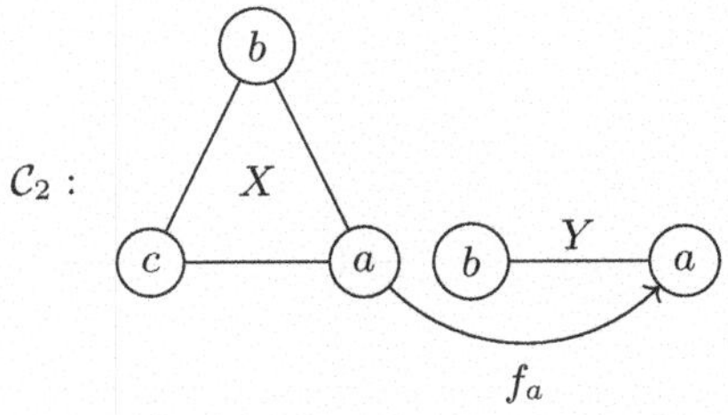

Fig. 2. Despite being in X, agent a thinks that the actual world is Y. Thus, a wrongly believes that c died.

In this paper, we present and interpret polychromatic simplicial complexes, i.e., complexes in which adjacent vertices may be of the same color, as simplicial models. Such models admit a notion of belief that satisfies the two conditions mentioned above. The belief studied in this work is based on the plausibility of states rather than on their possibility alone. That is, an agent believes a formula ϕ if and only if it is true in all states that it deems plausible enough. Since the actual global state need not be among them, the agent's beliefs might be wrong.

We start by defining a plausibility relation between the states based on the multiplicity of color within a state. If the color of an agent a has a lower or equal multiplicity in a state X than in a state Y, then a considers X to be at least as plausible as Y. If vertices are interpreted as belief states, then a possible reading of our relation is that an agent a considers worlds with fewer doxastic alternatives as more plausible. Since this relation is a wellfounded preorder, we can use the machinery of plausibility models [3,4] to define various notions of belief such as *safe belief* and *most plausible belief*.

Most plausible belief is a useful notion of belief. An agent a most plausibly believes ϕ if and only if ϕ is true in the worlds that it considers to be the most plausible ones. This kind of belief is appropriate for reasoning about distributed computing [7], where agents act based on guarantees that hold true with overwhelming probability, i.e., the states in which they hold are the most plausible ones. For example, when communicating over authenticated links, it is the most plausible case that if Alice receives a message m from Bob over such a link, then m was actually sent by Bob and not by an imposter. Our simplicial models can model this kind of belief while taking the topology of the model into account. The complex $\mathcal{C}_3$ in Fig. 3 depicts the same situation as $\mathcal{C}_2$ but without a belief function. When in X, a considers Y more plausible than X because its multiplicity in X is 2 and only 1 in Y. In this case, a considers Y the most plausible world. Further, since c is not present in Y, a believes that c died. Thus, an agent's most plausible belief can be false. It is also straightforward to verify that this notion of belief satisfies the principle of knowledge-yields-belief.

$\mathcal{C}_3$:

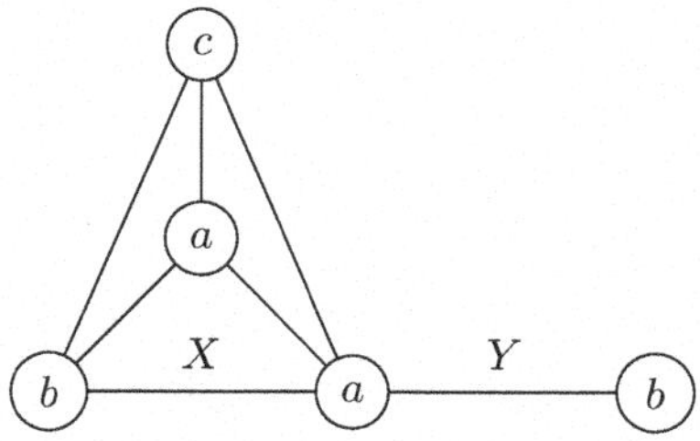

Fig. 3. A polychromatic simplicial complex that models the same situation as $\mathcal{C}_2$ but without a belief function.

We close this introduction by briefly comparing our method to another, yet unexplored, approach for modeling belief on simplicial structures mentioned in [10, Section 4.3.3]. The idea is to impose a distance measure by allowing edges between vertices with the same color. It is important, however, that these edges are not epistemically interpreted. The a-distance between two global states X and Y is the length of the shortest a-path between them. The shorter the path to a global state, the more plausible it becomes. For example, consider $\mathcal{C}_4$ in Fig. 4. When in X, a considers Y more plausible than Z. While this interpretation takes topological information into account, it does not model false beliefs because agents always consider the actual world as the most plausible one.

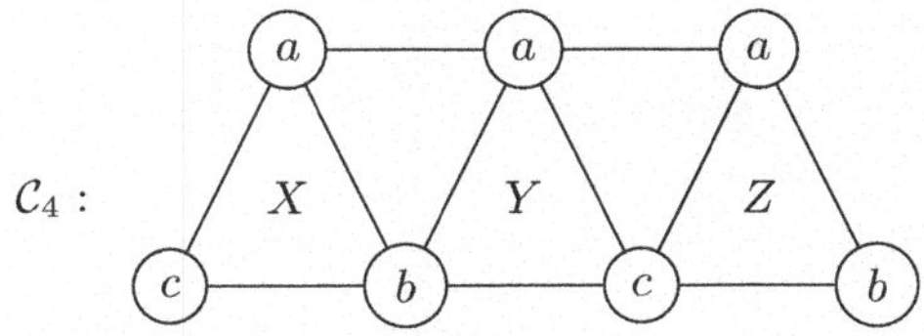

Fig. 4. Unicolored edges are used to represent a *distance* between worlds. For example, when in X, a considers Y to be more plausible than Z because the a-path from X to Y is shorter than the a-path from X to Z.

We give an overview of the theory of simplicial models in Sect. 2. In Sect. 3, we introduce polychromatic models and present different notions of belief on them. Next, we analyze in Sect. 4 knowledge and belief gain in our models. In Sect. 5, we discuss other possible semantics for our models, and, lastly, we conclude our work in Sect. 6.

2 Simplicial Knowledge

We quickly recall the standard definitions for distributed knowledge on simplicial complexes [11,16,17]. In the subsequent section, we will extend them to incorporate notions of belief.

Let Ag be the set of finitely many agents, and let Prop be a countable set of atomic propositions. We define the language of knowledge $\mathcal{L}_\mathcal{K}$ for $G \subseteq \mathsf{Ag}$ and $p \in \mathsf{Prop}$ inductively by the following grammar:

$$\phi ::= p \mid \neg\phi \mid \phi \wedge \phi \mid [\sim]_G \phi$$

The remaining Boolean connectives are defined as usual. In particular, we set $\phi \vee \psi := \neg(\neg\phi \wedge \neg\psi)$ and $\bot := p \wedge \neg p$ for some fixed $p \in \mathsf{Prop}$. We write $\mathsf{alive}(G)$ for $\neg[\sim]_G\bot$ and $\mathsf{dead}(G)$ for $[\sim]_G\bot$.

Definition 1. *Let $\mathcal{V}$ be a set of vertices. $C = (S, \mathcal{V})$ with $S \subseteq \mathsf{Pow}(\mathcal{V}) \setminus \{\emptyset\}$ is called a* simplicial complex *if*

$$\textit{for each } X \in S \textit{ and each } \emptyset \neq Y \subseteq X, \textit{ we have } Y \in S.$$

Remark 1. The standard definition of a simplicial complex (cf. [19]) requires that the singleton of each vertex is an element of S. Not doing so is purely cosmetic.

We call the elements of S *faces*. A face that is maximal under inclusion is called a *facet*. We denote the set of facets of C by $\mathcal{F}(C)$. A *coloring* is a mapping $\chi : \mathcal{V} \to \mathsf{Ag}$. A coloring is *proper* if it assigns a different value to each vertex within a face. We use $\chi(U)$ for the set $\{\chi(u) \mid u \in U\}$.

Definition 2. *Let* $C = (S, \mathcal{V})$ *be a simplicial complex. A* simplicial model *is a quadruple* $\mathcal{C} = (C, \chi, W, \ell)$ *where*

1. C *is a simplicial complex;*
2. $\chi : \mathcal{V} \to \mathsf{Ag}$ *is a proper coloring;*
3. $\mathcal{F}(C) \subseteq W \subseteq S$ *is a set of worlds;*
4. $\ell : W \to \mathsf{Pow}(\mathsf{Prop})$ *is a valuation.*

Remark 2. Definition 2 is not completely standard and is studied by Goubault et al. [15]. Moreover, it is worth mentioning that the valuation ℓ in Definition 2 assigns propositions to arbitrary faces instead of vertices only.

Given a simplicial model, a group of agents $G \subseteq \mathsf{Ag}$ cannot distinguish two worlds $X, Y \in W$, denoted by $X \sim_G Y$, if and only if $G \subseteq \chi(X \cap Y)$. We call $\sim_G$ the *epistemic indistinguishability relation.* If we need to indicate the simplicial complex $\mathcal{C}$ of which we consider the relation $X \sim_G Y$, we use the notation $X \sim_G^{\mathcal{C}} Y$. If G contains only a single agent a, we write $X \sim_a Y$ and $[\sim]_a$ instead of $X \sim_{\{a\}} Y$ and $[\sim]_{\{a\}}$, respectively.

It is worth noting that $[\sim]_G$ is the usual notion of distributed knowledge of a group G of agents, which is semantically given by the intersection of the individual indistinguishability relations of the group members (cf. [18]).

Definition 3. *For a simplicial model* $\mathcal{C} = (C, W, \chi, \ell)$*, a world* $X \in W$*, and a formula* $\phi \in \mathcal{L}_K$*, we define the relation* $\mathcal{C}, X \Vdash \phi$ *inductively by*

$$
\begin{array}{lll}
\mathcal{C}, X \Vdash p & \textit{iff} & p \in \ell(X) \\
\mathcal{C}, X \Vdash \neg\phi & \textit{iff} & \mathcal{C}, X \nVdash \phi \\
\mathcal{C}, X \Vdash \phi \wedge \psi & \textit{iff} & \mathcal{C}, X \Vdash \phi \textit{ and } \mathcal{C}, X \Vdash \psi \\
\mathcal{C}, X \Vdash [\sim]_G\phi & \textit{iff} & X \sim_G Y \textit{ implies } \mathcal{C}, Y \Vdash \phi \textit{ for all } Y \in W.
\end{array}
$$

We say that *agent a is alive in a world X* if $a \in \chi(X)$. The set of worlds in which a group $G \subseteq \mathsf{Ag}$ is alive is defined as

$$\mathsf{Alive}_{\mathcal{C}}(G) = \{X \in W \mid G \subseteq \chi(X)\}.$$

It is standard to show that $\sim_G$ is a partial equivalence relation. We have the following lemma.

Lemma 1. *Let* $\mathcal{C} = (C, \chi, W, \ell)$ *be a simplicial model. For each* $G \subseteq \mathsf{Ag}$*, the relation* $\sim_G$ *is an equivalence relation on* $\mathsf{Alive}_{\mathcal{C}}(G)$ *and empty otherwise.*

3 Simplicial Belief

We now drop the requirement that the coloring of a simplicial model must be proper. The resulting models are called polychromatic. We will define a well-founded preorder on the states of a polychromatic model, which will serve as a plausibility relation [3,4]. This makes it possible to interpret various notions of belief on simplicial models.

It is straightforward to verify that Lemma 1 does not hold for polychromatic models because $\sim_G$ need not be transitive. Indeed, consider the set of vertices $\{0, 1, 2, 3\}$ and a model that consists of the worlds

$$X := \{0, 1\}, \quad Y := \{1, 2\}, \quad \text{and} \quad Z = \{2, 3\}$$

with a coloring χ that assigns the same agent a to all vertices. We find that $X \sim_a Y$ and $Y \sim_a Z$, but not $X \sim_a Z$. In order to re-establish transitivity of $\sim_G$, we must require that for any three worlds $X, Y, Z \in W$ and any group of agents $G \subseteq \mathsf{Ag}$:

$$G \subseteq \chi(X \cap Y) \text{ and } G \subseteq \chi(Y \cap Z) \text{ implies } G \subseteq \chi(X \cap Z). \qquad (\star)$$

Definition 4. *A* polychromatic model *is a simplicial model where:*

1. *the coloring is not required to be proper;*
2. *condition* $(\star)$ *holds.*

Requiring condition $(\star)$ is like requiring a transitive accessibility relation in certain Kripke models.

The multiplicity, see Definition 5, of a color within a face induces for each agent a a wellfounded relation $\leq_a$ on worlds. We call this the (a priori) plausibility relation.

Definition 5. *Let* (C, χ, W, ℓ) *be a polychromatic model. We define the* multiplicity *of* $a \in \mathsf{Ag}$ *in a world* X *by*

$$m_a(X) = |\{v \in X \mid \chi(v) = a\}|$$

where $|\cdot|$ *denotes the cardinality of a set. Note that if agent* a *is alive in a world* X*, then* $m_a(X) \geq 1$*. For* $X, Y \in W$ *and* $a \in \mathsf{Ag}$*, we write*

$$X \leq_a Y \quad \textit{iff} \quad m_a(X) \leq m_a(Y).$$

Next, we introduce a local plausibility relation

$$\trianglelefteq_a \; := \; \leq_a \cap \sim_a,$$

which captures the agent's plausibility relation at a given state. Further, we write

$$X \geq_a Y \quad \text{iff} \quad m_a(X) \geq m_a(Y)$$

and we use $\trianglerighteq_a$ and $\triangleleft_a$ in the obvious way. The following lemma shows that the indistinguishability relation can be given in terms of the local plausibility relation.

Lemma 2. $\sim_a = \trianglelefteq_a \cup \trianglerighteq_a$.

Proof. Observing that $\leq_a$ is strongly connected and unfolding the definition yields

$$\sim_a = (\leq_a \cup \geq_a) \cap \sim_a = (\leq_a \cap \sim_a) \cup (\geq_a \cap \sim_a) = \trianglelefteq_a \cup \trianglerighteq_a.$$

□

From the relation $\trianglerighteq_a$, we get a corresponding modal operator $[\trianglerighteq]_a$, which is referred to in the literature as *safe belief* [4] (sometimes also as *defeasible knowledge* [2]). Our language of knowledge and belief $\mathcal{L}_{\mathcal{KB}}$ extends $\mathcal{L}_{\mathcal{K}}$ by the modal operator $[\trianglerighteq]_a$ for each agent $a \in \mathsf{Ag}$. It is inductively defined by as follows:

$$\phi ::= p \mid \neg\phi \mid \phi \wedge \phi \mid [\sim]_G \phi \mid [\trianglerighteq]_a \phi$$

where $p \in \mathsf{Prop}$ and $G \subseteq \mathsf{Ag}$. As usual, the dual of safe belief is defined as $\langle\trianglerighteq\rangle_a \varphi \equiv \neg[\trianglerighteq]_a \neg\varphi$.

Definition 6. *For a polychromatic model $\mathcal{C} = (C, \chi, W, \ell)$, a world $X \in W$, and a formula $\phi \in \mathcal{L}_{\mathcal{KB}}$, we define the relation $\mathcal{C}, X \Vdash \phi$ inductively by*

$\mathcal{C}, X \Vdash p$	*iff*	$p \in \ell(X)$
$\mathcal{C}, X \Vdash \neg\phi$	*iff*	$\mathcal{C}, X \nVdash \phi$
$\mathcal{C}, X \Vdash \phi \wedge \psi$	*iff*	$\mathcal{C}, X \Vdash \phi$ *and* $\mathcal{C}, X \Vdash \psi$
$\mathcal{C}, X \Vdash [\sim]_G \phi$	*iff*	$X \sim_G Y$ *implies* $\mathcal{C}, Y \Vdash \phi$, *for all* $Y \in W$
$\mathcal{C}, X \Vdash [\trianglerighteq]_a \phi$	*iff*	$X \trianglerighteq_a Y$ *implies* $\mathcal{C}, Y \Vdash \phi$, *for all* $Y \in W$.

The $[\trianglerighteq]_a$-modality satisfies the $\mathsf{S4.2}$ principles for alive agents.

Lemma 3. *The following formulas are valid:*

1. $[\trianglerighteq]_a(\phi \to \psi) \to ([\trianglerighteq]_a \phi \to [\trianglerighteq]_a \psi)$;
2. $\mathsf{alive}(a) \to ([\trianglerighteq]_a \phi \to \phi)$;
3. $[\trianglerighteq]_a \phi \to [\trianglerighteq]_a [\trianglerighteq]_a \phi$;
4. $\langle\trianglerighteq\rangle_a [\trianglerighteq]_a \phi \to [\trianglerighteq]_a \langle\trianglerighteq\rangle_a \phi$.

Proof. We will only show the last claim. Assume

$$\mathcal{C}, X \Vdash \langle\trianglerighteq\rangle_a [\trianglerighteq]_a \phi.$$

There exists Z with $X \trianglerighteq_a Z$ and

$$\mathcal{C}, Z \Vdash [\trianglerighteq]_a \phi. \tag{1}$$

From $X \trianglerighteq_a Z$ we get that a is alive in Z and thus $Z \trianglerighteq_a Z$. Therefore, by (1)

$$\mathcal{C}, Z \Vdash \phi. \tag{2}$$

Let Y be arbitrary with $X \trianglerighteq_a Y$. By ($\star$), we find $Y \sim_a Z$. Thus we get by Lemma 2, that $Z \trianglerighteq_a Y$ or $Y \trianglerighteq_a Z$. In the first case, we use (1) to obtain $\mathcal{C}, Y \Vdash \phi$. Further we get $Y \trianglerighteq_a Y$ from $X \trianglerighteq_a Y$, and thus

$$\mathcal{C}, Y \Vdash \langle \trianglerighteq \rangle_a \phi. \tag{3}$$

In the second case, (3) follows immediately from (2). Since Y was arbitrary with $X \trianglerighteq_a Y$, (3) implies $\mathcal{C}, X \Vdash [\trianglerighteq]_a \langle \trianglerighteq \rangle_a \phi$. □

As usual with plausibility models, not only can we define safe belief but also other notions of belief. We start by defining the set of most plausible worlds.

Definition 7. *Let $\mathcal{C} = (C, \chi, W, \ell)$ be a polychromatic model. For $X \in W$, we define*

$$\mathsf{Min}_{\trianglelefteq_a}(X) = \{Y \in W \mid Y \sim_a X \textit{ and } \nexists Z \in W.Z \lhd_a Y\}.$$

Observe that, in general, since $\leq_a$ is wellfounded, $\mathsf{Min}_{\trianglelefteq_a}(X) \neq \emptyset$ if agent a is alive in the world X.

The model $\mathcal{C}_5$ given in Fig. 5 shows a situation with two minimal worlds. Namely, we have $Y, Z \in \mathsf{Min}_{\trianglelefteq_a}(X)$. Further, the following relations hold: $Y \lhd_a X$, $Z \lhd_a X$, $Z \trianglelefteq_a Y$, and $Y \trianglelefteq_a Z$.

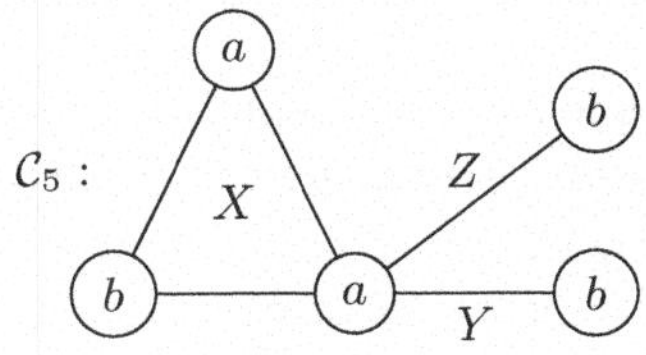

Fig. 5. A polychromatic model with two minimal worlds.

The following lemma about minimal worlds will be useful.

Lemma 4. *Let $\mathcal{C} = (C, \chi, W, \ell)$ be a polychromatic model, $X \in W$, and $Y \trianglelefteq_a X$. For each $Z \in \mathsf{Min}_{\trianglelefteq_a}(X)$, we have $Z \trianglelefteq_a Y$.*

Proof. From $Y \trianglelefteq_a X$ we get $X \sim_a Y$. Let $Z \in \mathsf{Min}_{\trianglelefteq_a}(X)$. By definition, this means

$$\nexists V \in W.V \lhd_a Z \tag{4}$$

and $X \sim_a Z$. Since $\sim_a$ is a partial equivalence relation, we get $Z \sim_a Y$. Suppose towards a contradiction that $Z \trianglelefteq_a Y$ does not hold. Since $Z \sim_a Y$ holds, we must have $Z \nleq_a Y$ and thus $Y <_a Z$. Therefore, $Y \lhd_a Z$. This contradicts (4), and we conclude $Z \trianglelefteq_a Y$. □

In the above lemma, we can let Y be X. Then we obtain the following instance:

$$Z \in \mathsf{Min}_{\trianglelefteq_a}(X) \quad \text{implies} \quad Z \trianglelefteq_a X. \tag{5}$$

We now include a new modality $\mathcal{B}_a$ for each agent a in our language $\mathcal{L}_{\mathcal{KB}}$. We extend the truth definition for $\mathcal{L}_{\mathcal{KB}}$ as follows.

Definition 8. *For a polychromatic model $\mathcal{C} = (C, \chi, W, \ell)$, a world $X \in W$, and a formula $\phi \in \mathcal{L}_{\mathcal{KB}}$, we define*

$$\mathcal{C}, X \Vdash \mathcal{B}_a\varphi \textit{ iff } \quad Y \in \mathsf{Min}_{\trianglelefteq_a}(X) \textit{ implies } \mathcal{C}, Y \Vdash \varphi \quad \textit{for all } Y \in W.$$

The modality $\mathcal{B}_a$ models agent a's (most plausible) belief. It is well-known that $\mathcal{B}_a$ can be expressed in terms of the $[\trianglerighteq]_a$-modality [4,22]. In our setting, we have the following theorem.

Theorem 1. *Let $\mathcal{C} = (C, \chi, W, \ell)$ be a polychromatic model, a an agent, and $X \in W$ such that a is alive in X. We find that*

$$\mathcal{C}, X \Vdash \mathcal{B}_a\varphi \quad \textit{if and only if} \quad \mathcal{C}, X \Vdash \langle\trianglerighteq\rangle_a[\trianglerighteq]_a\varphi.$$

Proof. For the direction from right to left, we assume $\mathcal{C}, X \Vdash \langle\trianglerighteq\rangle_a[\trianglerighteq]_a\varphi$. Thus, there exists Y with $X \trianglerighteq_a Y$ and $\mathcal{C}, Y \Vdash [\trianglerighteq]_a\varphi$. Now consider an arbitrary $Z \in \mathsf{Min}_{\trianglelefteq_a}(X)$. By Lemma 4 we find $Y \trianglerighteq_a Z$ and, therefore, $\mathcal{C}, Z \Vdash \varphi$. This yields $\mathcal{C}, X \Vdash \mathcal{B}_a\varphi$.

For the direction from left to right, we have $X \sim_a X$ since agent a is alive. Thus $\mathsf{Min}_{\trianglelefteq_a}(X)$ is non-empty and we let $Y \in \mathsf{Min}_{\trianglelefteq_a}(X)$. By (5), we obtain $Y \trianglelefteq_a X$. It remains to show $\mathcal{C}, Y \Vdash [\trianglerighteq]_a\varphi$. Let $Z \in W$ be such that $Z \trianglelefteq_a Y$. Then $Z \sim_a X$ by transitivity of $\sim_a$. Now we find $Z \in \mathsf{Min}_{\trianglelefteq_a}(X)$, for otherwise, we would find $V \in W$ with $V \triangleleft_a Z$, which yields $V \triangleleft_a Y$ and thus contradicts $Y \in \mathsf{Min}_{\trianglelefteq_a}(X)$. From $Z \in \mathsf{Min}_{\trianglelefteq_a}(X)$ and the assumption $\mathcal{C}, X \Vdash \mathcal{B}_a\varphi$, we get $\mathcal{C}, Z \Vdash \varphi$, which concludes the proof. □

Remark 3. A consequence of this theorem is that the properties of the $\mathcal{B}_a$-modality follow from properties of $[\trianglerighteq]_a$ such as the ones given in Lemma 3. For instance, we find that

$$\mathcal{B}_a\phi \wedge \mathcal{B}_a\psi \rightarrow \mathcal{B}_a(\phi \wedge \psi).$$

is valid.

Our model satisfies the knowledge yields belief principle. In particular, we have the following lemma.

Lemma 5. *Let $\mathcal{C} = (C, \chi, W, \ell)$ be a polychromatic model and $X \in W$. For any agent a and any formula φ, we have*

$$\mathcal{C}, X \Vdash [\sim]_a\varphi \rightarrow [\trianglerighteq]_a\varphi \quad \textit{and} \quad \mathcal{C}, X \Vdash [\trianglerighteq]_a\varphi \rightarrow \mathcal{B}_a\varphi.$$

Proof. For the first claim, assume

$$\mathcal{C}, X \Vdash [\sim]_a \varphi. \tag{6}$$

Let $X \trianglerighteq_a Y$, i.e., $X \sim_a Y$ and $X \geq_a Y$. By (6) we immediately get $\mathcal{C}, Y \Vdash \varphi$ and hence, $\mathcal{C}, X \Vdash [\trianglerighteq]_a \varphi$.

For the second claim, assume

$$\mathcal{C}, X \Vdash [\trianglerighteq]_a \varphi. \tag{7}$$

Let $Y \in \mathsf{Min}_{\trianglelefteq_a}(X)$ be arbitrary. Using (5), we obtain $Y \trianglelefteq_a X$. Now $\mathcal{C}, Y \Vdash \varphi$ follows immediately from (7), which yields $\mathcal{C}, X \Vdash \mathcal{B}_a \varphi$. □

As usual with preference-based semantics, our models are non-monotone. That is, agents may drop their beliefs when learning new information.

Example 1. Consider the set $\mathcal{V} = \{1, 2, 3, 4\}$ and let $C = (S, \mathcal{V})$ and $C^\psi = (S^\psi, \mathcal{V})$ be simplicial complexes given by

$$S = (\mathsf{Pow}(\{1,2,3\}) \cup \mathsf{Pow}(\{3,4\})) \setminus \{\emptyset\} \text{ and } S^\psi = \mathsf{Pow}(\{1,2,3\}) \setminus \{\emptyset\}.$$

Further, let $X = \{1, 2, 3\}$ and $Y = \{3, 4\}$. We define the models

$$\mathcal{C} = (C, \chi, \{X, Y\}, \ell\}) \text{ and } \mathcal{C}^\psi = (C^\psi, \chi, \{X\}, \ell^\psi).$$

The coloring χ is defined as follows

1. $\chi(1) = \chi(3) = a$;
2. $\chi(2) = \chi(4) = b$.

Figure 6 shows the colored complexes. We choose the valuation ℓ such that for some propositional formulas $\psi, \phi \in \mathcal{L}_{\mathcal{KB}}$:

$$\mathcal{C}, X \Vdash \neg\phi \wedge \psi \quad \text{and} \quad \mathcal{C},\ \mathrm{Y} \Vdash \phi \wedge \neg\psi$$

and we define $\ell^\psi(X) = \ell(X)$. The model $\mathcal{C}^\psi$ represents the situation after the agents in $\mathcal{C}$ learn that ψ is true. That is, it is the same as $\mathcal{C}$ but without the worlds where ψ is false. We observe that

$$\mathcal{C}, X \Vdash \mathcal{B}_a \phi \quad \text{and} \quad \mathcal{C}^\psi, X \nVdash \mathcal{B}_a \phi.$$

Hence, a only believes ϕ in X before it learns ψ. This is because removing worlds from $\mathcal{C}$ can result in a new world becoming a most plausible world.

A model is called *proper* if different worlds can be distinguished by at least one agent. Properly colored simplicial models are proper (cf. [15]). Formally, this is expressed as

$$\mathsf{alive}(G) \wedge \mathsf{dead}(G^c) \wedge \varphi \rightarrow [\sim]_G(\mathsf{dead}(G^c) \rightarrow \varphi) \tag{8}$$

being valid, where G^c stands for the complement of G. Lemma 6 shows that this is no longer true for polychromatic models.

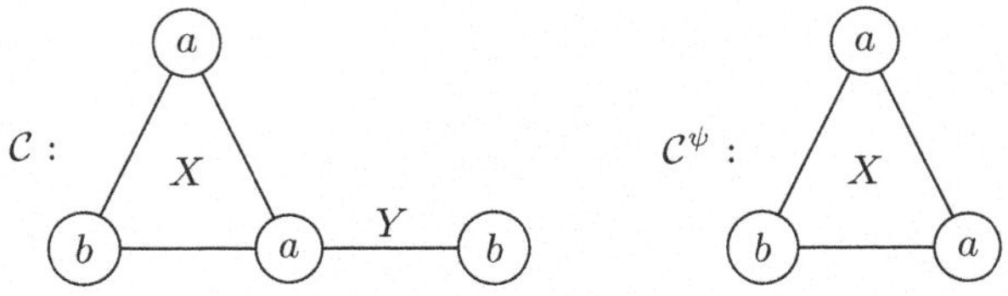

Fig. 6. The polychromatic complex $\mathcal{C}^{\psi}$ represents the state of affairs after the agents in $\mathcal{C}$ learn that ψ is true.

Lemma 6. *Polychromatic models are not proper, i.e.* (8) *is not valid.*

Proof. Consider the following counter-example. Let $C = (S, \mathcal{V})$ be the simplicial complex given by

$$\mathcal{V} = \{1, 2, 3\} \quad \text{and} \quad S = \mathsf{Pow}(\mathcal{V}) \setminus \{\emptyset\}.$$

Let a, b be agents. We set

$$\chi(1) := \chi(2) := a \quad \text{and} \quad \chi(3) := b.$$

The set of possible worlds W contains only the two worlds $X := \{1, 2, 3\}$ and $Y := \{1, 3\}$. The valuation is such that $\ell(X) := \{p\}$, and $\ell(Y) := \emptyset$.

Let $\mathcal{C} := (C, \chi, W, \ell)$. For $G = \{a, b\}$, we find that

$$\mathcal{C}, X \Vdash \mathsf{alive}(G) \wedge \mathsf{dead}(G^c) \wedge p.$$

However, we have $X \sim_G Y$ and $\mathcal{C}, Y \Vdash \mathsf{dead}(G^c) \wedge \neg p$. Hence

$$\mathcal{C}, X \nVdash [\sim]_G(\mathsf{dead}(G^c) \rightarrow p).$$

Therefore, (8) is not valid on polychromatic models. □

4 Knowledge Gain

An important result for epistemic simplicial models is that agents cannot gain new knowledge along morphisms of simplicial models. This property is crucial in distributed computing to show that certain morphisms do not exist, which implies that certain computation tasks are not solvable.

We adapt the notion of morphism from [16] to the setting where models contain a set of worlds. The fact that our models are polychromatic does not matter for the definition of morphism. As usual, given a function $f : U \rightarrow V$ and a set $W \subseteq U$, we let

$$f(W) := \{f(x) \mid x \in W\}.$$

Definition 9. *Let $C = (S, \mathcal{V})$ and $C' = (S', \mathcal{V}')$ be two simplicial complexes. A* simplicial map *from C to C' is a function $f : \mathcal{V} \rightarrow \mathcal{V}'$ such that if $X \in S$ then $f(X) \in S'$.*

Definition 10. *Let $C = (S, \mathcal{V})$ and $C' = (S', \mathcal{V}')$ be two simplicial complexes and let $\mathcal{C} = (C, \chi, W, \ell)$ and $\mathcal{C}' = (C', \chi', W', \ell')$ be two polychromatic models. A* morphism *from $\mathcal{C}$ to $\mathcal{C}'$ is a function f such that*

1. *f is a simplicial map from C to C';*
2. *$\chi'(f(v)) = \chi(v)$ for all $v \in \mathcal{V}$;*
3. *$f(X) \in W'$ for all $X \in W$;*
4. *$\ell'(f(X)) = \ell(X)$ for all $X \in W$.*

Morphisms respect the indistinguishability relation. We have the following lemma.

Lemma 7. *Let $C = (S, \mathcal{V})$ and $C' = (S', \mathcal{V}')$ be two simplicial complexes and let $\mathcal{C} = (C, \chi, W, \ell)$ and $\mathcal{C}' = (C', \chi', W', \ell')$ be two polychromatic models and f be a morphism from $\mathcal{C}$ to $\mathcal{C}'$. For $X, Y \in W$ we have*

$$X \sim_G^{\mathcal{C}} Y \quad \textit{implies} \quad f(X) \sim_G^{\mathcal{C}'} f(Y).$$

Proof. Assume $G \subseteq \chi(X \cap Y)$ and let a be an element of G. There exists $v \in \mathcal{V}$ with $v \in X$, $v \in Y$, and $\chi(v) = a$. We find that $f(v) \in f(X)$, $f(v) \in f(Y)$, and $\chi'(f(v)) = a$. Hence, $a \in \chi'(f(X) \cap f(Y))$. □

The positive formulas are the formulas of $\mathcal{L}_K$ where the operator $[\sim]_G$ occurs only in positive positions. Formally, we use the following definition.

Definition 11. *We consider the following grammar:*

$$\phi ::= p \mid \neg\phi \mid \phi \wedge \phi$$
$$\psi ::= \phi \mid \psi \wedge \psi \mid \psi \vee \psi \mid [\sim]_G \psi$$

where $p \in \mathsf{Prop}$ and $G \subseteq \mathsf{Ag}$. Formulas given by ψ are called positive formulas.

The result about no knowledge gain is standard [16]. Note that we are in a setting where agents may crash. However, since we assign atomic variables to the worlds (and not the local states of the agents), we can employ the usual formulation of positive formulas in the following theorem. Also, the fact that we have polychromatic models does not matter.

Theorem 2. *Let $\mathcal{C} = (C, \chi, W, \ell)$ and $\mathcal{C}' = (C', \chi', W', \ell')$ be two polychromatic models and f be a morphism from $\mathcal{C}$ to $\mathcal{C}'$. Further, let ψ be a positive formula. We find that*

$$\mathcal{C}', f(X) \Vdash \psi \quad \textit{implies} \quad \mathcal{C}, X \Vdash \psi$$

for all $X \in W$.

Proof. First we show that for a formula given by ϕ according to Definition 11, we have

$$\mathcal{C}', f(X) \Vdash \phi \quad \text{iff} \quad \mathcal{C}, X \Vdash \phi \tag{9}$$

for all $X \in W$. We proceed by induction on the structure of ϕ and distinguish:

1. ϕ is an atomic proposition p. Since f is a morphism, we have $\ell'(f(X)) = \ell(X)$. Thus $\mathcal{C}, f(X) \Vdash p$ iff $\mathcal{C}, X \Vdash p$.
2. ϕ is a negation or a conjunction. The claim follows immediately by I.H.

Now we proceed by induction on the structure of ψ (acc. to Definition 11) and assume $\mathcal{C}', f(X) \Vdash \psi$. We distinguish the following cases:

1. ψ is a formula ϕ (according to Definition 11). The claim follows from (9).
2. ψ is a conjunction or a disjunction. The claim follows immediately by the induction hypothesis.
3. ψ is of the form $[\sim]_G\psi'$. From $\mathcal{C}', f(X) \Vdash [\sim]_G\psi'$ we obtain that $f(X) \sim_G^{\mathcal{C}'} Y$ implies $\mathcal{C}', Y \Vdash \psi'$ for all $Y \in W'$.
   Let $Z \in W$ be such that $X \sim_G^{\mathcal{C}} Z$. Since f is a morphism, we find by Lemma 7 that
   $$f(X) \sim_G^{\mathcal{C}'} f(Z) \quad \text{and} \quad f(Z) \in W'.$$
   Thus, $\mathcal{C}', f(Z) \Vdash \psi'$. By I.H., we find $\mathcal{C}, Z \Vdash \psi'$. Since $Z \in W$ was arbitrary with $X \sim_G^{\mathcal{C}} Z$, we conclude $\mathcal{C}, X \Vdash [\sim]_G\psi'$.

□

The previous theorem only holds for knowledge but not for belief. That is, the operator $[\trianglerighteq]_a$ cannot be included in the class of positive formulas. We have the following lemma.

Lemma 8. *There are two polychromatic models $\mathcal{C}$ and $\mathcal{C}'$, a morphism f from $\mathcal{C}$ to $\mathcal{C}'$ and a world X of $\mathcal{C}$ such that for some agent a and $p \in \mathsf{Prop}$,*

$$\mathcal{C}', f(X) \Vdash [\trianglerighteq]_a p \quad \textit{but} \quad \mathcal{C}, X \nVdash [\trianglerighteq]_a p.$$

Proof. Consider the set $\mathcal{V} = \{1, 2, 3\}$. We let a be an agent and set

$$\chi(1) := \chi(2) := \chi(3) := a.$$

We let $S := \{\{1,2\}, \{2,3\}, \{1\}, \{2\}, \{3\}\}$ and $S' := \{\{1,2\}, \{1\}, \{2\}\}$. Hence $C = (S, \mathcal{V})$ and $C' = (S', \mathcal{V})$ are simplicial complexes. We let $X := \{1,2\}$, $Y := \{2,3\}$, and $Z := \{2\}$. Further, we set

$$W := \{X, Y\}, \qquad \ell(X) := \emptyset \quad \text{and} \quad \ell(Y) := \{p\}$$

and

$$W' := \{X, Z\}, \quad \ell'(X) := \emptyset \quad \text{and} \quad \ell'(Z) := \{p\}.$$

We define the polychromatic models

$$\mathcal{C} := (C, \chi, W, \ell) \quad \text{and} \quad \mathcal{C}' := (C', \chi, W', \ell')$$

and let f be such that $f(1) = 1$, $f(2) = 2$, and $f(3) = 2$. Obviously, f is a morphism from $\mathcal{C}$ to $\mathcal{C}'$ and we have $f(Y) = Z$. Finally, we note that

$$\mathcal{C}', f(Y) \Vdash [\trianglerighteq]_a p \quad \text{but} \quad \mathcal{C}, Y \nVdash [\trianglerighteq]_a p.$$

□

Besides the statement that belief gain is possible, this lemma could also be interpreted in such a way that condition ($\star$) is not strong enough or that we need a different notion of morphism, see also the discussion about simplicial sets in the next section.

5 Alternative Formalizations

The important feature of our approach to model belief with simplicial models is that agents are assigned a multiplicity within a world. Polychromatic models achieve this by allowing several vertices to represent the same agent. We have chosen this approach because it has the advantages of being very close to the original epistemic interpretation of simplicial complexes and having a simple geometric interpretation. Compared to [15], only the constraint of unique colors within a world has been relaxed. However, to guarantee that the indistinguishability relation is a partial equivalence relation, we have to impose condition ($\star$) on polychromatic models.

In this section, we present two alternative approaches to belief in simplicial models that are based on the idea of having repeated vertices instead of having several vertices with the same color. In these settings, the vertices of a world all have different colors, and thus condition ($\star$) is not needed.

We will only sketch the basic ideas and leave an elaborate presentation for future work. The first approach consists of using a version of simplicial complex that is based on multisets instead of sets. One could use the following definitions.

Let $\mathcal{V}$ be a set of vertices. A multiset over $\mathcal{V}$ is a function $M : \mathcal{V} \to \mathbb{N}$. The empty multi-set $\emptyset$ denotes the constant 0-function. A multiset N is a subset of a multiset M, in symbols as usual $N \subseteq M$, if we have $N(v) \leq M(v)$ for all $v \in \mathcal{V}$. The support of a multiset M is given by the set of its elements., i.e. $\mathsf{supp}(M) := \{v \mid M(v) > 0\}$.

Definition 12. *$C = (S, \mathcal{V})$ with S being a set of multisets over $\mathcal{V}$ is called a* multi-simplicial complex *if*

1. *S does not contain the empty set;*
2. *for each $X \in S$ and each $\emptyset \neq Y \subseteq X$, we have $Y \in S$.*

A *multi-simplicial model* is then defined like a simplicial model where a multi-simplicial complex is used instead of a simplicial complex. Since a world in a multi-simplicial model is a multiset, we can define the *multiplicity* of an agent a in a world X by

$$m_a(X) := \begin{cases} X(v) & \text{if there exists } v \text{ with } \chi(v) = a \text{ and } X(v) > 0, \\ 0 & \text{otherwise.} \end{cases}$$

Note that by the definition of multi-simplicial model, where the coloring must be proper,

$$\text{there is at most one } v \in \mathcal{V} \text{ with } \chi(v) = a \text{ and } X(v) > 0. \qquad (10)$$

Hence, $m_a(X)$ is well-defined.

We can employ the usual indistinguishability relation if we use the support of the worlds, meaning we consider the worlds as sets instead of multisets. For a group of agents G, we define

$$X \sim_G Y \quad \text{iff} \quad G \subseteq \chi(\mathsf{supp}(X) \cap \mathsf{supp}(Y))$$

Using (10), we immediately find that the so-defined relation $\sim_G$ is transitive. Thus, we do not need to additionally impose a condition like ($\star$). The disadvantage of multi-simplicial models is that they do not have a topological interpretation.

The second alternative approach, which does have a natural topological interpretation, is to work with simplicial sets, see, e.g. [12] for an introduction. For our very informal discussion here, it is sufficient to think of simplicial sets as simplices that may contain repeated vertices. Hence, the set of worlds of a simplicial set model may contain simplices as well as simplices with repeated vertices, which are called *degenerate* simplices.

It is a standard result [12, Prop. 4.8] that if Y is a degenerate simplex, then there is a unique non-degenerate simplex X such that $Y = s_{i_1} \cdots s_{i_k} X$ for some collection of degeneracy maps $s_{i_1}, \ldots, s_{i_k}$. Here, a degeneracy map creates one copy of a vertex. Given a degenerate simplex Y, we denote the unique non-degenerate simplex X of the above result by $\mathsf{nds}(Y)$. If Y is a non-degenerate simplex, we set $\mathsf{nds}(Y) := Y$.

Using this setting, we can define the indistinguishability relation for simplicial set models by

$$X \sim_G Y \quad \text{iff} \quad G \subseteq \chi(\mathsf{nds}(X) \cap \mathsf{nds}(Y)).$$

As before, $\sim_G$ is transitive, and condition ($\star$) is not needed. Moreover, we can define the multiplicity of a vertex v in a world Y as 1 plus the number of copies of v that the degeneracy maps introduce in the construction of Y from $\mathsf{nds}(Y)$.

Note that we have to be careful when we move from polychromatic models to simplicial set models. One of the reasons is that we have different notions of morphisms for the two models. The notion of morphism for polychromatic models (Definition 10) is based on simplicial maps that preserve the color. However, the multiplicity of an agent need not be preserved along morphisms for polychromatic models. The situation is different for simplicial set models. If we take the notion of morphism from simplicial sets and require that the color is preserved, then the multiplicity of an agent will also be preserved.

So far, simplicial sets have not been considered models for epistemic logic. However, we want to mention that semi-simplicial sets, which lie between simplicial complexes and simplicial sets, have been used to model novel notions of group knowledge [9,14].

6 Conclusion

We presented the first interpretation of belief on simplicial models that depends only on the topological structure without requiring additional machinery like belief functions. Our approach consists of dropping the requirement that the coloring must be proper and using the multiplicity of color within a face as an inverted plausibility measure.

The study of polychromatic models is still in its infancy. This paper presents first definitions, initial results, and key differences to simplical models such as the observations about non-proper models and belief gain.

Although we included a notion of group knowledge in our logic, we only considered individual belief. Obviously, notions of group belief provide an interesting topic for future research. A first definition could be

$$X \leq_G Y \text{ if and only if } \min\{m_a(X) \mid a \in G\} \leq \min\{m_a(Y) \mid a \in G\}$$

and then define the interpretation of the modalities $[\trianglerighteq]_G$ and $\mathcal{B}_G$ as in the individual case. Some principles of this notion of group belief are immediate, e.g. that group belief does not imply belief of subgroups (or individual belief). However, a detailed analysis of this approach should be performed and its relationship to existing approaches, e.g. [13,20], should be studied.

Another possible research direction for polychromatic models is to interpret them as a simplicial neighborhood semantics [10]. One approach would be to consider models in which simplices are either colored properly or unicolored. In Fig. 7, for instance, the simplices X_0, X_1, Y_0, Y_1 are properly colored (each vertex has a different color) whereas Z is unicolored (every vertex has the same color).

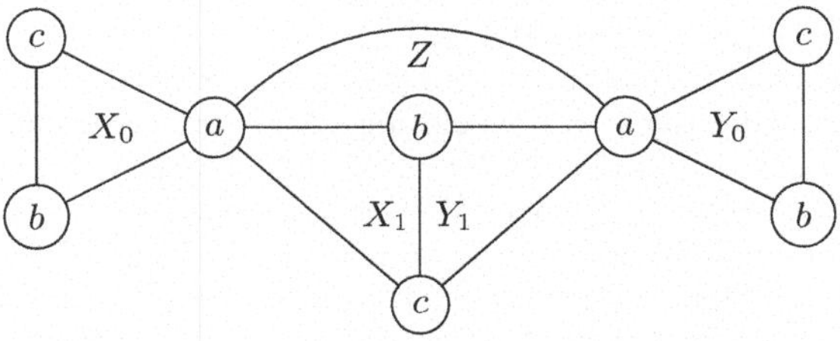

Fig. 7. Polychromatic complexes in which simplices are either properly colored or unicolored might be used for developing a simplicial neighborhood semantics.

Unicolored simplices can be used to define a neighborhood structure. The simplex X_0 has two a-neighborhoods $\{X_0, X_1\}$ and $\{Y_0, Y_1\}$ because agent a can reach Y_0 via the unicolored edge Z. It is straightforward to see that such neighborhoods will never intersect in a. The interpretation of knowledge is then standard: agent a knows φ if there is an a-neighborhood in which φ is true in all worlds.

Last but not least, let us mention that recently, a variant of topological semantics was introduced for polyhedra [1,5,6]. It will be interesting to investigate the similarities and differences between simplicial semantics and polyhedral semantics for modal logic.

Acknowledgments. We thank the anonymous reviewers for helpful suggestions and feedback. Special thanks goes to the organizers and participants of the Dagstuhl meeting *Epistemic and Topological Reasoning in Distributed Systems* [10], especially the working group on representing epistemic attitudes via simplicial complexes. We would also like to thank the anonymous reviewers of our AiML short paper [8] for many helpful comments.

This work has been funded by the Swiss National Science Foundation (SNSF). Christian Cachin and David Lehnherr are supported under SNSF grant agreement Nr. 219403 (Emerging Consensus) and Thomas Studer is supported under SNSF grant agreement Nr. 10000440 (Epistemic Group Attitudes).

Disclosure of Interests. The authors have no competing interests to declare that are relevant to the content of this article.

References

1. Adam-Day, S., Bezhanishvili, N., Gabelaia, D., Marra, V.: Polyhedral completeness of intermediate logics: the nerve criterion. J. Symb. Logic **89**(1), 342–382 (2024). https://doi.org/10.1017/jsl.2022.76
2. Baltag, A., Renne, B.: Dynamic epistemic logic. In: Zalta, E.N. (ed.) The Stanford Encyclopedia of Philosophy. Metaphysics Research Lab, Stanford University, Winter 2016 edn. (2016)
3. Baltag, A., Smets, S.: Dynamic belief revision over multi-agent plausibility models. In: Proceedings of the 7th Conference on Logic and the Foundations of Game and Decision (LOFT), pp. 11–24 (2006)
4. Baltag, A., Smets, S.: A qualitative theory of dynamic interactive belief revision. In: Bonanno, G., van der Hoek, W., Wooldridge, M. (eds.) Logic and the Foundations of Game and Decision Theory (LOFT7), pp. 13–60 (2008)
5. Bezhanishvili, N., Bussi, L., Ciancia, V., Fernández-Duque, D., Gabelaia, D.: Logics of polyhedral reachability. In: Advances in Modal Logic 2024 (2024)
6. Bezhanishvili, N., Marra, V., McNeill, D., Pedrini, A.: Tarski's theorem on intuitionistic logic, for polyhedra. Ann. Pure Appl. Logic **169**(5), 373–391 (2018). https://doi.org/10.1016/j.apal.2017.12.005
7. Cachin, C., Guerraoui, R., Rodrigues, L.E.T.: Introduction to Reliable and Secure Distributed Programming, 2nd edn. Springer (2011). https://doi.org/10.1007/978-3-642-15260-3
8. Cachin, C., Lehnherr, D., Studer, T.: Simplicial belief. In: Advances in Modal Logic 2024 – Short Papers, pp. 9–14 (2024)
9. Cachin, C., Lehnherr, D., Studer, T.: Synergistic knowledge. Theor. Comput. Sci. **1023**, 114902 (2025). https://doi.org/10.1016/J.TCS.2024.114902
10. Castañeda, A., van Ditmarsch, H., Kuznets, R., Moses, Y., Schmid, U.: Epistemic and topological reasoning in distributed systems (Dagstuhl Seminar 23272). Dagstuhl Reports **13**(7), 34–65 (2024). https://doi.org/10.4230/DagRep.13.7.34. https://drops.dagstuhl.de/entities/document/10.4230/DagRep.13.7.34
11. van Ditmarsch, H., Goubault, É., Ledent, J., Rajsbaum, S.: Knowledge and simplicial complexes. In: Lundgren, B., Nuñez Hernández, N.A. (eds.) Philosophy of Computing, pp. 1–50. Springer, Cham (2022)
12. Friedman, G.: Survey article: an elementary illustrated introduction to simplicial sets. Rocky Mountain J. Math. **42**(2), 353–423 (2012). https://doi.org/10.1216/RMJ-2012-42-2-353
13. Gaudou, B., Herzig, A., Longin, D., Lorini, E.: On modal logics of group belief. In: Herzig, A., Lorini, E. (eds.) The Cognitive Foundations of Group Attitudes and Social Interaction. SPS, vol. 5, pp. 75–106. Springer, Cham (2015). https://doi.org/10.1007/978-3-319-21732-1_4

14. Goubault, É., Kniazev, R., Ledent, J., Rajsbaum, S.: Semi-simplicial set models for distributed knowledge (2023). https://doi.org/10.1109/LICS56636.2023.10175737
15. Goubault, É., Kniazev, R., Ledent, J., Rajsbaum, S.: Simplicial models for the epistemic logic of faulty agents. Boletín de la Sociedad Matemática Mexicana **30**(3), 90 (2024). https://doi.org/10.1007/s40590-024-00656-x
16. Goubault, É., Ledent, J., Rajsbaum, S.: A simplicial complex model for dynamic epistemic logic to study distributed task computability. Inf. Comput. **278**, 104597 (2021). https://doi.org/10.1016/j.ic.2020.104597
17. Goubault, É., Ledent, J., Rajsbaum, S.: A simplicial model for $KB4_n$: epistemic logic with agents that may die. In: Berenbrink, P., Monmege, B. (eds.) 39th International Symposium on Theoretical Aspects of Computer Science. LIPIcs, vol. 219, pp. 33:1–33:20. Schloss Dagstuhl - Leibniz-Zentrum für Informatik (2022). https://doi.org/10.4230/LIPIcs.STACS.2022.33
18. Halpern, J.Y., Moses, Y.: Knowledge and common knowledge in a distributed environment. In: Kameda, T., Misra, J., Peters, J.G., Santoro, N. (eds.) Proceedings of the Third Annual ACM Symposium on Principles of Distributed Computing, pp. 50–61. ACM (1984). https://doi.org/10.1145/800222.806735
19. Herlihy, M., Kozlov, D.N., Rajsbaum, S.: Distributed Computing Through Combinatorial Topology. Morgan Kaufmann (2013). https://store.elsevier.com/product.jsp?isbn=9780124045781
20. Lindqvist, J., Velázquez-Quesada, F.R., Ågotnes, T.: Variations on distributed belief. CoRR abs/2408.10637 (2024). https://doi.org/10.48550/ARXIV.2408.10637
21. Randrianomentsoa, R., van Ditmarsch, H., Kuznets, R.: Impure simplicial complexes: complete axiomatization. Logical Methods Comput. Sci. **19**(4) (2023). https://doi.org/10.46298/lmcs-19(4:3)2023
22. Stalnaker, R.: On logics of knowledge and belief. Philos. Stud. **128**, 169–199 (2006)

When MIS and Maximal Matching are Easy in the Congested Clique

Keren Censor-Hillel[1], Tomer Even[1(✉)], Maxime Flin[2], and Magnús M. Halldórsson[2]

[1] Department of Computer Science, Technion, Haifa, Israel
ckeren@cs.technion.ac.il, tomer.even@campus.technion.ac.il
[2] Department of Computer Science, Reykjavik University, Reykjavik, Iceland
maximef@ru.is

Abstract. Two of the most fundamental distributed symmetry-breaking problems are that of finding a maximal independent set (MIS) and a maximal matching (MM) in a graph. It is a major open question whether these problems can be solved in constant rounds of the all-to-all communication model of Congested Clique, with $O(\log\log \Delta)$ being the best upper bound known (where Δ is the maximum degree).

We explore in this paper the boundary of the feasible, asking for *which graphs* we can solve the problems in constant rounds. We find that for several graph parameters, ranging from sparse to highly dense graphs, the problems do have a constant-round solution. In particular, we give algorithms that run in constant rounds when:

- the average degree is at most $d(G) \leq 2^{O(\sqrt{\log n})}$,
- the neighborhood independence number is at most $\beta(G) \leq 2^{O(\sqrt{\log n})}$, or
- the independence number is at most $\alpha(G) \leq |V(G)|/d(G)^{\mu}$, for any constant $\mu > 0$.

Further, we establish that these are tight bounds for the known methods, for all three parameters, suggesting that new ideas are needed for further progress.

Keywords: Distributed graph algorithms · Maximal independent set · Maximal matching · Congested clique

1 Introduction

Symmetry breaking is fundamental to distributed computing. Two of the most fundamental symmetry-breaking problems are that of computing a maximal

Due to space constraints, this paper takes a reduced form of the original paper, and the last section that contain a tight analysis for the running time of the state-of-the-art algorithms for MIS [33] and maximal matching [15] is omitted. A full version is available at https://arxiv.org/abs/2502.21031 and referred in [20].

U. Schmid and R. Kuznets (Eds.): SIROCCO 2025, LNCS 15671, pp. 194–210, 2025.
https://doi.org/10.1007/978-3-031-91736-3_12

independent set (MIS) and a maximal matching (MM) in the distributed network. We consider these problems in the Congested Clique model, which has a machine for each graph node and in each synchronous round, each machine can send a $O(\log n)$-bit message to every other node.

There is now a fairly good understanding of the complexity of these problems in the central model of locality, LOCAL. There is an algorithm that runs in $O(\log \Delta + (\log\log n)^{O(1)})$ rounds [26,31,36] that works even in the bandwidth-constrained CONGEST model and a lower bound of $\Omega\left(\min\left\{\frac{\log \Delta}{\log\log \Delta}, \sqrt{\frac{\log n}{\log\log n}}\right\}\right)$ [45] for the LOCAL model. Much less is known in the Congested Clique model.

Ghaffari [32] gave a $\tilde{O}(\sqrt{\log \Delta})$-round Congested Clique algorithm for MIS which beats the LOCAL lower bounds when Δ is small enough. It was followed by [33,43] which improved the complexity to $O(\log\log \Delta)$ by using a sparsification technique pioneered by [1] for correlation clustering in streaming. For MM, the state of the art is the $O(\log\log \Delta)$-round algorithm of Behnezhad, Hajiaghai and Harris [15] which implements a sparsification mechanism comparable to the one for MIS. Meanwhile, problems such as minimum spanning tree [42], $(\Delta + 1)$-vertex-coloring [21] or 2-ruling set [19] have $O(1)$-round algorithms in Congested Clique. The 2-ruling set problem in particular is a slight relaxation of the MIS problem which has a similar complexity in the LOCAL model [8,10].

Proving lower bounds in Congested Clique is known to be hard [24]. Interestingly, as the algorithm of [33] can be implemented in $O(\log\log \Delta)$ passes of streaming, a very recent streaming lower bound by Assadi, Konrad, Naidu and Sundaresan [4] shows that such algorithms cannot break the $O(\log\log n)$ barrier in general graphs. Contrary to MIS, there exist no such results for MM because it is trivial to compute in the streaming model. Unconditional lower bounds for MIS and MM are only known in the much weaker *broadcast congested clique*, where nodes must send the same message to everyone in a round. In this model, Assadi, Kol and Zhang [3] recently showed that any MIS or MM algorithm with $(\log n)^{O(1)}$ bandwidth needs $\Omega(\log\log n)$ rounds. Meanwhile, the best upper bound in this model is $O(\log n)$ rounds [48].

Turning back to Congested Clique, we change perspective and ask:

***When** can MIS and MM be solved in $O(1)$ rounds of Congested Clique?*

In other words, for what kind of graphs can we solve the problems fast, and for which ones are current techniques insufficient?

We consider both sparse and dense families of graphs, pinpointing properties that allow us to quickly converge toward a solution. On one hand, we focus on graphs of low-to-medium *average degree*, which subsumes other sparse graphs such as low-degree or low-arboricity graphs. On the other hand, we consider graphs with sublinear *independence number*, which are graphs that can be highly dense. Another related notion is *neighborhood independence number*, which is the largest number of independent nodes in the neighborhood of a given node.

Our Results. We summarize our algorithmic results succinctly as follows. Let n be the number of nodes in the graph G, $d(G)$ its average degree, $\alpha(G)$ the size of the largest independent set, and $\beta(G) = \max_{v \in V(G)} \alpha(G[N(v)])$ the neighborhood independence number. We say that a problem is *easy* if it can be solved in $O(1)$ rounds of Congested Clique (on the respective graph class.

> MIS and MM are easy on graphs of *average degree* at most $d(G) \leq 2^{O(\sqrt{\log n})}$.

The same holds in terms of neighborhood independence:

> MIS and MM are easy on graphs of *neighborhood independence* at most $\beta(G) \leq 2^{O(\sqrt{\log n})}$.

Additionally, we get bounds in terms of independence number.

> MIS and MM are easy on graphs with *independence number* at most $\alpha(G) \leq n/d(G)^{\mu}$, for any constant $\mu > 0$.

For larger ranges of these parameters, we get tradeoffs, with complexities as a function of the parameters: $O(\log(\log d(G)/\sqrt{\log n}) + 1))$, $O(\log(\log(\beta(G))/\sqrt{\log n})+1)$, and $O(\log 1/\mu+1)$, respectively. Note that classes of graph where those parameters are small have been studied and are known to appear in practice; see Sect. 1.2 for more details.

Our results are tight for all three graph parameters. More generally, we give a family of hard graphs for which current techniques require $\Omega(\log\log n)$ rounds. The parameters $\Delta, d(G), \beta(G), \alpha(G)$ in our hard graphs are such, that our bounds for them are tight given the known methods. Our construction provides concrete evidence of instances for which new approaches are needed.

1.1 Technical Overview

Upper Bounds. Our starting points for computing MIS and MM are the algorithms of [15,33] which take $O(\log\log(\Delta))$ rounds for MIS and MM, respectively. They both use degree reduction for sparsification.

For MIS, the algorithm samples a subset of vertices that induce a graph with $O(n)$ edges. One designated vertex collects the edges of this graph and locally computes an MIS S on it. To extend the set S into an MIS, the algorithm then proceeds by computing an MIS on $G[V \setminus \Gamma(S)]$, where $\Gamma(S)$ contains all vertices in S or with at least one neighbor in it (i.e., vertices that are *covered* by S). Computing an MIS on this residual graph $G[V \setminus \Gamma(S)]$ is an easier task as it has a maximal degree of $\tilde{O}(\sqrt{\Delta})$, w.h.p.. This algorithm is repeated $O(\log\log(\Delta))$ times to get a residual graph of a maximum degree $\text{poly}\log(n)$, at which point it switches to simulating a LOCAL algorithm by letting each vertex learn its $\log\log(\Delta)$ neighborhood.

Average Degree: As a warm-up to our main technical contribution, we observe that one can push the degree-reduction method beyond the parameter of the maximum degree, which can be replaced by the *average degree*. The key idea is to sample according to the average degree, rather than the maximum degree, and leverage a concentration bound that allows for dependencies. With efficient routing, this leads us to a constant-round algorithm when the average degree is at most $2^{O(\sqrt{\log n})}$, which we use next.

Independence Number: One of our main technical contributions is to bound the running time of the repeated degree reduction algorithm in terms of the *independence number* of the graph. To give a flavor of this, consider a graph G with n vertices, average degree $d(G)$, and an independence number as small as possible, i.e., $\alpha(G) = n/(d(G)+1)$. After one round of degree reduction, we get an independent set S and a residual graph $H = G[V \setminus \Gamma(S)]$, where the maximal degree in H is $\tilde{O}(\sqrt{d(G)})$. Note that since H is an induced subgraph of G, its independence number cannot be larger than that of G. We use Turán's bound to obtain the following bound: $\frac{n(H)}{2d(H)} \leq \alpha(H) \leq \alpha(G) \leq n/(d(G)+1)$. Further, we can bound the number of edges of H by multiplying both sides by $(d(H))^2$ to get that $|E(H)| = n(H) \cdot d(H)/2 \leq n \cdot \frac{(d(H))^2}{d(G)+1} \leq \tilde{O}(n)$, where the last inequality follows as $d(H) = \tilde{O}(\sqrt{d(G)})$. This, together with efficient routing, leads us to a constant-round algorithm, as the residual graph has average degree is at most $2^{O(\sqrt{\log n})}$.

This toy example was only for $\alpha(G) = n/(d+1)$. We extend it by showing a $O(\log(1/\mu)+1)$ round algorithm for MIS, where $\mu > 0$ is such that the independence number of G is $\alpha(G) = n/(d+1)^{\mu}$.

This framework also applies for MM. Consider the algorithm of [15]. It partitions the vertices of the graph into random sets, such that each set induces a graph with at most $O(n)$ edges. Then, a maximal matching is computed on each part. Let M' denote the union of those computed matchings. The algorithm then proceeds with a clean-up phase, in which vertices of high degree are matched. Let M denote the union of M' and the edges computed in the clean-up phase. The algorithm has the property that $G[V \setminus V(M)]$ has a maximal degree of $\tilde{O}(\Delta^{0.92})$.

As for the MIS case, we turn this into an algorithm which reduces the maximal degree in terms of the average degree and show that if the input graph has a small independence number then we can compute an MM faster.

Neighborhood Independence: Our second main technical contribution involves graphs with small neighborhood independence. A graph G has neighborhood independence $\beta(G)$ if the size of any independent set in the neighborhood of any vertex is at most $\beta(G)$. Note that $\beta(G)$ and $\alpha(G)$ are not comparable as $\beta(G) \leq \Delta(G)$, and $n/(\Delta(G)+1) \leq \alpha(G) \leq n\beta(G)/\delta(G)$.

Suppose a graph has a minimum degree of at least $(d(G))^{1/2}$, and that $\beta(G)$ is, say, at most $(d(G))^{1/4}$. Then we get that $\alpha(G) \leq n/(d(G))^{1/4}$, allowing us to use our framework for low independence number. This motivates us to find subgraphs for which this holds. However, a direct degree reduction is insufficient. The challenge here is that the minimum degree might be much smaller than the average degree.

Our approach for this case is to partition the vertices into log n degree classes where the i-th degree class, denoted by V_i, contains vertices of roughly the same degrees, and then use the fact that each $G[V_i]$ has relatively small independence number. Our final step here is to replace sampling of vertices with a uniform probability of $1/\sqrt{d(G)}$, by a *non-uniform* probability of $1/\sqrt{\deg_G(v)}$ (such sampling was applied for 2-ruling sets in [19]). We show that this algorithm has the property that within the residual graph, each node has at most $\tilde{O}(\sqrt{x})$ neighbors that had degrees in G in the range $[x, 2x]$. This allows us to show that after one iteration, we get a residual graph H, where $H[V_i]$ contains only $\tilde{O}(n\beta(G))$ edges w.h.p, and therefore so does H. We can therefore find an MIS on H in time $O(\log(\log(\beta(G))/\sqrt{\log n}) + 1)$ w.h.p..

For matching, our algorithm in terms of $\beta(G)$ is simpler. We partition the vertices into degree classes as before. In parallel, each class computes a matching on the edges inside the class, using the degree reduction algorithm. We claim that finding a maximal matching in each degree class takes only a constant number of rounds, after which, the residual graph H contains only $\tilde{O}(n\beta(G))$ edges w.h.p.. The crucial point here is as follows. If W is a set of unmatched vertices inside the same degree class, i.e., of degree between a and $2a$, then W is an independent set, from which we obtain that $|W| \leq n\beta(G)/a$. Therefore, these vertices contribute at most $n\beta(G)$ edges to H.

Tight Analysis for Existing Algorithms. We provide a hard distribution over graphs, on which the state-of-the-art methods require $\Omega(\log\log n)$ iterations to complete. Notably, the maximum degree in these graphs, the average degree, and the global- and neighborhood- independence numbers are all values for which our bounds with respect to them become tight given the known methods.[1] The construction of this hard distribution is omitted due to space considerations, but is available in the full version of the paper.

MIS: We provide a tight analysis for the running time of all algorithms that compute an MIS using the following framework: (i) samples a set of vertices S, such that the induced subgraph $G[S]$ has $O(n)$ edges w.h.p, (ii) computes a greedy MIS over $G[S]$, (iii) removes all covered vertices, and repeats. The high-level intuition behind our construction is as follows. We split the n vertices into roughly $\log\log n$ levels, and denote by k the number of vertices in each level. Within level i, we partition the vertices into $k^{1-1/4^i}$ equal-size clusters, each forming a clique. Note that if the algorithm sampled at least one vertex v from some cluster C, then a vertex from C joins the MIS, and all other vertices in C are covered. We say that a cluster is sampled if at least one of its vertices is sampled. Yet, if for a given cluster, no vertex gets sampled in a certain iteration, then none joins the independent set, then none of its vertices get covered. As the level number increases, so does its number of clusters, while their size decreases, making them less likely to be sampled.

The running time of an algorithm that samples vertices with some uniform probability might be $\Omega(\log\log n)$, but an algorithm that samples each vertex

[1] We do not know if this is the case for the construction of [4].

with probability $1/\sqrt{d_v}$, is likely to hit all clusters very quickly (plus it can be easily handled in a single iteration of a Luby approach).

To overcome this, we add random edges to the graph, obtaining a distribution over graphs. A random edge is added between two vertices in levels i, j for $i < j$ with probability that is roughly the inverse of the number of clusters in level i (up to poly $\log n$ factors). The obtained graph becomes nearly regular, and then it is unlikely that the smaller clusters (of the higher levels) are sampled, even if the sampling probability is based on the vertices degree. The random edges complicate the analysis as a vertex is covered when its cluster is sampled, or when a neighbor of the vertex in a different cluster is sampled. A crucial step in this analysis requires proving that w.h.p.at least $1 - o(1)$ fraction of the vertices in the r-th level are not covered before iteration r is over. Since the graph has roughly $\log \log n$ levels, we conclude that at least $\Omega(\log \log n)$ iterations are needed to cover all vertices. The full details of this proof appear in the full version of the paper.

MM: We also provide a tight analysis for the running time of all algorithms that compute an MM using the following framework: (i) randomly partitions the vertices into sets $(S_1, \ldots, S_k)$, such that for every i the induced subgraph $G[S_i]$ has $O(n)$ edges w.h.p, (ii) computes a greedy MM over each $G[S_i]$, (iii) invokes a clean-up phase that matches high-degree vertices, (iv) removes all covered vertices, and repeats. Our hard graph distribution for MM is only a slight modification of the one for MIS: The number of vertices in each level is increased by 4 as the level increases. This ensures that any maximal matching must "reveal" the last level of the graph. We also ensure that graphs taken from this distribution have a perfect matching, by setting the number of vertices in each cluster to be even.

A crucial step in the analysis of the running time of such algorithm requires proving that w.h.p.at least $\frac{2}{3} \cdot (1 - o(1))$ fraction of the vertices in the r-th level are not matched before iteration r is over. Using this property and the fact that 3/4 fraction of the vertices are in the last level, and 3/4 fraction of the remaining vertices are in the second-to-last level, we conclude that after $\ell/2$ iterations, where ℓ is the number of levels, at least $\frac{8(1-o(1))}{9}$ fraction of the vertices are not matched, and therefore the computed matching is not maximal, as any maximal matching must in a graph with a perfect matching must match at least half of the vertices. Therefore, the running time of the algorithm must be at least $\Omega(\log \log n)$ w.h.p..

1.2 Related Work

Algorithms in LOCAL and CONGEST. Maximal Independent Set and Maximal Matching have long been studied in the classical LOCAL and CONGEST models (see, e.g., [7,9,13,26,27,31,36,44,48,51,52]). Combining recent advances on local rounding [26,27] and network decomposition [35,36,52], authors of [34] gave a $\tilde{O}(\log^2 n)$-round state-of-the-art deterministic LOCAL upper bound. Using the shattering technique [13,31] this yields a $O(\log \Delta)+\tilde{O}(\log^2 \log n)$ round

LOCAL algorithm. As for LOCAL lower bounds, authors of [7] showed that any LOCAL algorithm that solves MIS or MM with probability at least $1-1/n$ needs $\Omega(\min\{\Delta, \frac{\log\log n}{\log\log\log n}\})$. Hence, local algorithms are nearly optimal.

Algorithms for Maximum Matching in Congested Clique. For matching, using different techniques, authors of [2,33] gave an algorithm to $(1+\varepsilon)$-approximate the integral maximum matching running in $O(\log\log n)$ rounds. Authors of [15] showed that a simple degree reduction algorithm could be implemented for matching, resulting in a $O(\log\log \Delta)$-round algorithm for maximal matching. A recent breakthrough of [28] shows that given a Congested Clique algorithm for maximal matching, one can boost the approximation factor to $1+\varepsilon$ by paying only a $(1/\varepsilon)^{O(1)}$ factor of overhead in the round complexity.

Algorithms for 2-Ruling Set. It is interesting to note that a weaker variant of the MIS called *2-ruling set* has also been studied in the Congested Clique [16,19, 38,41]. In this problem, we relax the domination requirement from MIS: every vertex should be within distance two from a node in the independent set. Very recently, Cambus, Kuhn, Pai and Uitto [19] showed this problem has $O(1)$-round algorithm. Deterministic algorithms for 2-ruling set were given by [38,50].

Sparse Graphs. Arboricity has been intensively studied in LOCAL and Congested Clique (see, e.g., [11,13,14,37]). Since the arboricity is bounded by the average-degree, our results show $O(1)$ round Congested Clique algorithms for MIS and MM in low arboricity graphs as well.

Small Independence Number. While there is no concrete characterization of graphs with small independence number, random graphs with degree d are known to have independence number $\approx 2n/d \log d$ (see, e.g., [17,29] for more precise bounds). In particular, for random graphs, the Turán bound is nearly-tight and our algorithm for low independence number ends in $O(1)$ rounds.

Bounded Neighborhood Independence. Graphs of bounded neighborhood independence capture a wide range of dense graphs. This includes line graphs, bounded-growth graphs (e.g., interval graphs [40] or disk graphs [39]). Graphs with $\beta+1$-bounded neighborhood independence are equivalent to β-claw-free graphs (i.e., that do not contain $K_{1,\beta+1}$). Such graphs have been extensively studied in structural graph theory (see, e.g., [23]). Motivated by applications to wireless networks, distributed algorithms for MIS and MM on graphs of bounded independence have also been studied (see, e.g., [12,30,39,46,53]). More recently, Assadi and Solomon [6] gave sublinear algorithms for MIS and MM. Milenković and Solomon [49] showed that one could compute $(1+\varepsilon)$-matching sparsifiers of size $O(n\beta/\varepsilon \log(1/\varepsilon))$ on graphs with β-bounded neighborhood-independence by simply having each vertex sample adjacent edges.

2 Preliminaries

Graph Theoretic Notation. For a graph H, let $d(H)$ be its average degree, $\Delta(H)$ its maximum degree, and $n(H) = |V(H)|$ its number of vertices. For a vertex v, let $N_H(v)$ its set of neighbors, and $\deg_H(v) = |N_H(v)|$ its degree. The hop-distance between u and v is $\mathrm{dist}_H(u,v)$. For a subset $S \subseteq V$, let $N_G(S) = \bigcup_{v \in S} N_G(v)$, $\Gamma_G(S) = S \cup N_G(S)$, and for sets $A, B \subseteq V(H)$, let $E_H(A,B) = \{e = (a,b) \in E(H) : a \in A, b \in B\}$.

Throughout this paper, the input graph is $G = (V(G), E(G))$, with $n = |V(G)|$ and $m = |E(G)|$. We denote the independence number of the graph by $\alpha(G)$. Let $\beta(G) \triangleq \max_{v \in V} \alpha(G[N(v)])$ denote the maximum independence number of an induced neighborhood subgraph.

We say that a vertex set A is *covered* by an independent set S if $A \subseteq \Gamma(S)$. The lexicographically first MIS (LFMIS) is defined with respect to an ordering $v_1, v_2, \ldots, v_n$ of the vertices as the outcome of the following greedy algorithm: beginning with an empty independent set S, loop through vertices in the given order and add a vertex to S if it is uncovered when reached by the algorithm. The lexicographically first MM (LFMM) is defined similarly, with respect to adding edges to the matching.

Lenzen's Routing Lemma [47]**.** The following is equivalent to the Congested Clique model: In every round each node can send (receive) $\{b_i\}_{i \in [n]}$ bits to (from) the i-th node, for any sequence $\{b_i\}_{i \in [n]}$ satisfying $\sum_{i=1}^{n} b_i = O(n \log n)$. In other words, any routing scheme in which no vertex sends or receives more than $O(n)$ messages can be performed in $O(1)$ rounds.

2.1 Opportunistic Routing

We leverage a powerful method of Chechik and Zhang [22] to learn r-hop neighborhoods in the Congested Clique using opportunistic routing. A deterministic variant of this method was presented in [18].

Proposition 1. (Opportunistic Routing, [22]**).** *Let $c \geq 1$ be a constant. Let G be an n-vertex graph of maximum degree Δ. For any r, possibly depending on n, such that*

$$\Delta^{2(r+1)}((r+1)\log \Delta + c \log n) \leq n \ , \tag{1}$$

there exists a constant round randomized Congested Clique algorithm such that, w.p. $1 - n^{-\Theta(c)}$, each vertex learns its r-hop neighborhood $E^r(v) = \{uw \in E, \mathrm{dist}_G(v,u) \leq r$ and $\mathrm{dist}_G(v,w) \leq r\}$.

As the output of a vertex v in a r-round LOCAL algorithm only depends on edges of $E^r(v)$ and identifiers of their endpoints, Proposition 1 implies that we can simulate any r-round LOCAL algorithm on graphs of maximum degree Δ provided that Eq. (1) is true. In this paper, we make use of this observation in the regime described by Proposition 2. This observation is of independent interest.

Proposition 2 (Simulating *LOCAL*Algorithms). *Consider any $r = o(\log n / \log\log n)$-round LOCAL algorithm for a problem with inputs of size $O(poly(\log n, \Delta))$ per node. Then the algorithm can be simulated on Congested Clique in $O(1)$ rounds for graphs of maximum degree $\Delta(G) \leq n^{1/(2r+2)}$.*

By using Ghaffari's algorithm [31] that runs in $O(\log \Delta + \log^3 \log n)$ rounds of LOCAL [31, 34] (both for MIS and MM), we immediately obtain from Proposition 2 the following fast algorithm for graphs of restricted maximum degree.

Corollary 1. *There is a $O(1)$-round Congested Clique algorithm for MIS (MM) on graphs with maximum degree $\Delta = 2^{\sqrt{\log n}/C}$ for some universal constant $C \geq 1$.*

3 Sparsification via Degree Reduction

The main technical contribution of this section is the following lemma. It is a twist on previous methods that reduce the maximum degree. We first state the previous result, and then present our modification.

Lemma 1 ([33] (MIS), [15] (MM)). *There is a $O(1)$-round Congested Clique algorithm that computes an independent set S (matching M) such that the residual graph $H = G[V \backslash \Gamma(S)]$ ($H = G[V \setminus M]$) has maximum degree $\Delta(H) = O(\sqrt{\Delta(G)} \log n)$.*

We give a twist to the method by bounding degrees in terms of average degree.

Lemma 2. *There is a $O(1)$-round Congested Clique algorithm that computes an independent set S (or a matching M) such that the residual graph $H = G[V \backslash \Gamma(S)]$ ($H = G[V \setminus M]$) has maximum degree $\Delta(H) = O(\sqrt{d(G)} \log n)$, w.h.p.*

The proof for MIS is given below. The proof for MM appears in the full version. Using this new lemma, combined with Corollary 1, we obtain one of our contributions.

Corollary 2. *There is a $O(1)$-round Congested Clique algorithm for MIS (MM) for graphs G with average degree $d(G) = 2^{O(\sqrt{\log n})}$, w.h.p.*

By applying Lemma 2 repeatedly, we obtain the following generalization. We refer readers to [43] or [33] for more details on this.

Proposition 3. *Let $r \geq 1$. There is an $O(r)$-round randomized Congested Clique algorithm that computes an independent set S (or a matching M) in the input graph G such that the residual graph $H \triangleq G[V \setminus \Gamma(S)]$ ($H = G[V \setminus M]$) has maximum degree $\Delta(H) = O(d(G)^{1/2^r} \log^2 n)$, w.h.p.*

Applying Proposition 3 with $r = O(\log(\log d(G))/\sqrt{\log n} + 1)$, followed by Corollary 2, we get the second contribution:

Theorem 1. *There is a randomized Congested Clique algorithm for MIS (or MM) running in $O(\log(\log d(G))/\sqrt{\log n} + 1)$ rounds, w.h.p.*

3.1 Proof of Lemma 2

The following simple $O(1)$-round algorithm has been well studied [1,5,33,43]:

Algorithm 1. OneShotMIS

Input: A graph $G = (V, E)$, with n vertices and a parameter p.
Output: An independent set S

1. Sample a subset of vertices V_F, where each vertex v is sampled independently w.p. p.
2. Let one designated vertex v collect the edges of the induced subgraph $F \triangleq G[V_F]$.
3. v locally computes the lexicographically first MIS in F. Let S be the resulting independent set.

We use the following well-known result.

Lemma 3 ([5, Lemma 2.2]). *Let $G = (V, E)$ be a graph and v one of its vertices. Let $S \leftarrow$ OneShotMIS(G, p). Let G' obtained from G by removing all edges incident to $\Gamma(S)$. Then, for any $t \geq 0$, we have* $\Pr[\deg_{G'}(v) \geq t] \leq e^{-tp}$.

This lemma implies a variant of our Lemma 2 in terms of *maximum degree*, by setting $p = 1/\sqrt{\Delta(G)}$. A proof that the call to OneShotMIS runs in $O(1)$ rounds appears in the full version of the paper.

We apply OneShotMIS with $p = 1/\sqrt{d(G)}$. Our main task is to bound from above the number of edges in the residual graph H.

Lemma 4. *Suppose $\Delta(G) = \tilde{O}(n^{1/4})$. After running OneShotMIS with $p = 1/\sqrt{d}$, the graph F contains at most $36n$ edges, w.h.p.*

Proof. For each $e \in E(G)$, let X_e be the random variable indicating if both endpoints of e were sampled in F. We have that the probability that $X_e = 1$ is $p^2 = 1/d$. Let $X = \sum_{e \in E(G)} X_e$ denote the number of edges in F. By linearity of expectation, $\mathbb{E}[X] = |E(G)|/d = 2n$. We show concentration using the method of bounded differences (see [25, Appendix A.2]). Every vertex joins F independently, and adding/removing a single vertex to F affects the number of edges by at most Δ. Thus, using the assumed bound on $\Delta(G)$, we get

$$\Pr[X \geq \mathbb{E}[X] + n] \leq \exp\left(-\frac{n^2}{n \cdot \Delta(G)^2}\right) \leq \exp\left(-\tilde{O}(n^{1/2})\right) \ll 1/\text{poly}(n) .$$

Proof (Proof of Lemma 2 for MIS). First apply Lemma 1 twice, obtaining a residual graph G' of maximum degree $\Delta(G') = O(\Delta(G)^{1/4} \log n) = \tilde{O}(n^{1/4})$. Then run OneShotMIS on G' with $p = 1/\sqrt{d(G')}$. By Lemma 4, the residual graph F contains $O(n)$ edges. We can then gather it at a single node, using Lenzen routing, and complete the MIS S on G' centrally. The bound on the maximum degree of the residual graph $H = G[V \setminus \Gamma(S)]$ follows from Lemma 3 with $t = 1/p \cdot \log n$ and $p = 1/\sqrt{d(H)}$, w.h.p.

4 Leveraging Independence for Sparsification

In this section, we provide fast algorithms for graphs of sublinear neighborhood-independence. We first derive a common sparsification lemma in Sect. 4.1, which is then applied to MIS in Sect. 4.2 and to MM in theorem 4. We first illustrate the proof strategy on graphs with low independence number.

Theorem 2. *Let G be a graph $\alpha(G) = n/d(G)^{\mu}$ for some $\mu = \mu(n) > 0$. There exists a randomized* Congested Clique *algorithm that, w.h.p, computes an MIS (MM) of G in $O(\log(2/\mu) + 1)$ rounds.*

Proof. If $d(G) < 1$, then G has $nd(G)/2 = O(n)$ edges and all edges can be gathered by a single node, using Lenzen routing, and complete the MIS S on G' centrally. Henceforth, assume $d(G) \geq 1$. Apply the algorithm of Proposition 3 with $r = \log(2/\mu)$, obtaining a residual graph H of maximum degree $\Delta(H) = O(d^{\mu/2} \log^2 n)$, w.h.p. By Turán bound and by assumption,

$$\frac{|V(H)|}{2\Delta(H)} \leq \alpha(H) \leq \alpha(G) \leq \frac{n}{d^{\mu}} . \tag{2}$$

By the degree bound on H, we can bound the number of edges in H by

$$|E(H)| \leq |V(H)|\Delta(H)/2 \leq n \cdot \frac{\Delta(H)^2}{d^{\mu}} = \tilde{O}(n) .$$

We can then compute the MIS (MM) of H in $O(1)$ rounds via Theorem 1.

4.1 Sparsification in Terms of Neighborhood Independence

The goal of this section is to prove Lemma 5. One should think of t as some small poly$(\log n)$.

Lemma 5. *Let $s \geq 1$ be an integer and $X \subseteq V$ be a subset of vertices such that for each $v \in X$, $\deg_{G[X]}(v) \leq s\sqrt{\deg_G(v)}$. Then, $G[X]$ contains $O(\beta(G)n \cdot s^2 \log n)$ edges.*

We also use the following standard result on the size of independent sets in β-bounded neighborhood independence graphs.

Lemma 6. *Let $\delta > 0$ and $A \subseteq V_G$ such that $\deg_G(v) \geq \delta$, for all $v \in A$. Then, $\alpha(G[A]) \leq n\beta(G)/\delta$.*

Proof. Let I be an independent set within A of size $\alpha = \alpha(G[A])$. By assumption, each $v \in I$ has at least δ incident edges (each necessarily incident only on a single node in I). On the other hand, by definition, a vertex can have at most $\beta(G)$ neighbors in I. The lemma follows as there are at most $n\beta(G)$ and at least $\alpha\delta$ edges with an endpoint in I, so $\alpha \leq n\beta(G)/\delta$.

Proof (Proof of Lemma 5). Split the vertices into degree classes as follows: for each $i \in \{0, 1, \ldots, \log n\}$, define: $d_i = 2^i$, and $V_i \triangleq \{v \in V \mid d_i \leq \deg_G(v) < d_{i+1}\}$. Fix $i \in [\log n]$ and consider the subset $X_i \triangleq X \cap V_i$ of X within V_i. We have

$$\frac{|X_i|}{2\,s\sqrt{d_i}} \leq \alpha(G[X_i]) \leq \alpha(G[V_i]) \leq \frac{n\beta(G)}{d_i} .$$

The first inequality follows by Turán's bound, by the assumed degree bound, and the last inequality follows by Lemma 6. Multiply both sides by $(2s\sqrt{d_i})^2$ to get

$$\left|E_{G[X]}(X_i)\right| \leq |X_i|\,(2\,s\sqrt{d_i}) \leq \frac{n\beta(G)}{d_i} \cdot (2\,s\sqrt{d_i})^2 = 4n\beta(G) \cdot s^2 .$$

Therefore, by summing over all degree classes, we get that $G[X]$ contains $O(n\beta(G) \cdot s^2 \log n)$ edges.

4.2 Maximal Independent Set

Theorem 3. *There is a randomized algorithm for MIS in Congested Clique that runs in $O(\log(\log(\beta(G))/\sqrt{\log n}) + 1)$ rounds, w.h.p..*

The uniform degree reduction (Proposition 3) does not run any faster than $\Theta(\log \log n)$ rounds on graphs of high average degree. Hence, we need a different argument for graphs of bounded neighborhood-independence. We show that sampling vertices with probability proportional to the square-root of their degrees yields good sparsification. By running the degree-reduction algorithm twice, we may assume that all degrees are $\tilde{O}(n^{1/4})$.

Algorithm 2. OneShotMISNonUniform

Input: A graph $G = (V, E)$, with n vertices and $\Delta(G) \leq \tilde{O}(n^{1/4})$.
Output: An independent set S

1. Sample a subset of vertices V_F, where each vertex v is sampled independently w.p. $p_v = 1/\sqrt{\deg_G(v)}$. Let a designated vertex w collect the edges of the induced subgraph $F \triangleq G[V_F]$.
2. The vertex w locally computes the LFMIS S on F and informs each vertex whether it belongs to S or not.
3. Apply the MIS algorithm of Theorem 1 on the graph $H \triangleq G[V \setminus \Gamma(S)]$.

Before proving Theorem 3, we introduce a refined version of the MIS sparsification lemma (Lemma 3) to capture non-uniform sampling.

Lemma 7. *Fix some n-vertex graph $G = (V, E)$ and parameters $p_v \in [0, 1)$ for each vertex $v \in V$. Let F be the random set of vertices obtained by picking each* *vertex $v \in V$ independently with probability p_v. Denote by S the independent set obtained by computing the LFMIS in F and let $H = G[V \setminus \Gamma(S)]$ be the residual graph. Let $p \in (0,1)$ and $X = \{u \in V \mid p_u \geq p\}$. Then, for $v \in V(H)$ and $t \geq 1$, the probability that v has at least t neighbors in X that are within H is bounded by*

$$\Pr[\deg_{H \cap X}(v) \geq t] \leq e^{-tp} .$$

Proof. Consider a vertex v. The algorithm can be described as running the following martingale process. Starting with an empty independent set S, iterate over vertices according to a fixed arbitrary order. When we reach a vertex $u \in N(v) \cap X$, if one of its neighbors (earlier in the ordering) already joined S, we do nothing; otherwise, we toss a p_u-coin and include u in the set S if it succeeds (i.e., w.p. p_u). We say a vertex is covered if it is in S or it has a neighbor in S.

For each neighbor $u \in X$ that v has in the residual graph, the following must have happened: "when we reached u, it was not covered and we failed the p_u-coin". Let A_i be the event that this event occurred at least i times. This is equivalent to the event $\deg_{H \cap X}(v) \geq i$ that v has residual degree in X of at least i. By the definition of conditional probability,

$$\Pr[A_{i+1}] = \Pr[A_{i+1} \cap A_i] = \Pr[A_i]\Pr[A_{i+1}|A_i] .$$

Also, $\Pr[A_{i+1}|A_i] \leq 1 - p$, as we only toss coins for nodes u in X, for which $p_u \geq p$. Hence,

$$\Pr[A_t] = \Pr[A_0] \prod_{i=1}^{t-1} \Pr[A_i|A_{i-1}] \leq (1-p)^t \leq e^{-tp} .$$

We can now prove Theorem 3.

Proof (Proof of Theorem 3). We analyze the decrease in degrees after OneShotMISNonUniform. In OneShotMISNonUniform, all vertices in $X = \cup_{j \leq i} V_j$ are sampled with probability at least $1/\sqrt{d_{i+1}}$. Hence, by Lemma 7 with $t = c \log n \cdot \sqrt{d_i}$ for some large enough constant c, we get that every vertex $v \in H$ has at most t neighbors in $H \cap X$, with probability at least $1 - n^{-c}$. Thus, w.h.p, we get $\Delta(H[X]) \leq c \log n \cdot \sqrt{d_i}$.

Next, we explain how to implement OneShotMISNonUniform in $O(1)$ rounds in Congested Clique. For this, we need to show that the induced graph F contains $O(n)$ edges with high probability. This follows from the assumption that $\Delta < n^{1/3}$ and a straightforward application of the method of bounded differences, see e.g., [19, Lemma 3]. By Lemma 5 with $s = O(\log n)$, the residual graph H contains $\tilde{O}(\beta(G)n)$ edges, and thus we can find an MIS on H in $O(\log(\log(\beta(G))/\sqrt{\log n}) + 1)$ rounds by Theorem 1.

Theorem 4. *There exists a randomized Congested Clique algorithm for Maximal Matching that runs in $O(\log(\log(\beta(G))/\sqrt{\log n}) + 1)$ rounds, w.h.p..*

The proof of Theorem 4 is omitted due to space constraints.

References

1. Ahn, K., Cormode, G., Guha, S., McGregor, A., Wirth, A.: Correlation clustering in data streams. In: International Conference on Machine Learning, pp. 2237–2246. PMLR (2015)
2. Assadi, S., Bateni, M., Bernstein, A., Mirrokni, V.S., Stein, C.: Coresets meet EDCS: algorithms for matching and vertex cover on massive graphs. In: Proceedings of the Thirtieth Annual ACM-SIAM Symposium on Discrete Algorithms, SODA 2019, San Diego, California, USA, 6–9 January 2019, pp. 1616–1635. SIAM (2019). https://doi.org/10.1137/1.9781611975482.98
3. Assadi, S., Kol, G., Zhang, Z.: Rounds vs communication tradeoffs for maximal independent sets. In: 63rd IEEE Annual Symposium on Foundations of Computer Science, FOCS 2022, Denver, CO, USA, 31 October–3 November 2022, pp. 1193–1204. IEEE (2022). https://doi.org/10.1109/FOCS54457.2022.00115
4. Assadi, S., Konrad, C., Naidu, K.K., Sundaresan, J.: O(log log n) passes is optimal for semi-streaming maximal independent set. In: Mohar, B., Shinkar, I., O'Donnell, R. (eds.) Proceedings of the 56th Annual ACM Symposium on Theory of Computing, STOC 2024, Vancouver, BC, Canada, 24–28 June 2024, pp. 847–858. ACM (2024). https://doi.org/10.1145/3618260.3649763
5. Assadi, S., Onak, K., Schieber, B., Solomon, S.: Fully dynamic maximal independent set with sublinear in n update time. In: Proceedings of the 50th Annual ACM SIGACT Symposium on Theory of Computing, pp. 815–826 (2018)
6. Assadi, S., Solomon, S.: When algorithms for maximal independent set and maximal matching run in sublinear time. In: 46th International Colloquium on Automata, Languages, and Programming, ICALP 2019, 9–12 July 2019, Patras, Greece. LIPIcs, vol. 132, pp. 17:1–17:17. Schloss Dagstuhl - Leibniz-Zentrum für Informatik (2019). https://doi.org/10.4230/LIPICS.ICALP.2019.17
7. Balliu, A., Brandt, S., Hirvonen, J., Olivetti, D., Rabie, M., Suomela, J.: Lower bounds for maximal matchings and maximal independent sets. J. ACM (JACM) **68**(5), 1–30 (2021)
8. Balliu, A., Brandt, S., Kuhn, F., Olivetti, D.: Distributed δ-coloring plays hide-and-seek. In: Leonardi, S., Gupta, A. (eds.) STOC 2022: 54th Annual ACM SIGACT Symposium on Theory of Computing, Rome, Italy, 20–24 June 2022, pp. 464–477. ACM (2022). https://doi.org/10.1145/3519935.3520027
9. Balliu, A., Brandt, S., Kuhn, F., Olivetti, D.: Distributed maximal matching and maximal independent set on hypergraphs. In: Proceedings of the 2023 Annual ACM-SIAM Symposium on Discrete Algorithms (SODA), pp. 2632–2676. SIAM (2023)
10. Balliu, A., Brandt, S., Olivetti, D.: Distributed lower bounds for ruling sets. SIAM J. Comput. **51**(1), 70–115 (2022). https://doi.org/10.1137/20M1381770
11. Barenboim, L., Elkin, M.: Sublogarithmic distributed MIS algorithm for sparse graphs using nash-williams decomposition. Distributed Comput. **22**(5–6), 363–379 (2010). https://doi.org/10.1007/S00446-009-0088-2
12. Barenboim, L., Elkin, M.: Distributed deterministic edge coloring using bounded neighborhood independence. Distributed Comput. **26**(5–6), 273–287 (2013). https://doi.org/10.1007/S00446-012-0167-7
13. Barenboim, L., Elkin, M., Pettie, S., Schneider, J.: The locality of distributed symmetry breaking. J. ACM (JACM) **63**(3), 1–45 (2016)
14. Barenboim, L., Khazanov, V.: Distributed symmetry-breaking algorithms for congested cliques. In: Fomin, F.V., Podolskii, V.V. (eds.) CSR 2018. LNCS, vol. 10846, pp. 41–52. Springer, Cham (2018). https://doi.org/10.1007/978-3-319-90530-3_5

15. Behnezhad, S., Hajiaghayi, M., Harris, D.G.: Exponentially faster massively parallel maximal matching. J. ACM **70**(5), 1–18 (2023)
16. Berns, A., Hegeman, J., Pemmaraju, S.V.: Super-fast distributed algorithms for metric facility location. In: Czumaj, A., Mehlhorn, K., Pitts, A., Wattenhofer, R. (eds.) ICALP 2012. LNCS, vol. 7392, pp. 428–439. Springer, Heidelberg (2012). https://doi.org/10.1007/978-3-642-31585-5_39
17. Bollobás, B., Erdös, P.: Cliques in random graphs. In: Mathematical Proceedings of the Cambridge Philosophical Society, vol. 80, pp. 419–427. Cambridge University Press (1976)
18. Bui, H.D., Chandra, S., Chang, Y.J., Dory, M., Leitersdorf, D.: Improved all-pairs approximate shortest paths in congested clique. In: Proceedings of the 43rd ACM Symposium on Principles of Distributed Computing (2024)
19. Cambus, M., Kuhn, F., Pai, S., Uitto, J.: Time and space optimal massively parallel algorithm for the 2-ruling set problem. arXiv preprint arXiv:2306.00432 (2023)
20. Censor-Hillel, K., Even, T., Flin, M., Halldórsson, M.M.: When MIS and maximal matching are easy in the congested clique (2025). https://arxiv.org/abs/2502.21031
21. Chang, Y., Fischer, M., Ghaffari, M., Uitto, J., Zheng, Y.: The complexity of $(\Delta+1)$ coloring in congested clique, massively parallel computation, and centralized local computation, pp. 471–480. ACM (2019)
22. Chechik, S., Zhang, T.: Constant-round near-optimal spanners in congested clique. In: Proceedings of the 2022 ACM Symposium on Principles of Distributed Computing, PODC 2022, pp. 325–334. Association for Computing Machinery (2022). https://doi.org/10.1145/3519270.3538439
23. Chudnovsky, M., Seymour, P.D.: The structure of claw-free graphs. In: Surveys in Combinatorics, 2005: Invited lectures from the Twentieth British Combinatorial Conference, Durham, UK, July 2005. London Mathematical Society Lecture Note Series, vol. 327, pp. 153–171. Cambridge University Press (2005)
24. Drucker, A., Kuhn, F., Oshman, R.: On the power of the congested clique model. In: ACM Symposium on Principles of Distributed Computing, PODC 2014, Paris, France, 15–18 July 2014, pp. 367–376. ACM (2014). https://doi.org/10.1145/2611462.2611493
25. Dubhashi, D.P., Panconesi, A.: Concentration of measure for the analysis of randomized algorithms. Cambridge University Press (2009)
26. Faour, S., Ghaffari, M., Grunau, C., Kuhn, F., Rozhon, V.: Local distributed rounding: generalized to MIS, matching, set cover, and beyond. In: Proceedings of the 2023 ACM-SIAM Symposium on Discrete Algorithms, SODA 2023, Florence, Italy, 22–25 January 2023, pp. 4409–4447. SIAM (2023). https://doi.org/10.1137/1.9781611977554.CH168
27. Fischer, M.: Improved deterministic distributed matching via rounding. Distrib. Comput. **33**(3–4), 279–291 (2020)
28. Fischer, M., Mitrovic, S., Uitto, J.: Deterministic $(1+\varepsilon)$-approximate maximum matching with poly$(1/\varepsilon)$ passes in the semi-streaming model and beyond. In: STOC 2022: 54th Annual ACM SIGACT Symposium on Theory of Computing, Rome, Italy, 20–24 June 2022, pp. 248–260. ACM (2022). https://doi.org/10.1145/3519935.3520039
29. Frieze, A.M.: On the independence number of random graphs. Discret. Math. **81**(2), 171–175 (1990)
30. Gfeller, B., Vicari, E.: A randomized distributed algorithm for the maximal independent set problem in growth-bounded graphs. In: Proceedings of the Twenty-Sixth Annual ACM Symposium on Principles of Distributed Computing, PODC

2007, Portland, Oregon, USA, 12–15 August 2007, pp. 53–60. ACM (2007). https://doi.org/10.1145/1281100.1281111
31. Ghaffari, M.: An improved distributed algorithm for maximal independent set. In: Proceedings of the Twenty-Seventh Annual ACM-SIAM Symposium on Discrete Algorithms, SODA 2016. SIAM (2016). https://doi.org/10.1137/1.9781611974331.CH20
32. Ghaffari, M.: Distributed MIS via all-to-all communication. In: Schiller, E.M., Schwarzmann, A.A. (eds.) Proceedings of the ACM Symposium on Principles of Distributed Computing, PODC 2017, Washington, DC, USA, 25–27 July 2017, pp. 141–149. ACM (2017). https://doi.org/10.1145/3087801.3087830
33. Ghaffari, M., Gouleakis, T., Konrad, C., Mitrović, S., Rubinfeld, R.: Improved massively parallel computation algorithms for mis, matching, and vertex cover. arXiv preprint arXiv:1802.08237 (2018)
34. Ghaffari, M., Grunau, C.: Faster deterministic distributed MIS and approximate matching. In: Proceedings of the 55th Annual ACM Symposium on Theory of Computing, STOC 2023, Orlando, FL, USA, 20–23 June 2023, pp. 1777–1790. ACM (2023). https://doi.org/10.1145/3564246.3585243
35. Ghaffari, M., Grunau, C., Haeupler, B., Ilchi, S., Rozhon, V.: Improved distributed network decomposition, hitting sets, and spanners, via derandomization. In: Proceedings of the 2023 ACM-SIAM Symposium on Discrete Algorithms, SODA 2023, Florence, Italy, 22–25 January 2023, pp. 2532–2566. SIAM (2023). https://doi.org/10.1137/1.9781611977554.CH97
36. Ghaffari, M., Grunau, C., Rozhoň, V.: Improved deterministic network decomposition. In: Proceedings of the 2021 ACM-SIAM Symposium on Discrete Algorithms (SODA), pp. 2904–2923. SIAM (2021)
37. Ghaffari, M., Sayyadi, A.: Distributed arboricity-dependent graph coloring via all-to-all communication. In: Baier, C., Chatzigiannakis, I., Flocchini, P., Leonardi, S. (eds.) 46th International Colloquium on Automata, Languages, and Programming, ICALP 2019, 9–12 July 2019, Patras, Greece. LIPIcs, vol. 132, pp. 142:1–142:14. Schloss Dagstuhl - Leibniz-Zentrum für Informatik (2019). https://doi.org/10.4230/LIPIcs.ICALP.2019.142
38. Giliberti, J., Parsaeian, Z.: Massively parallel ruling set made deterministic. In: Alistarh, D. (ed.) 38th International Symposium on Distributed Computing, DISC 2024, October 28 to November 1, 2024, Madrid, Spain. LIPIcs, vol. 319, pp. 29:1–29:21. Schloss Dagstuhl - Leibniz-Zentrum für Informatik (2024). https://doi.org/10.4230/LIPICS.DISC.2024.29
39. Halldórsson, M.M., Konrad, C.: Distributed large independent sets in one round on bounded-independence graphs. In: Moses, Y. (ed.) DISC 2015. LNCS, vol. 9363, pp. 559–572. Springer, Heidelberg (2015). https://doi.org/10.1007/978-3-662-48653-5_37
40. Halldórsson, M.M., Kortsarz, G., Shachnai, H.: Sum coloring interval and k-claw free graphs with application to scheduling dependent jobs. Algorithmica **37**(3), 187–209 (2003). https://doi.org/10.1007/S00453-003-1031-8
41. Hegeman, J.W., Pemmaraju, S.V., Sardeshmukh, V.B.: Near-constant-time distributed algorithms on a congested clique. In: Kuhn, F. (ed.) DISC 2014. LNCS, vol. 8784, pp. 514–530. Springer, Heidelberg (2014). https://doi.org/10.1007/978-3-662-45174-8_35
42. Jurdzinski, T., Nowicki, K.: MST in $O(1)$ rounds of congested clique. In: Czumaj, A. (ed.) Proceedings of the Twenty-Ninth Annual ACM-SIAM Symposium on Discrete Algorithms, SODA 2018, New Orleans, LA, USA, 7–10 January 2018, pp. 2620–2632. SIAM (2018). https://doi.org/10.1137/1.9781611975031.167

43. Konrad, C.: MIS in the congested clique model in o(log log Δ) rounds. CoRR abs/1802.07647 (2018). http://arxiv.org/abs/1802.07647
44. Kuhn, F., Moscibroda, T., Wattenhofer, R.: Local computation: lower and upper bounds. arXiv preprint arXiv:1011.5470 (2010)
45. Kuhn, F., Moscibroda, T., Wattenhofer, R.: Local computation: lower and upper bounds. J. ACM **63**(2), 17:1–17:44 (2016). https://doi.org/10.1145/2742012
46. Kuhn, F., Wattenhofer, R., Zollinger, A.: Ad hoc networks beyond unit disk graphs. Wirel. Networks **14**(5), 715–729 (2008). https://doi.org/10.1007/S11276-007-0045-6
47. Lenzen, C.: Optimal deterministic routing and sorting on the congested clique. In: Proceedings of the 2013 ACM Symposium on Principles of Distributed Computing, pp. 42–50 (2013)
48. Luby, M.: A simple parallel algorithm for the maximal independent set problem. SIAM J. Comput. **15**, 1036–1053 (1986)
49. Milenković, L., Solomon, S.: A unified sparsification approach for matching problems in graphs of bounded neighborhood independence. In: Proceedings of the 32nd ACM Symposium on Parallelism in Algorithms and Architectures, pp. 395–406 (2020)
50. Pai, S., Pemmaraju, S.V.: Brief announcement: deterministic massively parallel algorithms for ruling sets. In: Milani, A., Woelfel, P. (eds.) PODC 2022: ACM Symposium on Principles of Distributed Computing, Salerno, Italy, 25–29 July 2022, pp. 366–368. ACM (2022). https://doi.org/10.1145/3519270.3538472
51. Panconesi, A., Srinivasan, A.: On the complexity of distributed network decomposition. J. Algorithms **20**(2), 581–592 (1995)
52. Rozhon, V., Ghaffari, M.: Polylogarithmic-time deterministic network decomposition and distributed derandomization, pp. 350–363 (2020). https://doi.org/10.1145/3357713.3384298
53. Schneider, J., Wattenhofer, R.: An optimal maximal independent set algorithm for bounded-independence graphs. Distrib. Comput. **22**, 349–361 (2010)

Bankrupting DoS Attackers

Trisha Chakraborty[1], Abir Islam[2], Valerie King[3], Daniel Rayborn[4], Jared Saia[2(✉)], and Maxwell Young[4]

[1] Amazon Web Services, Minneapolis, MN 55401, USA
trichakr@amazon.com
[2] Department of Computer Science, University of New Mexico, Albuquerque, NM 87131, USA
saia@cs.unm.edu
[3] Department of Computer Science, University of Victoria, Victoria, BC V8P 5C2, Canada
val@cs.uvic.ca
[4] Department of Computer Science and Engineering, Mississippi State University, Mississippi State, MS 39762, USA
{dcr101,my325}@msstate.edu

Abstract. Can we make a denial-of-service attacker pay more than the server and honest clients? Consider a model where a server sees a stream of jobs sent by either honest clients or an adversary. The server sets a price for servicing each job with the aid of an *estimator*, which provides approximate statistical information about the distribution of previously occurring good jobs.

We describe and analyze pricing algorithms for the server under different models of synchrony, with total cost parameterized by the accuracy of the estimator. Given a reasonably accurate estimator, the algorithm's cost provably grows more slowly than the attacker's cost, as the attacker's cost grows large. Additionally, we prove a lower bound, showing that our pricing algorithm yields asymptotically tight results when the estimator is accurate within constant factors.

Keywords: Denial-of-service attacks · Resource burning · Distributed algorithms

1 Introduction

Since their first documented occurrence in 1974 [21], denial-of-service(DoS) attacks have evolved into a multi-billion dollar security dilemma [35]. The DoS problem has always been economic: success or failure is decided by who can marshal resources most effectively. In this paper, we provide a novel theoretical formulation and analysis of the DoS problem. We hope this will be a first step towards: (1) leveraging tools from theoretical computer science to solve a critical and persistent security problem; and (2) spurring new theoretical techniques and models; for example, in the same way that the theoretical focus on the paging problem eventually helped create the area of online algorithms.

U. Schmid and R. Kuznets (Eds.): SIROCCO 2025, LNCS 15671, pp. 211–227, 2025.
https://doi.org/10.1007/978-3-031-91736-3_13

It has long been known that fees can be used to curb DoS attacks [17,22,32, 39]. This fee may be in the form of computation: a client solves a computational puzzle or Captcha per request [16,23,27]; or bandwidth: a client sends multiple messages to the server per request [39,40,44]. Fees may be set to 0 in the absence of attack, so that in peaceful times, clients pay nothing.

In this paper, we design algorithms for setting the fee (the "price") in an online manner, assisted by an "estimator", which provides statistical information about the distribution of previously occurring good jobs in the stream. Our goal is to set prices so as to minimize the total algorithm's cost - that is, the cost to the server and honest clients - relative to the attacker's cost. To achieve this, we define an accuracy measure of the estimator, and determine how the algorithm's cost increases with this measure.

Due to space constraints, all proofs and detailed discussion of related work and references are included in [9].

1.1 The Dynamic Job Pricing Problem

Server and Clients. The ***server*** receives a stream of jobs. Each job is either ***good:*** sent by a client - to avoid redundancy, henceforth "clients" rather than "honest clients" - or ***bad:*** sent by the adversary (described below).

We assume a message-passing model of communication. The following happens in parallel for the server and clients.

- **Server:** The server uses the estimator to compute the current price. Jobs are serviced based on whether the attached fee meets this price. When the server receives a job request with an insufficient fee, it sends the current price back to the sender.
- **Clients:** Each client may send a job to the server with a fee.

Adversary. The adversary knows everything about: the distribution of good jobs, the estimator, and our algorithm. At any time, the adversary can send messages to the server with (bad) job requests with fees attached.

Costs. Our algorithmic costs are two-fold: (1) fees paid by clients; and (2) ***service costs*** incurred by the server. We assume a normalized service cost of 1 for servicing any job, whether good or bad. The cost to the adversary is the total cost of all fees paid to the server.

Estimator. The server has an estimator whose output may be used to set prices. The ***estimator*** estimates the number of good jobs generated in any contiguous time interval in the past. For any such interval, $\boldsymbol{\mathcal{I}}$, let $\boldsymbol{g(\mathcal{I})}$ be the number of good jobs in $\mathcal{I}$ and $\boldsymbol{\hat{g}}(\boldsymbol{\mathcal{I}})$ be the value returned by the estimator for interval $\mathcal{I}$. The estimator $\hat{g}$ is additive: for any two non-overlapping intervals, $\mathcal{I}_1$ and $\mathcal{I}_2$:

$$\hat{g}(\mathcal{I}_1 \cup \mathcal{I}_2) = \hat{g}(\mathcal{I}_1) + \hat{g}(\mathcal{I}_2).$$

For a given input, the ***estimation gap***, $\boldsymbol{\gamma}$, is the smallest number greater than or equal to 1, such that the following holds for all intervals, $\mathcal{I}$ in the input:

$$g(\mathcal{I})/\gamma - \gamma \leq \hat{g}(\mathcal{I}) \leq \gamma g(\mathcal{I}) + \gamma.$$

Communication Latency. We have two problem variants:

- **Zero-latency:** All messages are sent instantaneously. So, the clients always know the exact current price; and the fees attached to jobs sent to the server always arrive before any change in the current price, and thus are always serviced.
- **Asynchronous:** Communication is *asynchronous* [29]: there is some unknown maximum latency, $\boldsymbol{\Delta}$ seconds, for any message to be received after it is sent. This latency can also capture the maximum time it took any client to attach a fee. Also, there is some unknown maximum number of good jobs, $\boldsymbol{M}$, that can be generated over Δ seconds. Both Δ and M are known to the adversary, but not to the algorithm. Also, the adversary controls the timing of all messages subject to the above constraints.

(Lack of) Knowledge of the Servers and Clients. Other than the outputs of the estimator, the server and clients have no prior knowledge of the number of good jobs, the amount spent by the adversary, or the estimation gap γ. Recall that, in the latency model, Δ and M are also unknown.

1.2 Model Discussion

Estimator. Our estimator generalizes many DoS detection tools; summarized in detail in [9]. The estimator generalizes three detection techniques. First, for detection tools that **classify jobs:** the estimator can return the number of jobs classified as good in the input interval, perhaps adjusted based on the classifier error rates if these are known. Second, for detection tools that **detect attacks:** when there is no attack detected in the input interval, the estimator can return the number of jobs in the interval; when there is an attack detected, the estimator can return the expected number of good jobs based on past data before the attack. Finally, when good jobs arrive via a **Poisson, Weibull, Gamma, or similar stochastic process:** we describe an estimator, in Sect. 3, with estimation gap that is $O(\log g)$, with probability at least $1 - 1/g^k$ for any positive k, where g is the number of good jobs over the system lifetime.

Prices. Many DoS defenses incorporate prices for service. One such tool is **bandwidth-based pricing** [26,39,40,44]. Every message from a client incurs a resource cost to the client; a bandwidth price for a job is set by requiring the client to send a certain number of messages per job serviced. In the simplest case, the server services each request with some tunable probability. This sets an *expected* bandwidth cost for each job serviced. When this probability changes, it is shared with the clients. Speak-up [39,40] (402 cites, according to Google Scholar) and DOW [44] (112 cites) are popular DoS defenses that primarily use this approach. When these systems detect an attack, they notify clients that the service probability has decreased. If the attack continues, good and bad clients pay similar bandwidth costs, and also have similar rates of service; thus the total algorithmic cost is $\Omega(B)$.

Puzzle-based pricing defenses require clients to solve computational or proof-of-work (PoW) puzzles in order to receive service. Popular DoS defense systems using PoW include: Poseidon [48] (207 cites); Phalanx [14] (120 cites); Portcullis [33] (288 cites); and Puzzle Auctions [41] (304 cites). Other puzzle-based defenses use *Captchas* [37,38]: human-solvable puzzles. Kill-Bots [23] (572 cites) is one such system. In all of these puzzle-based systems, clients pay an amount during an attack that matches the attacker, and thus receive a fair share of job service. However, to the best of our knowledge, none of these defense systems have provable total cost of $o(B)$, where B is the amount spent by the attacker.

Asymptotic Advantage. Many DoS defenses **filter** out packets that are classified as bad [47]. But, even if filtering is 99.9% effective – and reported classification accuracy now hovers around 95% [15,31,46] – filtering alone still requires the defenders to spend $\Omega(B)$. Thus, standalone filtering will never give the defender an asymptotic advantage, although filtering can be combined with our results to reduce costs by constant factors.

Should DoS defenders care about an asymptotic advantage? Resource costs related to DoS attacks are increasingly liquid: attackers can rent bot-farms on the black market [34]; and defenders can provision additional resources from the cloud to service requests [3,45,49]. So, if a defender has access to a DoS defense algorithm with an asymptotic cost advantage, it seems feasible they should be able to outlast an attacker with a similar monetary budget.

1.3 Our Results

For a given input, let $\boldsymbol{g}$ be the number of good jobs, $\boldsymbol{B}$ be the total spent by the adversary, and γ be the estimation gap for the input. Our primary goal is to minimize the algorithmic cost—i.e. fees plus service costs—as a function of g, B, and γ

Our first result is for the algorithm, LINEAR, which assumes zero-latency.

Theorem 1. LINEAR *has total cost*

$$O\left(\gamma^{5/2}\sqrt{B\,(g+1)} + \gamma^3(g+1)\right).$$

Moreover, the server sends $O(g)$ messages to clients, and clients send a total of $O(g)$ messages to the server. Also, each good job is serviced after at most 3 *messages are exchanged.*

To build intuition for this result, consider the case where $\gamma = O(1)$. Then, the theorem states that the cost to LINEAR is $O(\sqrt{B(g+1)}+(g+1))$. Importantly, the cost to the algorithm grows asymptotically slower than B whenever $g = o(B)$. We note that the additive $O(g+1)$ term results from service costs due to good jobs, which must hold for any algorithm. Our result is actually better than stated in Theorem 1, when γ grows large; in that case the total cost is always bounded by $g^2 + B$, as will become clear in Sect. 2.

Finally, the number of messages sent from server to clients and from clients to server, and also the maximum number of messages exchanged to obtain service are all asymptotically optimal. To see this note that at least one message must be sent for each good job serviced.

What if there is latency? For our asynchronous variant, we design and analyze an algorithm LINEAR-POWER. Our main theorem for this case follows. The $\tilde{O}$ notation for the algorithmic cost hides terms that are logarithmic in B and γ.

Theorem 2. LINEAR-POWER *has total cost:*

$$\tilde{O}\left(\gamma^3\left(\sqrt{B+1}\min\left(g+1, M\sqrt{g+1}, M\sqrt{B+1}\right)+(g+1)\right)\right).$$

Moreover, the server sends $O(g\log(B+\gamma))$ *messages to clients, and clients sends a total of* $O(g\log(B+\gamma))$ *messages to the server. Also, each good job is serviced after at most* $4\log(\gamma(B+1))+3$ *messages are sent.*

To gain intuition for this result, assume that γ and M are both $O(1)$. Then the cost to LINEAR-POWER is $\tilde{O}(\sqrt{B+1}\ \min(\sqrt{g+1},\ \sqrt{B+1})+(g+1))$. This equals $\tilde{O}(\sqrt{B(g+1)}+(g+1))$, since when $B=O(g)$, the expression is $\tilde{O}(g+1)$; and otherwise the expression is $\tilde{O}(\sqrt{B(g+1)})$. Thus, when γ and M are constants, the cost of LINEAR-POWER matches that of LINEAR, up to logarithmic terms. For communication, the number of messages sent from the server to clients, clients to the server, and the number of messages exchanged until a job is serviced all increase by a logarithmic factor in B, when compared to LINEAR.

To complement these upper bounds, in Section C in [9], we also prove the following lower bound. This lower bound holds for our zero-latency variant, and so directly also holds for the harder, asynchronous variant. Let $\boldsymbol{n}$ be the total number of jobs.

Theorem 3. *Let* γ *be any positive integer;* g_0 *be any positive multiple of* γ*; and* n *any multiple of* γg_0*. Next, fix any randomized algorithm. Then, there is an input with* n *jobs,* g *of which are good for* $g\in\{\gamma g_0, g_0/\gamma\}$*; an estimator with gap* γ *on that input; and the randomized algorithm on that input has expected cost:*

$$\Omega\left(\sqrt{\gamma B(g+1)}+\gamma(g+1)\right).$$

When $\gamma=O(1)$, this lower bound asymptotically matches the cost for LINEAR, and is within logarithmic factors of the cost for LINEAR-POWER. Additionally, the lower bound allows for a fair amount of independence between the settings of γ, g and n for which the lower bound can be proven. For example, it holds both when g is small compared to n and also when g is large. Finally, note that the lower bound holds for randomized algorithms, even though both LINEAR and LINEAR-POWER are deterministic. Thus, randomization does not offer significant improvements for our problem.

Finally, we provide simulation results for LINEAR and LINEAR-POWER in our appendix. Our results suggest that the hidden constants in the asymptotic notation may be small, and that the scaling behavior for our algorithms' costs matches that predicted by theory.

1.4 Technical Overview

All our proofs are included in the appendix. Here, we give intuition for our technical results. As a warm up, we start with our easiest result: the zero-latency case. Then, we sketch how we build on this result in our asynchronous setting. Finally, we give intuition for our lower bound.

Zero-Latency. Our algorithm for the zero-latency problem is LINEAR. Here, we give a high-level overview of the algorithm and its analysis, see Sect. 2.1 for detailed pseudocode and Section A of [9] for a formal analysis.

LINEAR is conceptually simple. Time is partitioned into ***iterations***, which end whenever the estimator returns a value of at least 1 for the estimate of the number of good jobs seen in the iteration so far. The current price is set to s, where s is the number of jobs seen so far in the iteration.

Our analysis of LINEAR involves two key steps: (1) establishing upper and lower bounds on the number of iterations as a function of γ and g (Lemmas 1 and 3 of [9]); and (2) deriving an upper bound on the number of good jobs in any iteration (Lemma 2 [9]).

Next, using (1) and (2), we prove an upper bound on the cost of LINEAR. To build intuition, we consider a general class of current price setting algorithms where the charge to the s-th job in the iteration is s^α, for any constant $\alpha \geq 0$. We prove that $\alpha = 1$ gives the best algorithmic cost as a function of B, g and γ.

Our upper bound consists of four exchange arguments [24], which will be over the ordering of good and bad job requests. Our first two exchange arguments lower bound B and the next two upper bound the algorithm cost, $\boldsymbol{A}$. We begin by lower bounding B. To do so, our first exchange argument shows that B is minimized when, in each iteration, the bad jobs occur before the good. Our second exchange argument shows that, subject to this constraint, B is minimized when the bad jobs are distributed as evenly as possible across the iterations. From these two exchange arguments, a lower bound on B follows via an integral lower bound and basic algebra (see Lemma 4 [9]).

We next upper bound A. To do so, we make two observations: (1) within an iteration, A is maximized by putting all good jobs at the end; and (2) without loss of generality, we can assume that the iterations are sorted in decreasing order by the number of bad jobs they contain. Then, we use a third (simple) exchange argument to show that A is maximized when good jobs are "packed left": packed as much as possible in the earlier iterations, which have the higher number of bad jobs.

Finally, we use the fourth (harder) exchange argument to show that A is maximized when *bad* jobs are "packed left" when $\alpha \geq 1$, and are evenly spread when $\alpha < 1$. To show this, for an iteration i, we let $f(b_i)$ be the cost of the i-th iteration when there are b_i bad jobs at the start. Then, let $\delta(b_i) = f(b_i+1) - f(b_i)$ be the change in the cost of iteration i when we add 1 bad job. Then the change in cost when there is an exchange of one bad job from iteration k to iteration i for $k > i$ will be $\delta(b_i) - \delta(b_k - 1)$.

How do we show that $\delta(b_i) - \delta(b_k - 1)) \geq 0$? If we relax b_i so it can take on values in the real numbers, then $f(b_i)$ becomes a continuous, differentiable function. Thus, we can use the mean value theorem (MVT). In particular, the MVT shows that $\delta(b_i)$ equals $f'(x)$ at some value of $x \in [b_i, b_{i+1}]$. Then, upper and lower bounding the δ values reduces to bounding f' in the appropriate range. This is a simpler problem because f' is a monotonically non-decreasing (non-increasing) function when $\alpha \geq 1$ ($\alpha < 1$, respectively). Once we have this exchange argument in hand, we can calculate the algorithmic costs (see the proofs of Lemma 5 and Theorem 1 in [9] for details).

Asynchronous. If we naively use LINEAR for the asynchronous model, costs can be high. For simplicity, assume that $\gamma = \Theta(1)$ and $g = \Theta(M\sqrt{B})$. Then, the adversary can insert $\sqrt{B}$ bad jobs initially, causing the current price in LINEAR to increase $\sqrt{B}$ times. During this time $\Theta\left(\min\left(g, M\sqrt{B}\right)\right)$ good jobs join the system, all of them are returned without service repeatedly, as the current price increases, until the current price reaches its maximum value of $\Theta(\sqrt{B})$. Next, all $\Theta\left(\min(g, M\sqrt{B})\right)$ of these good jobs wind up paying a fee equal to $\Theta(\sqrt{B})$. Adding in the cost to service jobs and the cost for any additional good jobs that enter after the bad, we have a total cost for the LINEAR of $\Theta\left(\min\left(g+1, M\sqrt{B+1}\right)\sqrt{B+1} + g\right)$.

How can we improve this? One key idea is to reduce the number of times that the server increases the current price. Our second algorithm, LINEAR-POWER does just that: (1) the server sets the current price to the largest power of 2 less than the total number of jobs in the current iteration; and (2) clients double their fee whenever their job request is returned.

To compare LINEAR-POWER with LINEAR, suppose $\gamma = O(1)$. Then, it is not hard to see that in any iteration i where the adversary spends B_i, the maximum current price is $O(\sqrt{B_i})$ (see Lemma 8 in [9]). Thus, we can argue that any job is returned $O(\log B)$ times, and so pays a total of $O(\sqrt{B})$.

But the above argument gives a poor bound on fees paid by the clients, since it multiplies all such fees from the LINEAR analysis by a $\sqrt{B}$ factor. To do better, we need a tighter analysis for fees. But how do we proceed when a single job can be refused service over many iterations because it keeps learning the current price too late?

The first key idea is to partition time into epochs, for the purpose of analysis only. An ***epoch*** is a period of time ending with 2Δ seconds during which the current price never increases. Then, several facts follow. First, every good job is serviced within 2 epochs. Second, we can bound the total fees that all good jobs pay during epoch i to be essentially $O(M(\sqrt{B_{i-1}} + \sqrt{B_i}))$, where B_j is the amount spent by the adversary in epoch j (see Lemma 11 in [9]). Third, an epoch has overpaying jobs only if it or its preceding epoch spans the beginning of an iteration (see Lemma 10 in [9]). This last fact allows us to bound the total number of epochs that add to the overpayment amount using our previous bounds on the number of iterations. Putting all these facts together, we obtain a total cost for LINEAR-POWER that is only $\tilde{O}\big(\min\left(g+1, M\sqrt{g+1}, M\sqrt{B+1}\right)\sqrt{B+1} + g\big)$ (see the proof of Theorem 2 in [9]) versus the cost of LINEAR computed above of $\Theta(\min(g+1, M\sqrt{B+1})\sqrt{B+1} + g)$. Ignoring logarithmic factors, the cost of LINEAR-POWER may be less than LINEAR by a factor of $\sqrt{g+1}/M$. As an example, when $M = O(1)$ and $B = \Theta(g)$, LINEAR-POWER has cost $\tilde{O}(g)$, and LINEAR has cost $\Theta(g^{3/2})$.

Lower Bound. We prove a lower bound for the zero-latency case (Section C in [9]). This bound directly holds for our harder, asynchronous problem variant.

In our lower bound, for given values of γ, g_0 and n, we design an input distribution that (1) sets g to g_0/γ with probability 1/2, and sets g to γg_0 otherwise; (2) distributes g good jobs uniformly but randomly throughout the input; and (3) creates an estimator that effectively hides both the value of g and also the location of the good jobs, and always has estimation gap at most γ on the input (see Section C.1 and Lemma 16 in [9] for details).

The key technical challenge is to set up a matrix in order to use Yao's minimax principle [43] to prove a lower bound on the expected cost for any randomized algorithm. In particular, we need a game-theory payoff matrix where the rows are deterministic algorithms, and the columns are deterministic inputs. A natural approach is for the matrix entries to be the ratios $A/\sqrt{Bg}$, where A is the cost to our algorithm. Unfortunately, this fails to achieve our goal since this would yield $E[A/\sqrt{Bg}]$ and we need $E[A]/E[\sqrt{Bg}]$, which are not generally equal. Some lower bound proofs can circumvent this type of problem. For example, lower bound proofs on competitive ratios of online algorithms circumvent the problem because the offline cost does not vary with the choice of the deterministic algorithm, i.e. from row to row. So the expected offline cost can be factored out and computed separately (see [4], Chap. 5). Unfortunately, this does not work for us since our random variables A and B are interdependent, and vary across both rows and columns.

To address this challenge, we first set the matrix entries to $E(A)-\sqrt{E(B)g\gamma})$, and then prove that the minimax for such a payoff matrix has expected value at least 0. Then, we use Jensen's inequality and linearity of expectation to show that the minimax is also at least 0 when the matrix entries are $E(A - \sqrt{Bg\gamma})$ (see Lemma 17 in [9]). Next, using Yao's principle, we can show the maximin is nonnegative, and thereby obtain a key inequality for A and B. Once we have this inequality, we can use algebra to establish our lower bound for any randomized algorithm (see Theorem 3 of [9]).

We believe that this new approach – using Yao's principle on a matrix whose entries are $X - Y$ in order to bound the ratio $E(X)/E(Y)$, and thereby prove some target lower bound of $E(Y)$ – may have broader applications to proving lower bounds for resource competitive algorithms (see Sect. 1.6 in [9]).

2 Our Algorithms

In Sect. 2.1, we formally define the algorithm LINEAR; and in Sect. 2.2, we define the algorithm LINEAR-POWER. As described in Sect. 1.1, both of our algorithms make use of an estimator. So, in Sect. 3, we give an example of the type of estimator that might be used by these algorithms. Next, in Sect. 3.4, to build intuition, we give some example scenarios for our algorithms.

> **LINEAR**
>
> **Server:**
> Repeat forever:
>
> 1. If $\hat{g}(\mathcal{I}) \geq 1$, where $\mathcal{I} = (t, t']$, t is the start time of the current iteration, and t' is the current time, then begin a new iteration.
> 2. PRICE $\leftarrow s + 1$, where s is the number of jobs serviced so far in this iteration. If PRICE has changed, send the new value to all clients.
> 3. If the next job has a attached fee at least equal to PRICE then service it, otherwise do not service and reply with PRICE.
>
> **Client:** $x \leftarrow 1$.
> Until the job is serviced, do:
>
> 1. Send the server a job request with fee x.
> 2. $x \leftarrow$ most-recent price received from the server.

Fig. 1. Pseudocode for LINEAR.

2.1 Algorithm LINEAR

Our first algorithm, LINEAR, addresses our zero-latency problem variant. The server partitions the servicing of jobs into ***iterations***, which end when $\hat{g}(\mathcal{I}) \geq 1$, where $\mathcal{I}$ is the time interval in the current iteration. The current price, ***PRICE***, for the next job is set to $s+1$, where s is the number of jobs serviced so far in the iteration. Since there is no latency, good jobs will always send in a fee with value equal to PRICE, thus jobs with lower value solutions can be ignored. Pseudocode is given in Fig. 1. Again, because there is no latency, the client's action is just to either send in the job with attached fee equal to the current value of PRICE, or else drop the job. Every message is presumed to be received at a unique point in time; if not, the server can break ties arbitrarily.

2.2 Algorithm LINEAR-POWER

We next describe a second algorithm, LINEAR-POWER, that addresses the asynchronous variant of our problem; see Fig. 2 for the pseudocode.

Server. The server operates in ***iterations***, which end when $\hat{g}(\mathcal{I}) \geq 1$, where $\mathcal{I}$ is the time interval elapsed in the iteration.

In each iteration, the current price, PRICE, is set to $2^{\lfloor \log(s+1) \rfloor}$, where s is the number of jobs serviced so far in the iteration. If the fee attached to the next job is at least PRICE, the job is serviced. Otherwise, the job is not serviced, and the server sends a message with the value PRICE to the client that sent the job.

Client. Initially, each client sets a local x variable to 1. The client sends the server a fee of x with its job. When receiving a message from the server with a

LINEAR-POWER

Server:
Repeat forever:

1. If $\hat{g}(\mathcal{I}) \geq 1$, where $\mathcal{I} = (t, t']$, t is the start time of the current iteration, and t' is the current time, then begin a new iteration.
2. PRICE $\leftarrow 2^{\lfloor \lg(s+1) \rfloor}$, where s is the number of jobs serviced in this iteration.
3. If the next job has a fee at least equal to PRICE, then service it. Otherwise, do not service the job; instead send a message with the value PRICE to the client that sent the job.

Client: $x \leftarrow 1$.
Until the job is serviced, do:

1. Send the server a fee of x with the job.
2. $x \leftarrow$ maximum price ever received from the server.

Fig. 2. Pseudocode for LINEAR-POWER.

thresh value greater than x, it sets x to the value received, and re-sends the job with a fee of x, or chooses to drop the job.

3 An Estimator with Broad Application

Our main goal is to design algorithms that harness the power of estimators, rather than to create the estimators themselves. However, to convey the broad applicability of our results, we consider various distributions for the arrival of good jobs that correspond with real-world observations and established traffic models. We describe an example estimator, with the properties defined in Sect. 1.1, which we show has small γ for many types of distributions. We then illustrate executions of both LINEAR and LINEAR-POWER using this estimator.

3.1 Poisson Estimator

Assume good jobs are generated via a Poisson distribution with parameter λ, so that the expected interval length between good jobs is $1/\lambda$. Poisson job arrivals are one of the most commonly used statistical models, and are consistent with empirical findings [7,28].

For simplicity, we assume exact knowledge of the Poisson parameter λ, but a constant factor approximation suffices for the analysis below. For an interval $\mathcal{I}$, let $\ell(\mathcal{I})$ be the length of $\mathcal{I}$. Then, for any interval $\mathcal{I}$, set

$$\hat{g}(\mathcal{I}) = \ell(\mathcal{I})/\lambda$$

Bounding γ. We now show that w.h.p. in g, i.e. probability at least $1 - 1/g^k$ for any positive k, $\gamma = O(\log g)$ for the Poisson estimator. This holds for *all* distributions of bad jobs including: no bad jobs; more bad jobs than good; bad jobs uniformly distributed; and bad jobs that are clustered in any manner.

As a special case, consider when the adversary inserts bad jobs using the same Poisson distribution as the good jobs. Then, it is impossible to differentiate a good job from a bad job with better than a (fair) coin toss. Thus, the estimation gap remains $O(\log g)$ even though we can not classify jobs as good or bad. So, estimation can be easier than classification.

Analysis. To formally bound γ, consider any input distribution with g good jobs generated by the Poisson process. We show that w.h.p. in g, $\hat{g}$ ensures that $\gamma = O(\log g)$.

Recall that $g(\mathcal{I})$ is the number of good jobs generated during $\mathcal{I}$. We rely on the following fact which holds by properties of the Poisson process, Chernoff and union bounds. In the following w.h.p. means with probability of error that is polynomially small in g.

Fact 1. *For some constant C, w.h.p. for any interval $\mathcal{I}$:*

1. *If $g(\mathcal{I}) < C \ln g$, then $\ell(\mathcal{I}) = O((1/\lambda) \log g)$*
2. *If $g(\mathcal{I}) \geq C \ln g$, then $\ell(\mathcal{I}) = \Theta((1/\lambda) g(\mathcal{I}))$.*

We can now show the following simple fact.

Fact 2. *With high probability, the estimator $\hat{g}$ has estimation gap $\gamma = O(\log g)$.*

Proof. Consider two cases:

Case: $g(\mathcal{I}) < C \ln g$. Then, by Fact 1 (1), $\ell(\mathcal{I}) = O((1/\lambda) \log g)$, and so $\hat{g}(\mathcal{I}) = O(\log g)$. Thus, there exists some γ, $\gamma = O(\log g)$ such that

$$g(\mathcal{I}) - \gamma \leq \hat{g}(\mathcal{I}) \leq g(\mathcal{I}) + \gamma$$

Hence, the estimation gap on this type of interval is $O(\log g)$.

Case: $g(\mathcal{I}) \geq C \ln g$. Then by Fact 1 (2), w.h.p., $\ell(\mathcal{I}) = \Theta((1/\lambda) g(\mathcal{I}))$, and so $\hat{g}(\mathcal{I}) = \Theta(g(\mathcal{I}))$. So, there exists some γ, $\gamma = O(1)$ such that

$$(1/\gamma) g(\mathcal{I}) \leq \hat{g}(\mathcal{I}) \leq \gamma g(\mathcal{I})$$

Hence, the estimation gap on this type of interval is $O(1)$. We conclude that for this example input distribution, w.h.p., $\hat{g}$ is an estimator with $\gamma = O(\log g)$.

Estimating λ. A simple way to estimate λ is to use job arrival data from a similar day and time. To handle possible spikes in usage by clients, one may need to analyze data on current jobs, including: packet header information [8], geographic client location [18], timing of requests [10], and spikes in traffic rate [25].

3.2 An Estimator for Weibull, Gamma, and Distributions with Exponentially Bounded Moments

The $\gamma = O(\log g)$ result for the Poisson distribution can be generalized to arrival times given by other distributions whose moments do not grow super-exponentially.

In particular, for any $i \geq 1$, let X_i be a random variable giving the time elapsed between the $i-1$ and i-th arrival. Assume that the X_i values are independent and identically distributed. Then, let $Y_i = X_i - E(X_i)$, and assume that for some real positive L and every positive integer k,

$$E(|Y_i^k|) \leq \frac{1}{2}E(Y_i^2)L^{k-2}k!$$

In other words, the above says that the k-th moment of Y_i is bounded by an exponential in k. This is true for common arrival time processes such as the Weibull [6] and the gamma distributions [12] with fixed parameters.

Then set $\rho = 1/E(X_i)$, and we can use an estimator that for any interval $\mathcal{I}$, sets

$$\hat{g}(\mathcal{I}) = \rho\ell(\mathcal{I})$$

We can again show the following.

Fact 3. *With high probability, the estimator $\hat{g}$ has estimation gap $\gamma = O(\log g)$.*

To show this, we use Bernstein's inequality (specifically, the second inequality in [42]; see also [36]) to prove a result analogous to Fact 1, but with $1/\lambda$ replaced by μ. Then, the result follows directly.

3.3 Adversarial Queuing Theory Estimator

To capture a range of network settings, adversarial queuing theory (AQT) [5] parameterizes the behavior of packet arrivals by an injection rate, ρ where $0 < \rho \leq 1$, and a non-negative integer, b, that denotes burstiness. In AQT, a common arrival model is the ***leaky-bucket adversary*** [13] where, for any t consecutive slots, at most $\rho t + b$ packets may arrive, for some number b (for examples, see [1,2,11,19,20]). The physical metaphor is that there is a bucket of size b gallons. Water enters the bucket according to the (adversarial) arrival process and it leaves at a constant rate of ρ. The model requires that b be large enough so that the bucket never overflows.

We define a new variant, the ***rain-barrel model***, that both lower bounds and upper bounds the number of arriving good jobs. In particular, we require that: for any interval of t seconds, between $(1/b)\rho t - b$ and $b\rho t + b$ good jobs arrive. The physical metaphor is that there is a barrel of size b and an additional reservoir of b gallons that can be added to the barrel at any time. Water enters the barrel according to the (adversarial) arrival process and leaves at a tunable rate in the range $\rho \cdot [1/b, b]$. The new requirement is that there is always water available, but that the barrel never overflows.

In such a model, with knowledge of ρ, it is easy to design an estimator with estimation gap $\gamma = b$. In particular, again set $\hat{g}(\mathcal{I}) = \rho\,\ell(\mathcal{I})$. By the definition of the bounded bucket model, this estimator has estimation gap $\gamma = b$.

Notably, the value b need not be known a priori. Further, as with the previous estimators, using a constant factor approximation to ρ will not change the asymptotic value of γ.

3.4 Zero Latency and Asynchronous Example Runs

We now illustrate the execution of our zero-latency and asynchronous algorithms under Poisson arrivals, using our first estimator (Sect. 3.1); results are similar for our other two estimators.

Zero-Latency. First, suppose the estimator is perfectly accurate. Then, $\hat{g}(\mathcal{I})$ is 0 during an iteration until a good job arrives. Once a good job arrives, $\hat{g}(\mathcal{I})$ increases from 0 to 1, and a new iteration begins, starting with this new good job. The value PRICE is set back to 1, and the good job pays a fee of 1. Then, the cost of each good job is 1 and the cost of each bad job is the number of bad jobs received since the last good job was received. There are exactly g iterations. If b bad jobs are submitted, the adversary's costs are minimized when the bad jobs are evenly spread over the iterations, so the cost for each bad job ranges from 1 to b/g in each of the g iterations, for a total cost of $B = \Theta(b^2/g)$.

Now, consider the special case where the arrival of good jobs is a Poisson process with parameter λ, and we use the example estimator from Sect. 3.1. Figure 3 illustrates this scenario. Here, iterations are demarcated by multiples of $1/\lambda$ and PRICE is reset to 1 at the end of each iteration. Many bad jobs may be present over these iterations, and there may be more than 1 good job per iteration (as depicted).

The lengths of time between good jobs are independently distributed, exponential random variables with parameter λ (see [30], Chap. 8) that is, the probability that j good jobs occur in time t is $e^{(-t\lambda)}\frac{(t\lambda)^j}{j!}$. The adversary can decide where to place bad jobs. We now consider what happens with LINEAR. When using the example estimator from Sect. 3.1, for all $i \geq 1$, the i-th iteration ends at time i/λ, and the current price is reset to 1.

1. If there are no bad jobs, the expected cost of the i-th iteration to the algorithm is $O(X_i^2)$ where X_i is the number of good jobs in a period of length $1/\lambda$, and $\sum_i X_i = g$. Since $E[X_i^2] = \sum_j j^2 e^{(-\lambda/\lambda)}\frac{(\lambda/\lambda)^j}{j!}$, the expected cost to the algorithm for any iteration i is $(1/e)\sum_j j^2(1/j!) = O(1)$.
2. Fix any iteration i. By paying $\Omega(b_i^2)$, the adversary can post b_i bad jobs early on, thereby forcing each good job to pay a fee of at least $b_i + 1$. The algorithm also pays a service cost of $b_i + 1$.

Asynchronous. With communication latency between the server and the clients, jobs may submit multiple fees over time. If the submitted solution hardness is less than the current price, the job request is denied. When this happens,

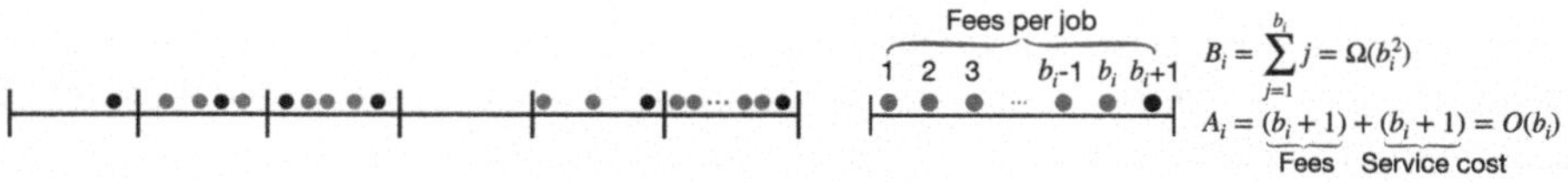

Fig. 3. (Left) LINEAR with iterations of length $1/\lambda$, using the estimator from the zero-latency example in Sect. 3.4. Black circles are good jobs, red circles are bad. Since the estimation gap is small, each iteration contains about 1 good job. (Right) Blowup of the last iteration, with b_i bad jobs followed by 1 good job. The adversary's cost is $B_i = \Omega(b_i^2)$ and the algorithm's cost is $A_i = O(b_i)$, as described in Sect. 3.4.

we say that the corresponding job is ***bounced*** (see Fig. 4). Alternatively, if the fee exceeds the price, then the job overpays.

Now, assume our example estimator, so that iterations end at the same time as the zero-latency case above. Further, assume a simple distribution for jobs where exactly 1 job is generated per iteration at some uniformly distributed time.

Note that M is determined by the maximum latency Δ, and λ. If the latency Δ is much larger than $1/\lambda$, then the adversary can delay messages so that $M = \Delta/\lambda$ good jobs, bounced from the previous Δ/λ iterations, are received in the current iteration. The first $M/2$ of these jobs are serviced, driving up the value of PRICE to at least $M/2$. During this time, the remaining jobs are bounced up to $\log(M/2)$ times. Thus, the algorithm's total cost is $\Omega((g/M)(M^2)) = \Omega(gM)$.

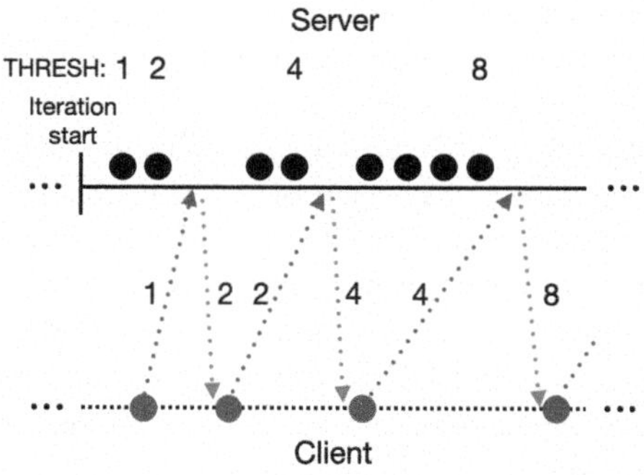

Fig. 4. The blue circle is a job that is repeatedly bounced, the black circles are other serviced jobs. Arrows are communication from the server to the client of the PRICE value, and from the client to the server of fees. The blue job is bounced 3 times and pays $1+2+4+8$. (Color figure online)

For these jobs $i = 2, ..., (M-1)$, the adversary can cause the i-th job to bounce $\lfloor \lg(M-i) \rfloor$ times with payments of 2^j, for $j = 0, ..., \lfloor \lg(M-i) \rfloor$ in the current iteration. Thus, the total cost to the algorithm is $\Omega((g/M)(M^2)) = \Omega(gM)$.

4 Conclusion and Future Work

We have provided two deterministic algorithms for DoS defense. Our algorithms dynamically adjust the server's prices based on feedback from an estimator which estimates the number of good jobs in any previously observed time interval. The accuracy of the estimator on the input, the number and temporal placement of the good and bad jobs, and the adversary's total spending are all unknown to the algorithm, but these all affect the algorithm's cost.

Critically, during a significant attack the cost to our algorithm is *asymptotically* less than the cost to the adversary. We believe this is a important property for deterring DoS attacks. We have also given a lower bound for randomized algorithms, showing that our algorithmic costs are asymptotically tight, whenever γ is a constant. We note that, while our algorithms are deterministic, our lower bound holds even for randomized algorithms.

Many interesting problems remain. Are there less restrictive properties for the estimator which suffice to achieve a good pricing strategy? Can we model both the clients and the server as selfish but rational agents? In particular, can we design a mechanism that results in a Nash equilibrium, and also provides good cost performance as a function of the adversarial cost?

Acknowledgements.. This material is based upon work supported by the National Science Foundation under grant numbers CNS-2210299 and CNS-2210300.

References

1. Anantharamu, L., Chlebus, B.S.: Broadcasting in ad hoc multiple access channels. Theoret. Comput. Sci. **584**, 155–176 (2015)
2. Anantharamu, L., Chlebus, B.S., Kowalski, D.R., Rokicki, M.A.: Deterministic broadcast on multiple access channels. In: IEEE INFOCOM, pp. 1–5. IEEE (2010)
3. AWS: AWS best practices for DDoS resiliency (2019). https://docs.aws.amazon.com/whitepapers/latest/aws-best-practices-ddos-resiliency/aws-best-practices-ddos-resiliency.pdf
4. Borodin, A., El-Yaniv, R.: Online Computation and Competitive Analysis. Cambridge University Press, Cambridge (2005)
5. Borodin, A., Kleinberg, J., Raghavan, P., Sudan, M., Williamson, D.P.: Adversarial queuing theory. J. ACM (JACM) **48**(1), 13–38 (2001)
6. Cao, J., Cleveland, W.S., Gao, Y., Jeffay, K., Smith, F.D., Weigle, M.: Stochastic models for generating synthetic HTTP source traffic. In: IEEE INFOCOM, vol. 3, pp. 1546–1557. IEEE (2004)
7. Cao, J., Cleveland, W.S., Lin, D., Sun, D.X.: Internet traffic tends toward Poisson and independent as the load increases. Nonlinear Estimation Classif. 83–109 (2003)
8. Carl, G., Kesidis, G., Brooks, R.R., Rai, S.: Denial-of-service attack-detection techniques. IEEE Internet Comput. **10**(1), 82–89 (2006)
9. Chakraborty, T., Islam, A., King, V., Rayborn, D., Saia, J., Young, M.: Bankrupting dos attackers despite uncertainty. arXiv preprint arXiv:2205.08287 (2022)
10. Chakraborty, T., Mitra, S., Mittal, S.: CAPoW: context-aware AI-assisted proof of work based DDoS defense. In: 20th International Conference on Security and Cryptography (SECRYPT), pp. 62–72 (2023)
11. Chlebus, B.S., Kowalski, D.R., Rokicki, M.A.: Maximum throughput of multiple access channels in adversarial environments. Distrib. Comput. **22**(2), 93–116 (2009)
12. Choi, H.K., Limb, J.O.: A behavioral model of web traffic. In: Seventh International Conference on Network Protocols, pp. 327–334. IEEE (1999)
13. Cruz, R.: A calculus for network delay. I. Network elements in isolation. IEEE Trans. Inf. Theory **37**(1), 114–131 (1991)

14. Dixon, C., Anderson, T.E., Krishnamurthy, A.: Phalanx: withstanding multimillion-node botnets. In: USENIX Symposium on Networked Systems Design and Implementation (NSDI), vol. 8, pp. 45–58 (2008)
15. Doriguzzi-Corin, R., Millar, S., Scott-Hayward, S., Martinez-del Rincon, J., Siracusa, D.: LUCID: a practical, lightweight deep learning solution for DDoS attack detection. IEEE Trans. Netw. Serv. Manage. **17**(2), 876–889 (2020)
16. Dwork, C., Naor, M.: Pricing via processing or combatting junk mail. In: Annual International Cryptology Conference, pp. 139–147. Springer (1992)
17. Dwork, C., Naor, M.: Pricing via processing or combatting junk mail. In: 12th Annual International Cryptology Conference on Advances in Cryptology, pp. 139–147 (1993)
18. Feng, W.C., Kaiser, E.: kaPoW webmail: effective disincentives against spam. In: Seventh Annual Collaboration, Electronic Messaging, Anti-abuse and Spam Conference (CEAS) (2010)
19. Garncarek, P., Jurdzinski, T., Kowalski, D.R.: Local queuing under contention. In: 32nd International Symposium on Distributed Computing (DISC), vol. 121, pp. 28:1–28:18 (2018)
20. Garncarek, P., Jurdziński, T., Kowalski, D.R.: Stable memoryless queuing under contention. In: 33rd International Symposium on Distributed Computing (DISC), vol. 146, pp. 17:1–17:16 (2019)
21. Inc., R.: DDoS attacks history (2017). https://www.radware.com/security/ddos-knowledge-center/ddos-chronicles/ddos-attacks-history
22. Juels, A., Brainard, J.: Client puzzles: a cryptographic countermeasure against connection depletion attacks. In: Network and Distributed System Security Symposium (NDSS), pp. 151–165 (1999)
23. Kandula, S., Katabi, D., Jacob, M., Berger, A.: Botz-4-sale: surviving organized DDoS attacks that mimic flash crowds. In: 2nd Conference on Symposium on Networked Systems Design & Implementation, vol. 2, pp. 287–300 (2005)
24. Kleinberg, J., Tardos, E.: Algorithm Design (Chapter 4). Pearson (2006)
25. Lakhina, A., Crovella, M., Diot, C.: Diagnosing network-wide traffic anomalies. ACM SIGCOMM Comput. Commun. Rev. **34**(4), 219–230 (2004)
26. Li, Q., et al.: A comprehensive survey on DDoS defense systems: new trends and challenges. Comput. Netw. 109895 (2023)
27. Liu, D., Camp, L.J.: Proof of work can work. In: 5th Workshop on the Economics of Information Security (WEIS) (2006)
28. Liu, Z., Niclausse, N., Jalpa-Villanueva, C.: Traffic model and performance evaluation of web servers. Perform. Eval. **46**(2–3), 77–100 (2001)
29. Lynch, N.A.: Distributed Algorithms. Elsevier, Amsterdam (1996)
30. Mitzenmacher, M., Upfal, E.: Probability and Computing: Randomization and Probabilistic Techniques in Algorithms and Data Analysis. Cambridge University Press, Cambridge (2017)
31. Ndibwile, J.D., Govardhan, A., Okada, K., Kadobayashi, Y.: Web server protection against application layer DDoS attacks using machine learning and traffic authentication. In: 39th IEEE Annual Computer Software and Applications Conference, pp. 261–267 (2015)
32. Noureddine, M.A., et al.: Revisiting client puzzles for state exhaustion attacks resilience. In: 49-th Annual IEEE/IFIP International Conference on Dependable Systems and Networks (DSN), pp. 617–629 (2019)
33. Parno, B., Wendlandt, D., Shi, E., Perrig, A., Maggs, B., Hu, Y.C.: Portcullis: protecting connection setup from denial-of-capability attacks. ACM SIGCOMM Comput. Commun. Rev. **37**(4), 289–300 (2007)

34. Somani, G., Gaur, M.S., Sanghi, D., Conti, M., Rajarajan, M., Buyya, R.: Combating DDoS attacks in the cloud: requirements, trends, and future directions. IEEE Cloud Comput. **4**(1), 22–32 (2017)
35. TechHQ: Technology and business: DDoS attacks cost US businesses 10bn per year (2019). https://techhq.com/2019/03/ddos-attacks-cost-us-businesses-10bn-per-year
36. Uspensky, J.V.: Introduction to Mathematical Probability. McCraw-Hill, New York (1937)
37. Von Ahn, L., Blum, M., Hopper, N.J., Langford, J.: CAPTCHA: using hard AI problems for security. In: International Conference on the Theory and Applications of Cryptographic Techniques, pp. 294–311. Springer (2003)
38. Von Ahn, L., Maurer, B., McMillen, C., Abraham, D., Blum, M.: reCAPTCHA: human-based character recognition via web security measures. Science **321**(5895), 1465–1468 (2008)
39. Walfish, M., Vutukuru, M., Balakrishnan, H., Karger, D., Shenker, S.: DDoS defense by offense. In: Conference on Applications, Technologies, Architectures, and Protocols for Computer Communications (SIGCOMM), pp. 303–314 (2006)
40. Walfish, M., Vutukuru, M., Balakrishnan, H., Karger, D., Shenker, S.: DDoS defense by offense. ACM Trans. Comput. Syst. (TOCS) **28**(1), 3 (2010)
41. Wang, X., Reiter, M.K.: Defending against denial-of-service attacks with puzzle auctions. In: Symposium on Security and Privacy, pp. 78–92. IEEE (2003)
42. Wikipedia: Bernstein inequalities (probability theory) (2024). https://en.wikipedia.org/wiki/Bernstein_inequalities_(probability_theory)
43. Yao, A.: Probabilistic computations: toward a unified measure of complexity. In: IEEE Symposium on Foundations of Computer Science (FOCS), pp. 222–227 (1977)
44. Yu, J., Li, Z., Chen, H., Chen, X.: A detection and offense mechanism to defend against application layer DDoS attacks. In: International Conference on Networking and Services (ICNS), p. 54. IEEE (2007)
45. Yuan, B., et al.: Minimizing financial cost of DDoS attack defense in clouds with fine-grained resource management. IEEE Trans. Netw. Sci. Eng. **7**(4), 2541–2554 (2020)
46. Yuan, X., Li, C., Li, X.: DeepDefense: identifying DDoS attack via deep learning. In: IEEE International Conference on Smart Computing (SMARTCOMP), pp. 1–8 (2017)
47. Zargar, S.T., Joshi, J., Tipper, D.: A survey of defense mechanisms against distributed denial of service (DDoS) flooding attacks. IEEE Commun. Surv. Tutorials **15**(4), 2046–2069 (2013)
48. Zhang, M., et al.: Poseidon: mitigating volumetric DDOS attacks with programmable switches. In: 27th Network and Distributed System Security Symposium (NDSS) (2020)
49. Zheng, S., Yang, X.: Dynashield: reducing the cost of DDoS defense using cloud services. In: 11th USENIX Workshop on Hot Topics in Cloud Computing (HotCloud 19) (2019)

Low-Distortion Clustering in Bounded Growth Graphs

Yi-Jun Chang[1(✉)], Varsha Dani[2], and Thomas P. Hayes[3]

[1] National University of Singapore, Singapore, Singapore
cyijun@nus.edu.sg
[2] Rochester Institute of Technology, Rochester, USA
varsha.dani@rit.edu
[3] University at Buffalo, Buffalo, USA
thayes2@buffalo.edu

Abstract. The well-known clustering algorithm of Miller, Peng, and Xu (SPAA 2013) is useful for many applications, including low-diameter decomposition and low-energy distributed algorithms. One nice property of their clustering, shown in previous work by Chang, Dani, Hayes, and Pettie (PODC 2020), is that distances in the cluster graph are rescaled versions of distances in the original graph, up to an $O(\log n)$ distortion factor and rounding issues. Minimizing this distortion factor is important for efficiency in computing the clustering, as well as in further applications, once the clustering has been constructed.

We prove that there exist graphs for which an $\Omega\left(\log^{1/3} n\right)$ distortion factor is necessary for any clustering. We also consider a class of nice graphs which we call *uniformly bounded independence* graphs. These include, for example, paths, lattice graphs, and "dense" unit disk graphs. For these graphs, we prove that clusterings of constant distortion always exist, and moreover, we give an efficient distributed algorithm to construct them. Our clustering algorithm is based on Voronoi cells centered at the vertices of a maximal independent set in a suitable power graph.

Applications of our new clustering include low-energy simulation of distributed algorithms in the LOCAL, CONGEST, and RADIO-CONGEST models, as well as efficient approximate solutions to distributed combinatorial optimization problems. We complement these results with matching or nearly matching lower bounds.

Keywords: Bounded independence · Energy complexity · Radio network

1 Introduction

Clustering, broadly speaking, refers to algorithms which partition the vertex set of a graph into subsets which induce connected subgraphs. In addition to

A full version of the paper is available at [14]. Y.-J. Chang is supported by the Ministry of Education, Singapore, under its Academic Research Fund Tier 1 (24-1323-A0001).

U. Schmid and R. Kuznets (Eds.): SIROCCO 2025, LNCS 15671, pp. 228–244, 2025.
https://doi.org/10.1007/978-3-031-91736-3_14

being a fundamental object of study in graph theory, clusterings have proven to be a useful tool in the design of graph algorithms, and also in distributed computation. Often, additional properties are desired, such as for the clusters to have small diameter, or for there to be few edges between clusters, or for the cluster graph to have certain properties, such as a bound on the degrees.

We consider the question of clusterings with low *distance distortion*, defined formally in Sect. 1.2. Informally, we would like clusterings for which distances in the base graph are approximately proportional to corresponding distances in the cluster graph; by *scale*, we mean the constant of proportionality, and by *distortion* we mean a particular quantification of "how approximately". Such clusterings have been studied before. Miller, Peng, and Xu [54] defined a clustering algorithm which was shown by Chang, Dani, Hayes, and Pettie [16] to have $O(\log n)$ distance distortion. They posed as an open question whether this is the best possible. In the present work, we present both positive and negative results on this question. On the one hand, we prove that there exist graphs and distance scales for which every clustering has distortion $\Omega\left(\log^{1/3} n\right)$; in particular, the lower bound applies to any family of regular graphs whose girth and diameter are both $\Theta(\log n)$. On the other hand, we show that, for a fairly broad class of "nice" graphs, including some that are frequently studied as models for distributed networks, it is possible to attain $O(1)$ distortion at all distance scales, and we give efficient distributed algorithms that achieve this.

1.1 Network Models

In the standard LOCAL model [53] of distributed computing, a network is modeled as a graph $G = (V, E)$ so that each vertex $v \in V$ is a computer and each edge $e \in E$ is a bidirectional communication link. Each vertex $v \in V$ is equipped with a distinct $O(\log n)$-bit identifier $\mathsf{ID}(v)$. The communication proceeds in synchronous rounds: In each round, each vertex $v \in V$ receives the messages sent from its neighbors, performs some arbitrary local computation, and sends a message of arbitrary length to each of its neighbors. The $\mathsf{CONGEST}$ model is a variant of the LOCAL model where the size of each message is $O(\log n)$ bits.

We also consider the extension of $\mathsf{CONGEST}$ to the more challenging *radio network* model. In the $\mathsf{RADIO\text{-}CONGEST}$ model, in each round of communication, each vertex can choose whether to transmit. In case a vertex chooses to transmit, it must transmit the same $O(\log n)$-bit message to all its neighbors. For each vertex v, it successfully receives a message from a neighbor u if u is the only neighbor of v that transmits a message in this round.

Throughout the paper, we write $n = |V|$ and $m = |E|$. Let $d(u, v)$ denote the graphical distance between vertices u and v in G. We also write $d_G(u, v)$ when disambiguation is needed. Let $\mathcal{B}_u(r)$ denote the ball of radius r around u in G.

1.2 Graph Clustering

In this paper, we focus on the distributed computation of graph clustering, where given an undirected graph $G = (V, E)$, the goal is to partition V into subsets, each of which induces a connected subgraph of G, satisfying some desired requirements. Throughout the paper, for any $v \in V$, we write $[v]$ to denote the cluster that contains v in the clustering under consideration.

Definition 1. *A* clustering *of a graph G is a partition of G into connected subgraphs, called* clusters. *Each cluster is associated with a canonical representative vertex, known as its* center, *as well as a BFS tree spanning the cluster and rooted at the center, where each vertex knows its level in the tree, its parent in the tree, and the cluster center.*

An equivalent way of specifying any clustering is to give a map $f : V \to V$, where, for every $v \in V$, $f(v)$ equals the center of the cluster containing v. We call this the *cluster center* map.

Graph clustering is a fundamental tool in distributed computing, with applications to many different kinds of distributed algorithms. In particular, clusterings that partition a graph into low-diameter components play a critical role in the complexity theory of local distributed graph problems [38]. There exist generic methods allowing us to obtain $(1 \pm \epsilon)$-approximate solutions to an arbitrary covering or packing integer linear program problem in the LOCAL model [18,38]. Moreover, it is known [38] that any *sequential local* polylogarithmic-round algorithm can be converted into a polylogarithmic-round distributed algorithm in the LOCAL model via such clustering.

Voronoi Clustering. For a given set S of *centers*, arguably the most natural clustering is Voronoi clustering, also known as Voronoi diagram or decomposition, or Dirichlet tesselation. For concreteness, we use the unique ID associated with each vertex to break ties in a canonical way.

Definition 2. *For $\emptyset \neq S \subseteq V$, the Voronoi clustering centered at S, which we denote* $\mathrm{Vor}(S)$ *is given by cluster center map*

$$f(v) = \arg\min\{d(s, v) : s \in S\}.$$

Equivalently, each vertex joins the cluster of the center closest to it. In case of a tie between several centers at minimal distance from v, we take $f(v)$ to be the one with the smallest ID.

A variation on this assigns additive weights to the various cluster centers.

Definition 3. *Let $\emptyset \neq S \subseteq V$, and let $W : S \to \mathbb{R}$. Then the associated* additively weighted Voronoi clustering *with weights W, which we denote* $\mathrm{Vor}(S, W)$, *is given by cluster center map*

$$f(v) = \arg\min\{d(s, v) - W(s) : s \in S\}.$$

Any ties are broken in favor of the center with the smallest ID, *as before.*

We emphasize that, with additive weights, the actual number of clusters may be smaller than $|S|$, since some element $u \in S$ may prefer another center $v \in S$ over itself. In this case, no vertex will select u as the center, as v is preferred compared with u by the triangle inequality.

Cluster Graph. For a given clustering $\mathcal{C}$, we will frequently be interested in the associated *cluster graph*, which is defined by the following quotient construction. Let V' be the set of clusters, and whenever an edge $\{v, w\}$ has its endpoints in two distinct clusters, $[v]$ and $[w]$, then E' contains edge $\{[v], [w]\}$. In other words, the cluster graph is the quotient graph $G/\sim$ where $\sim$ is the equivalence relation on vertices defined by the clustering $\mathcal{C}$.

1.3 Crossing Edges and Distance Distortion

Let $\mathcal{C}$ be a clustering, with associated cluster center map, f, and let $R \geq 1$. We are particularly interested in the following two metrics of how efficiently $\mathcal{C}$ represents graph G with respect to a given *scale factor* $R \geq 1$.

Few Crossing Edges. The first metric is the fraction of crossing edges, that is,

$$\frac{|E'|}{|E|},$$

where $E' = \{\{v, w\} \in E \colon f(v) \neq f(w)\}$ is the set of *crossing edges.* We would like most of the edges of the graph to have both endpoints within the same cluster. Ideally, the fraction of crossing edges should be $O(1/R)$, where R is the scale factor.

Low Distance Distortion. Our second goal is *low distortion*, which is, informally, the extent to which distances in the cluster graph are close to being distance in G, rescaled by the scale factor R. Intuitively, we would like the relationship $d \approx Rd'$ to hold, where $R \geq 1$ is the scale factor, d represents the graphical distance in the original graph G, and d' represents the graphical distance in the cluster graph G'. Since graphical distances are integers, there is a limit to what we can expect here, especially for nearby pairs of vertices. We resolve this issue by regularizing using an additive term before comparing them: $1 + (d/R) \approx 1 + d'$, which makes sense even when one or both of d, d' equals zero. This motivates the first part of the following definition. The second part of the definition requires that the *strong diameter* of the clusters should be small, capturing the intuition that the scale factor R is the "intended" cluster diameter bound.

Definition 4. *Let $(\mathcal{C}, R)$ be a clustering with scale factor R. Let d denote graphical distance in G, and d' denote graphical distance in the cluster graph. We define the* distance distortion *of $(\mathcal{C}, R)$ as the minimum $C \geq 1$ to satisfy the following conditions.*

1. *For every $v, w \in V$, we have*

$$\frac{1}{C} \leq \frac{1 + (d(v,w)/R)}{1 + d'([v],[w])} \leq C.$$

2. *The diameter of the subgraph induced by each cluster is at most CR.*

In Definition 4, we emphasize that Items 1 and 2 are logically independent. The *weak diameter* of a vertex subset $S \subseteq V$ is defined as $\max_{u \in S, v \in S} d_G(u, v)$. While Item 1 *does* imply that the weak diameter of each cluster is $O(CR)$, Item 1 *does not* imply an $O(CR)$ cluster diameter upper bound. For example, let G be the result of concatenating all the $n-1$ leaf vertices of an n-vertex star graph into a path P, and consider the clustering $\mathcal{C}$ where P forms a cluster and the star center forms a single-vertex cluster. Let $R = 1$. The clustering $(\mathcal{C}, R)$ satisfies Item 1 for $C = 3$, as $d(v, w) \in \{0, 1, 2\}$ and $d'([v], [w]) \in \{0, 1\}$ for all $v \in V$ and $w \in V$. This same clustering $(\mathcal{C}, R)$, however, does not satisfy Item 2 for any $C < n - 2$, as the diameter of P is $n - 2$.

1.4 MPX Clustering

Miller, Peng, and Xu [54] defined a particularly nice randomized algorithm which, for any input graph, produces a clustering that, with high probability, has both few crossing edges [54] and low distance distortion [16]. We will refer to this as "the MPX clustering algorithm.". Additionally, their algorithm can easily be implemented as a distributed algorithm, where the communication network is the graph to be decomposed.

Formally, the MPX clustering algorithm computes an additively weighted Voronoi clustering, where every vertex is a potential cluster center, and the weights are i.i.d. exponentially distributed with mean R. In other words, each vertex v independently samples a weight $W(v)$ from the exponential distribution with mean R, and then each vertex u joins a cluster "centered" on x, where x minimizes $d(u, x) - W(x)$. Using the terminology of Definition 3, for a given scale factor R, the MPX clustering is simply $\mathrm{Vor}(S, W)$ with $S = V$ and $W(v) \sim \mathsf{Exponential}(1/R)$.

A useful interpretation of the MPX algorithm is as follows. Each vertex v initiates a BFS tree rooted at v at time $t = -W(v)$. The BFS trees grow at the rate of one hop per unit of time. If the first BFS tree to reache v is rooted at u, then v joins the cluster of u.

Theorem 1 ([16,54]). *For any graph $G = (V, E)$ and any parameter $R \geq 1$, the MPX clustering algorithm produces a clustering for which the expected number of crossing edges is $O(m/R)$ and the distance distortion is $O(\log n)$ with probability $1 - 1/\operatorname{poly}(n)$.*

Proof. The bound on the number of crossing edges is from [54]. The bound on the distance distortion is from [16, Lemma 2.2] by setting $R = 1/\beta$.

MPX clustering has a wide range of applications in distributed computing: distributed approximation [7,11,18,30,31], distributed property testing [29,52], distributed spanner constructions [28,32], distributed expander decomposition construction [19,20], and radio network algorithms [15,16,22,24,44].

2 New Results

We set out to answer one unsolved question, namely: Is it possible to improve on the construction of MPX, and in particular to reduce its distance distortion? The importance of the question stems from the wide application of MPX. Notably, in the radio network algorithms by Chang, Dani, Hayes, and Pettie [16] and by Dani and Hayes [24], the distortion contributes significantly to their overall cost guarantees. Thus, improved distortion without sacrificing the other nice properties of MPX would immediately improve their results.

2.1 Distortion Lower Bound

The $O(\log n)$ distortion bound in Theorem 1 for analyzing the MPX clustering algorithm cannot be significantly improved, even for very nice graphs. For example, when G is a cycle, with probability $1 - 1/\operatorname{poly}(n)$, the MPX clustering with $R = \Theta(\log n/\log\log n)$ contains a cluster with diameter $\Omega(R\log n)$ and a sequence of $\Omega(\log n/\log\log n)$ consecutive single-vertex clusters. This means that, although the fraction of crossing edges is the optimal $\Theta(1/R)$, distances are somewhat distorted due to the non-uniformity of the clusters: At scale $R = \Theta(\log n/\log\log n)$, the distances can be *overestimated* by a distortion factor of $\Omega(\log n)$ and *underestimated* by a distortion factor of $\Omega(\log n/\log\log n)$. This naturally leads us to the question as to whether some other algorithm could find clusterings where the distortion is $O(1)$, or whether there exist graphs for which polylogarithmic distortion is unavoidable. We answer this question, negatively.

Theorem 2. *For every constant $d \geq 3$, there exist d-regular graphs whose girth and diameter are both $\Theta(\log n)$ such that for all $R = O\left(\log^{1/3} n\right)$, every clustering with scale factor R distorts distances by $\Omega(R)$.*

Indeed, Theorem 2 establishes that there are graphs and scale factors for which no clustering can have a distortion of $o\left(\log^{1/3} n\right)$, so now we know that polylogarithmic distortion is unavoidable. It remains an intriguing open question to determine the right exponent in $[1/3, 1]$ for the optimal distortion bound.

2.2 New Clustering

While for general graphs we came up with a negative answer, one could still hope that for some "nice" classes of graphs, it is possible to construct clusterings with constant distortion. For example, to obtain such clusterings for a cycle, it suffices to partition the cycle into paths of length $\Theta(R)$. This leads us to explore special graph classes for which such clusterings are possible. Our main contribution in this paper is a *simple* new clustering algorithm that completely meets *all* of our objectives: few crossing edges, constant distortion, and efficient distributed constructions, for a fairly big class of graphs.

Definition 5. *A graph $G = (V, E)$ is of* **bounded independence** *if there are constants $\gamma, k > 0$ such that for all $r > 0$ and $v \in V$, every independent set $S \subseteq \mathcal{B}_r(v)$ satisfies $|S| \leq \gamma r^k$.*

Bounded independence graphs, also known as bounded growth graphs in some works, have been extensively studied in the distributed setting [35,47,48, 56]. Notably, Schneider and Wattenhofer [56] presented a highly efficient MIS algorithm for such graphs in the CONGEST model, which will play a crucial role in our work. A primary motivation for investigating bounded independence graphs is that they cover *unit disk graphs*, which are commonly employed in modeling wireless networks for the study of distributed graph algorithms [2,10, 33,34,49].

Definition 6. *Fix $R > 0$. We write $G^{\leq R}$ to denote the graph with the same vertex set as G, but where there is an edge between u and v if $d_G(u, v) \leq R$. We call $G^{\leq R}$ the* **power graph** *of G with respect to parameter R.*

If G is a graph with bounded independence and R is a fixed *constant*, then $G^{\leq R}$ also has bounded independence. However, the actual independent set size bound gets much worse in the power graph. Indeed, if the polynomial bounding the size of independent subsets of r-balls in G is γr^k, then the corresponding bound for $G^{\leq R}$ is $\gamma_R r^k$, where $\gamma_R = \gamma R^k$, as r-balls in $G^{\leq R}$ are rR-balls in G. Since γ_R depends on the parameter R, this observation is useful only when R is small. We consider graphs whose power graphs $G^{\leq R}$ are of bounded independence *uniformly* for all R.

Definition 7. *A graph $G = (V, E)$ is of* ***uniformly*** **bounded independence** *if there exist constants $\gamma, k > 0$ such that for all $R > 0$, $G^{\leq R}$ is bounded independence with parameters γ, k.*

On the face of it, uniformly bounded independence seems like a very strong property and one might worry that it is impossible to satisfy. We show that various examples of interesting graph classes do exhibit this property, including paths, cycles, grids, lattices, and random geometric graphs, among others. Random geometric graphs [41] are widely used to model wireless sensor networks [43]. Distributed algorithms on grids have been extensively studied [3,8,9,42]. For these graphs, we show that clusterings with constant distortion are possible, and moreover, can be efficiently constructed by distributed algorithms.

Theorem 3. *Let G be a graph and $R \geq 1$ be a scale factor such that $G^{\leq R}$ has bounded independence. There exists a clustering of G with a constant distance distortion at scale R and an $O(1/R)$ fraction of crossing edges.*

If G is of uniformly bounded independence, then the condition of Theorem 3 is satisfied for any choice of the scale factor R. As we will later see, an application of low distortion clusterings is in low energy simulation of distributed algorithms. To achieve this we would like to follow the multi-scale clustering framework of Dani and Hayes [24]. In this framework, it is necessary to simultaneously build low-distortion clusterings at multiple scales that somehow interact well with each other. Thus, we are interested in graphs for which bounded independence carries through to the power graphs *uniformly at all scales.*

Our construction of the clustering of Theorem 3 is completely different from MPX: Instead of using randomized start times to determine cluster centers, which leads to large variations in cluster diameters, we first find a maximal independent set (MIS) in the power graph $G^{\leq R}$, and use those as cluster centers to partition G into Voronoi cells, which leads to far more uniform cluster diameters. Specifically, cluster diameters for MPX, even on really nice graphs like the square grid, range from 1 to $R \log n$, but cluster diameters for our construction are always $\Theta(R)$. The bounded independence property can be shown to imply low distortion for this clustering. To ensure that only $O(m/R)$ edges are crossing, we tweak this construction slightly using additive weights.

Although the above idea is almost as simple as the MPX algorithm, implementation is more difficult because we need to solve MIS to find the cluster centers. In the end, it can still be implemented efficiently thanks to excellent prior work on distributed MIS. As we will discuss, existing techniques for constructing maximal independent sets can be leveraged to convert the algorithm of Theorem 3 into an efficient distributed algorithm, specifically in the LOCAL model, which is essential for most of our algorithmic applications. For CONGEST and RADIO-CONGEST, further challenges are caused by the fact that we need a MIS for the power graph, $G^{\leq R}$, but the algorithm needs to run on the base graph G.

Given that MPX is already widely applied across various domains, our algorithm yields improved results for many problems for a large and practically useful class of graphs. The improvement from $O(\log n)$ to $O(1)$ is significant for problems whose complexity is already small. As we will later see, Theorem 3 allows us to obtain tight round complexity bounds and nearly tight energy complexity for many natural and well-studied distributed problems, such as $(1-\epsilon)$ approximation for the maximum independent set and maximum matching.

2.3 Distributed Computing Applications

We show that, for graphs G such that $G^{\leq R}$ is of bounded independence, very good low-distortion clusterings can be constructed efficiently.

Theorem 4. *Let $R \geq 1$. There is a randomized* LOCAL *algorithm which, when run on a graph G such that $G^{\leq R}$ is of bounded independence, produces a clustering at scale R that satisfies the following conditions in $O(R \log^* n)$ rounds.*

- *The distance distortion is constant, so the diameter of each cluster is $O(R)$.*
- *Each edge is a crossing edge with probability $O(1/R)$, so the clustering has an $O(1/R)$ fraction of crossing edges in expectation.*

Moreover, we show that the algorithm of Theorem 4 can be *derandomized* using a proper distance coloring with $O(1)$ colors and the method of conditional expectations.

Theorem 5. *Let $R \geq 1$. There is a deterministic* LOCAL *algorithm which, when run on a graph G such that $G^{\leq R}$ is of bounded independence, produces a clustering at scale R that satisfies the following conditions in $O(R \log^* n)$ rounds.*

- *The distance distortion is constant, so the diameter of each cluster is $O(R)$.*
- *The clustering has an $O(1/R)$ fraction of crossing edges.*

As we will later see, both the fraction of crossing edges $O(1/R)$ and the round complexity of the construction algorithm $O(R \log^* n)$ are asymptotically *optimal.*

Applications of our new clustering algorithms include low-energy simulation of distributed algorithms and efficient approximate solutions to distributed combinatorial optimization problems. We also complement these results with matching or nearly matching lower bounds.

While Theorem 5 seems to be strictly better than Theorem 4, as we will later see, some of our applications specifically require the property of Theorem 4 that each edge is a crossing edge with probability $O(1/R)$, so Theorems 4 and 5 are incomparable.

Low-Energy Simulation. In standard distributed computing one is often interested in round complexity, *i.e.*, the number of rounds taken by an algorithm, and message complexity, *i.e.*, the number of messages sent by vertices during the algorithm. While these measures are good descriptors of the complexity of a distributed algorithm, they do not account for the portion of the running time when a vertex is simply waiting for information from its neighbors. *Energy complexity*—also known as *awake complexity*—captures the idea that a vertex should not be charged for rounds when it is *predictably* not participating in the action of the algorithm. The predictability here is key: If a vertex knows it will not send or receive messages in certain rounds, then it can power down for those rounds of the algorithm. On the other hand, if it is powered down when it should be receiving a message, then the message is lost and could compromise the correctness of the algorithm. The goal, then, is to design distributed algorithms that enable vertices to have scheduled downtimes when they are far from the action in the algorithm, thereby saving energy.

We model energy usage as follows. In each round of the algorithm, a vertex v can choose to be asleep or awake. Every round where v chooses to be awake costs one unit of energy for v. A vertex can only participate in the algorithm, *i.e.*, communicate with other vertices, while it is awake. Notably, in the LOCAL model, an awake vertex may exchange complete information about its state with all of its awake neighbors. The setting where vertices are allowed to sleep is known as the *sleeping* model [21]. The energy usage of a vertex is its number of awake rounds. The energy complexity of an algorithm is the maximum energy usage of a vertex v among all $v \in V$.

Designing low-energy distributed algorithms for specific problems has been the subject of many prior works. The earlier research on this topic mostly focused on single-hop radio networks [6,17,46,50,55]. Recent works extended this line of research to multi-hop radio networks [13,15,16,23,24] as well as the LOCAL and CONGEST models [5,21,24,27,39,40].

Low-distortion clusterings have been used in the design of energy-efficient algorithms for breadth-first search [16] and generic algorithm simulation [24] in the radio network model. These applications critically rely on the low-distortion property of the clustering. It is not enough to know that good clusterings exist; we also need efficient algorithms to find them.

Our first application is to show that any LOCAL algorithm on a graph of uniformly bounded independence having round complexity $O(t \log^* n)$ can have its energy complexity reduced to $O(\log t \log^* n)$ without affecting its asymptotic round complexity. Informally, this implies that for *any* problem within this graph class, the energy complexity can be *exponentially* improved compared to the round complexity essentially for free!

Theorem 6. *Let $G = (V, E)$ be a graph with uniformly bounded independence. Let $t \geq 1$. Given any $t \log^* n$-round* LOCAL *algorithm $\mathcal{A}$, there is an algorithm $\mathcal{A}'$ that computes the same function as $\mathcal{A}$, in $O(t \log^* n)$ rounds and using $O(\log t \log^* n)$ energy per vertex.*

While Theorems 5 and 4 only requires constant distortion at scale R, Theorem 6 requires constant distortion uniformly at all scales because the algorithm is based on constructing low-distortion clusterings at *multiple scales*. The main proof idea behind Theorem 6 is similar to the hierarchical clustering in [16,24]. The idea is that within a cluster at a particular level, vertices know how far they are from the root of the cluster. This means that within the cluster, they can send and receive messages on a schedule, and sleep the rest of the time. Moreover, inter-cluster messages can also be arranged to be sent at predictable times. Therefore, a clustering at scale 2^{i+1} can be built energy-efficiently once we have a clustering at scale 2^i.

To achieve the generic low-energy simulation of Theorem 6, we construct a low-congestion overlapping clustering where $\mathcal{B}_t(v)$ is entirely contained in some cluster for every vertex $v \in V$. Leveraging the unlimited bandwidth of the LOCAL model, the BFS trees within each cluster enable an efficient low-energy simulation of any t-round LOCAL algorithm. Such clustering can be obtained through slight modifications to our construction of the clustering of Theorem 3 at scale t.

Extension to CONGEST *and* RADIO-CONGEST. Dani and Hayes [24] showed that, up to polylogarithmic factors in the round and energy complexity, there is a "best" generic method for reducing the energy cost of any algorithm in the RADIO-CONGEST model. Building on their work, we obtain an *improved* result for graphs with uniformly bounded independence by replacing their use of MPX clustering with our less distance-distorting clustering. The main challenge to realize this idea is to build the required multi-scale clustering in CONGEST and RADIO-CONGEST.

A key idea to address this challenge is the following modification to the multi-scale clustering algorithm: Instead of building the next-level clustering from scratch, we choose the new set of cluster centers to be a maximal subset of the existing cluster centers. While they do not form a maximal independent set in the power graph, we show that they are good enough for our purpose.

Combinatorial Optimization. Our clustering constructions can be used to give new fast approximation algorithms for certain combinatorial optimization problems.

It is well-known [7,11,18,30,31] that decomposing the communication network into low-diameter clusters with few crossing edges is useful for efficiently approximately solving various optimization problems. The high-level proof idea is to drop all edges between different clusters and solve the optimization problem on the resulting graph. Since each edge is a crossing edge with probability $O(1/R)$ in Theorem 6, a $1 - O(1/R)$ fraction of a fixed optimal solution survives the deletion of crossing edges, so the elimination of these edges does not change the optimum by much. After the deletion of crossing edges, the problem is now easy to solve because the graph is now decomposed into components of small diameter. Examples of combinatorial optimization problems that are amenable to this approach include maximum matching and maximum cut.

Moreover, such randomized algorithms can be derandomized using the *method of conditional expectations* to obtain deterministic algorithms attaining the same guarantees. The use of the method of conditional expectations in distributed graph algorithms was pioneered in [12] and subsequently used in many works [4,26,36,37]. The high-level idea is to set the values of the random variables sequentially to optimize the conditional expectation. To efficiently implement this approach as a distributed algorithm, we demonstrate that the value of many random variables can be fixed simultaneously in parallel.

Theorem 7. *The following holds for these two problems:* **Maximum Matching** *and* **Maximum Cut**. *Let G be a graph with uniformly bounded independence, and let $0 < \epsilon < 1$. There is a deterministic* LOCAL *algorithm that outputs a feasible solution whose size is at least $(1 - \epsilon)\,\mathsf{OPT}$, where OPT is the value of an optimal solution, with round complexity $O\left(\frac{1}{\epsilon}\log^* n\right)$ and energy complexity $O\left(\log\frac{1}{\epsilon}\log^* n\right)$.*

2.4 Bounded Growth Graphs

In this work, we also revisit the MPX clustering algorithm [54] and present a variant of it that achieves better guarantees in *bounded growth graphs*.

Definition 8. *A graph $G = (V, E)$ is of* **bounded growth** *if there are constants $\beta, k > 0$ such that for all $v \in V$, for all r, $|\mathcal{B}_r(v)| \le \beta r^k$.*

A bounded growth graph is automatically bounded independence, so bounded growth is a more stringent requirement than bounded independence. Cycles, grids, and lattices in dimension d are natural examples of bounded growth graphs. Distributed algorithms on bounded growth graphs have been studied in [1,57]. We show that in bounded growth graphs, a variant of the MPX algorithm can be used to construct clusterings with smaller maximum cluster diameters than the usual MPX clusters.

Theorem 8. *Given a parameter R, for any bounded growth graph $G = (V, E)$, there is an $O(R \log R)$-round algorithm in the* CONGEST *model that decomposes the vertex set V into clusters meeting the following properties.*

- *The subgraph induced by each cluster has diameter $O(R \log R)$.*
- *Each edge $e \in E$ is a crossing edge with probability $O(1/R)$.*

Theorem 8 is an improvement over the original MPX clustering, whose cluster diameter bound is $O(R \log n)$, as the maximum value of the additive weight $W(v)$ among all vertices $v \in V$ is most likely $\Theta(R \log n)$. Intuitively, the $O(\log n)$ factor is due to the *union bound* over n vertices in V.

The key idea behind our proof of Theorem 8 is to leverage the bounded growth property: Within any ball of radius $\Theta(R \log R)$, the number of vertices is polynomial in R, implying that the maximum value of the additive weight $W(v)$ among all vertices in the ball is likely to be $\Theta(R \log R)$. Intuitively, this implies that imposing a *cutoff* of $\Theta(R \log R)$ to cap the maximum value of all additive weights should have minimal impact on the analysis of the MPX algorithm on bounded growth graphs. Implementing the cutoff allows us to obtain a better cluster diameter bound. Using Theorem 8, we obtain improved results for approximating combinatorial optimization problems in bounded growth graphs.

Corollary 1. *The following holds for these two problems:* **Maximum Matching** *and* **Maximum Independent Set**. *Let G be a graph with bounded growth, and let $\epsilon = n^{-o(1)}$. There is a randomized* LOCAL *algorithm that outputs a feasible solution whose size is at least $(1-\epsilon)$ OPT with probability $1-1/\operatorname{poly}(n)$, where* OPT *is the value of an optimal solution, with round complexity $O\left(\frac{1}{\epsilon} \cdot \log \frac{1}{\epsilon}\right)$.*

2.5 Lower Bounds

We complement our upper bound results with matching or nearly matching lower bounds. The following lower bound was shown by Lenzen and Wattenhofer [51].

Theorem 9 ([51]). *Any deterministic algorithm that solves the $(1-\epsilon)$-approximate maximum independent set problem requires $\Omega\left(\frac{\log^* n}{\epsilon}\right)$ rounds in cycles in the* LOCAL *model.*

Theorem 10 immediately implies that the round complexity $O\left(\frac{\log^* n}{\epsilon}\right)$ in Theorem 7 is asymptotically optimal.

Moreover, Theorem 9 implies that any deterministic algorithm that removes at most ϵ fraction of the edges to decompose a cycle into connected components of diameter $O(1/\epsilon)$ must require $\Omega\left(\frac{\log^* n}{\epsilon}\right)$ rounds in the LOCAL model, as such a clustering allows us to construct a $(1-O(\epsilon))$-approximate maximum independent set of a cycle using an additional $O(1/\epsilon)$ rounds, as we explain below. Let each $O(1/\epsilon)$-diameter cluster locally compute a maximum independent set of this cluster, and then let S be the union of these independent sets. The size of S is at least $\lfloor n/2 \rfloor$. For each edge $e = \{u, v\}$ that crosses two clusters such that both u and v are in S, remove any one of them from S. The result is an independent set of size at least $\lfloor n/2 \rfloor - \epsilon n = (1 - O(\epsilon)) \cdot \mathsf{OPT}$, where $\mathsf{OPT} = \lfloor n/2 \rfloor$ is the size of a maximum independent set of a cycle.

As cycles are uniformly bounded independence graphs, by a change of variable $R = \Theta(1/\epsilon)$, the above argument implies that the round complexity $O(R \log^* n)$ of our deterministic clustering algorithm of Theorem 5 is asymptotically *optimal* in uniformly bounded independence graphs. The reason is that the fraction of crossing edges of any clustering of a cycle at scale R with constant distortion must be $O(1/R)$, since otherwise there is a path of length R intersecting $\omega(1)$ clusters, meaning that the distance distortion for the two endpoints of the path is $\omega(1)$.

We prove the following two theorems.

Theorem 10. *Any algorithm that solves the* $(1-\epsilon)$*-approximate maximum independent set problem with probability at least* 0.99 *requires* $\Omega\left(\log \frac{1}{\epsilon}\right)$ *energy in cycles in the* LOCAL *model.*

Theorem 10 immediately implies that the energy complexity $O\left(\log \frac{1}{\epsilon} \log^* n\right)$ in Theorem 7 is nearly optimal, up to a tiny $O(\log^* n)$ factor.

Moreover, Theorem 10 implies an $\Omega(\log t)$ lower bound for the low-energy simulation in Theorem 6. If the energy complexity of Theorem 6 can be improved to $o(\log t)$, then the improved low-energy simulation, combined with the approximate maximum independent set algorithm of Corollary 1, implies that a $(1-\epsilon)$-approximate maximum independent set of a cycle can be computed using $o\left(\log \frac{1}{\epsilon}\right)$ energy in LOCAL, contradicting the lower bound of Theorem 10.

Theorem 11. *Any algorithm that solves the* $(1-\epsilon)$*-approximate maximum independent set problem in expectation requires* $\Omega\left(\frac{1}{\epsilon} \cdot \log \frac{1}{\epsilon}\right)$ *rounds in bounded growth graphs in the* LOCAL *model.*

Theorem 11 implies that the round complexity of $O(R \log R)$ in Corollary 1 cannot be further improved. Consequently, our $O(R \log R)$-round variant of the MPX algorithm in bounded growth graphs stated in Theorem 8 attains the asymptotically optimal round complexity.

2.6 Related Work

Kuhn, Moscibroda, Nieberg, and Wattenhofer [47] designed distributed approximation algorithms for bounded independence graphs. Kuhn, Moscibroda, and Wattenhofer [48] showed that a maximal independent set can be computed in

$O(\log^* n)$ rounds in a class of bounded independence graphs with geometric information in the LOCAL model. The reliance on geometric information was removed in subsequent works [35,56]. Schneider and Wattenhofer [56] showed that for any bounded independence graphs, a maximal independent set can be computed in $O(\log^* n)$ rounds deterministically in the CONGEST model.

Bounded independence graphs arise naturally in geometric-based network models. In particular, bounded independence graphs have been studied in the signal-to-interference-plus-noise-ratio (SINR) wireless network model [45] and the radio network model [25].

Disclosure of Interests. The authors have no competing interests to declare that are relevant to the content of this article.

References

1. Abraham, I., Malkhi, D.: Name independent routing for growth bounded networks. In: Proceedings of the Seventeenth Annual ACM Symposium on Parallelism in Algorithms and Architectures, pp. 49–55 (2005)
2. Alzoubi, K.M., Wan, P.J., Frieder, O.: Message-optimal connected dominating sets in mobile ad hoc networks. In: Proceedings of the 3rd ACM International Symposium on Mobile Ad Hoc Networking & Computing, pp. 157–164 (2002)
3. Balliu, A., Hirvonen, J., Melnyk, D., Olivetti, D., Rybicki, J., Suomela, J.: Local mending. In: International Colloquium on Structural Information and Communication Complexity, pp. 1–20. Springer (2022)
4. Bamberger, P., Kuhn, F., Maus, Y.: Efficient deterministic distributed coloring with small bandwidth. In: Proceedings of the 39th Symposium on Principles of Distributed Computing (PODC), pp. 243–252 (2020)
5. Barenboim, L., Maimon, T.: Deterministic Logarithmic Completeness in the Distributed sleeping model. In: Gilbert, S. (ed.) 35th International Symposium on Distributed Computing (DISC). Leibniz International Proceedings in Informatics (LIPIcs), vol. 209, pp. 10:1–10:19. Schloss Dagstuhl – Leibniz-Zentrum für Informatik (2021)
6. Bender, M.A., Kopelowitz, T., Pettie, S., Young, M.: Contention resolution with log-logstar channel accesses. In: Proceedings of the 48th Annual ACM Symposium on Theory of Computing (STOC), pp. 499–508 (2016)
7. Bodlaender, M.H., Halldórsson, M.M., Konrad, C., Kuhn, F.: Brief announcement: local independent set approximation. In: Proceedings of the 2016 ACM Symposium on Principles of Distributed Computing, PODC 2016, pp. 93–95. ACM (2016)
8. Bonamy, M., Ouvrard, P., Rabie, M., Suomela, J., Uitto, J.: Distributed recoloring. In: Schmid, U., Widder, J. (eds.) 32nd International Symposium on Distributed Computing (DISC 2018). Leibniz International Proceedings in Informatics (LIPIcs), vol. 121, pp. 12:1–12:17. Schloss Dagstuhl – Leibniz-Zentrum für Informatik (2018)
9. Brandt, S., et al.: LCL problems on grids. In: Proceedings of the ACM Symposium on Principles of Distributed Computing (PODC), pp. 101–110 (2017)
10. Burkhart, M., Von Rickenbach, P., Wattenhofer, R., Zollinger, A.: Does topology control reduce interference? In: Proceedings of the 5th ACM International Symposium on Mobile Ad Hoc Networking and Computing, pp. 9–19 (2004)

11. Censor-Hillel, K., Levy, R., Shachnai, H.: Fast distributed approximation for max-cut. In: Fernández Anta, A., Jurdzinski, T., Mosteiro, M.A., Zhang, Y. (eds.) Algorithms for Sensor Systems, pp. 41–56. Springer, Cham (2017)
12. Censor-Hillel, K., Parter, M., Schwartzman, G.: Derandomizing local distributed algorithms under bandwidth restrictions. Distrib. Comput. 349–366 (2020). https://doi.org/10.1007/s00446-020-00376-1
13. Chang, Y.J.: The energy complexity of diameter and minimum cut computation in bounded-genus networks. In: International Colloquium on Structural Information and Communication Complexity, pp. 262–296. Springer (2023)
14. Chang, Y.J., Dani, V., Hayes, T.P.: Low-distortion clustering in bounded growth graphs. arXiv preprint arXiv:2405.05132 (2024)
15. Chang, Y.J., Dani, V., Hayes, T.P., He, Q., Li, W., Pettie, S.: The energy complexity of broadcast. In: Proceedings of the 2018 ACM Symposium on Principles of Distributed Computing, pp. 95–104 (2018)
16. Chang, Y.J., Dani, V., Hayes, T.P., Pettie, S.: The energy complexity of BFS in radio networks. In: Proceedings of the 39th Symposium on Principles of Distributed Computing (PODC), pp. 273–282. ACM (2020)
17. Chang, Y., Kopelowitz, T., Pettie, S., Wang, R., Zhan, W.: Exponential separations in the energy complexity of leader election. ACM Trans. Algorithms **15**(4) (2019)
18. Chang, Y.J., Li, Z.: The complexity of distributed approximation of packing and covering integer linear programs. In: Proceedings of the 2023 ACM Symposium on Principles of Distributed Computing (PODC), pp. 32–43 (2023)
19. Chang, Y.J., Pettie, S., Saranurak, T., Zhang, H.: Near-optimal distributed triangle enumeration via expander decompositions. J. ACM (JACM) **68**(3), 1–36 (2021)
20. Chang, Y.J., Saranurak, T.: Deterministic distributed expander decomposition and routing with applications in distributed derandomization. In: Proceedings of the 61st Annual IEEE Symposium on Foundations of Computer Science (FOCS), pp. 377–388 (2020)
21. Chatterjee, S., Gmyr, R., Pandurangan, G.: Sleeping is efficient: MIS in $O(1)$-rounds node-averaged awake complexity. In: Proceedings of the 39th Symposium on Principles of Distributed Computing (PODC), pp. 99–108. ACM (2020)
22. Czumaj, A., Davies, P.: Exploiting spontaneous transmissions for broadcasting and leader election in radio networks. In: Proceedings of the ACM Symposium on Principles of Distributed Computing (PODC), pp. 3–12 (2017)
23. Dani, V., Gupta, A., Hayes, T.P., Pettie, S.: Wake up and join me! An energy-efficient algorithm for maximal matching in radio networks. Distrib. Comput. **36**(3), 373–384 (2023)
24. Dani, V., Hayes, T.P.: How to wake up your neighbors: safe and nearly optimal generic energy conservation in radio networks. In: 36th International Symposium on Distributed Computing (DISC 2022), vol. 246, pp. 16:1–16:22. Schloss Dagstuhl-Leibniz-Zentrum für Informatik (2022)
25. Davies, P.: Uniting general-graph and geometric-based radio networks via independence number parametrization. In: Proceedings of the 2023 ACM Symposium on Principles of Distributed Computing, pp. 290–299 (2023)
26. Deurer, J., Kuhn, F., Maus, Y.: Deterministic distributed dominating set approximation in the congest model. In: Proceedings of the 2019 ACM Symposium on Principles of Distributed Computing (PODC), pp. 94–103 (2019)
27. Dufoulon, F., Moses Jr, W.K., Pandurangan, G.: Distributed MIS in $o(\log \log n)$ awake complexity. In: Proceedings of the 2023 ACM Symposium on Principles of Distributed Computing, pp. 135–145 (2023)

28. Elkin, M., Neiman, O.: Efficient algorithms for constructing very sparse spanners and emulators. ACM Trans. Algorithms (TALG) **15**(1), 1–29 (2018)
29. Even, G., et al.: Three Notes on Distributed Property Testing. In: Proceedings 31st International Symposium on Distributed Computing (DISC). Leibniz International Proceedings in Informatics (LIPIcs), vol. 91, pp. 15:1–15:30 (2017)
30. Faour, S., Fuchs, M., Kuhn, F.: Distributed CONGEST approximation of weighted vertex covers and matchings. In: Bramas, Q., Gramoli, V., Milani, A. (eds.) 25th International Conference on Principles of Distributed Systems (OPODIS 2021). Leibniz International Proceedings in Informatics (LIPIcs), vol. 217, pp. 17:1–17:20. Schloss Dagstuhl – Leibniz-Zentrum für Informatik (2022)
31. Faour, S., Kuhn, F.: Approximating bipartite minimum vertex cover in the CONGEST model. In: 24th International Conference on Principles of Distributed Systems (OPODIS 2020), vol. 184, pp. 29:1–29:16. Schloss Dagstuhl-Leibniz-Zentrum für Informatik (2021)
32. Forster, S., Grösbacher, M., de Vos, T.: An improved random shift algorithm for spanners and low diameter decompositions. In: Bramas, Q., Gramoli, V., Milani, A. (eds.) 25th International Conference on Principles of Distributed Systems (OPODIS 2021). Leibniz International Proceedings in Informatics (LIPIcs), vol. 217, pp. 16:1–16:17. Schloss Dagstuhl – Leibniz-Zentrum für Informatik (2022)
33. Funke, S., Kesselman, A., Meyer, U., Segal, M.: A simple improved distributed algorithm for minimum CDS in unit disk graphs. ACM Trans. Sens. Netw. (TOSN) **2**(3), 444–453 (2006)
34. Gao, J., Guibas, L.J., Hershberger, J., Zhang, L., Zhu, A.: Geometric spanner for routing in mobile networks. In: Proceedings of the 2nd ACM international symposium on Mobile ad hoc networking & computing, pp. 45–55 (2001)
35. Gfeller, B., Vicari, E.: A randomized distributed algorithm for the maximal independent set problem in growth-bounded graphs. In: Proceedings of the Twenty-Sixth Annual ACM Symposium on Principles of Distributed Computing (PODC), pp. 53–60 (2007)
36. Ghaffari, M., Kuhn, F.: Derandomizing distributed algorithms with small messages: spanners and dominating set. In: Schmid, U., Widder, J. (eds.) 32nd International Symposium on Distributed Computing (DISC 2018). Leibniz International Proceedings in Informatics (LIPIcs), vol. 121, pp. 29:1–29:17. Schloss Dagstuhl – Leibniz-Zentrum für Informatik (2018)
37. Ghaffari, M., Kuhn, F.: Deterministic distributed vertex coloring: simpler, faster, and without network decomposition. In: Proceedings of the IEEE 62nd Annual Symposium on Foundations of Computer Science (FOCS), pp. 1009–1020. IEEE (2022)
38. Ghaffari, M., Kuhn, F., Maus, Y., Uitto, J.: Deterministic distributed edge-coloring with fewer colors. In: Proceedings of the 50th Annual ACM SIGACT Symposium on Theory of Computing, STOC 2018, pp. 418–430. ACM (2018)
39. Ghaffari, M., Portmann, J.: Average awake complexity of MIS and matching. In: Proceedings of the 34th ACM Symposium on Parallelism in Algorithms and Architectures, pp. 45–55 (2022)
40. Ghaffari, M., Portmann, J.: Distributed MIS with low energy and time complexities. In: Proceedings of the 2023 ACM Symposium on Principles of Distributed Computing, pp. 146–156 (2023)
41. Gilbert, E.N.: Random plane networks. J. Soc. Ind. Appl. Math. **9**(4), 533–543 (1961)

42. Grebík, J., Rozhoň, V.: Local problems on grids from the perspective of distributed algorithms, finitary factors, and descriptive combinatorics. Adv. Math. **431**, 109241 (2023)
43. Haenggi, M., Andrews, J.G., Baccelli, F., Dousse, O., Franceschetti, M.: Stochastic geometry and random graphs for the analysis and design of wireless networks. IEEE J. Sel. Areas Commun. **27**(7), 1029–1046 (2009)
44. Haeupler, B., Wajc, D.: A faster distributed radio broadcast primitive. In: Proceedings of the 2016 ACM Symposium on Principles of Distributed Computing (PODC), pp. 361–370. ACM (2016)
45. Halldórsson, M.M., Holzer, S., Lynch, N.: A local broadcast layer for the SINR network model. In: Proceedings of the 2015 ACM Symposium on Principles of Distributed Computing (PODC), pp. 129–138 (2015)
46. Jurdziński, T., Kutyłowski, M., Zatopiański, J.: Efficient algorithms for leader election in radio networks. In: Proceedings of the 21st Annual ACM Symposium on Principles of Distributed Computing (PODC), pp. 51–57 (2002)
47. Kuhn, F., Moscibroda, T., Nieberg, T., Wattenhofer, R.: Local approximation schemes for ad hoc and sensor networks. In: Proceedings of the 3rd ACM Joint Workshop on Foundations of Mobile Computing (DIALM-POMC), pp. 97–103 (2005)
48. Kuhn, F., Moscibroda, T., Wattenhofer, R.: On the locality of bounded growth. In: Proceedings of the Twenty-Fourth Annual ACM Symposium on Principles of Distributed Computing (PODC), pp. 60–68 (2005)
49. Kuhn, F., Wattenhofer, R., Zollinger, A.: Worst-case optimal and average-case efficient geometric ad-hoc routing. In: Proceedings of the 4th ACM International Symposium on Mobile Ad Hoc Networking & Computing, pp. 267–278 (2003)
50. Lavault, C., Marckert, J.F., Ravelomanana, V.: Quasi-optimal energy-efficient leader election algorithms in radio networks. Inf. Comput. **205**(5), 679–693 (2007)
51. Lenzen, C., Wattenhofer, R.: Leveraging Linial's locality limit. In: Taubenfeld, G. (ed.) Distributed Computing, pp. 394–407. Springer, Berlin Heidelberg, Berlin, Heidelberg (2008)
52. Levi, R., Medina, M., Ron, D.: Property testing of planarity in the congest model. Distrib. Comput. **34**(1), 15–32 (2021)
53. Linial, N.: Locality in distributed graph algorithms. SIAM J. Comput. **21**(1), 193–201 (1992)
54. Miller, G.L., Peng, R., Xu, S.C.: Parallel graph decompositions using random shifts. In: Proceedings of the Twenty-Fifth Annual ACM Symposium on Parallelism in Algorithms and Architectures, SPAA 2013, pp. 196–203. ACM (2013)
55. Nakano, K., Olariu, S.: Randomized leader election protocols in radio networks with no collision detection. In: International Symposium on Algorithms and Computation, pp. 362–373. Springer (2000)
56. Schneider, J., Wattenhofer, R.: An optimal maximal independent set algorithm for bounded-independence graphs. Distrib. Comput. **22**(5–6), 349–361 (2010)
57. Schneider, J., Wattenhofer, R.: Distributed coloring depending on the chromatic number or the neighborhood growth. In: Structural Information and Communication Complexity: 18th International Colloquium, SIROCCO 2011, Gdańsk, Poland, 26–29 June 2011, pp. 246–257. Springer (2011)

Multimodal Search on a Line

Jared Coleman[1(✉)], Dmitry Ivanov[2], Evangelos Kranakis[2], Danny Krizanc[3], and Oscar Morales Ponce[4]

[1] Loyola Marymount University, Los Angeles, CA, USA
jared.coleman@lmu.edu
[2] Carleton University, Ottawa, ON, Canada
dimaivanov@cmail.carleton.ca, kranakis@scs.carleton.ca
[3] Wesleyan University, Middletown, CT, USA
dkrizanc@wesleyan.edu
[4] California State University, Long Beach, USA
Oscar.MoralesPonce@csulb.edu

Abstract. Inspired by the diverse set of technologies used in underground object detection and imaging, we introduce a novel multimodal linear search problem whereby a single searcher starts at the origin and must find a target that can only be detected when the searcher moves through its location using the correct of p possible search modes. The target's location, its distance d from the origin, and the correct search mode are all initially unknown to the searcher.

We prove tight upper and lower bounds on the competitive ratio for this problem. Specifically, we show that when p is odd, the optimal competitive ratio is given by $2p+3+\sqrt{8(p+1)}$, whereas when p is even, the optimal competitive ratio is given by c: the unique solution to $(c-1)^4 - 4p(c+1)^2(c-p-1) = 0$ in the interval $\left[2p+1+\sqrt{8p}, \infty\right)$. This solution c has the explicit bounds $2p+3+\sqrt{8(p-1)} \leq c \leq 2p+3+\sqrt{8p}$. The optimal algorithms we propose require the searcher to move infinitesimal distances and change directions infinitely many times within finite intervals. To better suit practical applications, we also propose an approximation algorithm with a competitive ratio of $c+\varepsilon$ (where c is the optimal competitive ratio and $\varepsilon > 0$ is an arbitrarily small constant). This algorithm involves the searcher moving finite distances and changing directions a finite number of times within any finite interval.

Keywords: Autonomous agent · Competitive ratio · Linear Search · Oblivious Mobile Target · Searcher

1 Introduction

Underground object detection and imaging is used in a wide variety of applications in archaeology, urban planning, environmental science, and more. Many different technologies, each with their own unique capabilities, advantages, and

E. Kranakis—Research supported in part by NSERC Discovery grant.

U. Schmid and R. Kuznets (Eds.): SIROCCO 2025, LNCS 15671, pp. 245–261, 2025.
https://doi.org/10.1007/978-3-031-91736-3_15

disadvantages, are used to detect different types of materials in different environments. Ground penetrating radars (GPR), for example, use radar pulses to image the subsurface and are used to locate pipes, cables, and other structures [18]. Electromagnetic induction (EMI) involves measuring electromagnetic fields to detect conductive materials (i.e., metal objects) [15]. Magnetometers measure variations in the earth's magnetic field caused by ferrous objects and Electrical Resistivity Tomography (ERT) involves measuring electrical resistivity to detect underground soil and groundwater characteristics [8].

Inspired by this diverse set of technologies, we study a novel variant of the classical linear search problem that features a new parameter p: the number of search modes available to the searcher, of which the target is only detectable in one.

1.1 Model, Notation, and Terminology

A point-like searcher starts at the origin of the real number line and can move with maximum speed 1. A stationary target is located an unknown distance $d \geq 1$ from the origin in an unknown direction. The searcher has p search modes and can only detect the target if the two become collocated while the searcher is in the correct search mode, which is not known in advance. The searcher can switch between modes and switch directions instantaneously, though it requires at least L time to search an interval of length L for every mode used. Note that for $p = 1$, this problem is equivalent to the standard linear search problem described in [5].

For a given algorithm $\mathcal{A}$, let $P_{\mathcal{A}}(t)$ denote the position of the searcher executing algorithm $\mathcal{A}$ at time $t \geq 0$. For simplicity and without loss of generality, we choose the positive direction to be that for which the searcher first travels a distance of 1 from the origin. Let $E_{\mathcal{A}}(x, t)$ be the event that the signed coordinate x has been explored in all p modes by the searcher using algorithm $\mathcal{A}$ at or before time t. We define $E_{\mathcal{A}}(x, t)$ to always be true if $|x| < 1$. Henceforth, we use the word "explored" to mean "explored in all p modes."

Our goal is to minimize the *competitive ratio* of our algorithms—essentially how many times longer the searcher takes to find the target compared to an omniscient searcher that knows the target's location and the correct search mode ahead of time. We define:

$$T_{\mathcal{A}}(x) = \inf\{t \geq 0 : E_{\mathcal{A}}(x, t)\}, \quad CR_{\mathcal{A}}(x) = \frac{1}{|x|} T_{\mathcal{A}}(x), \quad CR_{\mathcal{A}} = \sup_x CR_{\mathcal{A}}(x).$$

We consider only "successful" algorithms: those that yield finite exploration times for all coordinates x.

1.2 Related Work

Stochastic linear search was independently considered by Bellman and Beck in the 1960's (cf. [4,6]). Work in linear search dates back to the 1970s with

Beck and Newman who introduced the linear search problem and proposed an algorithm with optimal competitive ratio 9 [5]. Since then, many variants of the linear search problem have been studied and the competitive ratios of various algorithms have been analyzed [1,3,11,14].

The multimodal linear search problem we study is a generalization of linear search (when the number of modes $p = 1$, the two problems are equivalent). Despite the vast literature in search theory, work on multimodal search is sparse. A predator-prey model was introduced in [16] where the predator has two modes of searching: a fast mode where the predator has a low probability of detecting the prey when encountered and a slow mode where the predator has a high probability of detecting the prey. Similarly in [7], the authors study problems where the searcher moves at a faster speed at which the searcher cannot detect the target and a slower speed at which the searcher can detect the target (with certainty). The Beachcomber problem [10] extends this idea to a multi-agent setting where searchers must cooperate to find the target in as little time as possible. Another key similarity between these studies and ours is that, depending on the strategy the searcher uses, it may not detect the target when passing through its location. This is reminiscent of work in search with uncertain detection where the searcher may miss the target with some probability [1,13].

To the best of our knowledge, this work is the first to consider the multimodal linear search problem.

1.3 Outline and Results of the Paper

The rest of the paper is organized as follows. In Sect. 2, we propose an algorithm and prove an upper bound on its competitive ratio of $2p+3+\sqrt{8(p+1)}$ when p is odd and c (where c is the unique solution to $(c-1)^4 - 4p(c+1)^2(c-p-1) = 0$ in the interval $\left[2p+1+\sqrt{8p}, \infty\right))$ when p is even. This solution c has the explicit bounds $2p + 3 + \sqrt{8(p-1)} \leq c \leq 2p + 3 + \sqrt{8p}$. Then, in Sect. 3, we prove matching lower bounds on the competitive ratio for any algorithm, thus proving our upper bounds are tight. The algorithms we propose in Sect. 2 involve the searcher moving infinitesimal distances and changing directions infinitely many times within finite intervals. To better suit practical applications, we introduce in Sect. 4 modified algorithms (for both the odd and even cases) with a competitive ratio of $c + \varepsilon$ (where c is the optimal competitive ratio and $\varepsilon > 0$ is an arbitrarily small constant). This modified algorithm involves the searcher moving finite distances and changing directions a finite number of times within any finite interval. Numerical representations for the bounds proved in Sect. 2 are displayed in Table 1. Finally, in Sect. 5, we summarize our contributions and discuss potential avenues for future research. All proofs omitted due to space constraints can be found in the full version of the paper [9].

Table 1. Competitive ratio of multimodal search with p modes, for $p = 1, 2, \ldots, 16$.

Odd p		Even p	
p	Competitive Ratio	p	Competitive Ratio
1	9	2	10.27303
3	14.65685	4	16.08120
5	19.92820	6	21.43387
7	25	8	26.55911
9	29.94427	10	31.54214
11	34.79796	12	36.42569
13	39.58301	14	41.23468
15	44.31371	16	45.98516

2 Upper Bounds

In this section, we propose algorithms (one for odd p and one for even p) for solving the multimodal linear search problem and prove upper bounds on their competitive ratio.

2.1 Preliminaries

The algorithms that we present follow a certain principle: select an interval and explore it with all p modes by moving back and forth before proceeding. We refer to these intervals as "cells". The parity of p matters because the searcher will finish searching a cell on the opposite side of the cell that it started for odd p whereas it will finish at the same position that it started for even p. This can be beneficial or detrimental depending on where the searcher plans on going next. For odd p, ending on the opposite side facilitates searching multiple cells in a row but also means that the searcher will have to backtrack unproductively when returning to the origin. Meanwhile, for even p, the searcher finishes exploring the last cell while also making its way back to the origin; however, any prior cells must be traversed an extra time to reach the start of the next cell. Figure 1 depicts these advantages/disadvantages for the odd/even cases. This reflects a fundamental disparity between the two cases that is evident in our results.

We formalize this in the CELLSEARCH procedure, Algorithm 1, below. Note that we always begin cell search at the point closest to the origin.

Algorithm 1. (Cell Search)

procedure CELLSEARCH(x_{init}, δ, p)
  x_{init}, δ, p: current position, positive distance, number of search modes
  for $i \leftarrow 0, \ldots, p-1$ **do**
    travel to current position plus $(-1)^i \delta \cdot \text{sign}(x_{init})$ while searching in mode i
  endfor

Observe that splitting intervals into smaller cells enables points near the origin to be explored as soon as possible. Imagine a procedure that searches an interval by dividing it into a number of cells with a fixed size (except for the last cell). We call this procedure DiscreteThoroughSearch.

Algorithm 2. (Discrete Thorough Search)

procedure DiscreteThoroughSearch(x_{init}, δ, s, p)
  x_{init}: current position
  δ: positive distance
  s: cell size
  p: number of search modes
  $x \leftarrow 0$
  while $x < \delta$ **do**
    $\Delta x \leftarrow \min(s, \delta - x)$
    call CellSearch($x_{init} + x \cdot \text{sign}(x_{init})$, Δx, p)
    if p is even **then** travel to $x + \Delta x$ **endif**
    $x \leftarrow x + \Delta x$
  endwhile

There is no theoretical lower limit on the cell size s. In fact, as s approaches 0, DiscreteThoroughSearch essentially involves the agent exploring the interval in all search modes simultaneously at a slower speed $\frac{1}{p}$ if p is odd and $\frac{1}{p+1}$ if p is even. We permit this behavior and simply call it ThoroughSearch.

Algorithm 3. (Thorough Search)

procedure ThoroughSearch(x_{init}, δ, p)
  x_{init}: current position
  δ: positive distance
  p: number of search modes
  $u \leftarrow \frac{1}{p}$ **if** p **is odd else** $\frac{1}{p+1}$
  travel to $x_{init}+\delta \cdot \text{sign}(x_{init})$ at speed u while searching all modes simultaneously

2.2 Upper Bound for Odd P

For the case of odd p, we apply ThoroughSearch on intervals of exponentially growing size, alternating between the positive and negative directions. The result is Algorithm 4.

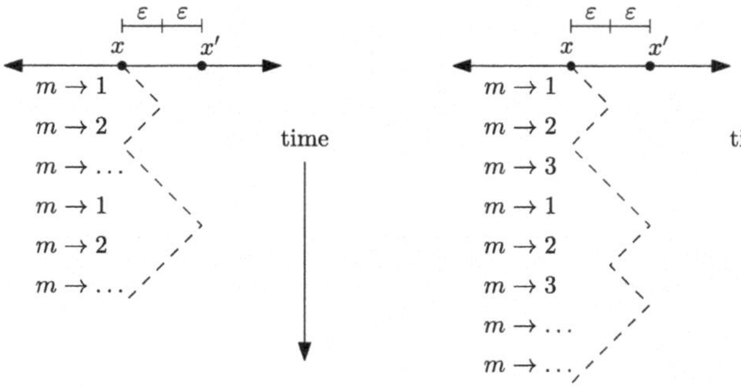

Fig. 1. The searcher explores the segment $[x, x']$ using DISCRETETHOROUGHSEARCH with two ε−width cells. The left figure depicts the case where $p = 2$ (even) and the right figure depicts the case where $p = 3$ (odd). The value of m represents the mode the searcher is operating in and $m \to \ldots$ is used to indicate that the searcher is traversing a cell that has already been fully explored.

Algorithm 4. (Multimodal search for odd p)

1: **input:** number of modes p and growth factor $a > 1$.
2: terminate upon detecting the target
3: **for** $i \leftarrow 0, 1, 2, ..., \infty$ **do** ▷ each iteration i represents a round
4: $\quad x \leftarrow (-1)^i a^{i-1}$
5: $\quad$ go to x
6: $\quad$ **call** THOROUGHSEARCH(x, $(a^{i+1} - a^{i-1})$, p)
7: $\quad$ return to the origin
8: **endfor**

Theorem 1. *Algorithm 4 satisfies* $CR_{\mathcal{A}} = 2p + 3 + \sqrt{8(p+1)}$ *with the growth factor* $a = 1 + \sqrt{\frac{2}{p+1}}$.

Proof. Algorithm 4 plays out in rounds numbered $i \geq 0$ (line 3). In each round i, the searcher explores the interval from a^{i-1} to a^{i+1} in direction $(-1)^i$ using THOROUGHSEARCH. Note that for round 0 only, some coordinates $|x| < 1$ are searched superfluously, though this does not significantly affect the upcoming analysis. Executing THOROUGHSEARCH takes $p(a^{i+1} - a^{i-1}) = pa^{i-1}(a^2 - 1)$ time and an additional $(a^{i+1} + a^{i-1}) = a^{i-1}(a^2 + 1)$ time is spent traveling back and forth from the origin. In total, completing a round i takes $t_i = a^{i-1}\left((p+1)(a^2 - 1) + 2\right)$ time. Completing all rounds from 0 to i takes $T_i = \frac{a^{i+1}-1}{a(a-1)}\left((p+1)(a^2 - 1) + 2\right)$ time. Coordinates that are explored on round i can be expressed in the form $(-1)^i a^{i-1}(1 + x)$ where $0 < x \leq a^2 - 1$. For rounds 0 and 1 only, x may be exactly 0; this does not affect the upcoming analysis. Each coordinate is explored at time $T_{i-1} + a^{i-1}(1 + px)$.

$$CR_{\mathcal{A}}(x) = \frac{T_{i-1} + a^{i-1}(1 + px)}{a^{i-1}(1 + x)} = p + \frac{T_{i-1} - (p-1)a^{i-1}}{a^{i-1}(1 + x)}$$

If $T_{i-1} \leq (p-1)a^{i-1}$, then the ratio is maximized by selecting $x = a^2 - 1$ and is no greater than p. Otherwise, the ratio is maximized by selecting x arbitrarily close to 0 (or simply equal to 0 in the case of the first two rounds). We now check the latter case for all rounds. What we will find is that the worst-case ratio is produced by arbitrarily late rounds and that it exceeds p, meaning that the former case does not produce the worst-case ratio.

$$\lim_{x\to 0^+} CR_{\mathcal{A}}(x) = 1 + \frac{T_{i-1}}{a^{i-1}} = 1 + \frac{1-a^{-i}}{a-1}\left((p+1)(a^2-1)+2\right)$$

$$\lim_{i\to\infty}\lim_{x\to 0^+} CR_{\mathcal{A}}(x) = 1 + \frac{1}{a-1}\left((p+1)(a^2-1)+2\right) = CR_{\mathcal{A}}$$

Selecting the growth factor $a = 1 + \sqrt{\frac{2}{p+1}}$ yields

$$CR_{\mathcal{A}} = 1 + (p+1)\left(2 + \sqrt{\frac{2}{p+1}}\right) + 2\left(\sqrt{\frac{2}{p+1}}\right)^{-1} = 2p + 3 + \sqrt{8(p+1)}$$

Indeed, this ratio is greater than p. This proves Theorem 1.

2.3 Upper Bound for Even P

In this section we give an algorithm with optimal competitive ratio for the case where p is even. The optimal ratio is given implicitly as a root of the polynomial $D_p(c) = (c-1)^4 - 4p(c+1)^2(c-p-1)$. By Theorem 3 (presented in Sect. 3 with other lower bounds), the competitive ratio for the even case must be at least $2p+1+\sqrt{8p}$ since the same problem with $p-1$ modes instead of p is strictly easier for the searcher. For this reason, we need only consider the function $D_p(c)$ for $c \geq 2p+1+\sqrt{8p}$. We derive the behavior of $D_p(c)$ on this interval in Lemma 1.

Lemma 1. *For even p, the function $D_p(c) = (c-1)^4 - 4p(c+1)^2(c-p-1)$ transitions from negative to positive exactly once within the range $\left[2p+1+\sqrt{8p}, \infty\right)$.*

Proof. Consider the polynomial $D_p(c)$ as well as its second and third derivatives.

$$\begin{aligned}
D_p(c) &= (c-1)^4 - 4p(c+1)^2(c-p-1) \\
&= c^4 - 4(p+1)c^3 + (4p(p-1)+6)c^2 + 4(2p-1)(p+1)c + (2p+1)^2 \\
D_p^{(2)}(c) &= 12c^2 - 24(p+1)c + 2(4p(p-1)+6) \\
&= 12c(c-2(p+1)) + 2(4p(p-1)+6) \\
D_p^{(3)}(c) &= 24c - 24(p+1)
\end{aligned}$$

We note that $D_p^{(3)}(c) > 0$ for all $c \geq 2p+1+\sqrt{8p} > p+1$. We also note that since $p \geq 2$, $D_p^{(2)}(c) > 0$ for all $c \geq 2p+1+\sqrt{8p} > 2(p+1)$. This shows that $D_p(c)$ is concave up for $c \geq 2p + 1 + \sqrt{8p}$. We now show that $D_p(2p + 1 + \sqrt{8p}) < 0$.

$$\begin{aligned} D_p(2p + 1 + \sqrt{8p}) &= (2p + \sqrt{8p})^4 - 4p(2p + 2 + \sqrt{8p})^2(p + \sqrt{8p}) \\ &= 16p^{\frac{3}{2}} \left(\sqrt{p}(\sqrt{p} + \sqrt{2})^4 - (p + 1 + \sqrt{2p})^2(\sqrt{p} + \sqrt{8})\right) \\ &= 16p^{\frac{3}{2}} \left(-p\sqrt{8} - 5\sqrt{p} - \sqrt{8}\right) < 0 \end{aligned}$$

Since $D_p(2p + 1 + \sqrt{8p}) < 0$ and $D_p(c)$ has the dominant term c^4, it must eventually become positive. Since it is concave up beyond $c = 2p + 1 + \sqrt{8p}$, it cannot switch signs more than once. This proves Lemma 1.

Again, our algorithm alternates searching intervals of growing size in each direction but in this case, a fraction r of the interval is searched using THOROUGHSEARCH and the remainder is treated as single cell. The result is Algorithm 5.

Algorithm 5. (Multimodal search for even p)

1: **input:** number of modes p, growth factor $a > 1$, and a coefficient $0 \leq r \leq 1$.
2: terminate at any point if collocated with target
3: **for** $i \leftarrow 0, 1, 2, ..., \infty$ **do** ▷ each iteration i represents a round
4: $\quad x_{start} \leftarrow (-1)^i a^{i-1}$
5: $\quad x_{end} \leftarrow (-1)^i a^{i+1}$
6: $\quad x_{mid} \leftarrow (1 - r)x_{start} + r x_{end}$
7: $\quad$ go to x_{start}
8: $\quad$ **call** THOROUGHSEARCH(x_{start}, $|x_{mid} - x_{start}|$, p)
9: $\quad$ **call** CELLSEARCH(x_{mid}, $|x_{end} - x_{mid}|$, p)
10: $\quad$ return to the origin
11: **endfor**

Theorem 2. *Algorithm 5 achieves a competitive ratio $CR_{\mathcal{A}} = c^*$ where c^* is the unique solution to $D_p(c) = 0$ in the range $\left[2p + 1 + \sqrt{8p}, \infty\right)$. The parameters a and r needed to do so are given in terms of the above value c^* as follows:*

$$a = \frac{(c^* - 1)^2}{2p(c^* + 1)}, \qquad r = \frac{\sqrt{(p(a^2 - 1) + 2)^2 + 8p(a^2 - 1)(a - 1)} - 2}{4(a^2 - 1)} - \frac{p}{4}.$$

Proof. Algorithm 5 plays out in rounds numbered $i \geq 0$ (line 3). Consider an arbitrary round i where the searcher goes in direction $(-1)^i$. The searcher begins by going to a^{i-1} which is at the boundary of everything searched thus far. The searcher then executes THOROUGHSEARCH to search in front of itself up to the point $m_i = a^{i-1} + r\left(a^{i+1} - a^{i-1}\right) = a^{i-1}\left(r(a^2 - 1) + 1\right)$. This will be dubbed phase 1. The remaining interval from m_i to a^{i+1} is searched by moving back and forth using a single call to CELLSEARCH. This will be dubbed phase 2. The searcher ends phase 2 at position m_i and then returns to the origin to end the round. Note that for round 0 only, some coordinates $|x| < 1$ are searched superfluously, though this does not significantly affect the upcoming analysis.

We define t_i to be the time taken to complete round i and $T_i = \sum_{j=0}^{i} t_j$. The above description implies that

$$\begin{aligned} t_i &= a^{i-1} + (p+1)(m_i - a^{i-1}) + p(a^{i+1} - m_i) + m_i \\ &= 2a^{i-1} + p\left(a^{i+1} - a^{i-1}\right) + 2r\left(a^{i+1} - a^{i-1}\right) \\ &= a^{i-1}\left(2 + (p+2r)\left(a^2 - 1\right)\right) \quad \text{and} \\ T_i &= \frac{a^{i+1} - 1}{a(a-1)}\left(2 + (p+2r)\left(a^2 - 1\right)\right). \end{aligned}$$

We now examine the competitive ratio of coordinates searched during phase 1. These are coordinates of the form $d_1(i,x) = (-1)^i a^{i-1}(1+x)$ for $x \in \left(0, a^2 - 1\right]$. For rounds 0 and 1 only, x may be exactly 0; this does not affect the upcoming analysis.

$$\begin{aligned} T_{\mathcal{A}}(d_1(i,x)) &= T_{i-1} + a^{i-1} + a^{i-1}(p+1)x \\ CR_{\mathcal{A}}(d_1(i,x)) &= 1 + \frac{T_{i-1} + a^{i-1}px}{a^{i-1}(1+x)} \\ &= 1 + \frac{\frac{a^i - 1}{a(a-1)}\left(2 + (p+2r)\left(a^2-1\right)\right) + a^{i-1}px}{a^{i-1}(1+x)} \\ &= 1 + \frac{\frac{1-a^{-i}}{a-1}\left(2 + (p+2r)\left(a^2-1\right)\right) + px}{1+x} \\ \lim_{i\to\infty} CR_{\mathcal{A}}(d_1(i,x)) &= 1 + \frac{\frac{2}{a-1} + (p+2r)(a+1) + px}{1+x} \\ &= 1 + p + \frac{\frac{2}{a-1} + (p+2r)a + 2r}{1+x} \end{aligned}$$

Since the numerator is positive, this value is maximized by minimizing x. We select x arbitrarily close to 0 (or simply equal to 0 in the case of the first two rounds).

$$\lim_{i\to\infty} \lim_{x\to 0^+} CR_{\mathcal{A}}(d_1(i,x)) = 1 + \frac{2}{a-1} + (p+2r)(a+1). \tag{2.1}$$

Note that this quantity is increasing with respect to r. Next, we consider coordinates searched during phase 2: $d_2(i,x) = (-1)^i x$ for $x \in \left(m_i, a^{i+1}\right]$.

$$T_{\mathcal{A}}(d_2(i,x)) = T_i - x, \quad CR_{\mathcal{A}}(d_2(i,x)) = \frac{T_i}{x} - 1.$$

This value is maximized as x approaches m_i.

$$\begin{aligned}
\lim_{x \to m_i^+} CR_{\mathcal{A}}(d_2(i,x)) &= \frac{T_i}{m_i} - 1 = \frac{\frac{a^{i+1}-1}{a(a-1)}\left(2 + (p+2r)\left(a^2-1\right)\right)}{a^{i-1}\left(r(a^2-1)+1\right)} - 1 \\
&= \frac{\frac{a-a^{-i}}{a-1}\left(2 + (p+2r)\left(a^2-1\right)\right)}{r(a^2-1)+1} - 1 \\
\lim_{i \to \infty} \lim_{x \to m_i^+} CR_{\mathcal{A}}(d_2(i,x)) &= \frac{\frac{a}{a-1}\left(2 + (p+2r)\left(a^2-1\right)\right)}{r(a^2-1)+1} - 1 \\
&= \frac{\frac{a}{a-1}p\left(a^2-1\right)}{r(a^2-1)+1} + \frac{2a}{a-1} - 1
\end{aligned}$$

Note that this quantity decreases with respect to r. Since the phase 1 ratio increases with respect to r while the phase 2 ratio decreases, the maximum of the two is minimized when they are equal.

$$\begin{aligned}
1 + \frac{2}{a-1} + (p+2r)(a+1) &= \frac{\frac{a}{a-1}p\left(a^2-1\right)}{r(a^2-1)+1} + \frac{2a}{a-1} - 1 \\
(p+2r)(a+1) &= \frac{\frac{a}{a-1}p\left(a^2-1\right)}{r(a^2-1)+1} \\
(p+2r)\left(r(a^2-1)+1\right) &= ap \\
2(a^2-1)r^2 + \left(p(a^2-1)+2\right)r + p(1-a) &= 0
\end{aligned}$$

$$r = \frac{\sqrt{\left(p(a^2-1)+2\right)^2 + 8p(a^2-1)(a-1)} - 2}{4(a^2-1)} - \frac{p}{4}$$

Substituting this into Eq. (2.1) yields

$$1 + \frac{p(a+1)}{2} + \frac{2 + \sqrt{\left(p(a^2-1)+2\right)^2 + 8p(a^2-1)(a-1)}}{2(a-1)}.$$

This function denotes the best competitive ratio achievable by Algorithm 5 for a given growth ratio a. Minimizing this function with respect to a yields the optimal parameters for Algorithm 5. Suppose that we desire some competitive ratio c:

$$\begin{aligned}
c = 1 + \frac{p(a+1)}{2} &+ \frac{2 + \sqrt{\left(p(a^2-1)+2\right)^2 + 8p(a^2-1)(a-1)}}{2(a-1)} \\
2(c-1)(a-1) - p(a^2-1) - 2 &= \sqrt{\left(p(a^2-1)+2\right)^2 + 8p(a^2-1)(a-1)}
\end{aligned}$$

For simplicity, let us define $f(a) = p(a^2 - 1) + 2$.

$$\begin{aligned}
2(c-1)(a-1) - f(a) &= \sqrt{f(a)^2 + 8p(a+1)(a-1)^2} \\
4(c-1)^2(a-1)^2 - 4(c-1)(a-1)f(a) + f(a)^2 &= f(a)^2 + 8p(a^2-1)(a-1) \\
(c-1)^2(a-1) - (c-1)\left(p(a^2-1)+2\right) &= 2p(a^2-1) \\
(c-1)^2(a-1) - p(c-1)(a^2-1) - 2(c-1) &= 2p(a^2-1) \\
(c-1)^2(a-1) - 2(c-1) &= p(c+1)(a^2-1) \\
(c-1)^2 a - (c-1)^2 - 2(c-1) &= p(c+1)a^2 - p(c+1) \\
p(c+1)a^2 - (c-1)^2 a + (c+1)(c-1-p) &= 0
\end{aligned}$$

Note that this quadratic has discriminant $D_p(c)$. We conclude that Algorithm 5 can attain competitive ratio c so long as $D_p(c) \geq 0$. According to Lemma 1, the lowest attainable ratio is c^*, the unique solution to $D_p(c) = 0$ in the range $\left[2p + 1 + \sqrt{8p}, \infty\right)$. The parameters to achieve this ratio are given by

$$a = \frac{(c^*-1)^2}{2p(c^*+1)}, \quad r = \frac{\sqrt{\left(p(a^2-1)+2\right)^2 + 8p(a^2-1)(a-1)} - 2}{4(a^2-1)} - \frac{p}{4}.$$

This concludes the proof of Theorem 2.

3 Lower Bounds

3.1 Preliminaries

When considering all possible algorithms, we permit infinitely precise motion such as that performed by ThoroughSearch, but in a more general form. To determine whether a pattern of motion is valid, we imagine a scenario where motion is restricted to small, discrete steps of size ε in which the searcher uses one search mode and moves a distance ε with speed 1 in either direction. A pattern of motion is valid if it can be approximated arbitrarily well in this restricted case as ε tends to 0.

We define several concepts that are integral to proving lower bounds for the competitive ratio of all algorithms. For a given algorithm $\mathcal{A}$, let

$$I^{\sigma}_{\mathcal{A}}(t) = (-1)^{\sigma} \sup\left\{x : x > 0, \forall 0 \leq x' \leq x \; E_{\mathcal{A}}\left((-1)^{\sigma}x', t\right)\right\}.$$

For a given t, the interval $\left(I^1_{\mathcal{A}}(t), I^0_{\mathcal{A}}(t)\right)$, hereafter referred to as the "central island," is the greatest-length contiguous interval around the origin such that all coordinates within the interval have been explored by algorithm $\mathcal{A}$ by time t. We are most interested in the length of the island; hence, the inclusion or exclusion of endpoints does not matter. Also, note that any integer values of σ may be used and only their parity is relevant. Even values represent the positive side of the line and odd values the negative side.

We define a function that extracts sequences of key values from an arbitrary algorithm. The existence of these key values and their properties will be used to set lower bounds on the performance of all algorithms. The function $X(\mathcal{A})$ takes an algorithm $\mathcal{A}$ and produces a tuple of sequences $\left((x_i)_{i=-2}^{\infty}, (t_i)_{i=0}^{\infty}\right)$ where the former is a sequence of signed coordinates and the latter is a sequence of time values.

To obtain the two sequences, we first define t_0 to be the time at which algorithm $\mathcal{A}$ first reaches $x = 1$. We also define $x_{-2} = 1$, $x_{-1} = -1$. Then, for each $i \geq 0$ in order of increasing i, we select t_{i+1} and x_i such that $t_{i+1} = \sup\left\{t : I_{\mathcal{A}}^{i+1}(t) = I_{\mathcal{A}}^{i+1}(t_i)\right\}$, $x_i = \lim_{t \to t_{i+1}^-} I_{\mathcal{A}}^i(t)$. We refer to the time interval $(t_i, t_{i+1}]$ as "period i" (not to be confused with the prior notion of rounds, though there do exist parallels between the two). During period i, the central island of explored points is grown in direction $(-1)^i$. Period i ends and period $i+1$ begins precisely when the searcher begins growing the island in the opposite direction. x_i is the signed endpoint of the central island in direction $(-1)^i$ at its furthest extent during period i.

The sequence $(x_i)_{i=-2}^{\infty}$ satisfies a number of key properties. By definition, $|x_i| > |x_{i-2}|$ for all $i \geq 0$. During each period, the searcher must at least traverse the interval $(-1, 1)$ in time 2 to start the next period. This means that a finite number of periods occur in any finite time. Furthermore, since we only consider "successful" algorithms, it must be true that $\lim_{i\to\infty} |x_i| = \infty$ as all finite distances have a finite exploration time.

Finally, we prove Lemma 2 which will be used in lower bound proofs. This lemma shows the impossibility of the existence of infinite sequences whose terms satisfy a certain inequality and is a modification of part of Lemma 2 from [5] that gives a more general result.

Lemma 2. *For $a > 0$, $b > 0$, and $a^2 - 4b < 0$, no positive, infinite sequence $(y_i)_{i=0}^{\infty}$ can satisfy $y_{i+2} \leq a y_{i+1} - b y_i$ for all i.*

3.2 Lower Bound for Odd P

Theorem 3. *For odd p, no algorithm $\mathcal{A}$ satisfies $CR_{\mathcal{A}} < 2p + 3 + \sqrt{8(p+1)}$.*

Proof. Consider an arbitrary algorithm $\mathcal{A}$. We obtain the sequences $(x_i)_{i=-2}^{\infty}$ and $(t_i)_{i=0}^{\infty}$ from $X(\mathcal{A})$ as described previously.

We start by establishing a lower bound on t_{i+2}: the time that it takes to complete all of the periods up to and including $i+1$. By the end of period $i+1$, the central island reaches length $|x_{i+1}| + |x_i|$. At $t = 0$, the island had length 2, meaning that at least $p\left(|x_{i+1}| + |x_i| - 2\right)$ time has been spent expanding the island across all of the periods before and including $i+1$. One may call this the "total productive time" as any further time is spent unproductively retreading old ground. We note that every period j includes a minimum amount of unproductive time. At the start of period j, the central island has length $|x_{j-1}|+|x_{j-2}|$, which must be retreaded during period j to start the next period. We call this the "base retreading time" of period j. The nature of odd p guarantees further retreading. The searcher cannot explore an interval using all p modes and return to the origin without retreading the interval in question at least once because that implies crossing something an odd number of times and returning to the starting side. The interval between x_{j-2} and x_j becomes fully explored by the end of period j, meaning that its length must have been retreaded at least once prior to the end of period j. The time taken to do so is not part of the base retreading time of period j or any prior period as the latter involves retreading ground on the origin-facing side of x_{j-2} while the former does not. We label this the "odd retreading time of period j". In summary, each period $0 \leq j \leq i+1$ contributes at least $|x_j| + |x_{j-1}|$ of unproductive time. In combination with the total productive time, this yields a lower bound on the time needed to complete all periods up to and including $i+1$. Let us use the shorthand $\hat{x}_i = \sum_{j=0}^{i} |x_j|$.

$$\begin{aligned} t_{i+2} &\geq p\left(|x_{i+1}| + |x_i|\right) + \sum_{j=0}^{i+1} \left(|x_j| + |x_{j-1}|\right) \\ &= (p+1)\left(|x_{i+1}| + |x_i|\right) + |x_i| + 2\hat{x}_{i-1} + |x_{-1}| \\ &\geq (p+1)\left(|x_{i+1}| + |x_i|\right) + |x_i| + 2\hat{x}_{i-1}. \end{aligned}$$

We will use this result to evaluate the performance of $\mathcal{A}$. Consider a period $i \geq 0$. Coordinates of the form $x_i + (-1)^i\varepsilon$ remain unexplored for certain, arbitrarily small $\varepsilon > 0$. Were this not true, such points would all be explored for sufficiently small values of ε, suggesting that the contiguous island extends further than x_i; however, this contradicts the definition of x_i. Such points are explored no sooner than period $i+2$. Generously, we assume that all coordinates explored during period $i+2$ are explored at the start of the period: t_{i+2}.

$$CR_{\mathcal{A}}\left(x_i + (-1)^i\varepsilon\right) \geq \frac{t_{i+2}}{|x_i| + \varepsilon}.$$

Since we can select arbitrarily small values of ε, it is in fact necessary that

$$CR_{\mathcal{A}}\left(x_i + (-1)^i\varepsilon\right) \geq \frac{t_{i+2}}{|x_i|} = 2p + 3 + \frac{(p+1)\left(|x_{i+1}| - |x_i|\right) + 2\hat{x}_{i-1}}{|x_i|}.$$

We now assume, for the sake of contradiction, that for some $c < \sqrt{8}$, for all $i \geq 0$,

$$\begin{aligned} CR_{\mathcal{A}}\left(x_i + (-1)^i\varepsilon\right) &\leq 2p + 3 + c\sqrt{p+1} \\ \frac{(p+1)\left(|x_{i+1}| - |x_i|\right) + 2\hat{x}_{i-1}}{|x_i|} &\leq c\sqrt{p+1} \\ (p+1)\left(|x_{i+1}| - |x_i|\right) + 2\hat{x}_{i-1} &\leq c\sqrt{p+1}|x_i| \\ (p+1)\left(\hat{x}_{i+1} - 2\hat{x}_i + \hat{x}_{i-1}\right) + 2\hat{x}_{i-1} &\leq c\sqrt{p+1}\left(\hat{x}_i - \hat{x}_{i-1}\right) \\ \hat{x}_{i+1} - 2\hat{x}_i + \hat{x}_{i-1} + \frac{2}{p+1}\hat{x}_{i-1} &\leq \frac{c}{\sqrt{p+1}}\hat{x}_i - \frac{c}{\sqrt{p+1}}\hat{x}_{i-1} \\ \hat{x}_{i+1} &\leq \left(\frac{c}{\sqrt{p+1}} + 2\right)\hat{x}_i - \left(\frac{2}{p+1} + \frac{c}{\sqrt{p+1}} + 1\right)\hat{x}_{i-1} \end{aligned}$$

We invoke Lemma 2 with $a = \frac{c}{\sqrt{p+1}} + 2$ and $b = \frac{2}{p+1} + \frac{c}{\sqrt{p+1}} + 1$. We verify that $a^2 - 4b < 0$:

$$0 > \left(\frac{c}{\sqrt{p+1}} + 2\right)^2 - 4\left(\frac{2}{p+1} + \frac{c}{\sqrt{p+1}} + 1\right)$$

$$0 > \frac{c^2}{p+1} + \frac{4c}{\sqrt{p+1}} + 4 - \frac{8}{p+1} - \frac{4c}{\sqrt{p+1}} - 4 \Rightarrow 0 > \frac{c^2-8}{p+1} \Rightarrow c < \sqrt{8}$$

By Lemma 2, a contradiction is reached as the positive sequence $(\hat{x}_i)_{i=0}^{\infty}$ cannot remain positive. We conclude that no algorithm $\mathcal{A}$ can satisfy $CR_{\mathcal{A}} < 2p + 3 + \sqrt{8(p+1)}$. This proves Theorem 3.

3.3 Lower Bound for Even P

Theorem 4. *For even p, no algorithm $\mathcal{A}$ can satisfy $CR_{\mathcal{A}} < c^*$ where c^* is the unique solution to $D_p(c) = 0$ in the range $[2p + 1 + \sqrt{8p}, \infty)$.*

Theorem 5. *The unique solution c^* to $D_p(c) = 0$ within the interval $[2p + 1 + \sqrt{8p}, \infty)$ is bounded as follows:* $2p+3+\sqrt{8(p-1)} \leq c^* \leq 2p+3+\sqrt{8p}$.

4 A Practical Algorithm

The optimal algorithms described thus far use ThoroughSearch, which requires constantly switching directions in a way that is infinitely precise. In this section, we show that a family of modified algorithms produces competitive ratios arbitrarily close to the optimal ones without the need for infinitesimal movements or infinitely many switches in direction. The algorithms in question replace all instances of ThoroughSearch with another procedure that we call CompliantThoroughSearch. This new procedure has five parameters. Three of the parameters are identical to those of ThoroughSearch, describing the interval that needs to be searched and the number of modes. The fourth parameter c represents the competitive ratio achieved by the algorithm that is subject to modification. The final parameter $\varepsilon > 0$ is chosen by the user and represents the amount by which CompliantThoroughSearch is permitted to deviate from competitive ratio c in its performance. Unlike DiscreteThoroughSearch, CompliantThoroughSearch partitions the searched interval into cells of unequal sizes (exponentially increasing). This is done strategically as large cells closer to the origin disproportionately affect the resulting competitive ratio. It should be noted that the way the interval is partitioned into cells does not affect the overall completion time of the procedure.

Algorithm 6. (Compliant Thorough Search)

procedure COMPLIANTTHOROUGHSEARCH(x_{init}, δ, p, c, ε)
  x_{init}: current position
  δ: positive distance
  c: competitive ratio
  ε: difference in competitive ratios
  p: number of search modes
  $p' \leftarrow p$ **if** p **is odd else** $p+1$
  $x \leftarrow 0$ ▷ current displacement from x_{init}
  $s \leftarrow \frac{\varepsilon |x_{init}|}{p'-1}$ ▷ cell width
  while $x < \delta$ **do**
    $\Delta x \leftarrow \min(s, \delta - x)$
    call CELLSEARCH($x_{init} + x \cdot \text{sign}(x_{init})$, Δx, p)
    if p **is even then** travel to $x + \Delta x$ **endif**
    $x \leftarrow x + \Delta x$
    $s \leftarrow s \cdot \frac{c+\varepsilon-1}{p'-1}$
  endwhile

Lemma 3. *Suppose that an algorithm $\mathcal{A}$ makes a call to* THOROUGHSEARCH *with parameters x_{init} and δ in order to explore the half-open interval composed of the set of coordinates $\{x : |x_{init}| < x \cdot sign\,(x_{init}) \leq |x_{init}| + \delta\}$. Suppose also that all coordinates x within this interval receive $CR_{\mathcal{A}}(x) \leq c$ for some c. Given any $\varepsilon > 0$, we can construct a modified algorithm $\mathcal{A}'$ in which the call to* THOROUGHSEARCH *is replaced with an analogous call to* COMPLIANTTHOROUGHSEARCH *which achieves $CR_{\mathcal{A}'}(x) \leq c + \varepsilon$ for the same coordinates x and which makes n calls to cell search where*

$$n = \left\lceil \left(\ln\left(\frac{c+\varepsilon-1}{p'-1}\right)\right)^{-1} \cdot \ln\left(1 + \frac{(c+\varepsilon-p')\delta}{\varepsilon |x_{init}|}\right) \right\rceil$$

and $p' = p$ for odd p while $p' = p + 1$ for even p.

Theorem 6. *For any odd $p \geq 3$, let $\mathcal{A}$ be an instance of Algorithm 4 with the optimal parameter $a = 1 + \sqrt{\frac{2}{p+1}}$ so that $CR_{\mathcal{A}} = 2p + 3 + \sqrt{8(p+1)}$. For any $\varepsilon > 0$, a modified algorithm $\mathcal{B}$ that replaces all calls to* THOROUGHSEARCH *with analogous calls to* COMPLIANTTHOROUGHSEARCH *with the additional parameters $c = CR_{\mathcal{A}}$ and ε achieves competitive ratio $CR_{\mathcal{B}} = CR_{\mathcal{A}} + \varepsilon$ while making $O\left(\log\left(\frac{p}{\varepsilon}\right)\right)$ calls to cell search per round.*

Theorem 7. *For any even p, let $\mathcal{A}$ be an instance of Algorithm 5 with parameters a and r such that it yields the optimal competitive ratio $CR_{\mathcal{A}}$ according to Theorem 2. For any $\varepsilon > 0$, a modified algorithm $\mathcal{B}$ that replaces all calls to* THOROUGHSEARCH *with analogous calls to* COMPLIANTTHOROUGHSEARCH *with the additional parameters $c = CR_{\mathcal{A}}$ and ε achieves competitive ratio $CR_{\mathcal{B}} = CR_{\mathcal{A}} + \varepsilon$ while making $O\left(\log\left(\frac{p}{\varepsilon}\right)\right)$ calls to cell search per round.*

5 Conclusion

In this paper, we introduced a new type of linear search, multimodal linear search, with p search modes, a generalization of the well-known linear search with $p = 1$. We proposed algorithms (one for the case of odd p and one for the case of even p) which we show achieve the optimal competitive ratio for multimodal search for an unknown target by a single searcher. As anticipated, our results generalize the well-known competitive ratio of 9 for the standard linear search problem (i.e., $p = 1$).

We also proposed modified algorithms (with competitive ratio $c + \varepsilon$ where c is the optimal competitive ratio) that involve the searcher moving finite distances and switching directions a finite number of times over any finite interval. This suggests that it may be interesting to study the multimodal search problem for the case where there is a cost associated with switching directions and/or search modes. In this variant, Algorithms 4 and 5 would not only be impractical, but have unbounded competitive ratio. Turn cost has been studied before [1][Section 8.4, p 132], [2,12,17] but, to the best of our knowledge, has never been studied in a multimodal context.

Many other interesting research questions remain open, including randomized algorithms, group search (with multiple cooperating searchers), fault tolerance of the searchers, as well as search on domains more general than the real line (e.g., a star of infinite rays).

References

1. Alpern, S., Gal, S.: The Theory of Search Games and Rendezvous. International Series in Operations Research and Management Science, vol. 55. Kluwer (2003)
2. Angelopoulos, S., Arsénio, D., Dürr, C.: Infinite linear programming and online searching with turn cost. Theor. Comput. Sci. **670**, 11–22 (2017)
3. Baeza-Yates, R.A., Schott, R.: Parallel searching in the plane. Comput. Geom. **5**, 143–154 (1995)
4. Beck, A.: On the linear search problem. Israel J. Math. **2**(4), 221–228 (1964)
5. Beck, A., Newman, D.J.: Yet more on the linear search problem. Israel J. Math. **8**(4), 419–429 (1970)
6. Bellman, R.: An optimal search. SIAM Rev. **5**(3), 274 (1963)
7. Bénichou, O., Loverdo, C., Moreau, M., Voituriez, R.: Intermittent search strategies. Rev. Mod. Phys. **83**(1), 81 (2011)
8. Budhu, M.: Soil Mechanics and Foundations. Wiley, Hoboken (2010)
9. Coleman, J.R., Ivanov, D., Kranakis, E., Krizanc, D., Morales-Ponce, O.: Multimodal search on a line. CoRR, abs/2502.07000 (2024)
10. Czyzowicz, J., Gasieniec, L., Georgiou, K., Kranakis, E., MacQuarrie, F.: The beachcombers' problem: walking and searching with mobile robots. Theor. Comput. Sci. **608**, 201–218 (2015)
11. Czyzowicz, J., Georgiou, K., Kranakis, E.: Group search and evacuation. In: Flocchini, P., Prencipe, G., Santoro, N. (eds.) Distributed Computing by Mobile Entities: Current Research in Moving and Computing, pp. 335–370. Springer, Cham (2019)

12. Demaine, E.D., Fekete, S.P., Gal, S.: Online searching with turn cost. Theor. Comput. Sci. **361**(2–3), 342–355 (2006)
13. Georgiou, K., Giachoudis, N., Kranakis, E.: Overcoming probabilistic faults in disoriented linear search. In: Rajsbaum, S., Balliu, A., Daymude, J.J., Olivetti, D. (eds.) Structural Information and Communication Complexity - 30th International Colloquium, SIROCCO 2023, Alcalá de Henares, Spain, 6–9 June 2023. Lecture Notes in Computer Science, vol. 13892, pp. 520–535. Springer (2023)
14. Ghosh, S.K., Klein, R.: Online algorithms for searching and exploration in the plane. Comput. Sci. Rev. **4**(4), 189–201 (2010)
15. Giancoli, D.C.: Physics: Principles with Applications. Prentice-Hall, Inc., Englewood Cliffs (1980)
16. Knoppien, P., Reddingius, J.: Predators with two modes of searching: a mathematical model. J. Theor. Biol. **114**(2), 273–301 (1985)
17. Lopez-Ortiz, A.: Searching in bounded and unbounded domains. Ph.D. thesis, Department of Computer Science, University of Waterloo (1996)
18. Srivastav, A., Nguyen, P., McConnell, M., Loparo, K.A., Mandal, S.: A highly digital multiantenna ground-penetrating radar (GPR) system. IEEE Trans. Instrum. Meas. **69**(10), 7422–7436 (2020)

Multi-agent Disk Inspection

James Conley and Konstantinos Georgiou(✉)

Department of Mathematics, Toronto Metropolitan University, Toronto, ON, Canada
{j1conley,konstantinos}@torontomu.ca

Abstract. We consider n unit-speed mobile agents initially positioned at the center of a unit disk, tasked with inspecting all points on the disk's perimeter. A perimeter point is considered covered if an agent positioned outside the disk's interior has unobstructed visibility of it, treating the disk itself as an obstacle. For $n = 1$, this problem is referred to as the shoreline problem with a known distance and was originally proposed as a more tractable variant of Bellman's famous lost-in-the-forest problem [14]. Isbell [44] derived an optimal trajectory that minimizes the worst-case inspection time for that problem. For $n \geq 2$ agents, worst-case optimal trajectories were shown in [1,29].

Our contributions are threefold. First, and as a warm-up, we extend Isbell's findings by deriving worst-case optimal trajectories addressing the partial inspection of a section of the disk, hence deriving an alternative proof of optimality for inspecting the disk with $n \geq 2$ agents. Second, we analyze the average-case inspection time, assuming a uniform distribution of perimeter points (equivalent to randomized inspection algorithms). Using spatial discretization and Nonlinear Programming (NLP), we propose feasible solutions to the continuous problem and evaluate their effectiveness compared to NLP solutions. Third, we establish Pareto-optimal bounds for the multi-objective problem of jointly minimizing the worst-case and average-case inspection times.

Keywords: Mobile Agents · Search · Disk · Shoreline Problem

1 Introduction

In 1955, Bellman [14] posed a seminal question in search theory: given a random point P within a region R, what trajectory minimizes the expected time to reach the boundary, and what is the maximum time required? This question laid the groundwork for search theory, inspiring subsequent research into optimal trajectories under uncertainty. Bellman also proposed studying specific cases, such as when R represents the area between two parallel lines separated by a known distance or when R is a semi-infinite plane with a known distance to

Due to space limitations, many of the proofs are omitted. An extended version of this work with all missing details is available on arXiv [23].

Research supported in part by NSERC Discovery grant and by the Toronto Metropolitan University Faculty of Science Dean's Research Fund.

U. Schmid and R. Kuznets (Eds.): SIROCCO 2025, LNCS 15671, pp. 262–280, 2025.
https://doi.org/10.1007/978-3-031-91736-3_16

the boundary. This problem also appears in Bellman's classic work on dynamic programming [15] (Problem 6, page 133), now recognized as the "lost-in-a-forest" problem. Over the decades, this question has inspired various modifications, with only a few specific cases fully resolved and many still open (see Sect. 1.1 for additional references). This question is a central reference in search theory [17] and has even appeared in the USSR Olympiad Problem Book [55]. Notably, the variant with two parallel lines is among the few examples that have been fully solved (see Problem 7, page 133, in [15]).

Bellman [14] introduced a notable variation of the problem as a simpler alternative to the lost-in-a-forest problem. This variation, solved by Isbell [44], is known as the *shoreline problem with known distance*. The problem seeks a minimum-length curve in $\mathbb{R}^2$ whose convex hull contains a disk of radius 1 centered at one endpoint. Alternatively, in modern mobile agent terms, it involves a unit-speed mobile agent tasked with finding a hidden infinite (shore)line at a known distance of 1, though its location is unknown. Isbell's solution provides the agent's optimal trajectory, minimizing the worst-case time to reach the shoreline.

The shoreline can also be interpreted as the visibility boundary for an agent tasked with inspecting (or patrolling) the perimeter of a unit disk, starting from its center. This application could represent inspecting the perimeter of a circular building for defects or intruders, where visibility of the exterior is possible only from outside the building. Specifically, a point on the disk is considered visible (inspected) if no convex combination of the agent's position and the point lies within the disk's interior, treating the disk as an obstacle to visibility. We refer to this task as the *1-agent inspection problem*, which Isbell resolved optimally by minimizing the worst-case inspection cost. Although originally posed as a theoretical question, Isbell's shoreline problem laid the groundwork for modern applications in operations research, particularly in autonomous navigation and notably in navy operations.

To the best of our knowledge, we are the first to consider the *multi-agent* variant of this problem. Our work explores both *worst-case* and *average-case* (equivalent to considering randomized search trajectories for) inspection times for n agents, with a focus on the *partial inspection problem*, where an agent is tasked to inspect only an arc of some fixed length on the unit disk. This generalization allows us to develop solutions for the multi-agent case, optimizing both worst-case and average-case costs. Our approach involves discretizing the problem, constructing feasible trajectories through Nonlinear Programming solutions, and quantifying inspection costs across discrete and continuous settings, thus enabling us also to address the multi-objective optimization of minimizing both worst-case and average-case costs.

1.1 Related Work

The study of search problems began in the 1950s and 1960s [12,16] and has since become a distinct field within operations research and theoretical computer science. Foundational surveys and texts [2–4,26,36] document its development. Early research focused on single-agent line search [8,46], while advances in robotics spurred interest in multi-agent variants [24,51].

In search problems, agents aim to locate hidden targets within a domain. Linear search problems, such as the classic cow-path problem, are fundamental and have been extensively studied for single [9] and multiple agents [21]. Recent work introduces variations minimizing weighted completion times [40] and adapts searches to complex domains such as rays [20], terrains [28], and graphs [5].

Two-dimensional search introduces added complexity, with early results for polygons [32] and disks [24], and expanded to include shapes like triangles, squares, ℓ_p disks, and regular polygons [10,27,37–39]. Multi-agent models in these contexts address coordination, speed [11], and fault tolerance [13,18,25]. The only related work addressing worst-case and average-case trade-offs in mobile agent search is in [22], examining single-agent search and evacuation in a unit disk.

The two-dimensional search problem most relevant to our work is Bellman's shoreline problem [15], which was first solved optimally in the worst-case sense for a single agent by Isbell [44], with the target shoreline at a known distance from the agent's starting point. Searching for a shoreline at a known distance with $n \geq 4$ agents was solved optimally in the worst-case sense in [1], while the cases of $n = 2, 3$ were treated in [29].

The more general lost-in-a-forest problem has been surveyed in [17,33]. Recent progress on the area include that for convex polygons in [41] and well as for general shapes using competitive analysis in [47]. Searching for a (shore)line without knowledge of its distance was introduced for a single robot in [8], with extensions for $n \geq 2$ robots explored in [6,7,9,45]. For $n = 1$, a logarithmic spiral achieves the best known competitive ratio of 13.81 [8,34], with lower bounds of 6.3972 unconditionally [7] and 12.5385 for cyclic trajectories [49]. For multiple robots, a double logarithmic spiral yields a competitive ratio of 5.2644 for $n = 2$ [7], and for $n \geq 3$, robots moving along rays achieve competitive ratio at most $1/\cos(\pi/n)$ [7], with matching lower bounds provided for $n \geq 4$ and $n = 3$ in [1] and [29], respectively. Variants where partial information is available yield varying competitive ratios based on distance, slope, or orientation knowledge [8,42–45].

Beyond shoreline search, strategies have been developed for circles [43] and other planar regions. Langetepe [48] proved that the spiral search is optimal for single-agent point search in two dimensions, while multi-agent and grid-based searches with memory constraints have been examined in [30,31,35,50]. Recent studies address point search in planar and geometric terrains [19,52] and explore trade-offs between search cost and information [53,54].

1.2 Roadmap

Section 2 presents the necessary preliminaries and result statements. Specifically, Sect. 2.1 formally introduces the problems studied and defines the notation, while Sect. 2.2 summarizes all results with appropriate quantification. The worst-case analysis begins in Sect. 3, with key inspection-related observations made and proved in Sect. 3.1. Section 3.2 then addresses the partial inspection problem, followed by the n-agent inspection problem for the worst case. The average-case problem is examined in Sect. 4. Specifically, Sect. 4.1 presents a discretized algorithm for the problem and quantifies its performance. This foundation allows us to address the partial continuous average inspection problem in Sect. 4.2. Finally, Sect. 5 explores the multi-objective problem.

Many proofs are omitted from this work due to space limitations. An extended version, with all proofs, can be found on arXiv [23].

2 Inspection Problems and New Contributions

2.1 Formal Definition and Some Auxiliary Problems

We formally define the main problems studied below, along with auxiliary problems whose solutions aid in deriving key results. All definitions are provided in this section for convenience.

n-Agent Disk Inspection Problem, $\mathcal{S}_n$: In this problem, n unit-speed agents (searchers) start from the center of a unit-radius disk. The agents can move freely within the plane of the disk, changing directions without added time costs. For a given trajectory, a point P on the disk's perimeter is said to be inspected if an agent passes through a point A in the plane such that no convex combination of A and P lies within the disk. The first time P is inspected by an agent is its *inspection time*, denoted by $\mathcal{I}(P)$. Feasible solutions to this problem are agents' trajectories that inspect every point on the disk's perimeter. In $\mathcal{S}_n$, the focus is on minimizing either the worst-case inspection time, $\sup_P \mathcal{I}(P)$, or the average-case inspection time, $\mathbb{E}_P \mathcal{I}(P)$, where the expectation is taken over a uniform distribution of P.

The problem of minimizing $\sup_P \mathcal{I}(P)$ for $\mathcal{S}_1$ was originally introduced and optimally solved in [44]. Also known as the *shoreline problem with known distance*, it seeks the shortest trajectory of a unit-speed agent to reach a hidden line located at a distance of 1 from the agent. Equivalently, the problem asks for the shortest continuous 1-dimensional curve, with one endpoint at the disk's center, whose convex hull encloses the disk. Minimizing $\sup_P \mathcal{I}(P)$ for $\mathcal{S}_n$ extends this description with n such curves.

For the problem of minimizing $\mathbb{E}_P \mathcal{I}(P)$ in $\mathcal{S}_n$, the relevant trajectories correspond to deterministic algorithms, evaluated by their average-case performance. Alternatively, these algorithms can be viewed as randomized, where the circle is uniformly rotated before execution. Under this randomized interpretation, the expected inspection time $\mathcal{I}(P)$ is identical for all points P and matches the expected performance of the deterministic algorithm under a uniformly random choice of P. Thus, minimizing $\mathbb{E}_P \mathcal{I}(P)$ can also be framed as minimizing the worst-case (expected) performance of a randomized algorithm.

Single-Agent Partial Disk Inspection Problem, $\mathcal{P}(c)$: Our approach to minimizing $\sup_P \mathcal{I}(P)$ for $\mathcal{S}_n$ is facilitated by examining the single-agent problem $\mathcal{P}(c)$, which involves inspecting an arc of length $c \in [0, 2\pi]$ on the disk's perimeter. Noting that in an optimal solution for $\mathcal{S}_n$, all agents should conclude their search simultaneously, each agent should inspect an arc of length $2\pi/n$. Thus, $\mathcal{S}_n$ is equivalent to $\mathcal{P}(2\pi/n)$.

Single-Agent Auxiliary Disk Inspection Problem, $\mathcal{A}(k, \theta, c)$: This problem is defined for parameters $k \in \mathbb{N} \cup \{\infty\}$, $\theta < \pi/2$, and $c \geq 2\theta$. It is inspired by optimal (worst-case performance) algorithms for $\mathcal{S}_n$. Here, a single-agent partial inspection problem is introduced where the agent's starting position is not at the origin but at a carefully selected point outside the disk's interior. This setup enables deriving results also for minimizing $\mathbb{E}_P\mathcal{I}(P)$ for $\mathcal{S}_n$.

The parameter k designates whether the inspection domain is discrete or continuous, with k specifying the number of discrete inspection points if applicable. The parameter c represents the range of points to be inspected, and θ specifies the agent's initial position. For simplicity, we consider a unit-radius disk centered at the origin in the 2D Cartesian plane, denoted O, with its perimeter parameterized as $\mathcal{P}_t := (\cos t, \sin t)$, where $t \in [0, 2\pi]$.

In $\mathcal{A}(k, \theta, c)$, the agent starts at $(1, \tan\theta)$. When $k = \infty$, the objective is to inspect points P_ϕ with $\phi \in [2\theta, c]$, noting that points P_ϕ in $[0, 2\theta]$ are already inspected due to the agent's initial placement, as we will prove later. When $k \in \mathbb{N}$, the goal is to inspect discrete points $P_i := \mathcal{P}_{\phi_i}$, where $\phi_i := 2\theta + \frac{c-2\theta}{k} i, \quad i = 0, \ldots, k$. Here, ϕ_i implicitly depends on the fixed values of k, θ, and c. As with $\mathcal{S}_n$, $\mathcal{A}(k, \theta, c)$ may aim to minimize either the worst-case or average-case inspection time. Later, we will connect and quantify solutions for $\mathcal{A}(\infty, \theta, c)$ and $\mathcal{A}(k, \theta, c)$, ultimately relating them to $\mathcal{S}_n$ for carefully selected values of k, θ, and c based on n.

Single-Agent Trade-Off Problem $\mathcal{T}(\lambda)$: With our results for $\mathcal{S}_n$ established, we consider the multi-objective optimization problem of minimizing both worst-case and average-case inspection times with a single agent. Alternatively, we approach bounding the Pareto-optimal set for these objectives by minimizing $\lambda \cdot \sup_P \mathcal{I}(P) + (1 - \lambda) \cdot \mathbb{E}_P\mathcal{I}(P)$, where $\lambda \in [0, 1]$. The trade-off problem $\mathcal{T}(\lambda)$ addresses real-life applications in which minimizing average performance is prioritized, subject to a manageable worst-case bound-such as cases with energy constraints that limit the operational duration of mobile agents.

2.2 Main Results Made Formal and Motivation

In this section we outline our main contributions. The motivating starting point for the present work is a result established by Isbell in 1957.

Theorem 1 ([44]). *The optimal worst case cost to $\mathcal{S}_1$ is $1+\sqrt{3}+\frac{7\pi}{6} \approx 6.39724$.*

We generalize the above result by proving, as a warm-up, the following.

Theorem 2. *The optimal worst case cost to $\mathcal{S}_n$ is $1 + \sqrt{3} + 2\pi/n - 5\pi/6$, if $n = 1, 2$, and $1/\cos(\pi/n)$, if $n \geq 3$.*

The results of Theorem 2 for $\mathcal{S}_n$, with $n \geq 2$, are also derived as corollaries from the lower bounds in [29] (for $n = 2, 3$) and in [1] for $n \geq 4$. These papers also address the search for a shoreline without knowledge of the distance from the starting point. However, some of the lower bound arguments in these works are independent of the shoreline's distance from the starting position, despite being based on competitive analysis. In our case, with known distance, we provide an alternative proof specifically designed for the problem where the distance is known. More significantly, we derive our results as corollaries of a solution to the more general problem of inspecting an arc of arbitrary length $c \in [0, 2\pi]$, a result not achieved in prior studies. Indeed, the proof of Theorem 2 follows directly from the broader Theorem 5, which establishes the optimal worst-case solution to the Partial Inspection Problem $\mathcal{P}(c)$ for every $c \in [0, 2\pi]$ and is of independent interest. In particular, for $c \geq 5\pi/6$ we show that an Isbell-type trajectory is the optimal solution to $\mathcal{P}(c)$, generalizing the optimal trajectory for $\mathcal{S}_1$ in [44] and the optimal trajectory for $\mathcal{S}_2$ in [29].

Our next main result pertains to upper bounds for the average-case inspection time to $\mathcal{S}_n$.

Theorem 3. *$\mathcal{S}_n$ can be solved in average-case inspection time* 3.5509015 *if* $n = 1$, 1.7946051 *if* $n = 2$, *and* $\frac{n}{2\pi} \log\left(\frac{1+\sin \pi/n}{1-\sin \pi/n}\right)$ *if* $n \geq 3$.

As before, our results are based on designing efficient average-cost inspection algorithms for the general problem $\mathcal{P}(c)$, $c \in [0, 2\pi]$, which may be of independent interest. The reader is referred to Fig. 1 for the derived trajectories minimizing the average-cost inspection time for $n = 1, 2$, corresponding to solutions for $\mathcal{P}(c)$ with $c = \pi$ and $c = 2\pi$.

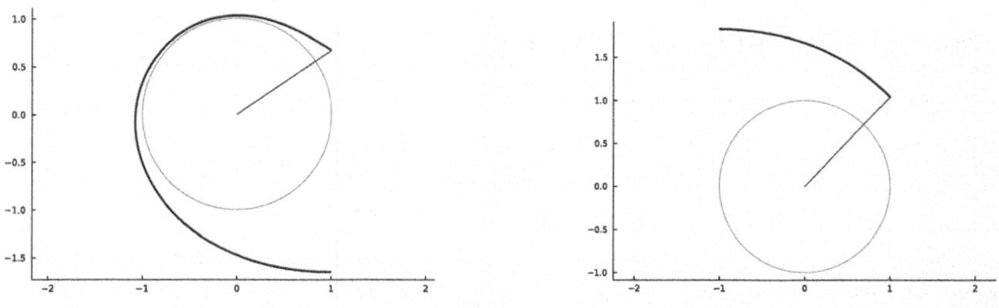

(a) The trajectory proving the average inspection upper bound to $\mathcal{S}_1$.

(b) The trajectory proving the average inspection upper bound to $\mathcal{S}_2$.

Fig. 1. Agent's trajectories of Theorem 3 for $n = 1, 2$.

Notably, the naive inspection algorithm for $\mathcal{S}_n$, where each agent inspects an arc of length $2\pi/n$ along the disk, yields an average inspection cost of $1 + \pi/n$, which is approximately 4.14159 and 2.5708 for $n = 1$ and $n = 2$, respectively. This improvement is substantial for all values of n, and asymptotically, our results show that the average inspection cost converges to $1 + \frac{\pi^2}{6n^2} + o\left(\frac{1}{n^3}\right)$.

In Sect. 5, we analyze the trade-offs between worst-case and average-case inspection costs for $\mathcal{S}_1$, focusing on minimizing the multi-objective function $\{\sup_P \mathcal{I}(P), \mathbb{E}_P\mathcal{I}(P)\}$. This is achieved by minimizing $\lambda \cdot \sup_P \mathcal{I}(P) + (1-\lambda) \cdot \mathbb{E}_P\mathcal{I}(P)$, for $\lambda \in [0,1]$. Theorem 2 (originally due to [44] for $\mathcal{S}_1$) provides the worst-case optimal inspection time for $\mathcal{S}_1$ as $1+\sqrt{3}+\frac{7\pi}{6} \approx 6.39724$, while the same trajectory has an average inspection cost of $\frac{5}{2\sqrt{3}} + \frac{91\pi}{144} + \frac{\log(6)}{2\pi} \approx 3.71386$, as shown in Lemma 6.

At the other end, Theorem 3 gives an upper bound on the average inspection cost at 3.5509015, corresponding to a trajectory length of approximately 6.8673829. We are thus motivated to establish upper bounds on average-case performance when the trajectory length varies between 6.39724 and 6.8673829.

To achieve this, we first determine the Pareto curve for the Isbell-type trajectories used in the proof of Theorem 1, parameterized by θ. For suitable values of θ, we compute in Lemma 6 both the worst-case and average-case costs for this family of algorithms, which serve as benchmarks for our subsequent results.

Finally, to improve the upper bounds for the multi-objective inspection problem, we extend our approach to minimize average-case inspection costs. This is accomplished using a discrete trajectory optimized through a Non-Linear Program with carefully selected parameters. The bounds for the Pareto curve of the multi-objective inspection problem are summarized in the following theorem.

Theorem 4. *Single Agent trade-off problem $\mathcal{T}(\lambda)$, where $\lambda \in [0,1]$ admits inspection trajectories with worst-case cost and average-case cost to $\mathcal{S}_1$ that are depicted in Fig. 2.*

3 Optimal Worst Case Inspection for $\mathcal{S}_n$

3.1 Some Preliminary Observations

To simplify the notation, we remind the reader of some existing abbreviations and introduce new ones. Throughout our analysis, we will work with a fixed $k \in \mathbb{N}$, and $\theta, c \in [0, 2\pi]$, where $\theta < \pi/2$ and $\theta \leq 2c$. For such fixed k, c, and θ, we define $\phi_i := 2\theta + \frac{c-2\theta}{k} i, \quad i = 0, \ldots, k$, and $P_i := \mathcal{P}_{\phi_i}, \quad i = 0, \ldots, k$ The values ϕ_i are chosen such that the intervals $[\phi_i, \phi_{i+1}]$ form a (non-disjoint) partition of the interval $[2\theta, c]$, for $i = 0, \ldots, k-1$. Additionally, the points P_i represent the endpoints of the corresponding arcs on the perimeter of the unit disk. For some $\phi \in [0, 2\pi]$, we also consider the parametric equation of the line $\mathcal{L}_\phi(t)$, which is tangent to the disk at the point $\mathcal{P}_\phi$: $\mathcal{L}_\phi(t) := \begin{pmatrix} \cos(\phi) \\ \sin(\phi) \end{pmatrix} + t \begin{pmatrix} \sin(\phi) \\ -\cos(\phi) \end{pmatrix}$. For the same fixed k, c, θ, we will also need to consider the tangent $\mathcal{L}_\phi(t)$ at $\phi = \phi_i$, hence it is convenient to introduce the abbreviation $L_i(t) := \mathcal{L}_{\phi_i}(t), \quad t \in \mathbb{R}, \quad i = 0, \ldots, k$.

The following lemma summarizes some important observations pertaining to conditions that guarantee when points on the perimeter of the disk are inspected.

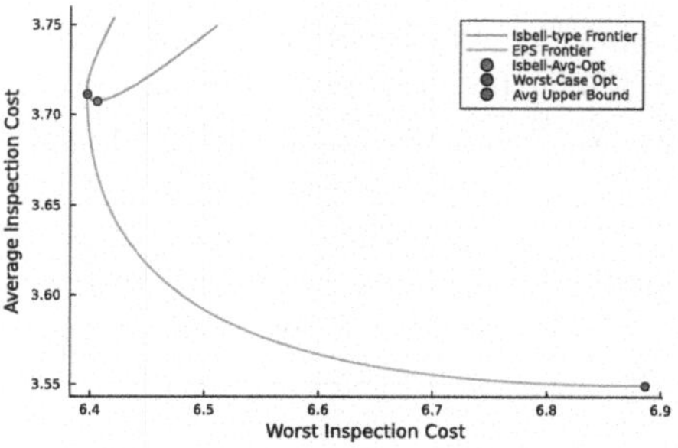

Fig. 2. Our Pareto upper bounds are depicted in orange and are achieved by the Extended-Poly-Segment (EPS) trajectory we define later. The blue line depicts the Pareto Frontier of the Isbell-type trajectories, parameterized by the initial deployment angle θ. The provable worst-case optimal solution and its corresponding average inspection performance are shown as the blue Pareto point. The red Pareto point corresponds to the performance of the Isbell-type trajectory that is average cost optimal. Finally, the green Pareto point is the performance of the EPS-type trajectory with the smallest average inspection cost. (Color figure online)

Lemma 1.

(a) An agent at point with polar coordinates (r, ϕ) and $r \geq 1$ inspects all points $\mathcal{P}_t$ that satisfy $|t - \phi| \leq \arccos(1/r)$.
(b) For every $\theta \in [0, \pi/2)$, an agent at point $(1, \tan(\theta))$ inspects all $\mathcal{P}_t$, with $t \in [0, 2\theta]$.
(c) An agent inspects point $\mathcal{P}_\phi$ if and only if it lies in the closed halfspace defined by $\mathcal{L}_\phi$ that does not contain the interior of the unit disk. In particular, an agent residing on any point of $\mathcal{L}_\phi$ inspects $\mathcal{P}_\phi$.
(d) Consider points $A = \mathcal{L}_\phi(t_1)$ and $B = \mathcal{L}_{\phi+\gamma}(t_2)$ satisfying $\gamma < \pi/2$, $t_1 \geq 0$ and $t_2 \geq \frac{1-\cos(\chi)}{\sin(\chi)}$. Then, for each $\phi \leq \chi \leq \chi + \gamma$, point $\mathcal{P}_\chi$ is inspected by some convex combination of A, B. In other words, the movement $A \to B$, inspects all points $\mathcal{P}_\chi$, with $\chi \in [\phi, \phi + \gamma]$.

By Lemma 1(c), $\mathcal{P}_\phi$ is inspected if an agent lies in the halfspace defined by $\mathcal{L}_\phi(t)$ that does not contain the disk. Therefore, unless an agent is not already in the halfspace of $\mathcal{L}_\phi(t)$ not containing the disk, $\mathcal{I}(\mathcal{P}_\phi)$ is also equal to the first time an agent hits line $\mathcal{L}_\phi(t)$.

3.2 Solving $\mathcal{S}_n$ Using the Partial Inspection Problem $\mathcal{P}(c)$

The following theorem establishes the optimal worst-case cost for solving the partial inspection problem $\mathcal{P}(c)$. Clearly, $\mathcal{P}(2\pi)$ is equivalent to the problem $\mathcal{S}_1$, and indeed, the case $c = 2\pi$ and the corresponding optimal result (see Theorem 1, first proved in [44]) are included in the statement below.

Theorem 5. *The optimal worst case inspection cost to $\mathcal{P}(c)$ is $\frac{1}{\cos(c/2)}$, if $0 \leq c \leq \frac{2\pi}{3}$, $1 - 2\cos(c)$, if $\frac{2\pi}{3} < c \leq \frac{5\pi}{6}$, and $1 + \sqrt{3} + c - \frac{5\pi}{6}$, if $\frac{5\pi}{6} < c \leq 2\pi$.*

To prove Theorem 5 for $\mathcal{P}(c)$, we identify the shortest curve, C, that inspects all points along the arc of length c. The proof distinguishes between two types of curves: Type-1 (Fig. 3a), which intersects the disk perimeter at one or two points and applies when $0 \le c \le 3\pi/2$, and Type-2 (Isbell-type) curves (Fig. 3b), which intersect the perimeter infinitely and apply when $c \ge \pi/2$. For each configuration, we derive conditions for minimal length based on convexity constraints around the disk, ensuring that C fully covers the arc while minimizing total distance.

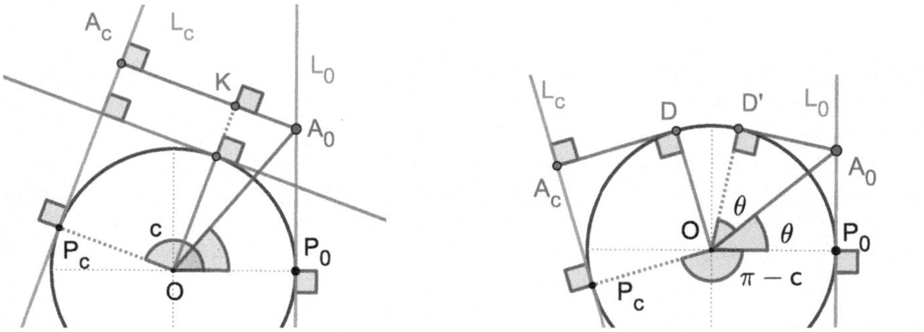

(a) Type-1 curves (in green) for solving $\mathcal{P}(c)$, applicable when $0 \le c \le 3\pi/2$ and $2\theta \ge c - \pi/2$.

(b) Type-2 (Isbell-type) curves (in green) for solving $\mathcal{P}(c)$, applicable when $\pi/2 \le c \le 2\pi$ and $2\theta \le c - \pi/2$.

Fig. 3. Trajectories achieving the worst-case optimal inspection cost to $\mathcal{P}(c)$ as per Theorem 5.

We note that the optimal worst case inspection cost to $\mathcal{P}(c)$ is continuous at $c = 2\pi/3$ and at $c = 5\pi/6$ with corresponding worst case costs equal to 2 and $1 + \sqrt{3}$, respectively. Moreover, and not surprisingly, it is easy to see that the cost is monotonically increasing in c.

As an immediate corollary we obtain the optimal worst case cost for the general inspection problem $\mathcal{S}_n$, as stated in Theorem 2.

Proof (of Theorem 2). Consider any partition of the perimeter of the disk that assigns points to inspect to the n agents. This defines a number of arcs covering the perimeter of the disk, each labeled according to the assigned agents. It is easy to see that the arcs in the partition can be rearranged (each agent still being assigned the same total portion of the disk) so that those with the same label are contiguous, without increasing the cost.

Hence, each agent needs to inspect a contiguous arc. Next, we claim that all agents inspect arcs of the same length. If not, we could slightly decrease the largest arc and slightly increase the smallest arc. From Theorem 5, the optimal cost for covering any arc of length c is monotonically increasing in c, and thus this adjustment would reduce the overall inspection cost.

We conclude that each agent inspects an arc of length $c = 2\pi/n$. The claim of the theorem follows from Theorem 5, which describes the worst-case optimal cost for inspecting an arc of length c, for every $c \in [0, 2\pi]$. □

4 Average-Case Inspection Upper Bounds for $\mathcal{S}_n$

4.1 Discrete Inspection Algorithms for $\mathcal{A}(k, \theta, c)$ and $\mathcal{P}(c)$

In this section, we introduce and analyze the trajectory (Algorithm) PolySegment for solving $\mathcal{A}(k, \theta, c)$ and the trajectory (Algorithm) ExtendedPolySegment, which, under certain conditions, will solve the 1-agent partial inspection problem $\mathcal{P}(c)$ and eventually the n-agent inspection problem $\mathcal{S}_n$.

Recall that $\theta \in [0, \pi/2)$, and $c \in [0, 2\pi]$, with $c \geq 2\theta$. We fix these constants and also fix $k \in \mathbb{N}$, i.e., for the description of the trajectory, we require k to be finite. The algorithm will be parameterized by $(t_1, \ldots, t_k)$, where $t_i \in \mathbb{R}$. Recall that in $\mathcal{A}(k, \theta, c)$, we need to inspect points P_i with angular values $\phi_i := 2\theta + \frac{c-2\theta}{k} i, \quad i = 0, \ldots, k$. Note that the remaining of the necessary notation was introduced earlier in Sect. 3.1.

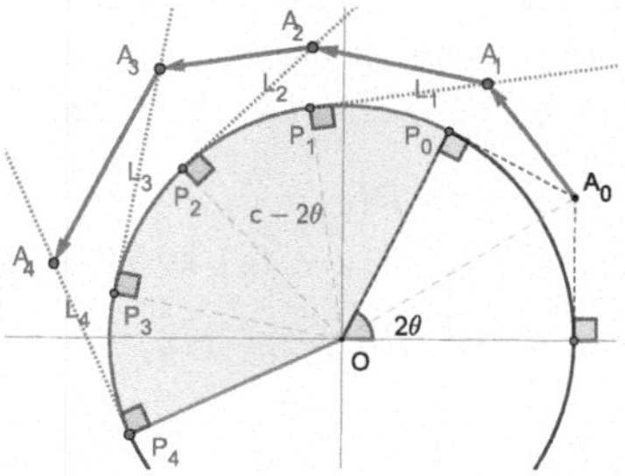

Fig. 4. Example with $k = 4$, showing $\mathrm{PS}(t_1, \ldots, t_4)$ as feasible for $\mathcal{A}(4, \theta, c)$. The light blue sector represents the disk segment of arc length $c - 2\theta$, which must be inspected. Tangent lines $L_i(t)$ (dotted red) are shown for $t \geq 0$. We define $A_0 = (1, \tan(\theta))$, where there exists t_0 such that $A_0 = \mathcal{L}_{\phi_0}(t_0)$ for $\phi_0 = 2\theta$. Points A_i follow as $A_i = L_i(t_i)$ with $t_i > 0$. The trajectory $A_0 \to A_1 \to A_2 \to A_3 \to A_4$ is feasible for both $\mathcal{A}(4, \theta, c)$ and $\mathcal{A}(\infty, \theta, c)$. Feasibility for $\mathcal{A}(\infty, \theta, c)$ is analyzed later, as $t_i > 0$ alone is insufficient. (Color figure online)

Algorithm 6 (*PolySegment* $(t_1, \ldots, t_k)$). *Given k, θ, c with $\theta < \pi/2$, $c \geq 2\theta$, we set $A_0 := (1, \tan(\theta))$, and $A_i := L_i(t_i)$. Then, the trajectory is defined by the movement between points $A_0 \to A_1 \to \ldots \to A_k$.*

For fixed k, θ, c, our goal will be to determine optimal values for $(t_1, \ldots, t_k)$ so as to minimize the average case performance to $\mathcal{A}(k, \theta, c)$. For the ease of notation, we will refer to the trajectory as $\mathrm{PS}(t_1, \ldots, t_k)$. See Fig. 4 for an illustration. Next we comment on the algorithms feasibility and compute its average case inspection cost.

Lemma 2. *For every $k \in \mathbb{N}$ and for all $t_i \in \mathbb{R}$, trajectory $\mathrm{PS}(t_1, \ldots, t_k)$ is feasible to $\mathcal{A}(k, \theta, c)$. Moreover, trajectory $\mathrm{PS}(t_1, \ldots, t_k)$ has worst case cost $\sum_{j=1}^{k} \|A_{j-1}A_j\|$ and average case cost $\frac{1}{k+1} \sum_{j=1}^{k} (k - j + 1)\|A_{j-1}A_j\|$.*

Trajectory PolySegment($t_1, \ldots, t_k$) will be used to derive upper bounds to $\mathcal{S}_n$. Below we extend the trajectory, and we quantify its performance assuming it is feasible for the continuous problem. In that direction, we present trajectory (algorithm) ExtendedPolySegment for solving $\mathcal{P}(c)$.

Algorithm 7 (*ExtendedPolySegment* $(\theta, t_1, \ldots, t_k)$). *Given $c \in [0, 2\pi], k \in \mathbb{N}$ and $\theta \le \{c/2, \pi/2 - \epsilon\}$ for some $\epsilon > 0$, we set $A_0 := (1, \tan(\theta))$, and $A_i := L_i(t_i)$.*[1] *Then, the trajectory of the agent is defined by the movement between points $O \to A_0 \to A_1 \to \ldots \to A_k$, where O is the center of the disk.*

For convenience, we introduce notation $\mathrm{EPS}(\theta, t_1, \ldots, t_k)$ to refer to the algorithm above. Next, we describe its performance on $\mathcal{P}(c)$ as a function of its performance on the discrete problem, assuming also that it is feasible for the problem. We will introduce the feasibility conditions later.

Lemma 3. *Let $k \in \mathbb{N}$ and $\theta \le \{c/2, \pi/2 - \epsilon\}$ for some $\epsilon > 0$. Suppose also that $\mathrm{PS}(t_1, \ldots, t_k)$*[2] *is feasible to $\mathcal{A}(\infty, \theta, c)$ with worst case cost and the average case inspection cost equal to s_1 and s_2, respectively. Then,*

(a) $\mathrm{EPS}(\theta, t_1, \ldots, t_k)$ is feasible to $\mathcal{P}(c)$.
(b) The worst case cost of $\mathrm{EPS}(\theta, t_1, \ldots, t_k)$ to $\mathcal{P}(c)$ is $\frac{1}{\cos(\theta)} + s_1$.
(c) The average case cost of $\mathrm{EPS}(\theta, t_1, \ldots, t_k)$ to $\mathcal{P}(c)$ is $\frac{1}{c} \log\left(\frac{1+\sin(\theta)}{1-\sin(\theta)}\right) + \left(1 - \frac{2\theta}{c}\right) s_2$.

We are now ready to introduce the feasibility conditions of the discrete inspection algorithm for the continuous inspection problems, as well as to relate the performance of the algorithm for the discrete and continuous inspection problems. As a reminder, the parameter k quantifies the refinement of the continuous space into a discrete set of points to be inspected. The value of k that we will eventually use is on the order of thousands, and the higher it is, the better our results for the continuous problem. The requirement in the lemma that $k \ge 5$ becomes apparent only in the proof, where we need the points to be inspected to be strictly less than $\pi/2$ apart.

Lemma 4. *Consider $k \ge 5 \in \mathbb{N}$ and $t_i \in \mathbb{R}$, satisfying $t_i \ge \frac{1-\cos\left(\frac{c-2\theta}{k}\right)}{\sin\left(\frac{c-2\theta}{k}\right)}, i = 1, \ldots, k$. Then, trajectory $\mathrm{PS}(t_1, \ldots, t_k)$ is feasible to $\mathcal{A}(\infty, \theta, c)$. Moreover, if s is the average case cost for the same trajectory to $\mathcal{A}(k, \theta, c)$, then the average case cost to $\mathcal{A}(\infty, \theta, c)$ is at most $\left(1 + \frac{1}{k}\right) s$.*

4.2 Upper Bounds to the Average Case Inspection of $\mathcal{P}(c)$ and $\mathcal{S}_n$

In this section we introduce a Non Linear Program (NLP-avg(k, c, ϵ)) for finding a feasible ExtendedPolySegment trajectory to $\mathcal{P}(c)$ and for minimizing its

[1] Recall that the definition of lines L_i is tailored to the fixed k, θ, c.
[2] Note that the trajectory points are tailored to the fixed values of k, θ, c.

average case performance. The algorithm also takes as input some $k \geq 5$ and $c \in [0, 2\pi]$, ensuring that the points to be inspected on the disk's perimeter are less than $\pi/2$ apart. In fact, the values of k we use are on the order of thousands, and the higher they are, the better our results for the continuous problem.

For fixed k, θ, and using the notation introduced at the beginning of Sect. 3.1, we write explicit formulas for points $A_0, A_1, \dots, A_k$ used by algorithm $\text{EPS}(\theta, t_1, \dots, t_k)$ (see the formal description in 7) with respect to the NLP variables $\theta \in \mathbb{R}$ and $t \in \mathbb{R}^k$. For this, we also recall that $A_i = L_{\phi_i}(t_i)$ for $i = 1, \dots, k$, and that $A_0 = (1, \tan(\theta))$.

Therefore, it is easy to see that $\|A_0A_1\|^2 = (-1 + c'_{\theta,k} + s'_{\theta,k}t_1)^2 + (-s'_{\theta,k} + c'_{\theta,k}t_1 + \tan(\theta))^2$, and $\|A_iA_{i+1}\|^2 = 2 - 2c_{\theta,k} + 2s_{\theta,k}t_i + t_i^2 - 2(s_{\theta,k} + c_{\theta,k}t_i)t_{i+1} + t_{i+1}^2$, where $c'_{\theta,k} = \cos((2(\pi + (-1+k)\theta))/k)$, $s'_{\theta,k} = \sin((2(\pi + (-1+k)\theta))/k)$, $c_{\theta,k} = \cos((2(\pi - \theta))/k)$, and $s_{\theta,k} = \sin((2(\pi - \theta))/k)$. We are now ready to write down the promised NLP, which, based on the observations above, is defined on the variables $t \in \mathbb{R}^k$ and $\theta \in \mathbb{R}$. The value of $\epsilon > 0$ is provided as part of the input and is chosen to be sufficiently small, ensuring that the expressions are well-defined over the domain of θ.

$$\min_{t \in \mathbb{R}^k, \theta \in \mathbb{R}} \frac{1}{c} \log\left(\frac{1+\sin(\theta)}{1-\sin(\theta)}\right) + \left(1 + \frac{1}{k}\right)\left(1 - \frac{2\theta}{c}\right)\left(\|OA_0\| + \sum_{i=0}^{k-1} \frac{k-i}{k+1}\|A_iA_{i+1}\|\right) \qquad (\text{NLP-avg}(k, c, \epsilon))$$

$$\text{s.t.} \quad \sin\left(\frac{c-2\theta}{k}\right) t_i \geq 1 - \cos\left(\frac{c-2\theta}{k}\right), \quad i = 1, \dots, k, \qquad (1)$$

$$0 \leq \theta \leq \min\left\{\frac{(1-\epsilon)c}{2}, \frac{\pi}{2} - \epsilon\right\}. \qquad (2)$$

Lemma 5. *Fix $\epsilon > 0$, small enough. For any $(\theta, t) \in \mathbb{R}^k$ feasible to (NLP-avg(k, c, ϵ)) with $k \geq 5$, $\text{EPS}(\theta, t_1, \dots, t_k)$ is a feasible inspection algorithm to $\mathcal{P}(c)$. Moreover the objective to the NLP quantifies the corresponding average case cost of the algorithm.*

Proof. For a solution $(\theta, t) \in \mathbb{R}^k$ to (NLP-avg(k, c, ϵ)), we observe that $t_i \geq (1 - \cos(\frac{c-2\theta}{k}))/\sin(\frac{c-2\theta}{k})$, for $i = 1, \dots, k$. By Lemma 4 the trajectory $\text{PS}(t_1, \dots, t_k)$ is feasible to $\mathcal{A}(\infty, \theta, c)$, and if its average case cost to $\mathcal{A}(k, \theta, c)$ equals s, then the average case cost to $\mathcal{A}(\infty, \theta, c)$ is at most $\left(1 + \frac{1}{k}\right)s$.

Then, by Lemma 3, we obtain that $\text{EPS}(\theta, t_1, \dots, t_k)$ is also feasible to $\mathcal{P}(c)$, with overall cost $\frac{1}{c}\log\left(\frac{1+\sin(\theta)}{1-\sin(\theta)}\right) + \left(1 + \frac{1}{k}\right)\left(1 - \frac{2\theta}{c}\right)s$. Finally, Lemma 2 computes quantity s, which results to the objective of (NLP-avg(k, c, ϵ)), as promised. □

Next we provide upper bounds to the average inspection cost for $\mathcal{P}(c)$.

Theorem 8. *$\mathcal{P}(c)$ has an average inspection cost solution shown in Fig. 5.*

For values of c from 0.01 to 2π with a step size of 0.01, we compute a numerical solution to (NLP-avg(k, c, ϵ)) using $k = 1000$ and $\epsilon = 10^{-3}$. By applying Lemma 5, we confirm local optimality of the solutions via Julia's JuMP and Ipopt, which employ an interior-point method. Notably, the constraint

$\theta \le \min\{(1-\epsilon)c/2, \pi/2-\epsilon\}$ is satisfied with slack for values $c \ge 2.39$, and the derived bounds are based on the numerical solutions to the NLP. For $c \le 2.39$, we report a closed formula, obtained by using $\theta = c/2$ in Lemma 3, part (c).

We conjecture that our reported average inspection times are optimal. Interestingly, from our numerical results, we see that the reported average inspection cost is eventually very close to be linear, i.e. it is well approximated by $0.5395026c + 0.1637722$ for $c \ge 4.3$. Some indicative and notable trajectories solving $\mathcal{P}(c)$ and proving Theorem 8 are those corresponding to $c = 2\pi$ and $c = \pi$, see Fig. 1 on page 6.

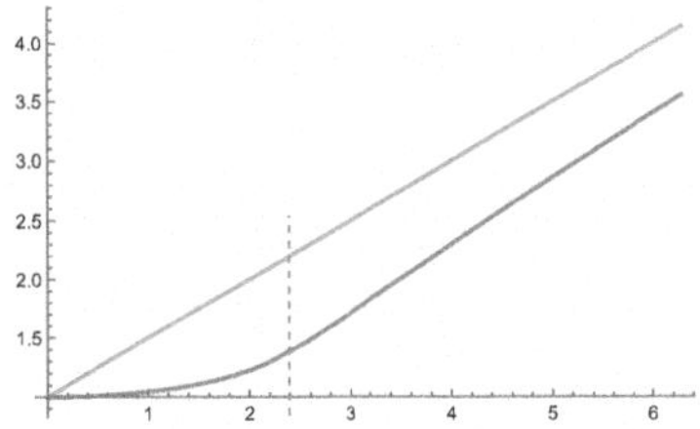

Fig. 5. The reported average inspection cost (in blue) of algorithm EPS$(\theta, t_1, \ldots, t_k)$ as per Theorem 8. For values $c \le 2.39$, our solution admits an analytic formula for the average cost. For comparison, we also display in yellow the performance $1 + c/2$ of the plain-vanilla randomized algorithm. (Color figure online)

We are now ready to prove Theorem 3.

Proof (of Theorem 3). The upper bounds on the average inspection cost for $\mathcal{S}_n$ are obtained from the upper bounds of $\mathcal{P}(c)$ in Theorem 8, by setting $c = 2\pi/n$. Since all agents are identical, each is assigned an arc of length $2\pi/n$ to inspect, and the n agents' trajectories are obtained by rotating one agent's trajectory by an angle of $2\pi/n$, $n-1$ times.

We note that for $n \ge 3$ agents, the length $c = c(n)$ of the arc that each agent inspects satisfies $c \le 2.39$, and as such the inspection trajectories we report have average cost $\frac{1}{c}\log\left(\frac{1+\sin(c/2)}{1-\sin(c/2)}\right)$, as per Lemma 3, part (c). More specifically, one agent's trajectory is given by the line segment OA, where $A = (1, \tan(c/2))$, and all other trajectories are rotations of this segment by $2\pi/n$.

It follows that only for $n = 1, 2$ are the trajectories more elaborate, requiring the numerical solution to (NLP-avg(k, c, ϵ)) for $c = 2\pi$ and $c = \pi$, and for large enough k. In that direction, we refine the results reported in Theorem 8, and we solve (NLP-avg(k, c, ϵ)) for $k = 2000$ inspection points to obtain slightly improved upper bounds, as shown in Table 3 on page 6.

In Fig. 1 on page 6, we depict the derived trajectory for $\mathcal{P}(2\pi)$ and $\mathcal{P}(\pi)$, corresponding to solutions to $\mathcal{S}_1$ and $\mathcal{S}_2$, respectively. Finally, the values returned by the NLP for the angle θ for $\mathcal{S}_1$ and $\mathcal{S}_2$ were $\theta = 0.5910554, 0.8054878$, respectively (while for $n \ge 3$, we use $\theta = \pi/n$ as explained before). $\square$

5 Worst/Average Case Inspection Trade-Offs to $\mathcal{S}_1$

The purpose of this section is twofold. First, we demonstrate that the upper bound we achieve for the average inspection cost of $\mathcal{S}_1$ using EPS$(\theta, t_1, \ldots, t_k)$ represents a significant improvement over previously known trajectories. Second, we show that our techniques also address the multi-objective optimization problem of simultaneously minimizing the worst-case and average-case inspection times. This is achieved by considering the trade-off problem $\mathcal{T}(\lambda)$, leading to the proof of one of our main contributions, Theorem 4, which establishes Pareto upper bounds for the multi-objective inspection problem.

5.1 Pareto Upper Bounds

We begin by considering the *Isbell-type* family of trajectories for $\mathcal{S}_1$, introduced in [44] as part of Theorem 1, parameterized by the deployment angle θ. Figure 3b shows this trajectory for $\mathcal{P}(c)$. This family of trajectories also appears in Theorem 5, where it establishes the optimal worst-case inspection solution for $\mathcal{P}(c)$ with $c \in [5\pi/6, 2\pi]$ (where $\mathcal{P}(2\pi)$ corresponds to $\mathcal{S}_1$).

In the *Isbell-type* trajectory with parameter $\theta \in [0, \pi/2)$, we define the point $A_0 = (1, \tan(\theta))$. The agent starts from the center of the disk and moves to A_0 (see Fig. 3b on page 9 and Fig. 6a on page 15). The agent then follows the tangent point $D' = \mathcal{P}_{2\theta}$ on the disk, searches the disk counterclockwise to point $D = \mathcal{P}_{3\pi/2}$, and finally moves along the projection to the tangent line at $\mathcal{P}_{2\pi}$, given by the line $x = 1$.

Lemma 6. *For every $\theta \in [0, \pi/2)$, the Isbell-type trajectory with deployment angle θ has the following inspection performance.*

(a) The worst case inspection equals $I_W(\theta) := 1/\cos(\theta) + \tan(\theta) + 3\pi/2 - 2\theta + 1$.
(b) The average case inspection cost $I_A(\theta)$ equals

$$\frac{1}{2\pi}\left(\log\left(\frac{1+\sin(\theta)}{1-\sin(\theta)}\right) + 2\theta^2 - 4\pi\theta + 2(\pi-\theta)\tan(\theta) + \frac{2(\pi-\theta)}{\cos(\theta)} + \frac{15\pi^2}{8} + \log 2\right)$$

The value of θ that minimizes the worst-case inspection cost $I_W(\theta)$, as described in Lemma 6, is $\theta_W = \pi/6 \approx 0.523599$ (see also Theorem 1 from [44], or our more general Theorem 2 for $n = 1$). For this value of θ_W, we have $I_W(\theta_W) = 1 + \sqrt{3} + \frac{7\pi}{6} \approx 6.39724$ and $I_A(\theta_W) = \frac{5}{2\sqrt{3}} + \frac{91\pi}{144} + \frac{\log(6)}{2\pi} \approx 3.71386$. Similarly, the value of θ that minimizes the average inspection cost $I_A(\theta)$, from Lemma 6, is $\theta_A = 0.592334$, which results in $I_W(\theta_A) \approx 6.406$, and $I_A(\theta_A) \approx 3.70737$. We expect the Pareto frontier of Isbell-type algorithms to be determined by $\theta \in [0.52359, 0.59233]$. However, for completeness, we plot the pairs $I(\theta) := (I_W(\theta), I_A(\theta))$ for all $\theta \in [0.4, 0.75]$ in Fig. 2, on page 8.

In Theorem 3, we achieved an average inspection cost as low as 3.5509015. We are ready to demonstrate how to significantly improve both the worst-case and average inspection costs across the entire Pareto frontier, effectively proving Theorem 4.

Proof (of Theorem 4). The upper bound to the Pareto frontier will be obtained by the family of algorithms EPS$(\theta, t_1, \ldots, t_k)$, with $k = 1000$, and for appropriate choices of parameters θ, t_i.

For this, we recall Lemma 3(b), from which for any feasible choice of the parameters, we obtain a formula $S_W(\theta, t_1, \ldots, t_k)$ for the worst case cost. Let also $S_A(\theta, t_1, \ldots, t_k)$ denote the objective of (NLP-avg(k, c, ϵ)), with $c = 2\pi$ and $\epsilon = 10^{-3}$. For all NLP solutions we obtain, the optimizer values for θ are bounded away from $\pi/2$ (in fact, they admit values in $[0.5, 0.6]$), and hence the value of ϵ does not affect the cost of the solution, rather only is there to justify its correctness. According to Lemma 5, $S_A(\theta, t_1, \ldots, t_k)$ correctly quantifies an upper bound to the average inspection cost to EPS$(\theta, t_1, \ldots, t_k)$, for any feasible solution $(\theta, t_1, \ldots, t_k)$ to the NLP.

Consequently, the new nonlinear program with objective $\lambda S_W(\theta, t_1, \ldots, t_k) + (1 - \lambda) S_A(\theta, t_1, \ldots, t_k)$ and subject to the same constraints as in (NLP-avg(k, c, ϵ)), correctly quantifies the objective for the trade-off problem $\mathcal{T}(\lambda)$.

For every λ from 0 to 1 and with step size 0.01, solving the new NLP we obtain a solution $(\theta(\lambda), t_1(\lambda), \ldots, t_k(\lambda))$, and we define $N_W(\lambda) := S_W(\theta(\lambda), t_1(\lambda), \ldots, t_k(\lambda))$, and $N_A(\lambda) := S_A(\theta(\lambda), t_1(\lambda), \ldots, t_k(\lambda))$. Thus, $N_W(\lambda)$ and $N_A(\lambda)$ accurately quantify the worst-case inspection cost and average inspection cost, respectively, for feasible EPS$(\theta(\lambda), t_1(\lambda), \ldots, t_k(\lambda))$. As expected, $\lambda = 1$ provides an approximation of the worst-case optimal inspection trajectory and a precise quantification of its average inspection cost. Indeed, $N_W(1) \approx 6.3972578$ and $N_A(1) \approx 3.71130923$ (compare these to $I_W(\theta_W)$ and $I_A(\theta_W)$ respectively of the Isbell-type algorithm, and note that due to the feasibility lower bound on the t_i's imposed in the NLP, the average inspection cost is slightly improved, whereas the worst case cost is slightly compromised). Similarly, the solution for $\lambda = 0$ yields the average inspection upper bound $N_A(0)$ from Theorem 3 for $\mathcal{P}(2\pi)$ (equivalently, $\mathcal{S}_1$) and a worst-case inspection cost of $N_W(0) = 6.8867382$. We conclude with Fig. 2 on page 8, where we depict the pair $(N_W(\lambda), N_A(\lambda))$ for $\lambda \in [0, 1]$. □

Trajectories that certify the reported average inspection cost upper bounds for $\mathcal{T}(\lambda)$ are depicted in Fig. 6, with $\lambda = 1$ corresponding to optimizing the worst-case cost, and for $\lambda = 2/3$ and $\lambda = 1/3$. Recall that the trajectory certifying the upper bound for $\mathcal{T}(0)$, which minimizes the average inspection cost, is shown in Fig. 1a on page 6.

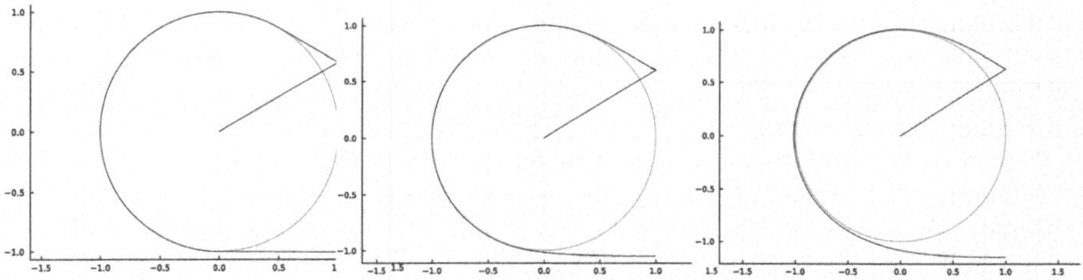

(a) The trajectory of the optimal $\mathrm{EPS}(\theta, t_1, \ldots, t_k)$ for solving $\mathcal{T}(1)$.

(b) The trajectory of the optimal $\mathrm{EPS}(\theta, t_1, \ldots, t_k)$ for solving $\mathcal{T}(2/3)$.

(c) The trajectory of the optimal $\mathrm{EPS}(\theta, t_1, \ldots, t_k)$ for solving $\mathcal{T}(1/3)$.

Fig. 6. Agent's trajectories for trade-off problem $\mathcal{T}(\lambda)$ of the proof of Theorem 3, for $\lambda = 1, 2/3, 1/3$, using algorithm $\mathrm{EPS}(\theta, t_1, \ldots, t_k)$ with $k = 1000$. The trajectory corresponding to $\lambda = 0$ (i.e. the best average performance of $\mathrm{EPS}(\theta, t_1, \ldots, t_k)$ solving $\mathcal{S}_1$) can be seen in Fig. 1a.

References

1. Acharjee, S., Georgiou, K., Kundu, S., Srinivasan, A.: Lower bounds for shoreline searching with 2 or more robots. In: 23rd OPODIS. LIPIcs, vol. 153, pp. 26:1–26:11. Schloss Dagstuhl - LZI (2019)
2. Ahlswede, R., Wegener, I.: Search Problems. Wiley, Hoboken (1987)
3. Alpern, S., Fokkink, R., Gasieniec, L., Lindelauf, R., Subrahmanian, V.S.: Search Theory. Springer, Heidelberg (2013)
4. Alpern, S., Gal, S.: The Theory of Search Games and Rendezvous, vol. 55. Springer, Heidelberg (2006)
5. Angelopoulos, S., Dürr, C., Lidbetter, T.: The expanding search ratio of a graph. Discret. Appl. Math. **260**, 51–65 (2019)
6. Baeza-Yates, R.: Searching: an algorithmic tour. Encycl. Comput. Sci. Technol. **37**, 331–359 (1997)
7. Baeza-Yates, R., Schott, R.: Parallel searching in the plane. Comput. Geom. **5**(3), 143–154 (1995)
8. Baeza-Yates, R.A., Culberson, J.C., Rawlins, G.J.E.: Searching with uncertainty. In: Scandinavian Workshop on Algorithm Theory, pp. 176–189. Springer (1988)
9. Baezayates, R.A., Culberson, J.C., Rawlins, G.: Searching in the plane. Inf. Comput. **106**(2), 234–252 (1993)
10. Bagheri, I., Narayanan, L., Opatrny, J.: Evacuation of equilateral triangles by mobile agents of limited communication range. In: Dressler, F., Scheideler, C. (eds.) ALGOSENSORS 2019. LNCS, vol. 11931, pp. 3–22. Springer (2019)
11. Bampas, E., et al.: Linear search by a pair of distinct-speed robots. Algorithmica **81**(1), 317–342 (2019)
12. Beck, A.: On the linear search problem. Israel J. Math. **2**(4), 221–228 (1964)
13. Behrouz, P., Konstantinidis, O., Leonardos, N., Pagourtzis, A., Papaioannou, I., Spyrakou, M.: Byzantine fault-tolerant protocols for (n, f)-evacuation from a circle. In: International Symposium on Algorithmics of Wireless Networks, pp. 87–100. Springer (2023)

14. Bellman, R.: Minimization problem. Bull. Amer. Math. Soc **62**(3), 270 (1956)
15. Bellman, R.: Dynamic Programming. Chapter IX. Princeton University Press, Princeton (1958)
16. Bellman, R.: An optimal search. SIAM Rev. **5**(3), 274 (1963)
17. Berzsenyi, G.: Lost in a forest (a problem area initiated by the late Richard E. Bellman). Quantum **41** (1995)
18. Bonato, A., Georgiou, K., MacRury, C., Prałat, P.: Algorithms for p-faulty search on a half-line. Algorithmica 1–30 (2022)
19. Bouchard, S., Dieudonné, Y., Pelc, A., Petit, F.: Deterministic treasure hunt in the plane with angular hints. In: 29th International Symposium on Algorithms and Computation, ISAAC 2018, vol. 123, pp. 48–1. Schloss Dagstuhl–Leibniz-Zentrum fuer Informatik (2018)
20. Brandt, S., Foerster, K.-T., Richner, B., Wattenhofer, R.: Wireless evacuation on m rays with k searchers. Theor. Comput. Sci. **811**, 56–69 (2020)
21. Chrobak, M., Gasieniec, L., Gorry, T., Martin, R.: Group search on the line. In: Italiano, G.F., Margaria-Steffen, T., Pokorný, J., Quisquater, J.-J., Wattenhofer, R. (eds.) SOFSEM. LNCS, vol. 8939, pp. 164–176. Springer (2015)
22. Chuangpishit, H., Georgiou, K., Sharma, P.: A multi-objective optimization problem on evacuating 2 robots from the disk in the face-to-face model; trade-offs between worst-case and average-case analysis. Information **11**(11), 506 (2020)
23. Conley, J., Georgiou, K.: Multi-agent disk inspection. arXiv preprint arXiv:2411.15391 (2024)
24. Czyzowicz, J., Gasieniec, L., Gorry, T., Kranakis, E., Martin, R., Pajak, D.: Evacuating robots via unknown exit in a disk. In: Kuhn, F. (eds.) DISC 2014. LNCS, vol. 8784, pp. 122–136. Springer (2014)
25. Czyzowicz, J., et al.: Evacuation from a disc in the presence of a faulty robot. In: Das, S., Tixeuil, S. (eds.) SIROCCO 2017. LNCS, vol. 10641, pp. 158–173. Springer (2017)
26. Czyzowicz, J., Georgiou, K., Kranakis, E.: Group search and evacuation. In: Flocchini, P., Prencipe, G., Santoro, N. (eds.) Distributed Computing by Mobile Entities; Current Research in Moving and Computing, chap. 14, pp. 335–370. Springer (2019)
27. Czyzowicz, J., Kranakis, E., Krizanc, D., Narayanan, L., Opatrny, J., Shende, S.M.: Wireless autonomous robot evacuation from equilateral triangles and squares. In: Papavassiliou, S., Ruehrup, S. (eds.) 14th International Conference, ADHOC-NOW. LNCS, vol. 9143, pp. 181–194. Springer (2015)
28. Czyzowicz, J., Kranakis, E., Krizanc, D., Narayanan, L., Opatrny, J., Shende, S.M.: Linear search with terrain-dependent speeds. In: Fotakis, D., Pagourtzis, A., Paschos, V.T. (eds.) CIAC 2017. LNCS, vol. 10236, pp. 430–441 (2017)
29. Dobrev, S., Královič, R., Pardubská, D.: Improved lower bounds for shoreline search. In: International Colloquium on Structural Information and Communication Complexity, pp. 80–90. Springer (2020)
30. Emek, Y., Langner, T., Stolz, D., Uitto, J., Wattenhofer, R.: How many ants does it take to find the food? Theoret. Comput. Sci. **608**, 255–267 (2015)
31. Emek, Y., Langner, T., Uitto, J., Wattenhofer, R.: Solving the ants problem with asynchronous finite state machines. In: Proceedings of International Colloquium on Automata, Languages, and Programming (ICALP). LNCS, vol. 8573, pp. 471–482 (2014)
32. Fekete, S.P., Gray, C., Kröller, A.: Evacuation of rectilinear polygons. In: Wu, W., Daescu, O. (eds.) COCOA 2010, Part I. LNCS, vol. 6508, pp. 21–30. Springer (2010)

33. Finch, S.R., Wetzel, J.E.: Lost in a forest. Am. Math. Mon. **111**(8), 645–654 (2004)
34. Finch, S.R., Zhu, L.-Y.: Searching for a shoreline. arXiv preprint math/0501123 (2005)
35. Fricke, G.M., Hecker, J.P., Griego, A.D., Tran, L.T., Moses, M.E.: A distributed deterministic spiral search algorithm for swarms. In: 2016 IEEE/RSJ International Conference on Intelligent Robots and Systems (IROS), pp. 4430–4436. IEEE (2016)
36. Gal, S.: Search Games. Wiley Encyclopedia of Operations Research and Management Science (2010)
37. Georgiou, K., Jang, W.: Triangle evacuation of 2 agents in the wireless model. In: Algorithmics of Wireless Networks: 18th International Symposium on Algorithmics of Wireless Networks, ALGOSENSORS 2022, Potsdam, Germany, 8–9 September 2022, pp. 77–90. Springer (2022)
38. Georgiou, K., Jones, C., Lucier, J.: Multi-agent search-type problems on polygons. In: International Conference on Current Trends in Theory and Practice of Computer Science, pp. 314–332. Springer (2025)
39. Georgiou, K., Leizerovich, S., Lucier, J., Kundu, S.: Evacuating from ℓ_p unit disks in the wireless model. Theoret. Comput. Sci. **944**, 113675 (2023)
40. Georgiou, K., Lucier, J.: Weighted group search on a line & implications to the priority evacuation problem. Theoret. Comput. Sci. **939**, 1–17 (2023)
41. Gibbs, P.: Bellman's escape problem for convex polygons (2016)
42. Gluss, B.: An alternative solution to the "lost at sea" problem. Naval Res. Logist. Q. **8**(1), 117–122 (1961)
43. Gluss, B.: The minimax path in a search for a circle in a plane. Naval Res. Logist. Q. **8**(4), 357–360 (1961)
44. Isbell, J.R.: An optimal search pattern. Naval Res. Logist. Q. **4**(4), 357–359 (1957)
45. Jeż, A., Łopuszański, J.: On the two-dimensional cow search problem. Inf. Process. Lett. **109**(11), 543–547 (2009)
46. Kleinberg, J.M.: On-line search in a simple polygon. In: SODA, vol. 94, pp. 8–15. Citeseer (1994)
47. Kübel, D., Langetepe, E.: On the approximation of shortest escape paths. Comput. Geom. **93**, 101709 (2021)
48. Langetepe, E.: On the optimality of spiral search. In: Proceedings of the twenty-first annual ACM-SIAM symposium on Discrete Algorithms, pp. 1–12. SIAM (2010)
49. Langetepe, E.: Searching for an axis-parallel shoreline. Theoret. Comput. Sci. **447**, 85–99 (2012)
50. Langner, T., Keller, B., Uitto, J., Wattenhofer, R.: Overcoming obstacles with ants. In: Anceaume, E., Cachin, C., Potop-Butucaru, M.G. (eds.) International Conference on Principles of Distributed Systems (OPODIS). LIPIcs, vol. 46, pp. 9:1–9:17. Schloss Dagstuhl - Leibniz-Zentrum fuer Informatik (2015)
51. Pattanayak, D., Ramesh, H., Mandal, P.S., Schmid, S.: Evacuating two robots from two unknown exits on the perimeter of a disk with wireless communication. In: Proceedings of the 19th International Conference on Distributed Computing and Networking, pp. 1–4 (2018)
52. Pelc, A.: Reaching a target in the plane with no information. Inf. Process. Lett. **140**, 13–17 (2018)

53. Pelc, A., Yadav, R.N.: Information complexity of treasure hunt in geometric terrains. arXiv preprint arXiv:1811.06823 (2018)
54. Pelc, A., Yadav, R.N.: Cost vs. information tradeoffs for treasure hunt in the plane. arXiv preprint arXiv:1902.06090 (2019)
55. Shklarsky, D.O., Chentzov, N.N., Yaglom, I.M.: The USSR Olympiad Problem Book: Selected Problems and Theorems of Elementary Mathematics, Part 2, vol. II. Nauka Publishers, Moscow (1962)

Pointer Chasing with Unlimited Interaction

Orr Fischer[1], Rotem Oshman[2], Adi Rosén[3], and Tal Roth[2(✉)]

[1] Bar Ilan University, Ramat Gan, Israel
fischeo@biu.ac.il
[2] Tel Aviv university, Tel Aviv-Yafo, Israel
roshman@tau.ac.il, roth1@mail.tau.ac.il
[3] IRIF, CNRS and Université Paris Cité, Paris, France
adiro@irif.fr

Abstract. Pointer-chasing is a central problem in two-party communication complexity: given input size n and a parameter k, the two players Alice and Bob are given functions $N_A, N_B : [n] \to [n]$, respectively, and their goal is to compute the value of p_k, where $p_0 = 1$, $p_1 = N_A(p_0)$, $p_2 = N_B(p_1) = N_B(N_A(p_0))$, $p_3 = N_A(p_2) = N_A(N_B(N_A(p_0)))$ and so on, applying N_A in even steps and N_B in odd steps, for a total of k steps. In some versions of the problem, the final output is not p_k itself, but rather some fixed function $f(p_k)$ of p_k. It is trivial to solve the problem using k communication rounds, with Alice speaking first, by simply "chasing the function" for k steps. Many works have studied the communication complexity of pointer chasing, although the focus has always been on protocols with $k-1$ communication rounds, or with k rounds where Bob (the "wrong player") speaks first. Many works have studied this setting giving sometimes tight or near-tight results.

In this paper we study the communication complexity of the pointer chasing problem when the interaction between the two players is unlimited, i.e., without any restriction on the number of rounds. Perhaps surprisingly, this question was not studied before, to the best of our knowledge. Our main result is that the trivial k-round protocol is nearly tight (even) when the number of rounds is not restricted: we give a lower bound of $\Omega(k \log(n/k))$ on the randomized communication complexity of the pointer chasing problem with unlimited interaction, and a somewhat stronger lower bound of $\Omega(k \log \log k)$ for protocols with zero error. When combined with prior work, our results also give a nearly-tight bound on the communication complexity of protocols using at most $k-1$ rounds, across all regimes of k; for $k > \sqrt{n}$ there was previously a significant gap between the upper and lower bound.

Keywords: Communication complexity · Pointer chasing · Lower bounds

U. Schmid and R. Kuznets (Eds.): SIROCCO 2025, LNCS 15671, pp. 281–296, 2025.
https://doi.org/10.1007/978-3-031-91736-3_17

1 Introduction

Pointer chasing is a natural and well-studied problem in communication complexity, which was used to demonstrate the inherent sequential nature of certain distributed tasks where the input is partitioned between a number of players. In particular, it was used to demonstrate that certain tasks must be solved "step by step" and cannot be parallelized; such tasks require a large number of back-and-forth communication rounds,[1] and therefore limiting the number of rounds of communication between two parties may have a dramatic effect on communication complexity. Pointer chasing has found many applications, including a proof for the monotone constant-depth hierarchy for Boolean circuits [22], and lower bounds in distributed computation [21], streaming algorithms [1,8,10], Matroid intersections [11], data structures [20,28], and more.

It is easiest to informally describe the pointer chasing problem using the terminology of graphs. For integer numbers n and $k \leq n$, in the two-party k-step pointer chasing problem, we have a directed bipartite graph $G = (U, W, E)$, where $|U| = |W| = n$, $U \cap W = \emptyset$, and the out-degree of all nodes is exactly 1. Starting from any node $u \in V$, there is a natural notion of a k-*walk* starting at that node, where we follow the outgoing edges for k steps, alternating between nodes in U and in W. In the k-step pointer chasing problem, Alice receives all the edges emanating from nodes in U and Bob receives all edges emanating from nodes in W, and the task is to find the endpoint v of a k-step walk starting from some fixed node $u \in U$ on Alice's side of the graph (or sometimes, to compute a function $f(v)$ of that node, for some function f that is fixed in advance and known to both players). The trivial protocol for k-step pointer chasing requires k rounds, with Alice speaking first: the players follow the k-walk by having each player in their turn announce the vertex to which the walk moves, with Alice speaking in odd turns (since she knows the edges going from left to right) and Bob speaking in even turns (since he knows the edges going from right to left). The total communication is $O(k \log n)$ bits. We remark that when $k = \Theta(n)$, the players can simply send each other their full inputs, and this requires one round of communication and $O(n \log n)$ bits.

The pointer chasing problem was first introduced by Papadimitriou and Sipser [23], who showed that for the case of $k = 2$, any one-round protocol requires exponentially more communication than the trivial two-round protocol. The communication complexity of (k-step) pointer chasing for $(k-1)$-round protocols, or for k-round protocols where Bob speaks first, was subsequently studied in a number of works [6,15,19,22,25,30]. At present, the state of the art for *deterministic* communication protocols is an upper bound of $O\left(n \log^{(k-1)} n + k \log n\right)$ [6] and a lower bound of $\Omega(n - k \log n)$ [22], which holds for any k. For *randomized* communication protocols, the best known upper bound is $O\left(\left(\frac{n}{k} + k\right) \log n\right)$ [22], and the best lower bound is $\Omega(n/k + k)$ [19]. Crucially, these lower bounds apply to protocols with *exactly* $k-1$ rounds where

[1] See Sect. 2 for a more formal definition of this and other notions in communication complexity.

Alice speaks first, or to protocols with *exactly* k rounds where Bob speaks first. See Sect. 1.1 for more details on these and other results.

Perhaps surprisingly, the communication complexity of the pointer chasing problem with *unlimited interaction*, i.e., without any upper or lower bound on the number of rounds, has, to the best of our knowledge, not been studied to date. In this work we give lower bounds on the communication complexity of the k-step pointer chasing problem when there is *no restriction on the number of rounds*. We give the following two main results:

- An $\Omega(k \log(n/k))$ lower bound on the expected communication of randomized protocols that have a constant but non-zero error probability, and
- An $\Omega(k \log\log k)$ lower bound on the expected communication of zero-error randomized protocols.

As a by-product of our results, we also get new, essentially tight, results for the communication complexity of the pointer chasing problem when the protocol is restricted to *at most* $k-1$ rounds and Alice speaks first, or *at most* k rounds and Bob speaks first, for certain regimes of k and n. Again, this is in contrast to all previous lower bounds results which were proved for the case when the protocol uses *exactly* this number of rounds. While this difference seems at first sight to be of no great consequence, some of the lower bounds mentioned above, such as those of [15] and [19], do rely on this fact, and their proofs do not hold for the more general case. Our results imply new, nearly-tight lower bounds for the regime where $k > \sqrt{n}$, closing a significant gap (when $k \gg \sqrt{n}$) between the upper bound of $O\left(\left(\frac{n}{k} + k\right) \log n\right)$ [22] and the lower bound of $\Omega(n/k)$ [19] (restated for the case of *at most* $k-1$ rounds; see Sect. 1.1 for more details).

Another by-product of our results is a negative answer to a question posed very recently [19]: it is conjectured in [19] that the upper bound of $O\left(\left(\frac{n}{k} + k\right) \log n\right)$ from [22] is not tight for $k = \omega(\log n)$, and the $\log n$ factor can be removed (this upper bound is indeed not tight for $k = o(\log n)$). The lower bound that we prove in Theorem 3 below shows that this is not the case for $k = \Theta\left(n^{\delta}\right)$, where $1/2 < \delta < 1$ is a constant: the factor of $\log n$ is inherent.

Our results are proved using two simple reductions: for non-zero error protocols we reduce from the OR-Index problem, and for zero-error protocols we reduce from Cycle (we review these problems in Sect. 2).

1.1 Related Work

The known upper and lower bounds on k-step pointer chasing are summarized in Tables 1 and 2 below. In the tables we list upper and lower bounds for deterministic protocols with worst-case error (listed as "deterministic" in the tables), for randomized protocols, and for deterministic protocols with *distributional error* (listed as "distributional" in the tables), where the protocol only needs to succeed with high probability over inputs drawn from some fixed distribution (in this case, typically the uniform distribution). Distributional lower bounds are the strongest, as they imply both deterministic worst-case lower bounds and, by Yao's principle, randomized lower bounds with worst-case error.

We emphasize that the lower bounds listed in Table 1 all apply only to protocol with *exactly* the number of rounds listed in the table. This may be confusing at first, as one might expect that a lower bound that applies to protocol with *exactly* R communication rounds would also apply to protocols with *at most* R communication rounds, but in fact this is not necessarily the case. The distinction is most significant when it comes to the lower bounds of [15] and [19], where the proofs make explicit use of the fact that the protocol has exactly some number of rounds. It is possible to adapt these proofs so that they apply to any protocol with *at most* k communication rounds, but this comes at the cost of k bits in the lower bound, which becomes $\Omega(n/k)$ in both cases. Therefore, when $k \gg \sqrt{n}$, the lower bounds of [15] and of [19] are significantly weaker than the lower bounds of $\tilde{\Omega}(k)$ that we prove in the present paper.

From a technical perspective, one reason that prior work has hit an obstacle at $k = \sqrt{n}$ is that most of it has worked with the *uniform input distribution*, where the outgoing edge of each node in the bipartite graph goes to a uniformly random node on the other side. It is not hard to see that under that distribution within the first $k \approx \sqrt{n}$ steps the walk enters a *cycle*, which means that the players do not need to "keep walking" and may instead reuse the information that they have already learned, without further communication. Thus, proving lower bounds that grow with k when $k \geq \sqrt{n}$ requires a new idea.

Table 1. Related work: lower bounds

Paper	lower bound	Comments
[23]	$\Omega(n)$	$k = 2$, deterministic one-way
[7]	$\Omega(n/k^2)$	deterministic, $k - 1$ rounds
[22]	$\Omega(n - k \log n)$	deterministic, k rounds (Bob speaks first)
	$\Omega((n/k^2) - k \log n)$	randomized, k rounds (Bob speaks first)
[25]	$\Omega(n \log^{(k-1)} n)$	constant k, deterministic, k-rounds (Bob speaks first)
[15]	$\Omega((n/k) + k)$	randomized, k rounds (Bob speaks first)
[30]	$\Omega((n/k) - k \log n)$	distributional, k rounds (Bob speaks first)
[19]	$\Omega((n/k) + k)$	distributional, $k - 1$ rounds

Table 2. Related work: upper bounds

Paper	Upper bound	Comments
Trivial protocol	$O(k \log n)$	k rounds (Alice speaks first), deterministic
[6]	$O\left(n \log^{(k-1)} n + k \log n\right)$	k rounds (Bob speaks first), deterministic
[22]	$O\left(\left(\frac{n}{k} + k\right) \log n\right)$	$k - 1$ rounds, randomized

1.2 Other Models

In addition to the standard two-party communication model, pointer chasing was also studied in the quantum communication model [12,13,15,16] and in the multiparty number-on-the-forehead model [2–6,9,14,18,22,29], where it found applications in lower bounds for set disjointness and circuit complexity, respectively.

1.3 Our Results

Our work focuses on showing $\widetilde{\Omega}(k)$ lower bounds on the randomized communication complexity of k-step pointer chasing for protocols with an unlimited number of rounds, which serves to show that the trivial k-round $O(k \log n)$-bit protocol described above is essentially tight. Of course, lower bounds with no restriction on rounds immediately imply lower bounds for any protocol with restricted rounds.

For protocols with constant, non-zero error, we show:

Theorem 1 (Informal). *Any randomized protocol that solves k-step pointer chasing with constant (non-zero) error must send $\Omega(k \log(n/k))$ bits in expectation, even for the weaker version of the problem where the goal is merely to compute some (non-constant) function $f(v)$ where v is the end of a k-walk starting from some fixed vertex.*

This lower bound is nearly tight, given the trivial k-round $O(k \log n)$-bit protocol. Moreover, the theorem shows that when $k \geq \sqrt{n}$, restricting the number of rounds to be less than k does not matter very much: even if we do not limit the number of rounds, a protocol will have to expend $\Omega(k \log(n/k))$ bits of communication to solve k-step pointer chasing, and when $k \geq \sqrt{n}$ this very nearly matches the $(k-1)$-round protocol from [22], which sends $O((n/k)\log n + k \log n) = O(k \log n)$ bits. This contrasts sharply with the case $k \ll \sqrt{n}$, where prior work has shown that $(k-1)$-round protocols have significantly higher communication complexity compared to k-round protocols.

Theorem 1 trivially implies a lower bound for protocols that are limited to at most $k-1$ communication rounds. In addition, as noted in Sect. 1.1, the lower bounds from [15,19] can be adapted to apply to protocols with at most $k-1$ rounds, yielding a lower bound of $\Omega(n/k)$. We can thus combine these two lower bounds to obtain the following:

Corollary 1 (Informal). *For any balanced function f (e.g., the parity function), any randomized constant-error protocol for k-step pointer chasing that uses at most $k-1$ rounds must send $\Omega((n/k) + k \log(n/k))$ bits in expectation.*

This nearly matches the $O\left((n/k + k)\log n\right)$ protocol of [22] across all regimes of k.

For *zero-error* randomized protocols, where the output must always be correct, we prove a slightly stronger lower bound:

Theorem 2. *Any zero-error randomized protocol that solves k-step pointer chasing must send $\Omega(k \log\log k)$ bits in expectation, even for the weaker version of the problem where the goal is merely to compute some (non-constant) function $f(v)$ where v is the end of a k-walk starting from some fixed vertex.*

2 Preliminaries

Throughout this paper, we denote $[n] = \{1, \dots, n\}$.

2.1 Two-Party Communication Complexity

In two-party communication complexity, there are two players, Alice and Bob, each with a private input X, Y (resp.). The goal is for one of the two parties, specified in advance, to output the value of some function $F(X, Y)$ of their inputs, using as little communication as possible. To that end, the parties engage in a *communication protocol*, where they communicate back-and-forth for some number of rounds, until eventually the party responsible for producing an output does so.[2]

A communication protocol may be deterministic, or it may be randomized; in this paper we consider *randomized public-coin* protocols, where both players have access to a shared uniformly-random string of arbitrarily large length. The *worst-case communication cost* of the protocol is the total number of bits exchanged between the players, in the worst case over inputs and random strings. The *expected communication cost* of the protocol is the worst-case over inputs X, Y of the expected number of total number of bits exchanged between the players (where the expectation is taken over the randomness).

A protocol is said to compute F with error $0 \leq \epsilon \leq 1$ if on any input X, Y, the probability that the protocol produces the correct output $F(X, Y)$ is at least $1 - \epsilon$ (where the probability is taken over the randomness). In the special case where $\epsilon = 0$, we refer to this protocol as a *randomized zero-error* protocol for F. The *randomized ϵ-error communication complexity* of F is the minimum expected communication cost of any protocol that computes F with error ϵ.

The *round complexity* of a protocol is the worst-case number of messages exchanged between the two parties, speaking in alternating order. Either Alice or Bob may speak first (this must be specified by the protocol).

[2] It is common to require that *both* parties learn the value of $F(X, Y)$, and this is without loss of generality (up to one bit of communication) when F is a Boolean function and the number of rounds is unrestricted: the party that learns $F(X, Y)$ can simply send it to the other party. However, when the number of rounds is restricted, requiring both parties to learn $F(X, Y)$ can significantly increase the communication required.

2.2 Pointer Chasing

The pointer chasing problem is defined by applying two functions, each given as input to one of Alice or Bob, in alternating order and for some fixed number of steps k.

This alternation is formally captured by the following definition:

Definition 1 (Walk functions). *Let $n \in \mathbb{N}$, and $N_A, N_B \in [n]^n$. For any $r \in \mathbb{N}$, we define functions* $\text{walk}_{A,r}, \text{walk}_{B,r} \in [n]^n$ *with respect to N_A, N_B via mutual recursion. For any $i \in [n]$:*

$$\text{walk}_{A,r}(i) = \begin{cases} i & r = 0 \\ \text{walk}_{B,r-1}(N_A(i)) & r > 0, \end{cases}$$

and symmetrically:

$$\text{walk}_{B,r}(i) = \begin{cases} i & r = 0 \\ \text{walk}_{A,r-1}(N_B(i)) & r > 0. \end{cases}$$

In order to solve the pointer chasing problem, the parties need to output the value $\text{walk}_{A,k}(1)$ reached after k steps starting from value 1, or more generally, compute a predetermined function f of this value:

Definition 2 (Pointer chasing problem). *Let $n, k \in \mathbb{N}$, and let $f : [n] \to \{0,1\}$. In the k-step* pointer chasing problem $\text{PC}^f_{n,k}$, *Alice and Bob are given functions $N_A, N_B \in [n]^n$, respectively, and both parties know the function f. Their goal is to compute*

$$\text{PC}^f_{n,k}(N_A, N_B) = f(\text{walk}_{A,k}(1)).$$

Throughout the paper, we consider a class of functions we refer to as *non-trivial*, which is a slightly stronger condition than non-constant.

Definition 3. $f : [n] \to \{0,1\}$ *is* non-trivial *if f is non-constant on $[n] \setminus \{1\}$, or in other words, if there exist $i, j \in [n] \setminus \{1\}$ such that $f(i) \neq f(j)$.*

Graph-Theoretic Formulation. For our purposes it is convenient to use an equivalent, graph-theoretic definition of the pointer chasing problem. Let $G = (L, R, E)$ be a balanced directed bipartite graph where $|L| = |R| = \{1, \ldots, n\}$, and each vertex has out-degree 1. If we think of N_A (and resp. N_B) as the edges going from L to R (resp. from R to L), then it is easy to see that $\text{walk}_{A,r}(1)$ (resp. $\text{walk}_{B,r}(1)$) returns the label of the vertex at the end of the unique r-step *walk* in G starting from the vertex labeled 1 on the left side L (resp. the right side R). Thus, we can think of $\text{PC}^f_{n,k}$ as the problem where Alice is given the edges N_A from left to right, Bob is given the edges N_B from right to left, and the goal is to compute the function f of the vertex reached by a length-k walk starting from vertex 1 on the left side.

2.3 The Problems Index and OR-Index

We use in our proofs a reduction from the OR-Index function defined below.

Definition 4 (The Index **problem** [17]**).** *Let* $x \in [r]$ *and* $y \in \{0,1\}^r$. *The* Index_r *function is defined as:*

$$\text{Index}_r(x, y) = y_x.$$

The OR-Index problem is simply the disjunction of m instances of the Index function:

Definition 5 (The OR-Index **problem).** *Let* $x \in [r]^m$, *and* $y \in (\{0,1\}^r)^m$. *The* $\text{OR-Index}_{r,m}$ *function is defined as*

$$\text{OR-Index}_{r,m}(x, y) = \bigvee_{i=1}^{m} (y_i)_{x_i}.$$

In [24] the following lower bound for OR-Index (which is called *blocky lopsided disjointness* in [24]) is given:

Lemma 1 ([24] Theorem 1.4, restated and simplified). *The randomized public-coin communication complexity of* $\text{OR-Index}_{r,m}$ *with error* $\frac{1}{9999}$ *is* $\Omega(m \log r)$.

2.4 The Cycle_n Problem

In [27], Raz and Spieker defined the following problem, which we refer to as the Cycle_n problem:

Definition 6. *Fix an integer* $n \in \mathbb{N}$ *and two vertex sets* U, W *of size* $|U| = |W| = n$. *In the* Cycle_n *problem, Alice and Bob are given perfect matchings* P_A, P_B *(resp.) in the complete bipartite graph on* $U \cup W$. *The goal is for the players to determine whether* $P_A \cup P_B$ *is a Hamiltonian cycle on* $U \cup W$.

Note that in the (multi)-graph $G = (U \cup W, P_A \cup P_B)$, each vertex has degree exactly 2, and therefore G is a non-empty collection of cycles (possibly including degenerate cycles of length 2). This characterization will be useful to us later.

Raz and Spieker showed in [27] (Theorem 1) that the nondeterministic communication complexity[3] of the Cycle_n problem is $\Omega(n \log \log n)$. Nondeterministic communication complexity is a *lower bound* on zero-error randomized communication complexity (see [17] Proposition 3.7), and the following statement immediately follows:

Lemma 2 (Corollary of [27] Theorem 1). *The randomized zero-error communication complexity of* Cycle_n *is* $\Omega(n \log \log n)$.

---

3 In *nondeterministic communication complexity*, there is a prover whose goal is to convince the two parties to output 1. We do not give a formal definition here, as it is not needed for our purposes. See [17,26] for the formal definition.

3 Lower Bounds

3.1 Lower Bound for Constant Error Randomized Protocols

In this section we prove our main lower bound on the communication complexity of k-step pointer chasing with unlimited interaction:

Theorem 3 (Formal statement of Theorem 1). *Let $k, n \in \mathbb{N}$ be such that $k \le n$, and let $f : [n] \to \{0,1\}$ be non-trivial. Then there is a constant $\epsilon \in (0,1)$ such that the randomized ϵ-error communication complexity of $\mathrm{PC}^f_{n,k}$ is $\Omega(k \log(n/k))$.*

If we take $k = O(n^\delta)$ for some constant $\delta \in (0,1)$, then the bound that we obtain from Theorem 3 is $\Omega(k \log n)$, which matches the naïve protocol described in Sect. 1.

Proof Overview. We show a reduction from $\text{OR-Index}_{r,m}$ to $\mathrm{PC}^f_{n,k}$, where we take r, m such that $n = \Theta(rm)$ and $k = \Theta(m)$. Since $\text{OR-Index}_{r,m}$ requires $\Omega(m \log r)$ bits of communication [24], this yields a lower bound of $\Omega(k \log(n/k))$ for $\mathrm{PC}^f_{n,k}$.

Recall that $\text{OR-Index}_{r,m}$ comprises m instances of Index_r, and the goal is to determine whether at least one of them evaluates to 1. For convenience, denote these instances $\text{Index}_r^{(1)}, \ldots, \text{Index}_r^{(m)}$, and the value of the ith instance as $\text{Index}_r^{(i)}(X,Y)$. We construct edge sets N_A, N_B such that $\mathrm{PC}^f_{n,k}(N_A, N_B) = 1$ iff there is at least one i such that $\text{Index}_r^{(i)}(X,Y) = 1$.

We describe the construction in graph-theoretic terms (see the explanation in Sect. 2 above). The graph that we construct is made up of r gadgets, each associated with one instance $\text{Index}_r^{(i)}$. In addition, the graph contains two special vertices $v_{\texttt{disj}}, v_{\texttt{int}}$, which are not part of any gadget. The gadgets are constructed such that for each $i = 1, \ldots, m$, if $\text{Index}_r^{(i)}(X,Y) = 1$ then a two-step walk on the ith gadget leads to vertex $v_{\texttt{int}}$; but if $\text{Index}_r^{(i)}(X,Y) = 0$, then the two-step walk on the ith gadget leads to the next gadget if $i < m$, or to vertex $v_{\texttt{disj}}$ if $i = m$ (i.e., this is the last gadget). Consequently, we can show that $\text{OR-Index}_{r,m}(X,Y) = 0$ if and only if $\text{walk}_{A,k}(1) = v_{\texttt{disj}}$, and that $\text{OR-Index}_{r,m}(X,Y) = 1$ if and only if $\text{walk}_{A,k}(1) = v_{\texttt{int}}$.

The Reduction. Fix parameters $m \ge 1, r \ge 2$. We construct a protocol for $\text{OR-Index}_{r,m}$ by reduction to $\mathrm{PC}^f_{n,k}$, where f is any non-trivial function, and $n, k \in \mathbb{N}$ satisfy

1. $rm + 2 \le n$, and
2. $2m \le k$.

Recall that in the OR-Index function, Alice receives $X \in [r]^m$ and Bob receives $Y \in (\{0,1\}^r)^m$. We construct a pointer-chasing instance as follows (Fig. 1). Let $v_{\texttt{disj}}, v_{\texttt{int}} \in [n] \setminus \{1\}$ be two vertices such that $f(v_{\texttt{disj}}) \ne f(v_{\texttt{int}})$.

(We know that such vertices exist, because f is non-trivial, i.e., non-constant on $[n] \setminus \{1\}$.) We assume w.l.o.g. that $v_{\texttt{disj}} = rm + 1$ and $v_{\texttt{int}} = rm + 2$, as otherwise we can re-order the vertices.

The edges N_A going from left to right in the pointer chasing instance are as follows.

- For any $i \in [m]$, we set $N_A(r(i-1)+1) = r(i-1) + x_i$.
- $N_A(v_{\texttt{disj}}) = v_{\texttt{disj}}$, and $N_A(v_{\texttt{int}}) = v_{\texttt{int}}$.
- For any vertex $s \in [n]$ such that $N_A(s)$ is not defined by the other cases, we set $N_A(s) = 1$ (this is an arbitrary choice, and any other value would work just as well).

The edges N_B going from right to left in the pointer chasing instance are as follows.

- For any $i \in [m]$ and $j \in [r]$:

$$N_B(r(i-1)+j) = \begin{cases} v_{\texttt{int}} & \text{if } (y_i)_j = 1 \\ ri+1 & \text{if } (y_i)_j = 0, \end{cases}$$

- $N_B(v_{\texttt{disj}}) = v_{\texttt{disj}}$, and $N_B(v_{\texttt{int}}) = v_{\texttt{int}}$.
- For any vertex $s \in [n]$ such that $N_B(s)$ is not defined by the other cases, we set $N_B(s) = 1$ (again, an arbitrary value).

Observe that Alice can construct N_A and Bob can construct N_B from their respective inputs without communication. Alice and Bob now solve the pointer-chasing instance (N_A, N_B) by calling a protocol Π for $\mathrm{PC}^f_{n,k}$, and return OR-Index$(x, y) = 0$ iff $\mathrm{PC}^f_{n,k}(N_A, N_B) = f(v_{\texttt{disj}})$.

Analysis. Let Π' be the protocol constructed above for OR-Index$_{r,m}$. Clearly, the communication complexity of Π' is the same as that of the protocol Π for $\mathrm{PC}^f_{n,k}$ (possibly plus one bit, depending on the party we want to output the function's value). We show that Π' succeeds whenever Π succeeds.

Lemma 3. *Let $i \in [m]$. Then a two-step walk starting from vertex $r(i-1)+1$ at the left side will end at vertex $ri+1$ if $(y_i)_{x_i} = 0$, and otherwise will end at the vertex $v_{\texttt{int}}$. More formally, for any $i \geq 1$,*

$$\mathrm{walk}_{A,2}(r(i-1)+1) = \begin{cases} ri+1 & \textit{if } (y_i)_{x_i} = 0 \\ v_{\texttt{int}} & \textit{otherwise}\ . \end{cases}$$

Proof. Follows immediately from the definition of $N_A(r(i-1)+1)$ and $N_B(r(i-1)+j)$.

By induction on the length of the walk, we can immediately deduce from Lemma 3:

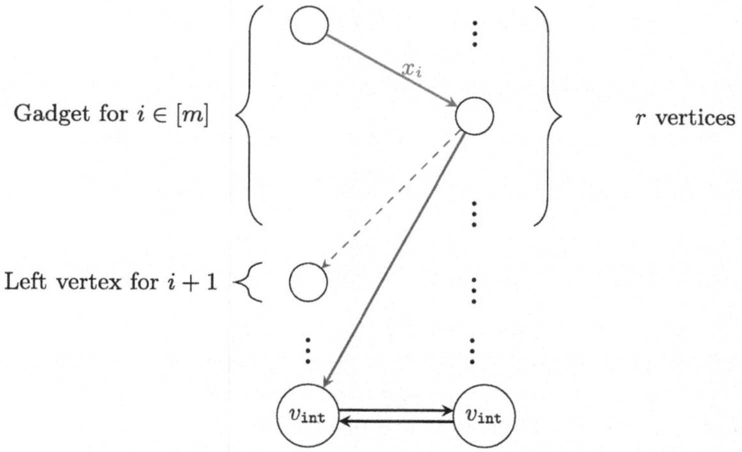

Fig. 1. The left vertex and x_i-th right vertex of the gadget for coordinate $i \in [m]$, together with the left vertex of the gadget for coordinate $i+1$, and $v_{\texttt{int}}$ vertices. Each gadget has one left vertex and r right vertices. The left vertex of the gadget for coordinate i is connected to the x_i-the right vertex. For any $j \in [r]$, the j-th right vertex of the gadget for coordinate i is connected to the $v_{\texttt{int}}$ vertex if $(y_i)_j = 1$, and otherwise to the left vertex of the gadget for coordinate $i+1$. The edges between the $v_{\texttt{int}}$ vertices always appear.

Corollary 2. *Let* $0 \leq \ell \leq m$ *be such that* $(y_i)_{x_i} = 0$ *for all* $1 \leq i \leq \ell$. *Then* $\text{walk}_{A,2\ell}(1) = r\ell + 1$.

The correctness of the reduction follows:

Corollary 3. *If* $\text{OR-Index}(x, y) = 0$, *then* $\text{walk}_{A,k}(1) = v_{\texttt{disj}}$, *and otherwise* $\text{walk}_{A,k}(1) = v_{\texttt{int}}$.

Proof. First, assume $\text{OR-Index}(x, y) = 0$. Then by Corollary 2 for $\ell = m$, we have that $\text{walk}_{A,m}(1) = rm + 1 = v_{\texttt{disj}}$. Since $r \leq k$, and $N_A(v_{\texttt{disj}}) = N_B(v_{\texttt{disj}}) = v_{\texttt{disj}}$, the claim follows.

Next, assume that $\text{OR-Index}(x, y) = 1$, and let $i \in [m]$ be the smallest index such that $(y_i)_{x_i} = 1$. Then by Corollary 2 for $\ell = i - 1$, we have that $\text{walk}_{A,2(i-1)}(1) = ri + 1$. Lemma 3 implies that $\text{walk}_{A,2(i-1)+2}(1) = v_{\texttt{int}}$. We note that $2(i-1) + 2 \leq m \leq k$. Since $N_A(v_{\texttt{int}}) = N_B(v_{\texttt{int}}) = v_{\texttt{int}}$, we conclude that $\text{walk}_{A,k}(1) = v_{\texttt{int}}$.

Taking $r = \Theta(n/k)$ and $m = \Theta(n)$, Theorem 3 now follows from the correctness of the reduction (Corollary 3), together with the lower bound of Lemma 1 on the communication complexity of $\text{OR-Index}_{r,m}$.

3.2 Lower Bound for Zero-Error Randomized Protocols

In this section we prove that any zero-error protocol solving the pointer chasing problem requires $\Omega(k \log \log k)$ bits of communication in expectation.

Theorem 4 (Formal statement of Theorem 2). *Let $n, k \in \mathbb{N}$ be such that $4k \leq n$, and let $f : [n] \to \{0,1\}$ be non-trivial. Then the randomized zero-error communication complexity of $\mathrm{PC}^f_{n,k}$ is $\Omega(k \log\log k)$.*

It is convenient to assume that f has the following properties.

Observation 1. *Without loss of generality, we may assume that*

(a) $|\{i \in [n] \mid f(i) = 0\}| \leq |\{i \in [n] \mid f(i) = 1\}|$, *and*
(b) $f(1) = 0$, *and* $f(i) = 1$ *for all* $2 \leq i \leq 2k$.

Proof. Property (a) is obtained by relabeling the output values so that 1 becomes the majority value. Given property (a), there are at least $2k$ indices such that $f(i) = 1$, and since f is non-trivial, there exists an index $j \in [n] \setminus \{1\}$ such that $f(j) = 0$. Then, using a single extra step, it is easy to see we can set $f(1) = 0$ (by using the first step to move to vertex j), and $f(i) = 1$ for $2 \leq i \leq 2k$ (by reordering the indices).

Proof Overview. We show that for any integer k, we can reduce the Cycle_k problem to the pointer-chasing problem $\mathrm{PC}^f_{n,2k'}$ for some $k < k' \leq 2k$ and for any $n \geq 4k$. We do so by first reducing Cycle_k to $\mathrm{Cycle}_{k'}$ for a prime $k < k' \leq 2k$, and then reducing $\mathrm{Cycle}_{k'}$ to $\mathrm{PC}^f_{n,2k'}$.

In any input to Cycle_n, each node lies on a single cycle. Thus, the first reduction can be trivially achieved by adding $2(k'-k)$ fresh vertices to the graph to artificially extend the cycle on which u_1 lies, so that its length is increased by $2(k'-k)$. As a result, if there is a Hamiltonian cycle, its length now becomes $2k+2(k'-k) = 2k'$ (recall that the original bipartite graph is over $2k$ vertices, k on each side); and if there is no Hamiltonian cycle, extending the cycle on which u_1 lies by adding fresh vertices will not create one.

The second reduction is done by constructing a pointer chasing instance that outputs 0 if a $2k'$-walk starting from u_1 returns to u_1, and outputs 1 otherwise. We then prove that since k' is prime, a $2k'$-walk from u_1 terminates at u_1 if and only if the cycle is Hamiltonian (excluding a case where u_i's cycle is of length 2, which can be handled trivially), and the theorem follows. As is clear from the formal definition of the reduction below, the reduction itself does not require the players to communicate.

The Reduction. Let $U = \{u_1, \dots, u_k\}, W = \{w_1, \dots, w_k\}$. Alice and Bob are given perfect matchings P_A, P_B between U and W, respectively. For $i \in [k]$, we denote by $P_A(i)$ the index of the unique vertex w_j neighboring u_i, i.e., $P_A(i) = j$ if $\{u_i, w_j\} \in P_A$. Similarly, for $w \in W$, we denote $P_B(j) = i$ if $\{u_i, w_j\} \in P_B$. Similar to the pointer chasing problem, we define a recursive notion of walk on the two matchings: for every $i \in [k]$ we denote

$$\mathrm{cwalk}_{A,r}(i) = \begin{cases} i & r = 0 \\ \mathrm{cwalk}_{B,r-1}(P_A(i)) & r > 0, \end{cases}$$

and symmetrically:

$$\text{cwalk}_{B,r}(i) = \begin{cases} i & r = 0 \\ \text{cwalk}_{A,r-1}(P_B(i)) & r > 0. \end{cases}$$

First, we reduce Cycle_k to $\text{Cycle}_{k'}$, where k' is the closest prime from above to k, i.e., $k < k' \leq 2k$. We do so by artificially increasing the length of the cycle on which u_1 lies by $2(k' - k) - 1$ edges. More formally:

- We add new vertices $x_1, \ldots, x_{k'-k}, y_1, \ldots, y_{k'-k}$ to obtain new vertex sets, of size k' each:

$$U' = \{u_1, \ldots, u_k\} \cup \{x_1, x_2, \ldots, x_{k'-k}\},$$
$$W' = \{u_1, \ldots, u_n\} \cup \{y_1, y_2, \ldots, y_{k'-k}\}.$$

- Alice removes the edge $\{u_1, w_{P_A(1)}\}$ from P_A.
- We add the path $(u_1, y_1, x_1, y_2, x_2, \ldots, x_{k'-k}, w_{P_A(1)})$ to the graph, by adding the odd edges to P_A and the even edges to P_B.

The new graph is a Hamiltonian cycle if and only if the original graph is a Hamiltonian cycle. Moreover, since we added $k' - k$ new vertices to each of the sets U, W to obtain U', W', the result is indeed an instance of $\text{Cycle}_{k'}$. This concludes the reduction from Cycle_k to $\text{Cycle}_{k'}$; observe that the reduction can be performed by Alice and Bob independently, with no communication.

Next, we reduce from $\text{Cycle}_{k'}$ to $\text{PC}^f_{n,2k'}$ for any $n \geq 2k'$. The players start by checking if the cycle that contains u_1 is of length 2. In order to do so, Alice first sends $P_A(1)$, and Bob sends $P_B(P_A(1))$, using $2\lceil \log k' \rceil$ of communication in total. If $P_B(P_A(1)) = 1$, the players conclude that $P_A \cup P_B$ is not a Hamiltonian cycle. Otherwise, the players can conclude that the cycle containing u_1 is of length more than 2.

Next, for $1 \leq i \leq k'$, define $N_A(i) = P_A(i)$ and $N_B(i) = P_B(i)$. For any $i > k'$, define $N_A(i), N_B(i)$ arbitrarily. In the next two lemmas, we show that the value of this instance of $\text{PC}^f_{n,2k'}$ is equal to the value of the $\text{Cycle}_{k'}$ problem, and hence to the original Cycle_k problem as well.

Lemma 4. *For any* $r \in \mathbb{N} \cup \{0\}$ *we have* $\text{cwalk}_{A,r}(1) = \text{walk}_{A,r}(1)$.

Proof. By induction on r. For $r = 0$, $\text{cwalk}_{A,0}(1) = \text{walk}_{A,0}(1) = 1$. Assume by induction on r that $\text{cwalk}_{A,r-1}(1) = \text{walk}_{A,r-1}(1)$. For even r,

$$\text{cwalk}_{A,r}(1) = P_B(\text{cwalk}_{A,r-1}(1)), \text{ and } \text{walk}_{A,r}(1) = N_B(\text{walk}_{A,r-1}(1)).$$

Since $\text{cwalk}_{A,r-1}(1) = \text{walk}_{A,r-1}(1)$, then in particular $\text{walk}_{A,r-1}(1) \in [k']$. By choice of N_B, $P_B(i) = N_B(i)$ for all $1 \leq i \leq k'$, and the claim follows. The case for odd r follows using a similar argument. □

Lemma 5. $P_A \cup P_B$ *is a Hamiltonian cycle on* $U' \cup W'$ *if and only if* $f(\text{walk}_{A,2k'}(1)) = 0$.

Proof. If $P_A \cup P_B$ is a Hamiltonian cycle (i.e., a cycle of length $2k'$), then a $2k'$-walk from u_1 terminates at u_1, or in other words $\text{cwalk}_{A,2k'}(1) = 1$. By Lemma 4, it holds that $\text{walk}_{A,2k'}(1) = \text{cwalk}_{A,2k'}(1) = 1$, and therefore by Property 1 we have $f(\text{walk}_{A,2k'}(1)) = 0$.

If $P_A \cup P_B$ is not a Hamiltonian cycle, then the cycle containing u_1 is of length 2ℓ for some $1 < \ell < k'$. Since k' is a prime number, ℓ does not divide k', implying that a $2k'$-walk from u_1 does not terminate at u_1. In other words, $\text{cwalk}_{A,2k'}(1) \in [k'] \setminus \{1\}$. By Lemma 4, it holds that $\text{walk}_{A,2k'}(1) = \text{cwalk}_{A,2k'}(1) \in [k'] \setminus \{1\}$, and therefore by Property 1, $f(\text{walk}_{A,2k'}(1)) = 1$. □

This completes the correctness of the reduction from $\text{Cycle}_{k'}$ to $\text{PC}^f_{n,2k'}$. Theorem 4 immediately follows, by Lemma 2.

Acknowledgments. Work by Orr Fischer was partially supported by the Israel Science Foundation (grant No. 1042/22 and 800/22). Rotem Oshman and Tal Roth are supported in part by the Israel Science Foundation, Grant No. 2801/20.

Disclosure of Interests. The authors have no competing interests to declare that are relevant to the content of this article.

References

1. Assadi, S., Chen, Y., Khanna, S.: Polynomial pass lower bounds for graph streaming algorithms. In: Proceedings of the 51st Annual ACM SIGACT Symposium on Theory of Computing, STOC, pp. 265–276 (2019). https://doi.org/10.1145/3313276.3316361
2. Brody, J.: The maximum communication complexity of multi-party pointer jumping. In: Proceedings of the 24th Annual IEEE Conference on Computational Complexity, CCC, pp. 379–386 (2009). https://doi.org/10.1109/CCC.2009.30
3. Brody, J., Chakrabarti, A.: Sublinear communication protocols for multi-party pointer jumping and a related lower bound. In: STACS 2008, 25th Annual Symposium on Theoretical Aspects of Computer Science, pp. 145–156 (2008). https://doi.org/10.4230/LIPIcs.STACS.2008.1341
4. Brody, J., Sanchez, M.: Dependent random graphs and multi-party pointer jumping. In: Approximation, Randomization, and Combinatorial Optimization. Algorithms and Techniques, APPROX/RANDOM, pp. 606–624 (2015). https://doi.org/10.4230/LIPIcs.APPROX-RANDOM.2015.606
5. Chakrabarti, A.: Lower bounds for multi-player pointer jumping. In: 22nd Annual IEEE Conference on Computational Complexity (CCC), pp. 33–45 (2007). https://doi.org/10.1109/CCC.2007.14
6. Damm, C., Jukna, S., Sgall, J.: Some bounds on multiparty communication complexity of pointer jumping. Comput. Complex. **7**(2), 109–127 (1998)
7. Duris, P., Galil, Z., Schnitger, G.: Lower bounds on communication complexity. In: Proceedings of the 16th Annual ACM Symposium on Theory of Computing (STOC), pp. 81–91. ACM (1984). https://doi.org/10.1145/800057.808668
8. Feigenbaum, J., Kannan, S., McGregor, A., Suri, S., Zhang, J.: Graph distances in the data-stream model. SIAM J. Comput. **38**(5), 1709–1727 (2008). https://doi.org/10.1137/070683155

9. Gronemeier, A.: NOF-multiparty information complexity bounds for pointer jumping. In: Mathematical Foundations of Computer Science, pp. 459–470 (2006). https://doi.org/10.1007/11821069_40
10. Guruswami, V., Onak, K.: Superlinear lower bounds for multipass graph processing. In: Proceedings of the 28th Conference on Computational Complexity, CCC, pp. 287–298 (2013). https://doi.org/10.1109/CCC.2013.37
11. Harvey, N.J.A.: Matroid intersection, pointer chasing, and young's seminormal representation of S_n. In: Proceedings of the Nineteenth Annual ACM-SIAM Symposium on Discrete Algorithms, SODA, pp. 542–549 (2008)
12. Jain, R., Radhakrishnan, J., Sen, P.: Privacy and interaction in quantum communication complexity and a theorem about the relative entropy of quantum states. In: 43rd Symposium on Foundations of Computer Science FOCS, pp. 429–438 (2002). https://doi.org/10.1109/SFCS.2002.1181967
13. Jain, R., Radhakrishnan, J., Sen, P.: The quantum communication complexity of the pointer chasing problem: The bit version. In: FST TCS Foundations of Software Technology and Theoretical Computer Science, pp. 218–229 (2002). https://doi.org/10.1007/3-540-36206-1_20
14. Jastrzebski, M.: On total communication complexity of collapsing protocols for pointer jumping problem. CoRR abs/1405.7596 (2014). http://arxiv.org/abs/1405.7596
15. Klauck, H.: On quantum and probabilistic communication: Las Vegas and one-way protocols. In: Proceedings of the Thirty-Second Annual ACM Symposium on Theory of Computing, pp. 644–651 (2000). https://doi.org/10.1145/335305.335396
16. Klauck, H., Nayak, A., Ta-Shma, A., Zuckerman, D.: Interaction in quantum communication and the complexity of set disjointness. In: Vitter, J.S., Spirakis, P.G., Yannakakis, M. (eds.) Proceedings on 33rd Annual ACM Symposium on Theory of Computing, pp. 124–133. ACM (2001). https://doi.org/10.1145/380752.380786
17. Kushilevitz, E., Nisan, N.: Communication Complexity. Cambridge University Press (1997)
18. Liang, H.: Optimal collapsing protocol for multiparty pointer jumping. Theory Comput. Syst. **54**(1), 13–23 (2014). https://doi.org/10.1007/s00224-013-9476-x
19. Mao, X., Yang, G., Zhang, J.: Gadgetless lifting beats round elimination: improved lower bounds for pointer chasing. In: 16th Innovations in Theoretical Computer Science Conference, ITCS. LIPIcs, vol. 325, pp. 75:1–75:14 (2025). https://doi.org/10.4230/LIPICS.ITCS.2025.75
20. Miltersen, P.B., Nisan, N., Safra, S., Wigderson, A.: On data structures and asymmetric communication complexity. J. Comput. Syst. Sci. **57**(1), 37–49 (1998). https://doi.org/10.1006/JCSS.1998.1577
21. Nanongkai, D., Sarma, A.D., Pandurangan, G.: A tight unconditional lower bound on distributed randomwalk computation. In: Proceedings of the 30th Annual ACM Symposium on Principles of Distributed Computing, PODC, pp. 257–266 (2011). https://doi.org/10.1145/1993806.1993853
22. Nisan, N., Wigderson, A.: Rounds in communication complexity revisited. SIAM J. Comput. **22**(1), 211–219 (1993)
23. Papadimitriou, C.H., Sipser, M.: Communication complexity. J. Comput. Syst. Sci. **28**(2), 260–269 (1984). https://doi.org/10.1016/0022-0000(84)90069-2
24. Patrascu, M.: Unifying the landscape of cell-probe lower bounds. SIAM J. Comput. **40**(3), 827–847 (2011). https://doi.org/10.1137/09075336X
25. Ponzio, S., Radhakrishnan, J., Venkatesh, S.: The communication complexity of pointer chasing: applications of entropy and sampling. In: Proceedings of the

Thirty-First Annual ACM Symposium on Theory of Computing, pp. 602–611 (1999). https://doi.org/10.1145/301250.301413
26. Rao, A., Yehudayoff, A.: Communication Complexity, and Applications. Cambridge University Press (2020)
27. Raz, R., Spieker, B.: On the “log rank”-conjecture in communication complexity. Comb. **15**(4), 567–588 (1995)
28. Sen, P., Venkatesh, S.: Lower bounds for predecessor searching in the cell probe model. J. Comput. Syst. Sci. **74**(3), 364–385 (2008). https://doi.org/10.1016/J.JCSS.2007.06.016
29. Viola, E., Wigderson, A.: One-way multiparty communication lower bound for pointer jumping with applications. Comb. 719–743 (2009). https://doi.org/10.1007/s00493-009-2667-z
30. Yehudayoff, A.: Pointer chasing via triangular discrimination. Comb. Probab. Comput. 485–494 (2020). https://doi.org/10.1017/S0963548320000085

Oblivious Robots Under Sequential Schedulers: Universal Pattern Formation

Paola Flocchini[1], Alfredo Navarra[2], Debasish Pattanayak[1], Francesco Piselli[2](✉), and Nicola Santoro[3]

[1] School of Electrical Engineering and Computer Science, University of Ottawa, Ottawa, Canada
paola.flocchini@uottawa.ca, drdebmath@gmail.com

[2] Department of Mathematics and Computer Science, University of Perugia, Perugia, Italy
alfredo.navarra@unipg.it, francesco.piselli@unifi.it

[3] School of Computer Science, Carleton University, Ottawa, Canada
santoro@scs.carleton.ca

Abstract. We study the computational power that oblivious robots operating in the plane have under *sequential* schedulers. We show that this power is much stronger than the obvious capacity these schedulers offer of breaking symmetry, and thus to create a leader. More precisely, we consider the class of *pattern formation* problems, and focus on the most general problem in this class, `Universal Pattern Formation` (`UPF`), which requires the robots to form *any* pattern given in input, starting from *any* initial configurations (where robots may occupy the same point). We first show that `UPF` is *unsolvable* under $\mathcal{FSYNC}$, even if the robots are endowed with additional strong capabilities (multiplicity detection, rigid movement, agreement on coordinate systems, presence of a unique leader). On the other hand, we prove that, except for point formation (`Gathering`), `UPF` is *solvable* under any sequential scheduler without any additional assumptions. We then turn our attention to the `Gathering` problem, and prove that weak multiplicity detection is necessary and sufficient for solvability under sequential schedulers. The obtained results show that the computational power of the robots under $\mathcal{FSYNC}$ (where `Gathering` is solvable without any multiplicity detection) and that under sequential schedulers are *orthogonal*.

1 Introduction

Distributed systems composed of a set $\mathcal{R}$ of autonomous mobile robots operating in Euclidean spaces, have been the object of rather extensive investigations in theoretical computer science since their introduction in [21]. Robots operate performing *Look-Compute-Move* cycles; in each cycle, a robot observes the location of the other robots (*Look*), executes its algorithm to determine a destination

Work funded in part by the Italian National Group for Scientific Computation INdAM - GNCS Project, CUP E53C24001950001, and by the Natural Sciences and Engineering Research Council of Canada, under Discovery Grants A2415 and 203254.

U. Schmid and R. Kuznets (Eds.): SIROCCO 2025, LNCS 15671, pp. 297–314, 2025.
https://doi.org/10.1007/978-3-031-91736-3_18

(*Compute*), and moves toward the computed destination (*Move*). Several formal models have been proposed, describing different levels of computation capabilities of the robots. In this paper we consider the standard model, $\mathcal{OBLOT}$, which describes the computationally weakest robots (e.g., see [9,16,22]). In this model, the robots are *anonymous*: they are indistinguishable by their appearance and do not have distinct identities; *disoriented*: they may not agree on a common coordinate system; *silent*: they have no means of direct communication; and they are *oblivious*: at the beginning of each cycle, a robot has no memory of past actions and computations.

As for other distributed systems, the main research objective is the determination of what is computable, under what conditions, and at what cost; the main goal is to identify the minimal capabilities needed by the robots for the system to be able to solve a given problem or perform an assigned task. Our focus is on deterministic computations. The feasibility of a task and the solvability of a problem are influenced by several specific factors. However, the most determinant one, concerns the nature of the timing of the robots actions imposed by the adversarial environment in which the robots operate.

Sequential Schedulers. The timing of when a robot is active and performs a cycle as well as the duration of each action within a cycle is determined by an adversary, called *scheduler*. There are two general classes of schedulers, the asynchronous and the synchronous ones; in both classes, all schedulers must respect the *fairness*: every robot is activated infinitely often.

In the *asynchronous* class (ASYNC), each robot may be activated at arbitrary times, independently of the others, and the duration of each operation in the cycle is finite but possibly unpredictable. In this class, the standard scheduler investigated in the literature is the one with unrestricted adversarial power, introduced in [11], called simply (*fully*) *asynchronous* and denoted by $\mathcal{ASYNC}$. A few other schedulers in this class have also been investigated (see, e.g., [3,14,17]).

In the *synchronous* class (SYNCH), time is logically divided into a sequence of *rounds*. In each round, the scheduler activates a non-empty subset of the robots, and they perform each phase of their *Look-Compute-Move* cycle simultaneously. The standard scheduler is the one which can generate all possible sequences, known as *semi-synchronous*, denoted by $\mathcal{SSYNC}$, introduced in [21]. At another end of the spectrum, there is the very popular *fully-synchronous* scheduler, denoted by $\mathcal{FSYNC}$, where all robots are activated in every round.

Interestingly, there are schedulers in SYNCH that allow robots to solve important problems unsolvable under $\mathcal{FSYNC}$. Such is the subclass of *sequential* schedulers, SEQUENTIAL $\subset$ SYNCH, where only a single robot is activated in each round. An immediate obvious computational advantage that such schedulers offer the robots is that they allow them to break symmetry (and thus, e.g., elect a leader), a task impossible to achieve under $\mathcal{FSYNC}$ when starting from a symmetric configuration. In spite of this remarkable fact, very little is known on the computational power of robots under sequential schedulers; in fact, unlike the interest shown on sequential schedulers in fields like self-stabilization and cellular automata, in the field of mobile robots on the plane they have been analyzed mainly in the area of fault-tolerance limited to the problems of gathering and scattering (see [7] for a recent survey).

In this paper we analyze the computational power of the robots under schedulers in SEQUENTIAL, focusing on the large and important class of *Pattern Formation* problems. In particular, we focus on the ability of robots to solve those problems under the scheduler that can generate all possible sequential activation sequences, henceforth denoted by $\mathcal{SEQ}$.

Pattern Formation. *Pattern formation* refers to the abstract problem of the robots rearranging themselves so that, at some point in time, the set of their locations in the plane (the *configuration*) satisfies some geometric predicate given in input (the *pattern*) and no longer move. Typically, the pattern is specified as a set of points $\hat{P}$ representing some geometric figure that the robots must form (irrespective to rotation, reflection, translation or scaling).

Due to their importance, the class of pattern formation problems has been extensively studied under a number of different assumptions. A very special role in this class is held by *point formation*; that is, the problem of forming the pattern consisting of a single point, i.e., $|\hat{P}| = 1$. It is relevant because it corresponds to the important coordination problem, extensively studied in the literature, called `Gathering` or `Rendezvous`, requiring all robots to meet at the same location, not fixed in advance (see, e.g., [5,18–20]). The class of pattern formation problems is quite large, not only because of the different types of geometric patterns that can be formed, many requiring special ad-hoc techniques for their formation (if at all possible), but also for wide range of more general research questions it poses. These include, for a specific adversarial scheduler: under what conditions can all patterns be formed, what patterns are formable from a given initial configuration, what patterns are formable from any initial configuration.

Clearly, to form a pattern, there must be at least as many robots as there are points in the pattern, i.e., $|\mathcal{R}| \geq |\hat{P}|$; we shall call this the *trivial assumption.* Most of the research has however relied on a variety of non-trivial assumptions, such as: the robots occupy distinct initial locations; the number of robots is equal to the number of points in the pattern, $|\mathcal{R}| = |\hat{P}|$; the robots have some agreement on coordinate systems or chirality (i.e., they agree on the clockwise and counterclockwise directions); one of the robots is visibly different from all the others (i.e., there is a *leader*), etc. (see, e.g., [1,2,4,10,12,15,21,22]).

In this paper we are interested in the most general problem in this class, `Universal Pattern Formation` (`UPF`), which requires the robots to form *every* pattern given in input starting from *any* initial configurations, regardless of their number, of the locations they initially occupy, and of the number of points in the pattern, and we are interested in satisfying only the trivial assumption with the minimal robots capabilities.

Contributions. In this paper we bring to light the very strong computational power that robots have under the SEQUENTIAL class of schedulers in regards to pattern formation problems.

We do so by focusing on the general `UPF` problem. We prove that this problem is *unsolvable* under $\mathcal{FSYNC}$ (and, thus, under $\mathcal{SSYNC}$), even if the robots are empowered with strong multiplicity detection (i.e., robots can detect the exact number of robots occupying the same location), their movements are rigid (i.e., they always reach their destination regardless of the distance), have total agreement on the coordinate system, and there is a unique leader robot (i.e., distinguishable from all others), see Sect. 3.

On the other hand, in Sect. 4 we show that, except for point formation, the UPF problem is *solvable* under any sequential schedulers without any additional assumptions on the computational capabilities of the robots, namely even if they are unable of multiplicity detection, their movements are non-rigid, they have no agreement on coordinate system nor on chirality, and are all identical. The proof is constructive as we present a solution protocol. The corresponding proof of its correctness along with the analysis on its time complexity can be found in [8]. These results indicate the strong impact that the presence of a sequential scheduler can have on the computational power of the robots, allowing them to solve, with minimal capabilities, problems unsolvable even under $\mathcal{FSYNC}$ with stronger power and a leader.

Then, in Sect. 5, we turn our attention to the point formation or Gathering problem. This is known to be unsolvable under $\mathcal{SSYNC}$ even with strong multiplicity detection. In contrast, we prove that, under any sequential scheduler, the problem becomes solvable with weak multiplicity detection (i.e., a robot can detect only whether a point is occupied by one or more than one robot). We also prove that, without weak multiplicity detection, the problem, while solvable under $\mathcal{FSYNC}$, is unsolvable under $\mathcal{SEQ}$. The obtained results prove that the computational power of the robots under $\mathcal{SEQ}$ and that under $\mathcal{FSYNC}$ are *orthogonal*, i.e., incomparable.

2 Model and Terminology

We consider the standard $\mathcal{OBLOT}$ model of distributed systems of autonomous mobile robots (e.g., see [9,13]).

Robots. The system is composed of a set $\mathcal{R} = \{r_1, r_2, \ldots, r_n\}$ of computational *robots*, moving and operating in the Euclidean plane $\mathbb{R}^2$. Viewed as points, the robots can move freely and continuously in the plane. A robot is endowed with motorial and computational abilities, a private coordinate system of which it perceives itself at the origin, and sensorial devices that allow it to observe the positions of the other robots with respect to its local coordinate system. When active, a robot performs a *Look-Compute-Move* cycle:

1. *Look:* The robot obtains an instantaneous snapshot of the positions occupied by the robots expressed within a private coordinate system;
2. *Compute:* The robot executes its built-in deterministic algorithm, the same for all robots, using the snapshot as input; the result of the computation is a destination point;
3. *Move:* The robot moves toward the computed destination. If the destination is the current location, the robot stays still.

The duration of each cycle is finite. Robots are *autonomous*: they operate without a central control or external supervision; *identical*: they are indistinguishable by their appearance and do not have distinct identities that can be used during the computation; *disoriented*: they may not agree on a common coordinate system nor on chirality (a clockwise orientation of the plane); *silent*: they have no means of direct communication of information to other robots, so any communication occurs in a totally implicit manner, by moving and by observing the positions of the other robots; and they are *oblivious*: at the beginning of each cycle, a robot has no memory of past actions and computations, so the computation is based solely on what determined in the current cycle.

Movements are said to be: *rigid*, if the robots always reach their destination in the same round they started moving; *non-rigid*, if they may be unpredictably stopped by an adversary whose only limitation is the existence of $\delta > 0$, unknown to the robots, such that if the destination is at distance at most δ, the robot will reach it, otherwise it will move at least δ toward the destination.

More than one robot can be at the same point at the same time; such a point is said to be a *multiplicity*. The robots are said to be capable of *multiplicity detection* if they can determine whether or not points are multiplicities; multiplicity detection is said to be *strong* if a robot can determine the exact number of robots located at a point, *weak* otherwise.

Schedulers. The timing of when a robot becomes active as well as the duration of *Compute* and *Move* phases within each cycle, and the rate of motion within each *Move* phase, is viewed as determined by an adversarial *scheduler*. The shortest period of time in which all the robots have finished at least one cycle is called *epoch*. There are two classes of schedulers, the asynchronous and the synchronous ones.

In the *asynchronous* class (ASYNCH), each robot may be activated at arbitrary times, independently of the others, the duration of each *Compute* and *Move* phase is finite but unpredictable, and might be different in different cycles. Traditionally, the scheduler in this class with unrestricted adversarial power is simply called asynchronous and denoted by $\mathcal{ASYNC}$.

In the *synchronous* class (SYNCH), time is logically divided into *rounds*. In each round, the scheduler activates a non-empty subset of the robots, and they perform each phase of their *Look-Compute-Move* cycle simultaneously, in perfect synchrony. Clearly, SYNCH $\subset$ ASYNCH.

Given a synchronous scheduler $\mathcal{S} \in$ SYNCH and a set of robots $\mathcal{R}$, an *activation sequence* of $\mathcal{R}$ by (or, under) $\mathcal{S}$ is an infinite sequence $E = \langle R_1, R_2, \ldots, R_i, \ldots \rangle$, where $R_i \subseteq \mathcal{R}$ denotes the set of robots activated in round i, satisfying the *fairness constraint*: $(\forall r \in \mathcal{R},\ \exists i : r \in R_i)$ **and** $(\forall i \geq 1, r \in R_i \Rightarrow \exists j > i : r \in R_j)$.

Let $\mathcal{E}(\mathcal{S}, \mathcal{R})$ denote the set of all activation sequences of $\mathcal{R}$ by $\mathcal{S}$. We shall denote by $\mathcal{SSYNC}$ the synchronous scheduler, traditionally called *semi-synchronous*, without any other restriction. The special *fully-synchronous* scheduler $\mathcal{FSYNC}$, which activates every robot in every round (i.e., a round coincides with an epoch) corresponds to further restricting the activation sequences by: $\forall i \geq 1, R_i = \mathcal{R}$.

The class of *sequential* schedulers SEQUENTIAL $\subset$ SYNCH is defined by the restriction: $\forall i \geq 1, |R_i| = 1$, that is, only one robot is activated in each round. We shall denote by $\mathcal{SEQ}$ the sequential scheduler without any other restriction. We remind that the fairness constraint is common to every scheduler.

Pattern Formation Problems. A *configuration* $C(t)$ is the *multiset* of the locations of the robots at time t. Let $Q(t) = \{q_1, q_2, \ldots, q_m\}$ be the corresponding *set* of unique points occupied by the robots at time t, where $m \leq n$.

Since robots can only observe the positions of others and move, in the $\mathcal{OBLOT}$ model, a *problem* to be solved is typically expressed in terms of a *temporal geometric predicate*, which the configuration formed by the robot positions must satisfy from some time on.

A *pattern* $\hat{P} = \{\hat{p}_1, \hat{p}_2, \ldots, \hat{p}_k\} \subset \mathbb{R}^2$ is a set of distinct points. A pattern P' is said to be *isomorphic* to a pattern P'' if P'' can be obtained by a combination of translation, rotation, reflection and scaling of pattern P'. We shall denote by $\mathcal{B}(\hat{P})$ the (infinite) set of patterns isomorphic to P. We say the robots have *formed* a pattern $\hat{P}$ at time t, if the configuration $C(t)$ is such that $Q(t) \in \mathcal{B}(\hat{P})$.

The class of *Pattern Formation* problems is quite large, not only because of the different types of geometric patterns that can be formed, many types requiring special ad-hoc techniques for their formation (if at all possible), but also for the scope of the requirements.

The most general problem in this class is `Universal Pattern Formation` (`UPF`) that requires the robots, executing the same algorithm and starting from any arbitrary initial configuration $C(0)$, to form any arbitrary pattern $\hat{P}$, given as its sole input and no longer move.

It is defined by the predicate:

$$\begin{aligned} \texttt{UPF} \equiv\ & \forall \hat{P} \subset \mathbb{R}^2, \forall n \geq |\hat{P}|, \forall C \in \mathbb{R}^2_n, \\ & (C(0) = C) \Rightarrow \exists t \geq 0 : (Q(t) \in \mathcal{B}(\hat{P})) \textbf{ and } (\forall t' > t, C(t') = C(t)). \end{aligned}$$

A particular instance is the problem of forming the pattern consisting of a single point, i.e., $|\hat{P}| = 1$. This problem is special because it corresponds to the important coordination problem, called `Gathering` or `Rendezvous`, of having all robots meet at the same location, not fixed in advance; it is defined by the predicate:

$$\begin{aligned} \texttt{Gather} \equiv\ & \forall n \in \mathbb{Z}^+, \forall C \in \mathbb{R}^2_n, \\ & (C(0) = C) \Rightarrow \exists t \geq 0 : (|Q(t)| = 1) \textbf{ and } (\forall t' > t, Q(t') = Q(t)), \end{aligned}$$

where $\mathbb{R}^2_n$ denotes all possible set of n points.

Due to its particular nature, the `Gathering` problem has been treated in the literature differently from all other cases of pattern formation where $|\hat{P}| > 1$. Also in this paper we shall make the same distinction, and refer to the case $|\hat{P}| > 1$ of `UPF` as `UPF*`.

3 Impossibilities for UPF Under $\mathcal{FSYNC}$

In this section, we show that UPF is unsolvable under the fully-synchronous scheduler $\mathcal{FSYNC}$, and, thus, under $\mathcal{ASYNC}$ and $\mathcal{SSYNC}$.

Theorem 1. UPF *is unsolvable under* $\mathcal{FSYNC}$ *even if the robots have strong multiplicity detection, share the same coordinate system, their movements are rigid, and one of them is distinguishable from all others (i.e., there is a leader).*

Proof. By contradiction, let $\mathcal{A}$ be a solution protocol for UPF. Consider an initial configuration $C(0)$ where there is a multiplicity composed by non-leader robots and $|Q(0)| < |\hat{P}|$. Consider an execution of $\mathcal{A}$. Under $\mathcal{FSYNC}$, since all robots share the same coordinate system, the non-leader robots composing the multiplicity have the same view in the *Look* phase; thus, they will compute the same destination and reach it at the end of the cycle because of the rigid movement, continuing to be co-located. In other words, during the execution of $\mathcal{A}$, separate groups of non-leader robots may become co-located, but it is impossible for co-located non-leader robots to separate; as for the leader robot, in each round, it might become co-located with non-leader robots, or remain alone.

In other words, in this execution, $|Q(t)| \leq |Q(0)|$ for all $t > 0$. Since $|Q(0)| < |\hat{P}|$, it follows that the robots will never be able to form $\hat{P}$, contradicting the correctness of $\mathcal{A}$. □

On the other hand, it is known that the special case of Gathering is solvable under $\mathcal{FSYNC}$ without any additional robot capability [6].

4 Possibility for UPF* Under $\mathcal{SEQ}$

In this section, we study the UPF problem under the $\mathcal{SEQ}$ scheduler in the case $|\hat{P}| > 1$ (i.e., UPF*) where the robots have no additional capabilities. We do so by presenting a solution algorithm under the $\mathcal{SEQ}$ scheduler when the robots have no multiplicity detection, no agreement on coordinate systems nor chirality, their movements are non-rigid, $|\mathcal{R}|$ is unknown, and $|\hat{P}| \geq 5$. The case where $|\hat{P}| < 5$ will be addressed in the subsequent section.

We start by introducing some important terminology and then we present the algorithm.

4.1 Terminology and Preliminaries

The *Smallest Enclosing Circle* of $C(t)$ is the circle of smallest radius such that every point of $Q(t)$ lies on the circle or in its interior; it is denoted by $SEC(Q(t))$. Let $\rho(t)$ and $c(Q(t))$ denote the radius and the center of $SEC(Q(t))$, respectively. In the following, should no ambiguity arise, the temporal indicator t shall be omitted. Furthermore, $c(Q)$ will be also denoted simply by O.

Let $S \subset \mathbb{R}^2$ be a finite set of $l > 2$ points where no two points are *co-radial*,[1] and let $c(S)$ be the center of $SEC(S)$.

The positions of the points of S, in the clockwise order, around $c(S)$ induce a cyclic order ψ among the points in S. Let $\psi(S) = (s_1, s_2, \ldots, s_l)$ be the corresponding cyclic sequence of the points, with s_1 chosen arbitrarily, and let us denote by $\theta(s_i, s_{i+1})$ the angle, with respect to $c(S)$, between the two consecutive points s_i and s_{i+1}.

Let $\theta(S) = \{\theta(s_i, s_{i+1}), \theta(s_{i+1}, s_{i+2}), \ldots, \theta(s_{i-1}, s_i)\}$ be an *angle sequence* starting at s_i induced on S.

We now consider a set S' containing only robot positions Q or both Q and pattern points $\hat{P}$, suitably overlapped in such a way that $SEC(S') = SEC(Q) = SEC(\hat{P})$. Furthermore, if there are co-radial points, then only the most external robot position is taken in S', if any, otherwise the most external pattern point is considered, and all other points are discarded. In this way, S' does not contain co-radial points. Still we assume $|S'| = l' > 2$. An angle sequence $\theta(S') = (\theta_1, \ldots, \theta_l)$ is a *leader angular sequence*, denoted by $\lambda(Q)$ or $\lambda(Q \cup \hat{P})$, if *(i)* $\theta_1 < \min\{\theta_2, \ldots, \theta_l\}$, and *(ii)* the two adjacent points s_i and s_{i+1} defining θ_1 belong to Q, with s_i inside $SEC(Q)$ and s_{i+1} on $SEC(Q)$.

Without loss of generality, if S' admits a leader angular sequence, then the rotational direction to overlap the radius containing s_i with the radius containing s_{i+1} spanning angle θ_1, is considered to be clockwise. Therefore, a configuration $C(t)$ (possibly overlapped with a pattern $\hat{P}$), inducing the set S' that admits a leader angular sequence, is asymmetric (apart for possible multiplicities) and all the robots can agree on the chirality of the system.

When S' admits a leader angular sequence, we define the *"radiangular"* distance between two points as follows. Let A and B be two points that make the clockwise (according to the leader angular sequence) angle $\theta = \angle AOB$ at O, the center of the SEC. The radiangular distance is defined by the tuple *(i)* $(\theta, |\overline{OA}| + |\overline{OB}|)$, if $\theta > 0$, *(ii)* $(0, |\overline{AB}|)$, otherwise. When A and B are not co-radial, the path of a robot from A to B consists of moving along $\overline{AO}$ to reach the center O, then along $\overline{OB}$ to reach B. When A and B are co-radial, the path is simply $\overline{AB}$. In both cases, the defined path is called a *radial detour.*

Let us now consider the set of pattern points $\hat{P}$, with $|\hat{P}| \geq 5$. Two pattern points p_i and p_{i+1} are said to be *adjacent* if they are co-radial and no other pattern point is in between them, or there is no other pattern point inside the sector defined by angle $\theta(\hat{p}_i, \hat{p}_{i+1})$ calculated in the clockwise order, and there are no other pattern points on the segments $\overline{O\hat{p}_i}$ and $\overline{O\hat{p}_{i+1}}$. Let $\alpha^{(i)}(\hat{P}) = \{(\theta(\hat{p}_i, \hat{p}_{i+1}), |\overline{O\hat{p}_i}|, |\overline{O\hat{p}_{i+1}}|), (\theta(\hat{p}_{i+1}, \hat{p}_{i+2}), |\overline{O\hat{p}_{i+1}}|, |\overline{O\hat{p}_{i+2}}|), \ldots, (\theta(\hat{p}_{i-1}, \hat{p}_i), |\overline{O\hat{p}_{i-1}}|, |\overline{O\hat{p}_i}|)\}$ be the set of triples determining what we call the *clockwise pattern sequence* induced on $\hat{P}$ starting at an arbitrary $\hat{p}_i$. Similarly, let $\beta^{(i)}(\hat{P}) = \{(\theta(\hat{p}_i, \hat{p}_{i-1}), |\overline{O\hat{p}_i}|, |\overline{O\hat{p}_{i-1}}|), (\theta(\hat{p}_{i-1}, \hat{p}_{i-2}), |\overline{O\hat{p}_{i-1}}|, |\overline{O\hat{p}_{i-2}}|), \ldots, (\theta(\hat{p}_{i+1}, \hat{p}_i), |\overline{O\hat{p}_{i+1}}|, |\overline{O\hat{p}_i}|)\}$ be the *counter-clockwise pattern sequence.* We shall denote by $\hat{\mu}$ the lexicographical smallest of all pattern sequences $\alpha^{(i)}(\hat{P})$ and

[1] Two points of a set S are co-radial if they reside on the same radius of the $SEC(S)$.

$\beta^{(i)}(\hat{P})$, $1 \leq i \leq k$. Note that, if $\hat{P}$ admits symmetries, then there is more than one index determining the same $\hat{\mu}$.

4.2 Strategy in Case $|\hat{P}| \geq 5$

When $|\hat{P}| \geq 5$, the strategy followed by the robots ideally consists of four consecutive stages: *Initialization*, *Leader Configuration*, *Partial Pattern Formation*, and *Finalization.*

The scope of the *Initialization* stage is to create a configuration C, if not already available, where the number $|Q|$ of distinct locations occupied by the robots is at least as large as the number $|\hat{P}|$ of points in the pattern, i.e., $|Q| \geq k = |\hat{P}|$. This is achieved through Procedure SEPARATE, invoked and executed as long as $|Q| < k$; its execution by an activated robot consists of the robot choosing as its destination: (i) an unoccupied point located along its own x-axis at its own unit distance, if this path does not contain other robots; or (ii) the point located along its own x-axis, at half of the distance from the closest robot along such a direction. After the activation of at most $k-1$ distinct robots, there will be at least k distinct locations occupied by the robots. Once $|Q| \geq k$, or if it was already as such from the initial configuration, the activated robot considers the pattern at a scale such that $SEC(\hat{P})$ overlaps $SEC(Q)$, i.e., their respective radii have the same length.

In the *Leader Configuration* stage, the goal is to create a specific map, called joint configuration, that shows both the points occupied by the robots and the points of the pattern. Informally, this is done as follows. After *Initialization*, if the current configuration does not concern the special cases managed by Procedure LAST, Procedure OVERLAP is executed to verify whether the robots can agree on the same joint configuration. If such an agreement is not possible, then Procedure LEADER is executed to create a leader configuration. To do so, the activated robot moves to the center of the configuration (if not yet occupied), provided it has a free path to the center and it is not responsible for maintaining the $SEC(Q)$. Then, the robot at the center moves toward a point to create the smallest unique angle determining a leader angular sequence. Once Procedure OVERLAP returns a joint configuration, then we check if it is a leader configuration. If this is not the case, Procedure LEADER is executed to create a leader angular sequence in the joint configuration.

Subsequently, in the *Partial Pattern Formation* stage, the pattern points are filled sequentially by the robots; this is accomplished through Procedure OCCUPY. This involves the movement of suitable robots along radial detours that avoids passing over other robots. Ideally, this stage lasts as long as only one robot position does not coincide with a pattern point.

In the *Finalization* stage, Procedure LAST is invoked when $|Q| \geq k$, any pattern point $\hat{p}_i$, $1 \leq i < k$ is occupied, and any robot position not located on a pattern point, is located at $\overline{O\hat{p}_k}$, i.e., the segment from O to the lowest priority pattern point $\hat{p}_k$. If the robot running Procedure LAST is located at $\overline{O\hat{p}_k}$, it moves toward $\hat{p}_k$. Moreover, whenever the activated robot detects that the current configuration matches the pattern irrespective to rotation, reflection, and translation,[2] it does not move, i.e., the pattern is formed.

Our solution to the Pattern Formation problem is provided by means of Algorithm SeqPF, which in turn exploits five procedures, determining the four stages described above: Separate (in *Initialization*), Overlap and Leader (in *Leader Configuration*), Occupy (in *Partial Pattern Formation*) and Last (in *Finalization*); they are individually presented and described next, whereas the pseudocode can be found in [8].

Algorithm SeqPF**.** In this section, we describe the overall structure of the resolution algorithm SeqPF from the point of view of the robot executing it. An activated robot r, obtains a snapshot Q in the *Look* phase, and then it executes Algorithm SeqPF in its *Compute* phase. Let $q \in Q$ be the position of robot r.

The first operation the robot performs is to check if the number $|Q|$ of robots currently on distinct locations is sufficient to form the pattern. If $|Q| < k$, then it executes Procedure Separate, to determine an appropriate destination point, and terminates this execution. Otherwise, it proceeds to the next step and the first thing it does is to scale the pattern $\hat{P}$ in such a way that the radius of $SEC(\hat{P})$ has the same length ρ of the radius of $SEC(Q)$.

Whenever $|Q| \geq k$, the activated robot checks if the current configuration Q falls in one of the special cases, i.e., if it is already equivalent to the target pattern or the robot positions not coinciding with a specific pattern point are all located at $\overline{O\hat{p}_k}$. In the former case, i.e., if the configuration already satisfies the pattern, no robot moves. In the latter, Procedure Last is invoked to finalize the formation of the pattern.

If it is not a special case, i.e., Procedure Last returns *null*, then the activated robot computes a joint configuration, denoted by Γ, following Procedure Overlap to properly overlap the pattern on the current configuration.

If a joint configuration cannot be computed, i.e., $\Gamma = null$, then there must be at least a robot that can move toward O (provided it is not responsible for $SEC(Q)$ and it has a free radial path) by means of Procedure Leader, or there must be a robot located at O. In this latter case, again Procedure Leader is executed to create a leader configuration with a leader angular sequence.

Once Q admits a leader angular sequence, the activated robot overlaps Q and $\hat{P}$ in a specific way creating the joint configuration Γ, and then it checks if Γ admits a leader angular sequence. If not, Procedure Leader is executed to create a leader angular sequence in Γ. Otherwise, Procedure Occupy is executed. The goal of this procedure is to move a specific robot to the highest priority empty pattern point p_l.

Once all robot positions but the lowest priority one coincide with the highest priority pattern points, then this is one of the special cases described before.

Procedure Separate**.** This procedure consists of a very simple operation. The activated robot r computes a point q' along its own x-axis at one unit distance.

[2] The scale has been fixed to the dimension of $SEC(Q)$.

If $\overline{qq'}$ does not contain other robots, r selects as its destination q'. Otherwise, r computes the position of its closest robot q'' along its own x-axis, and selects as its destination the midpoint of $\overline{qq''}$.

Procedure LAST. As previously described, Procedure LAST is executed to manage some special cases. Once the activated robot r has verified that $|Q| \geq k$, it starts with checking if the current configuration already is the target pattern, i.e., $Q \in \mathcal{B}(\hat{P})$. Robot r checks this by rotating and reflecting the pattern as follows. Let $\hat{p}_1$ be a pattern point on $SEC(\hat{P})$. Consider all the robot positions $(q_1, \dots, q_s) = SEC(Q) \cap Q$ in the cyclic order of some $\psi(S')$. Let q_i be a robot position in $(q_1, \dots, q_s)$. The robot r overlaps the pattern point $\hat{p}_1$ with each of the robot positions q_i "one by one" by reflecting the pattern (across the radius $\overline{O\hat{p}_1}$). If one of these transformations exactly overlaps all the points of Q with all the points of $\hat{P}$, it follows that $Q \in \mathcal{B}(\hat{P})$, and r does not move. In the worst case, if all the robot positions are on the $SEC(Q)$, then r has to check at most $2k$ transformations (k rotations and k reflections).

If $|Q| \geq k$ and $Q \notin \mathcal{B}(\hat{P})$, the activated robot checks if the configuration falls into the special cases "last one". These cases occur when pattern points $\hat{p}_i$, $1 \leq i < k$, are occupied and the only robot positions not on pattern points are located at $\overline{O\hat{p}_k}$, with $\hat{p}_k$ being the lowest priority pattern point. The activated robot checks at most $2k + 2$ transformations.

The lowest priority pattern point $\hat{p}_k$ is defined by any lexicographical smallest pattern sequence $\hat{\mu}$. Once a robot located at $\overline{O\hat{p}_k}$ is activated, it moves toward $\hat{p}_k$, if not yet there.

Procedure OVERLAP. When the activated robot r determines that $|Q| \geq k$ and that the configuration does not concern the special cases managed by Procedure LAST (i.e., call LAST() returns *null*), it checks if Q has a leader angular sequence $\lambda(Q)$. If this is the case, r proceeds to overlap the pattern points on the current configuration and subsequently achieve a joint configuration Γ.

Let $\nabla = \{Cir_1, Cir_2, \dots, Cir_\sigma\}$ be the set of all concentric circles, in the decreasing order of their radii, containing points of $\hat{P}$, and let ρ_j be the radius of Cir_j, $1 \leq j \leq \sigma$; thus $Cir_1 = SEC(\hat{P})$ and $\rho_1 = \rho$. The objective of the robot is to rotate and/or reflect $\hat{P}$ until some specific points in $SEC(Q)$ and $SEC(\hat{P})$ are co-located, i.e., they precisely overlap. This is done as follows.

Let $\hat{\mu}$ be the lexicographical smallest pattern sequence of $\hat{P}$ and $\psi(\hat{P})$ be the corresponding cyclic order of points, $\hat{p}_i$ be the first pattern point on the $SEC(\hat{P})$ in the cyclic order $\psi(\hat{P})$. Let q_j be the second robot position determining θ_1 in $\lambda(Q)$. The activated robot proceeds to overlap Q and $\hat{P}$ so that q_j overlaps with $\hat{p}_i$ and the order induced by $\hat{\mu}$ follows the clockwise direction induced by $\lambda(Q)$.

Procedure LEADER. As input, Procedure LEADER can get: a configuration Q with no leader angular sequence if $\Gamma = null$ or, a joint configuration Γ if Procedure OVERLAP has been executed from a configuration Q admitting a leader angular sequence.

The objective of LEADER is to first send a robot to O if O is empty, and then to create a leader configuration. If O is empty, then the robots that can move are the ones not responsible for the $SEC(Q)$ and that have a path with no other robot positions toward O, or if there is a joint configuration Γ, the robot that also cannot move is the one responsible for the smallest angle θ_1 if it has no other co-radial robots.

If, instead, O is occupied by a robot, once such robot is activated, it computes ξ, the smallest angle in the input configuration, and moves to u' such that $|\overline{u'O}| = \rho/2$ and for a given $q_j \in SEC(Q)$, it forms $\angle u'Oq_j = \xi/3$.

Procedure Occupy. Procedure Occupy deals with the movement of robots to reach pattern points. Given a joint configuration Γ with leader angular sequence, the ranking of the pattern points can be established. A robot is then chosen as the one to move, i.e., the *walker*, provided it satisfies some conditions. The walker robot moves to the target pattern point via a radial detour. Recall that u' and q_j are the robot positions defining the smallest angle θ_1 with q_j on the $SEC(\Gamma)$. Let q_j be located at A. Now $\overline{OA}$ is the radius. Consider points $A_2, \ldots, A_\sigma$ to be the projections of A on the circles $Cir_2, Cir_3, \ldots, Cir_\sigma$, respectively. For each Cir_i, the pattern points are ranked in the increasing order starting from A_i and moving following the clockwise direction induced by $\lambda(\Gamma)$ until we reach A_i again. The rank of pattern points in Cir_i is smaller than Cir_{i+1} for any $i = 1, \ldots, \theta - 1$. For Cir_1, the rank follows a different strategy. This is required in order to maintain the original SEC during the execution of the algorithm. By Procedure Overlap, we already have a robot at q_j, which is also the highest priority pattern point. We start at q_j as p_1. Consider p_1' as the antipodal point of p_1, positioned on the SEC. If there is a pattern point at p_1', then that becomes p_2. Otherwise, consider the two closest pattern points to p_1' positioned on the SEC, going from p_1' to p_1, one in the counterclockwise direction and the other one in the clockwise direction. They respectively become p_2 and p_3. The remaining pattern points on Cir_1 are assigned in the increasing order. Notice that, p_1 is the smallest rank pattern point and p_k is the highest rank pattern point. Point p_k is on Cir_σ.

Defining the Walker. The walker is a robot in the current configuration that will move to occupy an unoccupied pattern point. Before we define the walker, we define an order among the robots. The order among robots uses the radiangular distance from p_1 and we say that a robot is *free* if it is not located at a pattern point. Let r' and r'' be two robots that have radiangular distance from p_1 equal to (θ', x') and (θ'', x''), respectively. We say that r' has higher priority than r'', if either (i) $\theta' < \theta''$, (ii) $\theta' = \theta'' \neq 0$ and $x' > x''$, or (iii) $\theta' = \theta'' = 0$ and $x' < x''$. In other words, a robot that is closer in angle to the radius $\overline{Op_1}$ has higher priority and when two (or more) robots are on the same radius, the robot that is farther from the center has higher priority.

Now, consider the radial detour from the highest priority free robot to the highest priority empty pattern point, and the radius ρ_l where such a point lies. The following could happen:

- If the radial detour and ρ_l do not contain other robots, then the highest priority free robot becomes the walker;

- If the radial detour contains another robot and ρ_l does not contain other robots, the robot with smallest radiangular distance from the target pattern point, along the computed radial detour, becomes the walker;
- Otherwise, if ρ_l contains other robots, compute the path from the highest priority free robot in ρ_l to p_l: if this path is empty, then the free robot becomes the walker, otherwise, in the computed path, the robot closest to p_l becomes the walker.

Notice that the walker has always a free path to the target pattern point and that under these restrictions, the internal robot defining θ_1 becomes the last walker since it has the largest radiangular distance from p_1.

Also, notice that once the pattern points fixing the $SEC(\hat{P})$ are occupied by robots, the remaining robots on the $SEC(Q)$ are free to move, since they will not be necessary to maintain the SEC anymore.

Movement of a Robot. The movement of a robot always follows a radial detour. If a robot r is determined as the walker and its target pattern point is located on a different radius from the one where it is currently located, r first moves to the center, and then moves toward its target pattern point from the center. When r is co-radial with its target pattern point, it moves directly to that point. By definition, the walker does not have any other robot on the radial detour, and hence it can move to its target pattern point unobstructed.

4.3 Strategy in Case $|\hat{P}| \leq 4$

In this section, we consider the case $|\hat{P}| \leq 4$. We remind also that in UPF* $|\hat{P}| > 1$. Similarly to SeqPF, the strategy to solve this case follows four stages: *Initialization*, *Unique Maximum*, *Equalization*, and *Finalization*. An activated robot r does the following:

1. If $|Q| < |\hat{P}|$, r executes Procedure Separate to make $|Q| \geq |\hat{P}|$ (*Initialization* stage);
2. If Q does not admit a unique pair of points defining the maximum distance among any pair of points in Q, then the first activated robot r located at one of the points that define the current maximum distance, makes a move so as to enlarge its maximum distance (*Unique Maximum* stage);
3. Once Q admits a unique maximum distance defined by the pair $\{q_1, q_2\}$, then:
   - If $|Q| > |\hat{P}|$ then any robot not in $\{q_1, q_2\}$ moves toward the closest point among q_1 and q_2 (*Equalization* stage);
   - Else (i.e., $|Q| = |\hat{P}|$), r suitably overlaps $\hat{P}$ with Q so as to make the pair $\{q_1, q_2\}$ coincident with one of the pairs $\{\hat{p}_i, \hat{p}_j\}$ of $\hat{P}$ that define the maximum distance in $\hat{P}$. The *Finalization* stage can then start to finalize the formation of the pattern.

In the *Finalization* stage, when $|Q| = |\hat{P}|$ and Q admits a unique maximum distance defined by the pair $\{q_1, q_2\}$, the overlap is computed as follows. Without loss of generality, let $\hat{p}_1$ and $\hat{p}_2$ be two pattern points (not necessarily unique) that define the maximum distance in $\hat{P}$. The activated robot r overlaps the pattern by scaling it in such a way that $\overline{q_1 q_2}$ and $\overline{\hat{p}_1 \hat{p}_2}$ coincide. The pattern is considered at this scale. Now, if the activated robot r is located at q_1 or q_2, it does not move. Consider r not located at q_1 or q_2. There are multiple ways to overlap $\hat{P}$ with Q. Since there are at most four points in the pattern, there can be at most three pairs in $\hat{P}$ that can admit the maximum distance (three vertices of an equilateral triangle). Hence, the total number of distinct overlaps with rotation and reflection is at most 12. The activated robot computes the pattern with smallest distance deviation, i.e., the sum of distances between pattern points and robot positions. If there are two overlaps of the pattern that have the same smallest distance deviation, then r can consider one of them indiscriminately.

We define the distance deviation between two sets of points in Q and P (the corresponding overlap) as follows. Let (q_1, q_2) be the unique maximum distance pair, and let (p_1, p_2) be the corresponding maximum distance pair in the pattern. The distance deviation is given by:

1. $|\overline{p_3 q_3}|$ if $|Q| = |\hat{P}| = 3$;
2. $\min(|\overline{p_3 q_3}| + |\overline{p_4 q_4}|, |\overline{p_4 q_3}| + |\overline{p_3 q_4}|)$ if $|Q| = |\hat{P}| = 4$.

Now, for each unoccupied pattern point, r determines the closest robot position, and moves toward the unoccupied pattern point from which it is the closest. Notice that, it does not move toward the empty pattern point that it is closest to, since there maybe another robot that is closer to that pattern point.

5 Gathering Under Sequential Schedulers

In the previous section, we have seen how to solve UPF* (i.e., for $|\hat{P}| > 1$) under $\mathcal{SEQ}$. We now consider the case $|\hat{P}| = 1$, that is, Gathering. Recall that this problem is called Rendezvous when there are only two robots (i.e., $|\mathcal{R}| = 2$). In the literature, Gathering has been extensively studied. However, one of the main assumptions that has been considered was to start from initial configurations without multiplicities. Despite such an assumption, still some known results can be easily extended to the more general Gathering studied here, where initial configurations can admit multiplicities.

For instance, in $\mathcal{FSYNC}$, Gathering (and hence Rendezvous as well) can be easily solved by the Go_to_the_center_of_SEC algorithm (see, e.g., [6]) that makes robots move toward the center of the current SEC.

Concerning $\mathcal{SSYNC}$, it is known that Rendezvous, and hence Gathering, are unsolvable (see, e.g., [22]).

In $\mathcal{SEQ}$ (actually in any sequential scheduler), Rendezvous can be easily solved by making the active robot move toward the position occupied by the other robot. Although the moving robot may not reach the destination, due to the non-rigid movements, still the distance between the two robots decreases of at least δ. Hence, as soon as the distance becomes smaller than or equal to δ, we are ensured that Rendezvous is accomplished. Without any additional robot capability, we can prove that this result however does not extend to Gathering.

In fact, we show that, under a sequential scheduler, `Gathering` of more than two robots cannot be solved in general.

Theorem 2. *`Gathering` is unsolvable under a sequential scheduler.*

Proof. By contradiction, let $\mathcal{A}$ be an algorithm that solves `Gathering` under a sequential scheduler $\mathcal{S} \in$ SEQUENTIAL.

Consider a configuration $C(t)$ with $|\mathcal{R}| > 2$ and let $Q(t)$ be the corresponding set of distinct points. A robot r activated at time t decides its destination based on the observed set of points $Q(t)$; as a result, there are only three possible outcomes: (Act_1) not moving; (Act_2) moving to a point occupied by another robot; (Act_3) moving to a point not occupied by another robot. Since the robots have no multiplicity detection capabilities, this decision is taken by a robot without knowing whether or not there are other robots located at its own position. We shall examine now the execution of $\mathcal{A}$ starting from $Q(t)$.

Consider first the case where $|Q(t)| > 2$. Because of the assumed correctness of $\mathcal{A}$, and since at most one robot is moving in a round, there must exist a time $t' \geq t$ where $|Q(t')| = 2$.

Consider then the case $|Q(t)| = 2$. Let A and B be the two points of $Q(t)$; since $|Q(t)| = 2 < |\mathcal{R}|$, the configuration must have at least one multiplicity, say A. Consider the scenario where at any time $t' \geq t$, as long as $|Q(t')| = 2$, all robots co-located at A (resp. B) have the same coordinate system, which is reflexive to that of those co-located at B (resp. A). In this scenario, called *mirror*, any activated robot makes the same decision as long as $|Q(t')| = 2$.

Observe that, regardless of where the activated robot r resides, the decision according to $\mathcal{A}$ cannot be Act_1, i.e., not moving. Otherwise, in the execution of $\mathcal{A}$ in the mirror scenario, no activated robot will ever move; that is, $\forall t' > t$, $Q(t') = Q(t) = 2$ and `Gathering` remains unsolved, contradicting the assumed correctness of $\mathcal{A}$.

Moreover, the decision by a robot according to $\mathcal{A}$ cannot be Act_2, i.e., move to B. Consider in fact the execution of $\mathcal{A}$ in the mirror scenario where the scheduler activates sequentially all robots but one (say a) from A, then sequentially all the robots originally in B, and finally a. In so doing, the reached configuration is another mirror configuration, where the positions of the robots have now switched. Observe that, at any time t' during this process, $Q(t') = 2$. From this moment on, the scheduler continues to perform these switches. Thus, again $\forall t > 0$, $Q(t) = Q(0) = 2$, and `Gathering` remains unsolved, contradicting the assumed correctness of $\mathcal{A}$.

It remains to analyze Act_3, i.e., moving to a point distinct from A and B, say C; if the activated robot located in either A or B is part of a multiplicity, then $|Q(t+1)| = 3$, otherwise $|Q(t+1)| = 2$. In other words, in any execution of $\mathcal{A}$, if $|Q(t)| = 2$, then $|Q(t+1)| \geq 2$.

Summarizing, $\forall t \geq 0$, $|Q(t)| > 1$; that is, in any execution of $\mathcal{A}$ the robots never gather, contradicting the assumed correctness of $\mathcal{A}$. □

The next result shows that `Gathering` is solvable under any scheduler in $\mathcal{SEQ}$ when robots are empowered with weak multiplicity detection.

Theorem 3. `Gathering` *is solvable with weak multiplicity detection under any sequential scheduler.*

Sketch Proof. Algorithm SeqGathering solves `Gathering` by maintaining a unique multiplicity point and moving other robots sequentially to that multiplicity point. If there is no multiplicity, then the activated robot creates a multiplicity. If there is more than one multiplicity at time t, then the robots move to unoccupied points to separate such that at some time $t' > t$, there is only one multiplicity. Then for all $t'' > t'$, that multiplicity is maintained. □

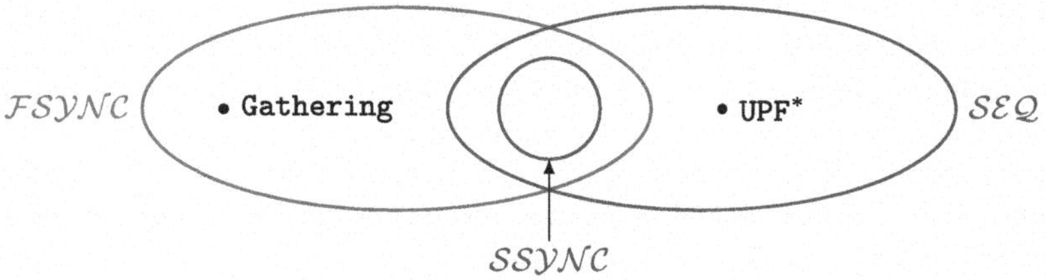

Fig. 1. Relationship between the set of problems solvable under $\mathcal{FSYNC}$, $\mathcal{SEQ}$, and $\mathcal{SSYNC}$, without any additional assumptions.

By contrast, the addition of multiplicity detection is not enough under most synchronous schedulers, as shown by the following known result:

Lemma 1. `Gathering` *is unsolvable with multiplicity detection under* $\mathcal{SSYNC}$.

Proof. From [22], `Rendezvous` is unsolvable under $\mathcal{SSYNC}$. Such a result can be easily extended to the case where robots are empowered with weak multiplicity detection. It is sufficient to start from a configuration with two multiplicities. Similarly, the result can be extended to the case where robots are empowered with the strong multiplicity detection. In this case, it is sufficient to start from a configuration with two multiplicities composed by the same number of robots. □

6 Conclusion

We have studied the computational power of oblivious robots operating in the plane under *sequential* schedulers. Focusing on the fundamental class of pattern formation problems, we investigated the computational power of the robots under $\mathcal{SEQ}$, with respect to

the other two synchronous schedulers studied in the literature: $\mathcal{FSYNC}$ and $\mathcal{SSYNC}$. Specifically, we considered the resolution of the following problems: `UPF`, the most general problem in the class; `Gathering`, or point-formation; and `UPF`* = `UPF` \ `Gathering`.

We have shown that, under any sequential scheduler, oblivious robots can solve: UPF*, without any additional assumption; Gathering and, hence, UPF, with the only addition of weak multiplicity detection.

In contrast, we proved that UPF* is unsolvable under $\mathcal{FSYNC}$ (and hence under $\mathcal{SSYNC}$ and $\mathcal{ASYNC}$) even by assuming strong multiplicity detection, a leader, rigid movements, and orientation.

Since it is known from the literature that Gathering is solvable under $\mathcal{FSYNC}$ but not under $\mathcal{SSYNC}$, then we conclude that $\mathcal{SEQ}$ and $\mathcal{FSYNC}$ are orthogonal in general, whereas the computational power of the robots under $\mathcal{SEQ}$ is stronger than under $\mathcal{SSYNC}$, see Fig. 1.

References

1. Cicerone, S., Di Stefano, G., Navarra, A.: Asynchronous arbitrary pattern formation: the effects of a rigorous approach. Dist. Comput. **32**(2), 91–132 (2019)
2. Cicerone, S., Di Stefano, G., Navarra, A.: Embedded pattern formation by asynchronous robots without chirality. Dist. Comput. **32**(4), 291–315 (2019)
3. Cicerone, S., Di Stefano, G., Navarra, A.: Semi-asynchronous: a new scheduler in distributed computing. IEEE Access **9**, 41540–41557 (2021)
4. Cicerone, S., Di Stefano, G., Navarra, A.: Solving the pattern formation by mobile robots with chirality. IEEE Access **9**, 88177–88204 (2021)
5. Cieliebak, M., Flocchini, P., Prencipe, G., Santoro, N.: Distributed computing by mobile robots: gathering. SIAM J. Comput. **41**(4), 829–879 (2012)
6. Cohen, R., Peleg, D.: Convergence properties of the gravitational algorithm in asynchronous robot systems. SIAM J. Comput. **34**(6), 1516–1528 (2005)
7. Défago, X., Potop-Butucaru, M., Tixeuil, S.: Fault-tolerant mobile robots. Chapter 10 of [13], pp. 234–251 (2019)
8. Flocchini, P., Navarra, A., Pattanayak, D., Piselli, F., Santoro, N.: Oblivious robots under sequential schedulers: universal pattern formation. arXiv preprint arXiv:2412.10733 (2024)
9. Flocchini, P., Prencipe, G., Santoro, N.: Distributed computing by oblivious mobile robots. Synthesis Lect. Distrib. Comput. Theory (2012)
10. Flocchini, P., Prencipe, G., Santoro, N., Viglietta, G.: Distributed computing by mobile robots: uniform circle formation. Dist. Comput. **30**(6), 413–457 (2017)
11. Flocchini, P., Prencipe, G., Santoro, N., Widmayer, P.: Hard tasks for weak robots. In: Proceedings of 10th International Symposium on Algorithms and Computation (ISAAC), pp. 93–102 (1999)
12. Flocchini, P., Prencipe, G., Santoro, N., Widmayer, P.: Arbitrary pattern formation by asynchronous, anonymous, oblivious robots. Theor. Comput. Sci. **407**(1), 412–447 (2008)
13. Flocchini, P., Prencipe, G., Santoro, N. (eds.): Distributed Computing by Mobile Entities. Springer (2019)
14. Flocchini, P., Santoro, N., Sudo, Y., Wada, K.: On asynchrony, memory, and communication: separations and landscapes. In: Proceedings of 27th International Conference on Principles of Distributed Systems (OPODIS), pp. 28:1–28:23 (2023)
15. Hahn, C., Harbig, J., Kling, P.: Forming large patterns with local robots in the OBLOT model. In: Proceedings of 3rd Symposium on Algorithmic Foundations of Dynamic Networks (SAND) (2024)

16. Izumi, T., et al.: The gathering problem for two oblivious robots with unreliable compasses. SIAM J. Comput. **41**(1), 26–46 (2012)
17. Kirkpatrick, D., Kostitsyna, I., Navarra, A., Prencipe, G., Santoro, N.: On the power of bounded asynchrony. Dist. Comput. **37**, 1–30 (2024)
18. Pattanayak, D., Augustine, J., Mandal, P.S.: Randomized gathering of asynchronous mobile robots. Theor. Comput. Sci. **858**, 64–80 (2021)
19. Pattanayak, D., Mondal, K., Ramesh, H., Mandal, P.S.: Gathering of mobile robots with weak multiplicity detection in presence of crash-faults. J. Parallel Distrib. Comput. **123**, 145–155 (2019)
20. Prencipe, G.: Impossibility of gathering by a set of autonomous mobile robots. Theor. Comput. Sci. **384**(2–3), 222–231 (2007)
21. Suzuki, I., Yamashita, M.: Distributed anonymous mobile robots: formation of geometric patterns. SIAM J. Comput. **28**(4), 1347–1363 (1999)
22. Yamashita, M., Suzuki, I.: Characterizing geometric patterns formable by oblivious anonymous mobile robots. Theor. Comput. Sci. **411**, 2433–2453 (2010)

On the Dynamical Hierarchy in Gathering Protocols with Circulant Topologies

Raphael Gerlach(✉), Sören von der Gracht, and Michael Dellnitz

Paderborn University, Warburger Straße 100, 33098 Paderborn, Germany
{raphael.gerlach,soeren.von.der.gracht}@uni-paderborn.de, dellnitz@upb.de

Abstract. In this article, we investigate the convergence behavior of two classes of gathering protocols with fixed circulant topologies using tools from dynamical systems. Given a fixed number of mobile entities moving in the Euclidean plane, we model a gathering protocol as a system of (linear) ordinary differential equations whose equilibria are exactly all possible gathering points. For a circulant topology, we derive a decomposition of the state space into stable invariant subspaces with different convergence rates. This decomposition is identical for every linear circulant gathering protocol. Only the convergence rates depend on the weights in the interaction graph. In the second part, we consider a normalized nonlinear version of the equation of motion that is obtained by scaling the speed of each entity. Again, we find a similar decomposition of the state space that is based on our findings in the linear case. In both situations, we also consider visibility preservation properties.

Keywords: Mobile agents and autonomous distributed systems · Gathering of mobile agents and robots · Circulant communication topology · Dynamical hierarchy in convergence rates

1 Introduction

This article applies dynamical systems theory to the analysis of swarms of mobile robots, which are widely studied under the headline of *distributed computing* (e.g. [15,19]). This precise and robust mathematical framework can be exploited to formally prove statements about the collective behavior but also to inform design choices for algorithms, which recently has been exemplified in the investigation of symmetry properties of robot swarms performing Near-Gathering [16].

Gathering: The problem of interest in this article is the *gathering problem* in which the robots are supposed to converge to a single, not predefined, point. The only capabilities the robots have are observing other robots' positions, performing computations in their local memory, and moving. The strategy that each robot pursues is called a *protocol.* While most studies of the gathering problem employ discrete time (e.g. [4,5,7,10,14]), some protocols using continuous time have been proposed as well [9,26]. The latter framework is the focus of this manuscript. Gathering is an instance of famous problems known

U. Schmid and R. Kuznets (Eds.): SIROCCO 2025, LNCS 15671, pp. 315–332, 2025.
https://doi.org/10.1007/978-3-031-91736-3_19

under terms such as *synchronization, agreement problems, rendezvous problems, consensus algorithms*, and likely many more. For example, in synchronization literature one commonly studies Laplacian (or Laplacian like) dynamics of coupled oscillators. Very general internal dynamics and coupling functions are possible and synchronization is not restricted to equilibria (see e.g. [13,20,32] and references therein). Very powerful synchronization results have also been derived in the control theory literature under the headline of *distributed consensus control/algorithms* or *asymptotic agreement problems*. Asymptotic consensus was investigated for abstract variables, time varying interaction topologies, communication delay, and even defective agents (e.g. [6,28–31]).

The Circulant Multi-Agent System Model (CMAS): We consider a swarm of $N \in \mathbb{N}$ autonomous, mobile robots modeled as deterministic points in the 2-dimensional plane. The robots are *oblivious* (have no memory), disoriented (own local self-centered coordinate system without agreement on direction and chirality) and to have a fixed (ordered) list of *interaction partners*. This list is realized as a fixed *interaction graph*. We assume the interaction graph, respectively its adjacency matrix, to be *circulant* which allows all robots to execute the same algorithm (*homogeneous*). This interaction topology can be seen as a generalization of a fixed closed chain. We emphasize, that the robots are *not anonymous* as they can distinguish their interaction partners. Interaction consists of the observation of the partner's positions which are used for the internal computations. These are not necessarily symmetric. Finally, we assume that the robots have a limited, constant *viewing range*, such that they cannot perceive anything beyond that range. Robots move according to a *time continuous fully-synchronous* $\mathcal{F}$SYNC model, i.e., each robot continuously adjusts its speed and direction by synchronously performing LOOK-COMPUTE-MOVE (LCM) rounds permanently and instantaneously.

Our Contribution: We propose two *classes* of protocols called *Circulant Laplacian Dynamics* and *Nonlinear Scaling of Circulant Laplacian Dynamics*. Protocols from the first class solve the gathering problem using the CMAS model via convergence in infinite time. The scaled version facilitates finite time gathering as well. For their respective analyses, we apply the mathematical theory of dynamical systems which allows for a profound understanding of the collective dynamics in terms of fine grained gathering rates. We unveil a foliation of the space of all robots' positions into dynamically invariant subspaces in which the configurations gather with different speeds (Theorems 3 and 4). This foliation is independent of the precise protocol. It allows to identify gathering rates for specific initial configurations and to decompose arbitrary initial configurations into components with different gathering rates (Corollary 1). Moreover, we show that visibility is preserved under the dynamics (cf. Theorem 2 and Proposition 2)

Example 1. Consider $N = 7$ mobile robots $\{0, \ldots, 6\}$ running the N-BUG protocol, i.e., if $z_i(t) \in \mathbb{R}^2$ is the position of robot i at time t the i-th robot is influenced by the $i+1$-th robot and will move towards it. Theorems 3 and 4 show that the configuration in Fig. 1(a) has the slowest gathering rate, while the configuration in Fig. 1(b) has the fastest gathering rate. Further intermediate configurations are unveiled as well.

(a) Slowest gathering configuration (b) Fastest gathering configuration

Fig. 1. Illustration of the gathering hierarchy exhibited by $N = 7$ robots running the N-BUG protocol. Blue dots indicate positions of each robot. Gray arrows symbolize the interaction relation, and *not* the movement direction. (Color figure online)

Further Related Work: The fixed circulant interaction topology serves mainly as an academic example. Nonetheless, it can be used in the theoretical worst case run-time analysis of distributed strategies to form communication *chains* between stationary relays (cf. [3,11,24,25]).

Furthermore, we would like to mention that comparable results are known in the synchronization literature for coupled oscillators. There, dynamical hierarchy corresponds to *normal modes* or *eigenmodes* which can unveil communities that synchronize faster than others (see for example [1,27]). Moreover, overall convergence speed to synchrony is commonly related to spectral properties of the Laplacian or system matrix, which has been done in particular for circulant interaction topologies (e.g. [22,23,31,33]). However, to the best of our knowledge a similarly detailed exploitation of the full spectral decomposition of phase space and the interpretation in terms of a dynamical hierarchy of configurations in positional variables has not been done before.

Outline: The article is organized as follows: In Sect. 2, we gather notation to formalize the underlying interaction structure by a weighted graph. In Sect. 3, we propose two classes of protocols (*Circulant Laplacian Dynamics (CLD)* and *Nonlinear Scaling of Circulant Laplacian Dynamics (NSCLD)*). We also present the mathematical dynamical systems framework for their analysis. The main dynamical analysis is carried out in Sect. 4 for the linear class and in Sect. 5 for the nonlinear class. In Sect. 4.3, the hierarchy of different gathering rates is proven, which is extended in Sect. 5 to the nonlinear setting. Visibility preservation is shown in Sect. 4.2 and Sect. 5. Finally, in Sect. 6 we end this article with a conclusion and a brief outlook on future research. Details of technical mathematical proofs can be found in the Appendix of the extended version of this article [17].

2 Circulant Interaction Topology

We begin by introducing some notation from algebraic graph theory. As we assume the interaction structure to be fixed for all times independent of the robots' positions, we can encode interactions between robots by a (directed) *graph* $G = (V, E)$ that we call *interaction graph.* The *vertices* are the robots $V = \{0, \dots, N-1\}$ and an *edge* $e = (j, i) \in E \subseteq V \times V$ represents the fact that robot i uses the position of robot j in its internal computations. We assume the interaction graph to be *weighted*: each edge $e = (j, i) \in E$ is assigned the *weight* $w_{i,j} \in \mathbb{R}$ and $w_{i,j} = 0$ if and only if $e = (j, i) \notin E$. A negative weight corresponds to the partner's position *reflected at the origin* and would force a robot to intentionally *move away* from that partner (cf. Sect. 3 below). Hence, we consider non-negative weights only from now on. A commonly considered special case is when the interaction graph is *symmetric* or *undirected*, i.e., the weight matrix $W \in \mathbb{R}^{N\times N}$ is symmetric. The interaction graph is said to be *connected* (or *weakly connected*) if for any two vertices $j, i \in \{0, \dots, N-1\}$ there is an *undirected path* from j to i. If there is always a *directed path* the graph is said to be *strongly connected.*

The interaction graph is *circulant* if and only if there are integers $0 \le s_1 < \dots < s_k \le N-1$ such that $(j, i) \in E$ if and only if $j = i + s_r \bmod N$ for an $r \in \{1, \dots, k\}$. Therein $i+s \bmod N = i+s$ if $i+s < N$ and $i+s-N$ if $i+s \ge N$. Circulant interaction structure implies that the weight matrix $W \in \mathbb{R}^{N\times N}$ is a *circulant matrix* of the form

$$W = \operatorname{circ}(w_0, w_1, \dots, w_{N-1}) := \begin{pmatrix} w_0 & w_1 & \cdots & w_{N-2} & w_{N-1} \\ w_{N-1} & w_0 & w_1 & & w_{N-2} \\ \vdots & w_{N-1} & w_0 & \ddots & \vdots \\ w_2 & & \ddots & \ddots & w_1 \\ w_1 & w_2 & \cdots & w_{N-1} & w_0 \end{pmatrix} \in \mathbb{R}^{N\times N}.$$

We say the vector $w = (w_0, \dots, w_{N-1}) \in \mathbb{R}^N$ *generates* the circulant matrix W. For a symmetric circulant weight matrix W, this implies $w_{N-i} = w_i i = 1, \dots, N-1)$ and W has fewer degrees of freedom, i.e., W is determined by only $\lfloor \frac{N}{2} \rfloor + 1$ elements $w_0, \dots, w_{\lfloor \frac{N}{2} \rfloor}$. For a circulant graph, weak and strong connectivity coincide (cf. [12, Corollary 1]). Hence, we drop the distinction.

3 Two Classes of Protocols

We consider two classes of protocols: *Circulant Laplacian Dynamics (CLD)* and *Nonlinear Scaling of Circulant Laplacian Dynamics (NSCLD)*. In both cases, a robot i continuously and instantaneously adapts its velocity vector according to the following rules corresponding to the LCM phases: Robot i

LOOK. observes the positions of the robots j such that $e = (j, i) \in E$ in relative coordinates within its vision range,

COMPUTE. computes a target point as a linear combination of the observed positions where the coefficients are given by the weights of the interaction graph, and

MOVE. adapts its velocity vector as the connecting vector between its own position and the target point. In an NSCLD protocol, this vector is rescaled.

For the dynamical analysis, we take the stance of an external observer describing the dynamics in global coordinates. We denote by $z_i(t) = (x_i(t), y_i(t)) \in \mathbb{R}^2$ the position of robot $i \in \{0, \ldots, N-1\}$ at time $t \in \mathbb{R}$. Typically, the time argument will be omitted for notational brevity. The collection of all robots' positions is called a *configuration*. The LCM steps of a CLD protocol described above are mathematically modeled by the *linear* ordinary differential equations

$$\dot{z}_i = -z_i + \sum_{j=0}^{N-1} w_j z_{i+j}, \tag{1}$$

where robot indices are counted mod N, i.e., $z_{(N-1)+1} = z_0$ and $z_{0-1} = z_{N-1}$.

CLD protocols satisfying the linear equations (1) are idealized. They do not generally prevent unbounded velocity and realize gathering only in the limit $t \to \infty$, as we will see below in Sect. 4. Both restrictions are at least problematic for the robot model. However, in commonly investigated protocols robots compute their *direction* as a linear combination of the (relative) positions of their neighbors (as in (1)) and move with bounded speed until they are gathered. This requires rescaling the velocity vector in the MOVE step of an NSCLD protocol. The modification is realized by a *nonlinear* function $\mathcal{N} : \mathbb{R}^2 \to \mathbb{R}^2$ with $\mathcal{N}(0) = 0$ of the form

$$\mathcal{N}(x, y) = n(\|(x, y)\|_2) \begin{pmatrix} x \\ y \end{pmatrix} \tag{2}$$

for some non-negative function $n : \mathbb{R} \to \mathbb{R}_{\geq 0}$. This function *scales* the length of the original velocity vector, and therefore the robot's speed, but does not change the direction. Here, $\|\cdot\|$ denotes the Euclidean norm. Other norms are possible. In fact, the theoretical results below can be adapted without much effort. The mathematical model is adapted to the *nonlinear* ordinary differential equations

$$\dot{z}_i = \mathcal{N}\left(-z_i + \sum_{j=0}^{N-1} w_j z_{i+j}\right). \tag{3}$$

A common approach is robots moving with constant speed one which requires

$$n(\|(x, y)\|_2) = \|(x, y)\|_2^{-1}. \tag{4}$$

Remark 1. Both mathematical formulations (1) and (3) do not incorporate the finite viewing range of robots explicitly. Instead, we analyze the frameworks under the hypothetical assumption of unlimited viewing range and prove in hindsight that interaction partners do not lose sight of each other when they started suitably close. This approach is similar to commonly studied multi-agent systems in which the communication graph is all-to-all connected and communication failure corresponds to vanishing weights. Examples include flocking models [8], bounded confidence opinion dynamics [21], or consensus algorithms [28].

In Sects. 4 and 5, we derive conditions under which a protocol from the two classes CLD and NSCLD solves the gathering problem. To make the notion precise, we state

Definition 1. *A protocol is called* gathering, *if the solution with any initial configuration converges to a* gathering point $z_i = z^* = (x^*, y^*) \in \mathbb{R}^2$ *for all* i *for* $t \to \infty$ *and if the system is in equilibrium in any such gathering point, i.e.,* $\dot{z}_i = 0$ *for all* i. *The subspace of all gathering configurations is denoted by*

$$V_0 = \{ Z^* = (x^*, \dots, x^*, y^*, \dots, y^*) \mid x^*, y^* \in \mathbb{R} \} \subseteq \mathbb{R}^{2N}.$$

Example 2. As a running example, we consider the N-BUG problem [34]. Here, each robot i is only influenced by its first neighbor to the right, which yields a connected circulant interaction structure of the form $W = \mathrm{circ}(0, w_1, 0 \dots, 0)$ with $w_1 \neq 0$.

4 Analysis of the Circulant Laplacian Dynamical Model

4.1 Gathering

In this section we study the convergence properties of CLD protocols using tools from dynamical systems theory. By Definition 1 it can readily be seen, that a necessary condition for a linear strategy (1) to be gathering is $z^* = \sum_{j=0}^{N-1} w_j z^*$, which is satisfied if and only if the weights satisfy

$$\sum_{j=0}^{N-1} w_j = 1. \tag{5}$$

In this situation, we say that the weights, the weight matrix, and the linear strategy are *consistent*. Under this condition, the right hand side of (1) is a linear combination of the *relative* positions $z_i - z_j$. In particular, the CMAS model is able to perform computations of the form (1).

For the analysis we collect all robots' positions in a single vector $Z = (X, Y) \in \mathbb{R}^{2N}$, where $X = (x_0, \dots, x_{N-1}) \in \mathbb{R}^N$ denotes the x-coordinates of the N robots and $Y = (y_0, \dots, y_{N-1}) \in \mathbb{R}^N$ the y-coordinates. This allows us to rewrite (1) as

$$\dot{Z} = (-\mathbf{I}_{2N} + \mathbf{W})\, Z \text{ with } \mathbf{W} := \mathbf{I}_2 \otimes W = \left(\begin{array}{c|c} W & \mathbf{0} \\ \hline \mathbf{0} & W \end{array}\right) \in \mathbb{R}^{2N \times 2N}, \tag{6}$$

where $\mathbf{I}_k \in \mathbb{R}^{k \times k}$ denotes the identity matrix and $\mathbf{I}_2 \otimes W$ the Kronecker product.

The behavior of linear systems of ordinary differential equations is well understood. In the particular case of a circulant interaction graph, the classification of gathering protocols follows from Theorem 3.2 in [2], Theorem 4 in [31], and consistency condition (5) which we saw to be necessary for a gathering protocol. In fact, under the assumption of consistency, the system matrix $-\mathbf{I}_{2N}+\mathbf{W}$ in (6) equals $\mathbf{I}_2 \otimes L$, where $L \in \mathbb{R}^{N\times N}$ is the graph Laplacian matrix of the weighted interaction graph so that these results apply.

Theorem 1. *For a* CLD *protocol with weight matrix* $W = \text{circ}(w_0, \dots, w_{N-1})$, *where* $w_i \geq 0$ $(i = 0, \dots, N-1)$, *the following are equivalent:*

(i) The protocol is gathering.
(ii) The interaction graph is connected and the weight matrix is consistent.

Seminal works as [2,28–31] and many others have classified gathering protocols (consensus control strategies in their context) also in more general settings.

Example 3. Consider the N-BUG problem introduced in the running Example 2. For this protocol, Theorem 1 yields that it is gathering if and only if its generating weight vector $w \in \mathbb{R}^N$ has the form $w = (0, 1, 0, \dots, 0)^T \in \mathbb{R}^N$.

4.2 Preservation of Visibility

So far in our analysis, we have ignored the fact that robots can only use local information specified by their finite viewing range $\mathcal{C} > 0$. While synchronization results under similar constraints are available (e.g. [29]), it remains largely unclear if visibility of different agents with a limited vision radius can be preserved under the dynamics. However, we may prove that this issue does not occur with CLD protocols *which solve the gathering problem.* In fact, Theorem 2 below shows that no robot loses sight of any of its interaction partners if initially all their interaction partners are within viewing range. The proof crucially depends on the time continuity of the model. It uses the consistency of the weight matrix of a gathering circulant protocol to show that the two maximally distant robots cannot move further away from each other. Note that robots can leave the convex hull of the configuration, however, the swarms diameter shrinks sufficiently fast, such that loss of sight is impossible. For more details, we refer to [17].

Theorem 2. *Consider a gathering* CLD *protocol with circulant weight matrix* $W \in \mathbb{R}^{N\times N}$. *Let* $z_0(0), \dots, z_{N-1}(0) \in \mathbb{R}^2$ *be a valid initial configuration, that is,* $\|z_i(0) - z_j(0)\| \leq \mathcal{C}$ *for all* $(j, i) \in E$. *Then, visibility is preserved under the dynamics, i.e.,*

$$\|z_i(t) - z_j(t)\| \leq \mathcal{C} \quad \text{for all } t \geq 0 \text{ and } (j, i) \in E.$$

4.3 Hierarchy of Convergence Rates

Theorem 1 provides a simple condition for CLD protocols to be gathering. For general linear systems such as (1) it is well known that spectral properties of the system matrix determine the longtime behavior. For a circulant matrix all eigenvalues and eigenvectors can be analytically computed as follows (e.g. [18]).

Proposition 1. *Let* $W = \operatorname{circ}(w_0, \dots, w_{N-1}) \in \mathbb{R}^{N \times N}$ *be a circulant matrix. Then its eigenvectors* $v_j \in \mathbb{C}$ $(j = 0, \dots, N-1)$ *are given by*

$$v_j = (1, \omega^j, \omega^{2j}, \dots, \omega^{(N-1)j})^T \in \mathbb{C}^N, \tag{7}$$

where $\omega = \exp\left(\frac{2\pi \mathrm{i}}{N}\right)$ *is a primitive* N*-th root of unity and its eigenvalues are*

$$\lambda_j = \sum_{i=0}^{N-1} w_i \omega^{ij} \in \mathbb{C} \text{ for } j = 0, \dots, N-1. \tag{8}$$

Remark 2. (a) For $j = 0$, we obtain the eigenvector $v_0 = (1, \dots, 1)^T \in \mathbb{R}^N$ with corresponding (simple) eigenvalue $\lambda_0 = \sum_{i=0}^{N-1} w_i$, which has to be one for the protocol to be gathering according to Theorem 1.

(b) Since $W \in \mathbb{R}^{N \times N}$ has real-valued entries, we immediately obtain $\lambda_j = \overline{\lambda_{N-j}}$ for $j = 1, \dots, N-1$. For even $N = 2k$ this implies $\lambda_k \in \mathbb{R}$. Moreover, if $W \in \mathbb{R}^{N \times N}$ is symmetric, this reduces to $\lambda_j = \lambda_{N-j}$ for $j = 1, \dots, N-1$, and the eigenvalues $\lambda_1, \dots, \lambda_{\lfloor \frac{N-1}{2} \rfloor} \in \mathbb{R}$ are duplicated.

(c) By (7), all eigenvectors $v_j \in \mathbb{C}^N$ are independent of the generating vector $w \in \mathbb{R}^N$, i.e., the given CLD protocol. In Fig. 2 we visualize the eigenvectors $v_j \in \mathbb{C}$, $j \neq 0$, in the complex plane for some $N \in \mathbb{N}$.

For the upcoming analysis, we derive solutions of (6) in closed form, which allows us to hierarchically rank them according to their speed of convergence. By Proposition 1 a circulant matrix W is diagonalizable, which yields the *fundamental solutions* of (6) to have the simple form

$$Z_j^x(t) = e^{(-1+\lambda_j)t} \begin{pmatrix} v_j \\ 0 \end{pmatrix} \text{ and } Z_j^y(t) = e^{(-1+\lambda_j)t} \begin{pmatrix} 0 \\ v_j \end{pmatrix}, \quad j = 0, \dots, N-1.$$

However, for $j \neq 0$, the eigenvector v_j and its corresponding eigenvalue λ are, in general, complex-valued, which gives little possibility for interpretation in terms of configurations of $N \in \mathbb{N}$ robots in the two-dimensional Euclidean plane.

We propose a real-valued basis as follows. Let $N = 2k$ be even or $N = 2k+1$ be odd. For $j = 0, \dots, k$ we define

$$V_j = \operatorname{span}\left(\begin{pmatrix} \Re(v_j) \\ -\Im(v_j) \end{pmatrix}, \begin{pmatrix} \Im(v_j) \\ \Re(v_j) \end{pmatrix}, \begin{pmatrix} -\Im(v_j) \\ \Re(v_j) \end{pmatrix}, \begin{pmatrix} \Re(v_j) \\ \Im(v_j) \end{pmatrix} \right), \tag{9}$$

as the subspaces spanned by linear combinations of real- and imaginary parts of the eigenvector $v_j \in \mathbb{C}$. The collection of all $V_j \subseteq \mathbb{R}^{2N}$ decompose the state space as $\mathbb{R}^{2N} = \bigoplus_{j=0}^{k} V_j$ since the basis vectors are linearly independent.

The first three basis vectors in (9) (considered as points/robots in the Euclidean plane) are reflections of the last one. In this sense, we say $v_j \in \mathbb{C}^N$, resp. the configuration $(\Re(v_j), \Im(v_j))^T \in \mathbb{R}^{2N}$, *generates* the subspace $V_j \subseteq \mathbb{R}^{2N}$. In particular, the x-coordinate of the i-th robot is given by the real part of the i-the entry of $v_j \in \mathbb{C}^N$, while its y-coordinate is the corresponding imaginary

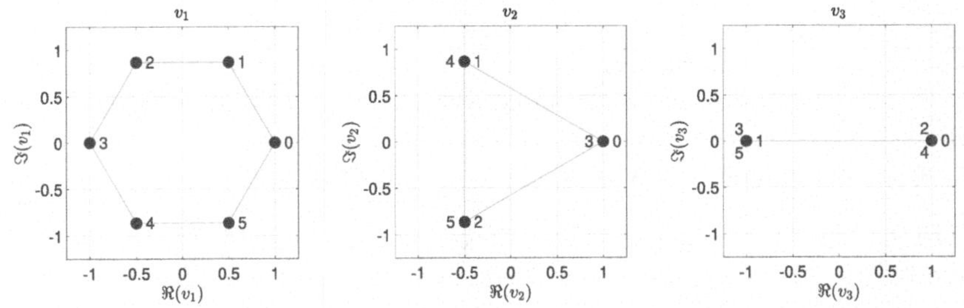

(a) $N = 6$

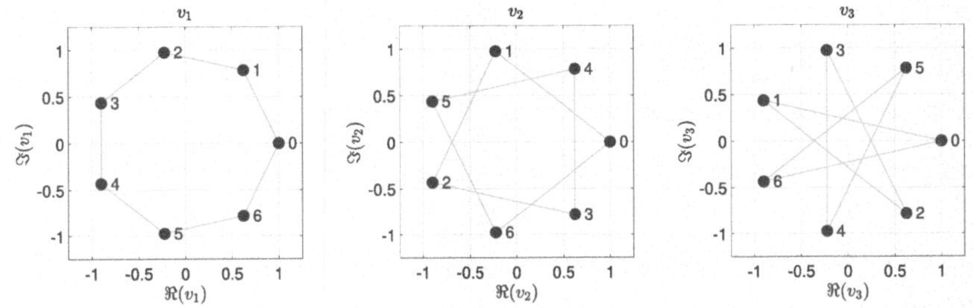

(b) $N = 7$

Fig. 2. Visualization of the complex-valued eigenvectors $v_j \in \mathbb{C}^N$, $j \neq 0$, of a circulant matrix in the complex plane for (a) $N = 6$ and (b) $N = 7$. Here, the number i next to the blue dots correspond to i-th entry of $v_j \in \mathbb{C}$. The connections in gray between the points illustrate the (cyclically) ascending index $i = 0, 1, \ldots, N-1$. In particular, it does not indicate any dynamical influences according to an underlying interaction graph. Note that, by definition all points are on the unit circle and thus form a regular N-gon (see (7)). For increasing j, if $N = 2k$ is even, the N-gon 'degenerates' since some entries are the same, while for $N = 2k+1$ only the ordering changes. Later on, we assign to every entry of $v_j \in \mathbb{C}$ its corresponding robot such that the dots also represent the position $x_i = \Re((v_j)_i)$ and $y_i = \Im((v_j)_i)$ of the i-th robot in the Euclidean plane (by changing the axis labels accordingly). Hence, the *generating* configuration of the stable invariant subspace $V_j \subseteq \mathbb{R}^{2N}$ is also visualized (cf. (9)). (Color figure online)

part. Hence, the generating configuration $(\Re(v_j), \Im(v_j))^T \in \mathbb{R}^{2N}$ of $V_j \subseteq \mathbb{R}^{2N}$ can be visualized as in Fig. 2 by changing the axis labels to x and y.

On each $V_j \subseteq \mathbb{R}^{2N}$ the matrix $\mathbf{W} \in \mathbb{R}^{2N \times 2N}$ in (6) acts as a block-diagonal matrix of two identical blocks $\Lambda_j \in \mathbb{R}^{2\times 2}$, i.e.,

$$\mathbf{W} V_j = V_j \begin{pmatrix} \Lambda_j & \\ & \Lambda_j \end{pmatrix} \text{ with } \Lambda_j = \begin{pmatrix} \Re(\lambda_j) & \Im(\lambda_j) \\ -\Im(\lambda_j) & \Re(\lambda_j) \end{pmatrix}. \tag{10}$$

This is due to the fact, that the eigenvalue pair $\lambda_j \in \mathbb{C}$ and $\overline{\lambda_j} = \lambda_{N-j}$ act as a matrix of the form $\Lambda_j \in \mathbb{R}^{2\times 2}$. Here, we abuse notation and also consider V_j to be the matrix containing the basis vectors of V_j (9). Since $\mathbf{W} \in \mathbb{R}^{2N\times 2N}$ maps each $V_j \subseteq \mathbb{R}^{2N}$ into itself, any initial configuration starting in one V_j will remain therein for all times $t > 0$. Moreover, the corresponding eigenvalues have real part less than 1, which implies that solutions will furthermore converge to 0 with the exponential decay rate $-1 + \Re(\lambda_j)$ in t (cf. [17]). The *smaller* $\Re(\lambda_j) < 1$ is, the *faster* the solution converges. In particular, if $\Re(\lambda_s) < \Re(\lambda_j)$ for all $j \neq s$, the convergence rate of any configuration in $V_s \subseteq \mathbb{R}^{2N}$ is faster than any configuration in any other $V_j \subseteq \mathbb{R}^{2N}$ for $0 \neq j \neq s$.

Remark 3. If $v_j \in \mathbb{C}^N$ has only real-valued entries, a suitable basis of $V_j \subseteq \mathbb{R}^{2N}$ is already given by the first (or last) two spanning vectors and (10) reduces to $\mathbf{W}V_j = V_j\Lambda_j$. This is the case when (1) $j = 0$ or (2) $N = 2k$ and $j = k$. As it complicates the notation, we continue with the general case below. However, similar arguments can be made in this situation and we refer to [17] for details.

In order to dynamically rank the subspaces $V_j \subseteq \mathbb{R}^{2N}$ and obtain a dynamical hierarchy of gathering speeds, we define the following.

Definition 2. *Consider a gathering* CLD *protocol with weight matrix* $W \in \mathbb{R}^{N\times N}$*, eigenvalues* $\lambda_j \in \mathbb{C}$ *and invariant subspaces* V_j *as in* (9).

(a) $V_0 \subseteq \mathbb{R}^{2N}$ *denotes the* 2*-dimensional subspace of all gathering points (cf. Definition 1 and Remark 2 (a)).*
(b) For $j \neq 0$, $V_j \subseteq \mathbb{R}^{2N}$ *is called* stable invariant subspace *with* convergence rate $\Re(\lambda_j) < 1$ *(since the* CLD *protocol is gathering).*
(c) If $\Re(\lambda_s) < \Re(\lambda_j)$ *for all* $j \neq s$*, the subspace* $V_s \subseteq \mathbb{R}^{2N}$ *is called* strong stable invariant subspace.

We summarize the results found above in the following theorem and state that the $2N$-dimensional state space of a gathering CLD protocol can be hierarchically decomposed. For more details of the proof, we refer to [17].

Theorem 3. *For a gathering* CLD *protocol of* $N \in \mathbb{N}$ *robots with weight matrix* $W \in \mathbb{R}^{N\times N}$ *there exist a family of stable invariant subspaces* $(V_j)_{j=1}^k \subseteq \mathbb{R}^{2N}$ *with convergence rates* $\Re(\lambda_j) < 1$ *and a* 2*-dimensional subspace* $V_0 \subseteq \mathbb{R}^{2N}$ *of gathering points such that*

$$\mathbb{R}^{2N} = V_0 \oplus \left(\bigoplus_{j=1}^{k} V_j\right), \quad \text{where either} \tag{11}$$

(i) every subspace $V_j \subseteq \mathbb{R}^{2N}$ *is* 4*-dimensional, if* $N = 2k+1$ *is odd,*
(ii) or, if $N = 2k$ *is even, the subspace* $V_k \subseteq \mathbb{R}^{2N}$ *is* 2*-dimensional, whereas the remaining* $V_j \subseteq \mathbb{R}^{2N}$, $j \neq k$*, are* 4*-dimensional.*

Each subspace $V_j \subseteq \mathbb{R}^{2N}$ *is spanned by a generating eigenvector* $v_j \in \mathbb{C}^N$ *of the weight matrix* $W \in \mathbb{R}^{N\times N}$ *in the sense of* (9)*. By Remark 2 (c) this decomposition in* (11) *is identical for every gathering* CLD *protocol. Only the explicit values of the convergence rates* $\Re(\lambda_j)$ *depend on the protocol itself.*

Example 4. We apply Theorem 3 to the running Example 2. To this end, we apply the eigenvalue formula (8) to $w = (0, 1, 0, \ldots, 0)^T \in \mathbb{R}^N$ (cf. Example 3):

$$\lambda_j = \omega^j = \cos\left(\frac{2\pi j}{N}\right) + \mathbf{i}\sin\left(\frac{2\pi j}{N}\right) \tag{12}$$

and thus $\Re(\lambda_j) = \cos\left(\frac{2\pi j}{N}\right)$ for $j = 0, 1, \ldots, N-1$. These convergence rates are plotted in Fig. 3 for some choices of $N \in \mathbb{N}$.

Note that the convergence rates are strictly decreasing in j (up to k), i.e., $\Re(\lambda_j) > \Re(\lambda_{j+1})$ for $j < k$. Hence, the stable subspaces $V_j \subseteq \mathbb{R}^{2N}$ ($j \neq 0$) are hierarchically ordered such that any configuration $Z \in V_j$ converges faster to a gathering point than any other configuration $\widetilde{Z} \in V_i$ for $i < j$. In particular, this model has a strong stable subspace $V_k \subseteq \mathbb{R}^{2N}$. It is 4-dimensional with convergence rate $\Re(\lambda_k) = \cos\left(\frac{2\pi k}{2k+1}\right)$ for $N = 2k+1$ odd and 2-dimensional with $\Re(\lambda_k) = -1$ for $N = 2k$ even. The corresponding generating configuration is illustrated in the right panels of Fig. 2.

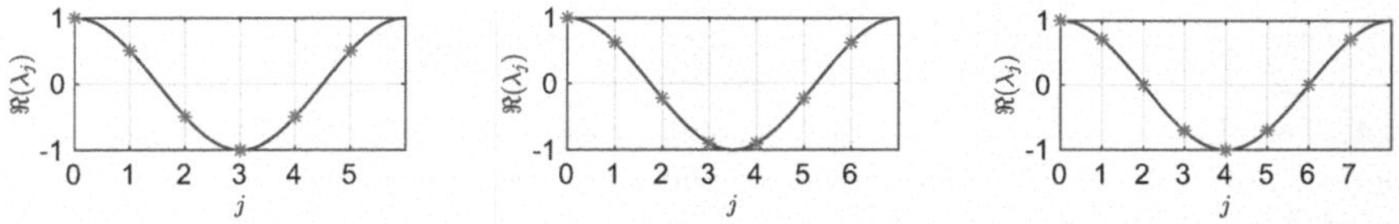

(a) $N = 6$ (b) $N = 7$ (c) $N = 8$

Fig. 3. Illustration of the convergence rates $\Re(\lambda_j)$ of the N-BUG problem for $N \in \{6, 7, 8\}$ by red stars. For reference, the cosine curve is plotted in blue. By Remark 2 (b) the convergence rates $\Re(\lambda_j)$ are symmetrically distributed.

As an arbitrary initial configuration $Z \in \mathbb{R}^{2N}$ can be written as a linear combination of the basis vectors of all $V_j \subseteq \mathbb{R}^{2N}$ and the CLD protocol is linear, Theorem 3 implies that for $j \neq 0$ every part in $V_j \subseteq \mathbb{R}^{2N}$ vanishes as time proceeds and only the gathering point in $V_0 \subseteq \mathbb{R}^{2N}$ remains.

Corollary 1. *Let $Z(0) = (X(0), Y(0)) \in \mathbb{R}^{2N}$ be an initial configuration and the state space $\mathbb{R}^{2N}$ be decomposed as in (11). Then the solution $Z(t) \in \mathbb{R}^{2N}$ of the* CLD *protocol (6) with initial condition $Z(0) \in \mathbb{R}^{2N}$ can be written as*

$$Z(t) = Z^* + \sum_{j=1}^{k} \alpha_j(t)\ \Xi_j(t), \text{ where} \tag{13}$$

(i) $Z^* = (X^*, Y^*) = (x^*, \ldots, x^*, y^*, \ldots, y^*)$ is the final gathering point of $Z \in \mathbb{R}^{2N}$. In particular, $x^* = \frac{1}{N}\sum_{i=0}^{N-1} X_i(0)$ and $y^* = \frac{1}{N}\sum_{i=0}^{N-1} Y_i(0)$.
(ii) $\alpha_j(t) = \exp((-1 + \Re(\lambda_j))t) \in \mathbb{R}$ is the exponentially decaying coefficient corresponding to $\Xi_j(t) \in V_j$.

(iii) by abusing notation, $\Xi_j(t) = V_j\beta_j(t) \in V_j$ *for a time-dependent vector* $\beta_j(t) \in \mathbb{R}^{\dim V_j}$ *with constant norm, i.e.,* $\|\beta_j(t)\|_2 = \|\beta_j(0)\|_2$ *for all* $t \geq 0$.

Example 5. We apply Corollary 1 to the N-BUG problem. For $N = 7$, we show in Fig. 4a a random initial configuration $Z(0) \in \mathbb{R}^{2N}$ as well as its final gathering point $Z^* \in \mathbb{R}^{2N}$. The corresponding exponentially decaying coefficients $\alpha_j(t) \in \mathbb{R}$ are plotted in Fig. 4b. The initial decomposition into $\Xi_j(0) \in V_j$ is visualized in Fig. 5a. The coefficients $\beta_j(t) \in \mathbb{R}^{\dim V_j}$ are shown in Fig. 5b.

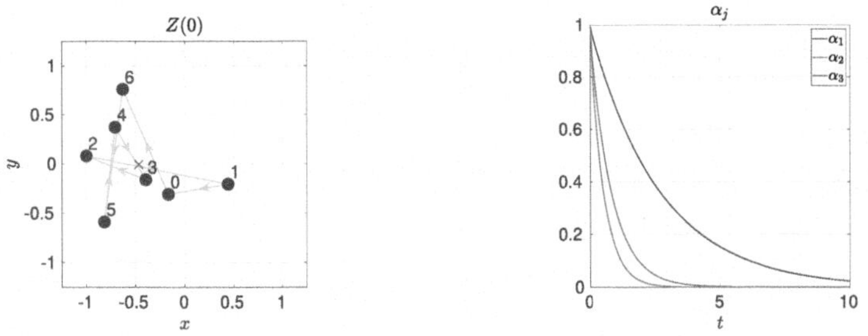

(a) $Z(0) \in \mathbb{R}^{2N}$ and $Z^* \in \mathbb{R}^{2N}$ (b) Trajectory of $\alpha_j(t) \in \mathbb{R}$

Fig. 4. Visualization of a random initial configuration $Z(0) \in \mathbb{R}^{2N}$ (blue dots) and its gathering point $Z^* \in \mathbb{R}^{2N}$ (red cross) in the Euclidean plane for $N = 7$ in (a). Its corresponding decaying coefficients $\alpha_j(t) \in \mathbb{R}$ are shown in (b). Note that $\alpha_j(t) \in \mathbb{R}$ decreases faster for increasing j, which illustrates the dynamical hierarchy discussed in Example 4. (Color figure online)

5 Analysis of the Nonlinear Scaled Circulant Laplacian Dynamics

In this section, we study the dynamical properties of NSCLD protocols of the form (3) by extending Theorem 3 to the nonlinear case. Reconsider the subspaces $V_j \subseteq \mathbb{R}^{2N}$ for $j \neq 0$ that are invariant for CLD protocols. Due to their block structure in (10), these can be further decomposed into $V_j = V_j^+ \oplus V_j^-$ with

$$V_j^+ = \operatorname{span}\left(\begin{pmatrix} \Re(v_j) \\ -\Im(v_j) \end{pmatrix}, \begin{pmatrix} \Im(v_j) \\ \Re(v_j) \end{pmatrix}\right) \text{ and } V_j^- = \operatorname{span}\left(\begin{pmatrix} -\Im(v_j) \\ \Re(v_j) \end{pmatrix}, \begin{pmatrix} \Re(v_j) \\ \Im(v_j) \end{pmatrix}\right)$$

on which we have $WV_j^{\pm} = V_j^{\pm}\Lambda_j$, where $V_j^{\pm}$ denotes either V_j^+ or V_j^-.

We evaluate the right-hand side of (3) for $Z \in V_j^{\pm}$. As computing $\mathcal{N}$, resp. n, in (3) involves evaluating the 'individual' length of each direction $(x_i, y_i) \in \mathbb{R}^2$ we define for $Z = (X, Y) \in \mathbb{R}^{2N}$ the length of the i-th direction by

$$\operatorname{norm_i}(Z) = \|(x_i, y_i)\|_2 \,. \tag{14}$$

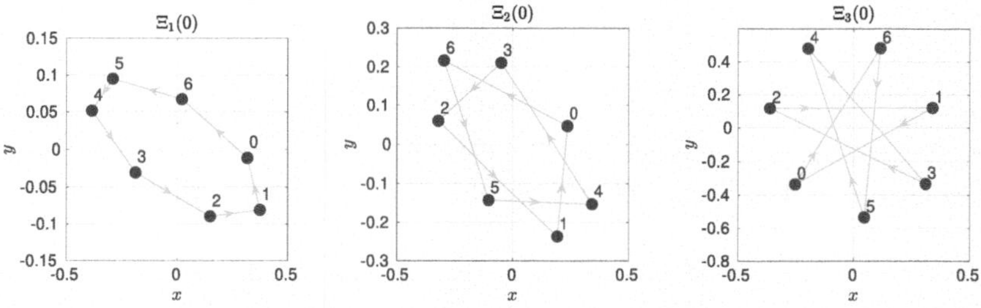

(a) Initial decomposition of $Z(0)$ into $\Xi_j(0) \in V_j$

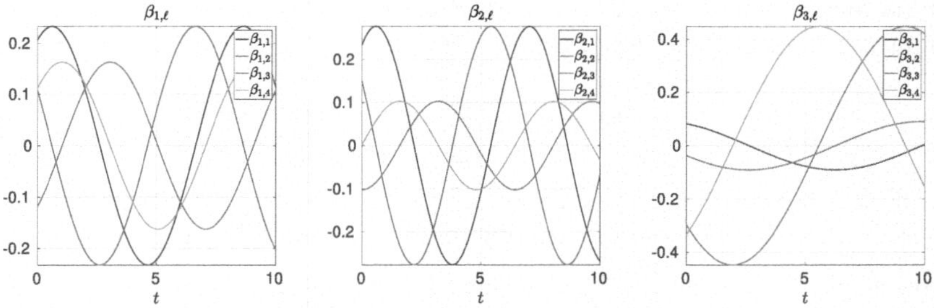

(b) Trajectory of $\beta_j(t) \in \mathbb{R}^{\dim V_j}$ such that $\Xi_j(t) = V_j\beta_j(t) \in V_j$

Fig. 5. Illustration of the decomposition (13) and its individual dynamics. Note that $\beta_j(t) \in \mathbb{R}^{\dim V_j}$ only consists of sinusoidal curves since the decaying part is contained in $\alpha_j(t) \in \mathbb{R}$. In particular, we have $\|\beta_j(t)\|_2 = \|\beta_j(0)\|_2$ for all $t \geq 0$.

A short computation shows that $\mathrm{norm_i}$ has a useful representation when writing $Z \in V_j^{\pm}$ as a linear combination of the basis vectors (see [17]). Since the entries of v_j are roots of unity we have

$$Z = V_j^{\pm}\begin{pmatrix} a \\ b \end{pmatrix} = a\begin{pmatrix} \Re(v_j) \\ \mp\Im(v_j) \end{pmatrix} + b\begin{pmatrix} \pm\Im(v_j) \\ \Re(v_j) \end{pmatrix} \implies \mathrm{norm_i}(Z) = \|(a,b)\|_2,$$

which only depends on the coefficients $(a,b) \in \mathbb{R}^2$ and, in particular, *not* on the index i. What remains to consider is how multiplication of $Z \in V_j^{\pm}$ by $-\mathbf{I}_{2N} + \mathbf{W}$ affects $\mathrm{norm_i}$. To this end, we compute

$$\mathrm{norm_i}\left((-\mathbf{I}_{2N} + \mathbf{W})\,Z\right) = \left\|(-\mathbf{I}_2 + \Lambda_j)\begin{pmatrix} a \\ b \end{pmatrix}\right\|_2 = |-1+\lambda_j|\,\|(a,b)\|_2,$$

since $V_j^{\pm} \subseteq \mathbb{R}^{2N}$ is invariant under the linear dynamics and $\Lambda_j \in \mathbb{R}^{2\times 2}$ acts on $\mathbb{R}^2$ as the multiplication by $\lambda_j \in \mathbb{C}$. Again, this expression only depends on the eigenvalue $\lambda_j \in \mathbb{C}$ and the coefficients $(a,b) \in \mathbb{R}^2$.

As the length of each individual direction is the same for $Z \in V_j^{\pm}$, we can rewrite the scaling in (3) as the multiplication with a single non-negative scalar $n(|-1+\lambda_j|\,\|(a,b)\|_2) \in \mathbb{R}$. This yields that the dynamics of an NSCLD protocol leaves the subspaces $V_j^{\pm}$ invariant and (3) can be written as

$$\dot{Z} = n(|-1+\lambda_j| \, \|(a,b)\|_2) \, (-\mathbf{I}_{2N} + \mathbf{W}) \, Z, \quad Z \in V_j^{\pm}. \tag{15}$$

Thus, the nonlinear dynamics restricted to $V_j^{\pm} \subseteq \mathbb{R}^{2N}$ reduces to a nonlinear uniform scaling of the linear dynamics. In (a,b)-coordinates, this becomes

$$\dot{\begin{pmatrix} a \\ b \end{pmatrix}} = n(|-1+\lambda_j| \, \|(a,b)\|_2) \, (-\mathbf{I}_2 + \Lambda_j) \begin{pmatrix} a \\ b \end{pmatrix}. \tag{16}$$

To derive an analogue of the convergence rate for NSCLD protocols, we consider the dynamical behavior of $\|(a,b)\|_2$ on $V_j^{\pm}$ for $j \neq 0$. Using the chain rule in combination with (16) its dynamics is given by

$$\left\| \dot{\begin{pmatrix} a \\ b \end{pmatrix}} \right\|_2 = (-1 + \Re(\lambda_j)) n(|-1+\lambda_j| \, \|(a,b)\|) \left\| \begin{pmatrix} a \\ b \end{pmatrix} \right\|_2 < 0, \tag{17}$$

since $\Re(\lambda_j) < 1$ for $j \neq 0$ (cf. Theorem 3) and n is assumed to be non-negative (cf. [17] for details). We summarize our findings in the following theorem.

Theorem 4. *For an* NSCLD *protocol with underlying gathering* CLD *protocol of $N \in \mathbb{R}$ robots with weight matrix $W \in \mathbb{R}^{N \times N}$ there exist two families of stable invariant subspaces $\left(V_j^+\right)_{j=1}^k \subseteq \mathbb{R}^{2N}$ and $\left(V_j^-\right)_{j=1}^k \subseteq \mathbb{R}^{2N}$ assigned with convergence rate $\Re(\lambda_j) < 1$, as well as a 2-dimensional subspace $V_0 \subseteq \mathbb{R}^{2N}$ of gathering points such that*

$$\mathbb{R}^{2N} = V_0 \oplus \left(\bigoplus_{j=1}^{k} V_j^+ \oplus V_j^- \right), \quad \textit{where either} \tag{18}$$

(i) every subspace $V_j^{\pm} \subseteq \mathbb{R}^{2N}$ is 2-dimensional, if $N = 2k+1$ is odd,
(ii) or, if $N = 2k$ is even, the subspace $V_k^{\pm} \subseteq \mathbb{R}^{2N}$ is 1-dimensional, whereas the remaining $V_j^{\pm} \subseteq \mathbb{R}^{2N}$, $j \neq k$, are 2-dimensional.

In each of the spaces $V_j^{\pm}$, gathering properties are described by (17)*. Again, this decomposition is identical for every* NSCLD *protocol.*

In addition to that, it can be shown that visibility is preserved in $V_j^{\pm}$. More precisely, Proposition 2 below states that interacting robots do not lose sight of each other, if the initial configuration was in any of the subspaces $V_j^{\pm}$ and all robots had all their interaction partners within viewing range. The statement is a consequence of the specific equation of motion on $V_j^{\pm} \subseteq \mathbb{R}^{2N}$ given by (15) and an analogous estimates as in the proof of Theorem 2 (details in [17]).

Proposition 2. *Consider an* NSCLD *protocol with underlying gathering* CLD *protocol of $N \in \mathbb{R}$ robots with weight matrix $W \in \mathbb{R}^{N \times N}$ and decomposition* (18)*. Let $z_0(0), \ldots, z_{N-1}(0) \in \mathbb{R}^2$ be a valid initial configuration in one of the subspaces $V_j^{\pm} \subseteq \mathbb{R}^{2N}$, that is, $\|z_p(0) - z_q(0)\| \leq C$ for all $(q,p) \in E$. Then, visibility is preserved under the dynamics, i.e.,*

$$\|z_p(t) - z_q(t)\| \leq C \quad \textit{for all } t \geq 0 \textit{ and } (q,p) \in E.$$

Finally, we discuss the implication of Theorem 4 for the non-smooth scaling $n(\|(x,y)\|_2 = \|(x,y)\|_2^{-1}$ in comparison to no scaling, i.e., the linear case. Non-smooth scaling causes $Z \in V_j^{\pm} \subseteq \mathbb{R}^{2N}$ to gather in *finite time* depending on the eigenvalue $\lambda_j \in \mathbb{C}$ and the 'size' of Z. For the proof we refer to [17].

Proposition 3. *An* NSCLD *protocol with* $n(\|(x,y)\|_2) = \|(x,y)\|_2^{-1}$ *achieves gathering of initial configurations* $Z_0 \in V_j^{\pm} \subseteq \mathbb{R}^{2N}$ $(j \neq 0)$ *in finite time that, for any* i, *is given by*

$$t_{gather,j} = -\frac{|-1+\lambda_j|}{-1+\Re(\lambda_j)}\text{norm}_i(Z_0) = -\frac{|-1+\lambda_j|}{-1+\Re(\lambda_j)}\left\|\begin{pmatrix}a_0\\b_0\end{pmatrix}\right\|_2 > 0.$$

In contrast, without scaling, i.e., purely linear dynamics, (17) becomes

$$\left\|\dot{\begin{pmatrix}a\\b\end{pmatrix}}\right\|_2 = (-1+\Re(\lambda_j))\left\|\begin{pmatrix}a\\b\end{pmatrix}\right\|_2.$$

Convergence is exponential with decay rate $-1+\Re(\lambda_j)$. However, the origin is never actually reached, meaning that gathering is not possible in finite time.

Example 6. According to (12) and Proposition 3 the gathering times for the N-BUG problem with scaling $n(\|(x,y)\|_2) = \|(x,y)\|_2^{-1}$ are given by

$$t_{\text{gather},j} = -\frac{|-1+\lambda_j|}{-1+\Re(\lambda_j)} = -\frac{\left|-1+\cos\left(\frac{2\pi j}{N}\right)+\mathbf{i}\sin\left(\frac{2\pi j}{N}\right)\right|}{-1+\cos\left(\frac{2\pi j}{N}\right)},$$

when a normalized initial condition $Z \in V_j^{\pm}$ is considered, i.e., $\|(a_0,b_0)\|_2 = 1$. For $N = 2k$ we have the real eigenvalue $\lambda_k = -1$ corresponding to the *strong* stable subspace $V_k^{\pm} \subseteq \mathbb{R}^{2N}$ and the *fastest* gathering time is $t_{\text{gather},k} = 1$.

6 Conclusion and Outlook

In this article, we investigate gathering properties of CLD and NSCLD protocols and derived a decomposition of the swarm state spaces into stable invariant subspaces which represent a dynamical hierarchy in terms convergence speed. The decomposition is independent of the explicit protocol, whereas only the gathering rates depend on the weights in the interaction graph.

We illustrate our findings on the N-BUG problem which is in the class of our considered protocols. Certainly, there are more protocols that can be described by CLD and/or NSCLD models such as the GO-TO-THE-MIDDLE and the GO-TO-THE-AVERAGE protocol. The applications of our findings to these protocols can be found in the Appendix of the extended version of this article [17].

We consider fixed *circulant* interaction structure. However, our results can be adapted to *general* fixed interaction graphs. Here, the eigendecomposition is still crucial for the dynamical analysis and stable invariant subspaces given by eigenspaces with different gathering rates can be found. Although, the explicit structure of this decomposition is generally unclear as the spectral decomposition of an arbitrary weight matrix cannot be analytically derived.

In future research, the general case could be considered, where the interaction structure is not fixed, but rather depends on the *visibility graph*, i.e., $e = (j, i) \in E$ if and only if robots i and j are sufficiently close. Here, a well-known model is the $\mathcal{OBLOT}$ model [15], where robots are *anonymous* in contrast to our CMAS model. In this situation the interaction graph is only changing finitely often at discrete times and we expect that the eigendecomposition (in between each topology change) remains a useful tool for the dynamical analysis.

Finally, in this work we consider a time-continuous robot model. However, a similar dynamical analysis is possible for a time-discrete model as well. In particular, the mathematical theory of linear discrete dynamical systems is just as well developed. In this context, it is an interesting open question if the results can even be extended to *asynchronous* time models.

Acknowledgments. The authors would like to express their special gratitude to Friedhelm Meyer auf der Heide, Jannik Castenow, and Jonas Harbig who provided invaluable inspiration and counsel for the distributed computing side of the project and took part in the development of some of the research ideas presented in this manuscript in numerous spirited and productive discussions.

This work is funded by the Deutsche Forschungsgemeinschaft (DFG, German Research Foundation) – Project number 453112019.

Disclosure of Interests. The authors have no competing interests to declare.

References

1. Arenas, A., Díaz-Guilera, A., Pérez-Vicente, C.J.: Synchronization reveals topological scales in complex networks. Phys. Rev. Lett. **96**(11), 114102 (2006). https://doi.org/10.1103/PhysRevLett.96.114102
2. Beard, R., Stepanyan, V.: Information consensus in distributed multiple vehicle coordinated control. In: 42nd IEEE International Conference on Decision and Control (IEEE Cat. No. 03CH37475), vol. 2, pp. 2029–2034 (2003). https://doi.org/10.1109/CDC.2003.1272913
3. Brandes, P., Degener, B., Kempkes, B., Meyer auf der Heide, F.: Energy-efficient strategies for building short chains of mobile robots locally. Theor. Comput. Sci. **509**, 97–112 (2013). https://doi.org/10.1016/j.tcs.2012.10.056, pII: S0304397512009887
4. Castenow, J., Fischer, M., Harbig, J., Jung, D., Meyer auf der Heide, F.: Gathering anonymous, oblivious robots on a grid. Theor. Comput. Sci. **815**, 289–309 (2020)
5. Castenow, J., Harbig, J., Jung, D., Knollmann, T., Meyer auf der Heide, F.: Gathering a euclidean closed chain of robots in linear time and improved algorithms for chain-formation. Theor. Comput. Sci. **939**, 261–291 (2023). https://doi.org/10.1016/j.tcs.2022.10.031
6. Chen, Y., Lu, J., Yu, X., Hill, D.J.: Multi-agent systems with dynamical topologies: consensus and applications. IEEE Circuits Syst. Mag. **13**(3), 21–34 (2013). https://doi.org/10.1109/MCAS.2013.2271443

7. Cohen, R., Peleg, D.: Convergence properties of the gravitational algorithm in asynchronous robot systems. SIAM J. Comput. **34**(6), 1516–1528 (2005). https://doi.org/10.1137/S0097539704446475
8. Cucker, F., Smale, S.: Emergent behavior in flocks. IEEE Trans. Autom. Control **52**(5), 852–862 (2007). https://doi.org/10.1109/TAC.2007.895842
9. Degener, B., Kempkes, B., Kling, P., Meyer auf der Heide, F.: Linear and competitive strategies for continuous robot formation problems. TOPC **2**(1), 2:1–2:18 (2015). https://doi.org/10.1145/2742341
10. Degener, B., Kempkes, B., Langner, T., Meyer auf der Heide, F., Pietrzyk, P., Wattenhofer, R.: A tight runtime bound for synchronous gathering of autonomous robots with limited visibility. In: Proceedings of the 23rd ACM Symposium on Parallelism in Algorithms and Architectures, SPAA, pp. 139–148 (2011). https://doi.org/10.1145/1989493.1989515
11. Degener, B., Kempkes, B., Meyer auf der Heide, F.: Energy-awareness in self-organising robotic exploration teams. In: Müller-Schloer, C., Schmeck, H., Ungerer, T. (eds.) Organic Computing — A Paradigm Shift for Complex Systems, pp. 531–543. Springer Basel (2011). https://doi.org/10.1007/978-3-0348-0130-0_35
12. van Doorn, E.A.: Connectivity of circulant digraphs. J. Graph Theory **10**(1), 9–14 (1986). https://doi.org/10.1002/jgt.3190100103
13. Dörfler, F., Bullo, F.: Synchronization in complex networks of phase oscillators: a survey. Automatica **50**(6), 1539–1564 (2014). https://doi.org/10.1016/j.automatica.2014.04.012
14. Flocchini, P.: Gathering. In: Distributed Computing by Mobile Entities, Current Research in Moving and Computing, pp. 63–82. Springer, Cham (2019)
15. Flocchini, P., Prencipe, G., Santoro, N.: Moving and computing models: robots. In: Flocchini, P., Prencipe, G., Santoro, N. (eds.) Distributed Computing by Mobile Entities, Current Research in Moving and Computing, Lecture Notes in Computer Science, vol. 11340, pp. 3–14. Springer, Cham (2019). https://doi.org/10.1007/978-3-030-11072-7_1
16. Gerlach, R., von der Gracht, S., Hahn, C., Harbig, J., Kling, P.: Symmetry preservation in swarms of oblivious robots with limited visibility. In: Bonomi, S., Galletta, L., Rivière, E., Schiavoni, V. (eds.) 28th International Conference on Principles of Distributed Systems (OPODIS 2024). Leibniz International Proceedings in Informatics (LIPIcs), vol. 324, pp. 13:1–13:28. Schloss Dagstuhl – Leibniz-Zentrum für Informatik, Dagstuhl, Germany (2025). https://doi.org/10.4230/LIPIcs.OPODIS.2024.13
17. Gerlach, R., von der Gracht, S., Dellnitz, M.: On the dynamical hierarchy in gathering protocols with circulant topologies (2024). https://arxiv.org/abs/2305.06632
18. Gray, R.M.: Toeplitz and circulant matrices: a review. Found. Trends® Commun. Inf. Theory **2**(3), 155–239 (2005). https://doi.org/10.1561/0100000006
19. Hamann, H.: Swarm Robotics - A Formal Approach. Springer, Cham (2018)
20. Heagy, J.F., Carroll, T.L., Pecora, L.M.: Synchronous chaos in coupled oscillator systems. Rev. Mod. Phys. **50**(3), 1874–1885 (1994). https://doi.org/10.1103/physreve.50.1874
21. Hegselmann, R., Krause, U.: Opinion dynamics and bounded confidence. J. Artif. Societies Soc. Simul. **5**(3) (2002). https://www.jasss.org/5/3/2.html
22. Iqbal, M., Leth, J., Ngo, T.D.: Cartesian product-based hierarchical scheme for multi-agent systems. Automatica **88**, 70–75 (2018). https://doi.org/10.1016/j.automatica.2017.11.009

23. Irofti, D., Atay, F.M.: On the delay margin for consensus in directed networks of anticipatory agents. IFAC-PapersOnLine **49**(10), 206–211 (2016). https://doi.org/10.1016/j.ifacol.2016.07.530
24. Kling, P.: Unifying the analysis of communication chain strategies. Master's thesis, Paderborn University (2010)
25. Kling, P., Meyer auf der Heide, F.: Convergence of local communication chain strategies via linear transformations. In: Meyer auf der Heide, F. (ed.) Proceedings of the 23rd ACM Symposium on Parallelism in Algorithms and Architectures, p. 159. ACM Conferences. ACM, New York (2011). https://doi.org/10.1145/1989493.1989517
26. Kling, P., Meyer auf der Heide, F.: Continuous protocols for swarm robotics. In: Flocchini, P., Prencipe, G., Santoro, N. (eds.) Distributed Computing by Mobile Entities: Current Research in Moving and Computing, pp. 317–334. Springer, Cham (2019). https://doi.org/10.1007/978-3-030-11072-7_13
27. McGraw, P.N., Menzinger, M.: Laplacian spectra as a diagnostic tool for network structure and dynamics. Phys. Rev. E Stat. Nonlinear Soft Matter Phys. **77**(3 Pt 1), 031102 (2008). https://doi.org/10.1103/PhysRevE.77.031102
28. Moreau, L.: Stability of continuous-time distributed consensus algorithms. In: 2004 43rd IEEE Conference on Decision and Control, vol. 4, pp. 3998–4003. IEEE Operations Center, Piscataway (2004). https://doi.org/10.1109/CDC.2004.1429377
29. Olfati-Saber, R., Murray, R.M.: Agreement problems in networks with directed graphs and switching topology. In: Zhu, J.J. (ed.) Proceedings / 42nd IEEE Conference on Decision and Control, pp. 4126–4132. IEEE Service Center, Piscataway (2003). https://doi.org/10.1109/CDC.2003.1271796
30. Olfati-Saber, R., Murray, R.M.: Consensus protocols for networks of dynamic agents. In: Proceedings of the 2003 American Control Conference, ACC, pp. 951–956. IEEE Service Center, Piscataway (2003). https://doi.org/10.1109/ACC.2003.1239709
31. Olfati-Saber, R., Murray, R.M.: Consensus problems in networks of agents with switching topology and time-delays. IEEE Trans. Autom. Control **49**(9), 1520–1533 (2004). https://doi.org/10.1109/TAC.2004.834113
32. Pikovskij, A., Rosenblum, M., Kurths, J.: Synchronization: A Universal Concept in Nonlinear Sciences. Cambridge Nonlinear Science Series, vol. 12. Cambridge University Press, Cambridge, 1st paperback edn., repr edn. (2003)
33. Rubido, N.: Energy Transmission and Synchronization in Complex Networks: Mathematical Principles. Springer Theses Series. Springer, Cham (2016). https://doi.org/10.1007/978-3-319-22216-5
34. Watton, A., Kydon, D.W.: Analytical aspects of the N-bug problem. Am. J. Phys. **37**(2), 220–221 (1969). https://doi.org/10.1119/1.1975458

Low-Bandwidth Matrix Multiplication: Faster Algorithms and More General Forms of Sparsity

Chetan Gupta[1], Janne H. Korhonen[2], Jan Studený[3], Jukka Suomela[3(✉)], and Hossein Vahidi[3]

[1] IIT Roorkee, Roorkee 247667, Uttarakhand, India
chetan.gupta@cs.iitr.ac.in
[2] Helsinki, Finland
janne.h.korhonen@gmail.com
[3] Aalto University, Konemiehentie 2, 02150 Espoo, Finland
{jan.studeny,jukka.suomela,hossein.vahidi}@aalto.fi

Abstract. In prior work, Gupta et al. (SPAA 2022) presented a distributed algorithm for multiplying sparse $n \times n$ matrices, using n computers. They assumed that the input matrices are *uniformly sparse*—there are at most d non-zeros in each row and column—and the task is to compute a uniformly sparse part of the product matrix. The sparsity structure is globally known in advance (this is the *supported* setting). As input, each computer receives one row of each input matrix, and each computer needs to output one row of the product matrix. In each communication round each computer can send and receive one $O(\log n)$-bit message. Their algorithm solves this task in $O(d^{1.907})$ rounds, while the trivial bound is $O(d^2)$. We improve on the prior work in two dimensions: First, we show that we can solve the same task faster, in only $O(d^{1.832})$ rounds. Second, we explore what happens when matrices are not uniformly sparse. We consider the following alternative notions of sparsity: row-sparse matrices (at most d non-zeros per row), column-sparse matrices, matrices with bounded degeneracy (we can recursively delete a row or column with at most d non-zeros), average-sparse matrices (at most dn non-zeros in total), and general matrices. We present a near-complete classification of the complexity of matrix multiplication in the above setting for all combinations of these notions of sparsity.

Keywords: distributed algorithms · low-bandwidth model · matrix multiplication

1 Introduction

In this work we study sparse matrix multiplication in a distributed setting, for the low-bandwidth model (which is closely related to the node-capacitated clique model). Our work improves over the prior work by [13] in two ways: we present a faster algorithm, and we study more general forms of sparsity.

U. Schmid and R. Kuznets (Eds.): SIROCCO 2025, LNCS 15671, pp. 333–349, 2025.
https://doi.org/10.1007/978-3-031-91736-3_20

Setting and Prior Work. The task is to compute the matrix product $X = AB$ for $n \times n$ matrices using a network of n computers. Initially, A and B are evenly distributed among computers. Computation proceeds in rounds, and in one round each computer can send one $O(\log n)$-bit message to another computer and receive one message from another computer; we will assume that the elements of A, B, X, and all intermediate values of the multiplication fit in one message. Eventually each computer has to report its own part of the product matrix X. How many communication rounds are needed to solve the task?

The trivial solution for *dense* matrices takes $O(n^2)$ rounds: everyone sends all information to computer number 1, which solves the task locally and then distributes the solution to other computers. However, we can do better: for matrix multiplication over semirings there is an algorithm that runs in $O(n^{4/3})$ rounds, and for matrix multiplication over fields there is an algorithm that runs in $O(n^{2-2/\omega})$ rounds, where ω is the exponent of centralized matrix multiplication [3]. By plugging in the latest value of $\omega < 2.371552$ [23], we obtain $O(n^{1.157})$ rounds.

The key question is how much better we can do when we multiply *sparse* matrices. Let us first look at the case of *uniformly sparse* matrices, with at most d nonzero elements in each row and column; in particular, we are interested in the case where $d \ll n$. We also assume that we are only interested in a uniformly sparse part of the product matrix X. Now it is natural to assume that each computer initially holds one row of A and one row of B, and it needs to know one row of X. We assume the *supported* setting: the sparsity structures of A, B, and X are known in advance, meaning that computers know which entries are nonzero, while the values of the nonzero elements are revealed at run time. There are two algorithms from prior work that are applicable in this setting: for moderately large values of d we can use the algorithm by [2], which runs in $O(dn^{1/3})$ rounds, while for small values of d the fastest algorithm is due to [13], and the round complexity is $O(d^{1.927})$ for semirings and $O(d^{1.907})$ for fields; see Table 1 for an overview.

Contribution 1: Faster Algorithm. Our first contribution is improvements in the running time: we design a faster algorithm that solves the case of semirings in $O(d^{1.867})$ rounds and the case of fields in $O(d^{1.832})$ rounds.

While there are no non-trivial unconditional lower bounds, we point out that $O(d^{1.333})$ for semirings or $O(d^{1.156})$ for fields would imply major breakthroughs for dense matrix multiplication (by simply plugging in $d = n$). In the following figure, we illustrate the progress we make towards these milestones:

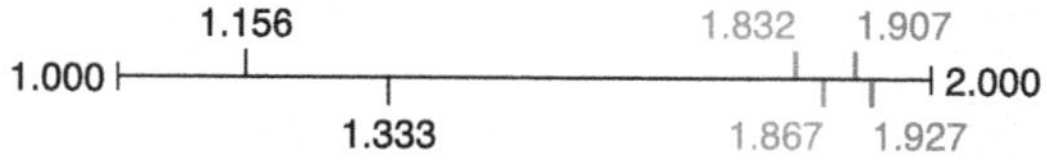

The algorithm from prior work [13] is based on the idea of *processing triangles.* Here a *triangle* is a triple $\{i, j, k\}$ such that A_{ij} and B_{jk} are nonzeros and X_{ik} is one of the elements of interest, and we say that we *process* the triangle if we add the product $A_{ij}B_{jk}$ to the sum X_{ik}. Once all such triangles are processed,

Table 1. Complexity of distributed sparse matrix multiplication

Semirings	Fields	Reference
$O(n^2)$	$O(n^2)$	trivial
$O(n^{4/3})$	$O(n^{1.157})$	[3, 23]
$O(dn^{1/3})$	$O(dn^{1/3})$	[2]
$O(d^2)$	$O(d^2)$	trivial, [13]
$O(d^{1.927})$	$O(d^{1.907})$	[13]
$O(d^{1.867})$	$O(d^{1.832})$	this work, Theorem 2

we have computed all elements of interest in the product matrix. There are two strategies that [13] uses for processing triangles:

1. If there are many triangles in total, we can find a dense cluster of them, we can interpret such a cluster as a tiny instance of dense matrix multiplication, and we can apply dense matrix multiplication algorithms to batch-process such clusters.
2. If there are few triangles in total, we can afford to process the triangles one by one.

They show that suitably applying the first strategy leaves us with few triangles in total, to which we can then apply the second strategy. Our improvements are in the second phase. Trivially, if we have a uniformly sparse instance, there are at most d^2n triangles in total, and we can process them in a naive manner in $O(d^2)$ rounds. The second phase of [13] shows that if the number of triangles is $d^{2-\varepsilon}n$, one can process them in $O(d^{2-\varepsilon/2})$ rounds. Here the key obstacle is the expression $\varepsilon/2$ in the exponent: even if we eliminate many triangles in the first phase, the second phase is still relatively expensive. We design a new algorithm that is able to process $d^{2-\varepsilon}n$ triangles in $O(d^{2-\varepsilon})$ rounds. By combining this with the general strategy of [13] and optimizing the trade-off between using time in the first phase vs. using time in the second phase, we obtain the new round complexities $O(d^{1.867})$ and $O(d^{1.832})$—the difference between semirings and fields is in the dense matrix multiplication routine that we use in the first phase.

Contribution 2: Beyond Uniform Sparsity. So far we have followed the lead of [13] and discussed *uniformly sparse* matrices (i.e., each row and column has at most d nonzeros). This is a rather restrictive notion of sparsity—can we extend prior work to more general notions of sparsity?

Perhaps the most natural candidate to consider would be *average sparsity*: we would simply assume that the total number of nonzeros in A and B and the total number of elements of interest in X is at most dn. Unfortunately, it turns out that this notion of sparsity is way too general, and we cannot expect fast algorithms for this case: in our full version [14] we show that if we could solve average-sparse matrix multiplication for $n \times n$ matrices using n computers in time that is independent of n or only mildly depends on n (say, polylogarithmic in n), it would imply major breakthroughs in our understanding of dense matrix multiplication.

In summary, in order to have any real hope of pushing the complexity of sparse matrix multiplication down to $O(d^{1.832})$ or even something more modest like $O(d^2 + \log n)$, we need to explore some *intermediate notions of sparsity*. In this work we make use of the following notions of sparsity:

- $\mathsf{US}(d)$ = uniformly sparse: at most d nonzeros per row and column.
- $\mathsf{RS}(d)$ = row-sparse: at most d nonzeros per row.
- $\mathsf{CS}(d)$ = column-sparse: at most d nonzeros per column.
- $\mathsf{BD}(d)$ = bounded degeneracy: we can recursively eliminate the matrix so that at each step we delete a row or column with at most d nonzeros.
- $\mathsf{AS}(d)$ = average-sparse: at most dn nonzeros in total.
- GM = general matrices.

We will omit (d) when it is clear from the context, and we will use e.g. expressions such as $\mathsf{US} \times \mathsf{BD} = \mathsf{AS}$ to refer to the task of computing $AB = X$ such that $A \in \mathsf{US}(d)$, $B \in \mathsf{BD}(d)$, and $\hat{X} \in \mathsf{AS}(d)$; here we write $\hat{X}$ for the matrix that indicates which elements of the product X we are interested in.

Family BD may at first look rather unusual in the context of linear algebra. However, if we interpret a matrix A as a bipartite graph G (with an edge $\{i, j\}$ whenever A_{ij} is nonzero), then $A \in \mathsf{BD}(d)$ corresponds to the familiar graph-theoretic notion that G is d-degenerate. This is a widely used notion of sparsity in the graph-theoretic setting, and also closely connected with other notions of sparsity such as bounded arboricity.

We also point out that any matrix $A \in \mathsf{BD}(d)$ can be written as a sum $A = X + Y$ such that $X \in RS(d)$ and $Y \in CS(d)$; to see this, eliminate A by deleting sparse rows or sparse columns, and put each sparse row in X and put each sparse column in Y. In particular, matrix multiplication $\mathsf{BD} \times \mathsf{BD}$ decomposes into operations of the form $\mathsf{CS} \times \mathsf{CS}$, $\mathsf{RS} \times \mathsf{CS}$, $\mathsf{CS} \times \mathsf{RS}$, and $\mathsf{RS} \times \mathsf{RS}$. In summary, we have

$$\mathsf{US} \subseteq \begin{Bmatrix} \mathsf{RS} \\ \mathsf{CS} \end{Bmatrix} \subseteq \mathsf{BD} \subseteq \mathsf{AS} \subseteq \mathsf{GM}.$$

As we will see, all of our algorithmic results are symmetric w.r.t. the three matrices. To help present such results, we will introduce the following shorthand notation. Let X, Y, and Z be families of matrices. We write $[\mathsf{X} : \mathsf{Y} : \mathsf{Z}]$ to refer to the following six operation: computing $\mathsf{X} \times \mathsf{Y} = \mathsf{Z}$, $\mathsf{Y} \times \mathsf{X} = \mathsf{Z}$, $\mathsf{X} \times \mathsf{Z} = \mathsf{Y}$, $\mathsf{Z} \times \mathsf{X} = \mathsf{Y}$, $\mathsf{Y} \times \mathsf{Z} = \mathsf{X}$, and $\mathsf{Z} \times \mathsf{Y} = \mathsf{X}$.

Now equipped with this notation, we can rephrase the main result from the prior work [13]: they show that $[\mathsf{US} : \mathsf{US} : \mathsf{US}]$ can be solved in $O(d^{1.927})$ rounds for semirings and in $O(d^{1.907})$ rounds for fields. On the other hand, in our full version [14] we argue why it is unlikely to extend this all the way to $[\mathsf{AS} : \mathsf{AS} : \mathsf{AS}]$. The key question is what happens between the two extremes, US and AS? In particular, could we solve $[\mathsf{BD} : \mathsf{BD} : \mathsf{BD}]$ efficiently?

In this work we present a near-complete classification of the complexity of matrix multiplication for all possible combinations of classes US, BD, AS, and GM. We show that almost all cases fall into one of these classes:

1. We have a trivial lower bound of $\Omega(d)$ rounds from [13] and a new upper bound of $O(d^{1.832})$ rounds from Theorem 2. For example, [US : US : AS] falls in this class.
2. We have a new lower bound of $\Omega(\log n)$ rounds from in addition to the lower bound of $\Omega(d)$ rounds from [13], and we show a near-matching upper bound of $O(d^2 + \log n)$ rounds in Theorems 3 and 4. For example, [BD : BD : BD] falls in this class.
3. We have a new lower bound of $\Omega(\sqrt{n})$ rounds in our full version [14] and the general upper bound of $O(n^{1.157})$ rounds from [3,23] that holds for any matrix. For example, [BD : BD : GM] falls in this class.
4. We have a conditional lower bound: a fast algorithm would imply major improvements in *dense* matrix multiplication. For example, [AS : AS : AS] falls in this class. We refer to the full version of this paper [14] for the details.

There is one outlier in our current classification: [US : US : GM] admits a trivial algorithm that runs in $O(d^4)$ rounds, but we do not know if it can be solved in $O(d^{1.832})$ rounds.

Our results are summarized in Table 2. The table summarizes the results for the case of semirings, but analogous results hold for the case of fields—the only difference is that the exponent 1.867 is replaced with 1.832 and we have got $1 \le \lambda \le 2 - 2/\omega < 1.157$.

In summary, three qualitatively different regions emerge: (i) matrices that can be multiplied in $f(d)$ rounds for some f, independently of n; (ii) matrices that can be multiplied in $f(d) + O(\log n)$ rounds for some f but not in $f(d) + o(\log n)$ rounds for any f; and (iii) matrices that can be multiplied in $n^{O(1)}$ rounds but not in $f(d) + n^{o(1)}$ rounds for any f.

Key Conceptual Message: Role of Bounded Degeneracy. Our work highlights the role that the class of bounded-degeneracy matrices plays in the theory and practice of distributed matrix multiplication. While the class of average-sparse matrices is arguably more natural, it is also too broad to admit fast algorithms, even if we are in the supported setting. On the other hand, the class of uniformly-sparse matrices familiar from prior work can be far too restrictive for many applications. Our work suggests a promising new notion of sparsity in this context: matrices with bounded degeneracy. First, such matrices naturally arise especially in graph-theoretic applications, and they also contain as special cases the important classes of row-sparse and column-sparse matrices. Second, our work shows that bounded-degeneracy matrices admit very efficient algorithms in the low-bandwidth setting (and as a corollary, also in any setting that is at least as strong as this model), at least if we know the structure of the matrices in advance (Theorem 4). We believe this can open a new research direction that studies the complexity of matrix multiplication across different distributed and parallel settings for bounded-degeneracy matrices.

Table 2. Summary of results (for semirings). Here $1 \leq \lambda \leq 4/3$ is the infimum of exponents λ such that dense matrix multiplication for semirings can be done in $O(n^\lambda)$ rounds. The lower bounds marked with † only hold for certain permutations of the matrix families. In the full version of this paper [14] we present our lower bounds in detail.

Sparsity	Upper bound		Lower bound	
[US : US : US] ... [US : US : AS]	$O(d^{1.867})$	Theorem 2	$\Omega(d^\lambda)$	trivial
[US : US : GM]	$O(d^4)$	trivial	$\Omega(d^\lambda)$	trivial
[US : BD : BD] ... [US : AS : GM]	$O(d^2 + \log n)$	Theorem 3	$\Omega(d^\lambda)$, $\Omega(\log n)$	full version
[BD : BD : BD] ... [BD : AS : AS]	$O(d^2 + \log n)$	Theorem 4	$\Omega(d^\lambda)$, $\Omega(\log n)$	full version
[US : GM : GM] ... [GM : GM : GM]	$O(n^{4/3})$		$\Omega(\sqrt{n})$ †	full version
[BD : BD : GM] ... [GM : GM : GM]	$O(n^{4/3})$		$\Omega(\sqrt{n})$ †	full version
[AS : AS : AS] ... [GM : GM : GM]	$O(n^{4/3})$		$\Omega(n^{(\lambda-1)/2})$	full version

Related Work and Applications. Our model of computing is the low-bandwidth model; this is closely related to the model known as *node-capacitated clique* or *node-congested clique* in the literature [1]—in the low-bandwidth model each computer can send and receive one message, while the node-capacitated clique model is usually defined so that each computer can send and receive $O(\log n)$ messages per round. Both the low-bandwidth model and the node-capacitated clique model can be interpreted as variants of the *congested clique* model [18]. Indeed, any algorithm that runs in $T(n)$ rounds in the congested clique model can be simulated in $nT(n)$ rounds in the low-bandwidth model, and for many problems (such as the dense matrix multiplication) such a simulation also results in the fastest known algorithms for the low-bandwidth model. Both the low-bandwidth model and the congested clique model can be interpreted as special cases of the classic *bulk synchronous parallel model* [22].

Matrix multiplication in this context has been primarily studied in the congested clique model [2,3,5,17]. However, as observed by [13], the congested clique model is poorly-suited for the study of sparse matrix multiplication: one can only conclude that for sufficiently sparse matrices, the problem is solvable in $O(1)$ rounds, and one cannot explore more fine-grained differences between different algorithms.

Similar to [13], we work in the *supported* version of the low-bandwidth model. In general, the supported model [11,12,21] refers to a setting in which the structure of the input is known in advance (we can do arbitrary *preprocessing* based on the structure of the input), while the specific instance is revealed at run time. In the case of graph problems, we can see an unweighted graph G in advance, while the adversary reveals a weighted version G' of G (a special case being a 0/1-weighted or $1/\infty$-weighted graph, which can be interpreted as a subgraph of G). For matrix multiplication, we know the sparsity structure in advance (i.e., which elements are potentially nonzero), while the adversary reveals the concrete values at run time.

Our logarithmic lower bounds build on the classic results in the CREW PRAM model by Cook, Dwork and Reischuk [8] and Dietzfelbinger, Kutylowski and Reischuk [9]. As observed by e.g. Roughgarden, Vassilvitskii and Wang [20], such lower bounds can be applied also in more modern settings.

One of the main applications that we have in mind here is triangle detection, which has been extensively studied in the distributed setting [4,6,7,10,13,15,16,19]. If we can multiply matrices, we can also easily detect triangles in a graph. Moreover, [US : US : US] corresponds to triangle detection in a bounded-degree graph, while e.g. [AS : AS : AS] corresponds to triangle detection in a sparse graph.

Open Questions for Future Work. In this work we initiated the study of different notions of sparsity in the context of low-bandwidth matrix multiplication. This work also gives rise to a number of questions for future work.

First, our running times—$O(d^{1.867})$ rounds for semirings and $O(d^{1.832})$ for fields—are still far from the conditional lower bounds $\Omega(d^{1.333})$ and $\Omega(d^{1.156})$, respectively. Pushing the running time further down is a major challenge for future work. If we follow the two-phase approach of [13], our present work essentially makes the second phase optimal; the main hope for future improvements now lies in the first phase of the algorithm.

Our algorithms are designed for the supported model. Eliminating the knowledge of the support (i.e., knowledge of the sparsity structure) is a major challenge for future work.

Finally, there are some gaps in Table 2. For example, we do not know if [US : US : GM] is strictly harder than [US : US : US]. Finally, we do not know if $O(d^2 + \log n)$ can be further improved to, say, $O(d^{1.867} + \log n)$.

2 Preliminaries

We work in the low-bandwidth model. There are n computers. Initially each computer holds its own part of A and B, and eventually each computer has to report its own part of X. When we study sparse matrices with at most dn nonzeros, we assume that each computer holds at most d elements, while when we study dense matrices, we assume that each computer holds at most n elements.

For our algorithms, it does not matter how the input and output is distributed among the computers—with an additional $O(d)$ time we can permute the input and output as appropriate. Our lower bounds hold for any fixed distribution of input and output that may also depend on the support.

Supported Model and Indicator Matrices. When we study matrix multiplication in the supported model, we assume that we know in advance indicator matrices $\hat{A}$, $\hat{B}$, and $\hat{X}$ that encode the structure of our instance: $\hat{A}_{ij} = 0$ implies $A_{ij} = 0$, $\hat{B}_{jk} = 0$ implies $B_{jk} = 0$, and $\hat{X}_{ik} = 0$ indicates that we do not need to compute the value of X_{ik}. When we make assumptions on sparsity, our assumptions refer to the sparsity of $\hat{A}$, $\hat{B}$, and $\hat{X}$. For example, when we study $\mathsf{US} \times \mathsf{BD} = \mathsf{AS}$, we assume that $\hat{A} \in \mathsf{US}(d)$, $\hat{B} \in \mathsf{BD}(d)$, and $\hat{X} \in \mathsf{AS}(d)$. It then also follows that $A \in \mathsf{US}(d)$ and $B \in \mathsf{BD}(d)$.

Tripartite Graph and Triangles. It will be convenient to assume that our matrices are indexed with indices from three disjoint sets I, J, and K, each of size n. For example, elements of matrix A are indexed with A_{ij} where $i \in I$ and $j \in J$. Following [13], we write $\hat{\mathcal{T}}$ for the set of all *triangles*, which are triples $\{i, j, k\}$ with $i \in I$, $j \in J$, and $k \in K$ such that $\hat{A}_{ij} \neq 0$, $\hat{B}_{jk} \neq 0$, and $\hat{X}_{ik} \neq 0$.

For a collection of triangles $\mathcal{T}$, we write $G(\mathcal{T})$ for the tripartite graph $G(\mathcal{T}) = (V, E)$, where the set of nodes is $V = I \cup J \cup K$ and there is an edge $\{u, v\} \in E$ if there is a triangle $T \in \hat{\mathcal{T}}$ with $\{u, v\} \subseteq T$. For a set of nodes $U \subseteq V$, we write $\mathcal{T}[U] = \{T \in \mathcal{T} : T \subseteq U\}$ for the set of triangles induced by U.

We say that we *process* a triangle $\{i, j, k\}$ if we have added the product $A_{ij}B_{jk}$ to the sum X_{ik}. The key observation is that processing all triangles is exactly equivalent to computing all values of interest in the product matrix $X = AB$.

Clusters and Clusterings. We say that $U \subseteq V$ is a *cluster* if $U = I' \cup J' \cup K'$ for $I' \subseteq I$, $J' \subseteq J$, $K' \subseteq K$, $|I'| = d$, $|J'| = d$, and $|K'| = d$. We say that a collection of triangles $\mathcal{P}$ is *clustered* if there are disjoint clusters $U_1, \ldots, U_t$ such that $\mathcal{P} = \mathcal{P}[U_1] \cup \cdots \cup \mathcal{P}[U_t]$.

A clustered collection of triangles can be processed efficiently by applying an algorithm for dense matrix multiplication in parallel to each cluster [13]. By applying the strategy of [3] and the latest value of ω from [23], we obtain:

Lemma 1. *A clustered instance of matrix multiplication can be solved in $O(d^{4/3})$ rounds over semirings, and in $O(d^{1.156671})$ rounds over fields.*

3 Handling Few Triangles Fast

We will start by a general result that will form the foundation for all of our upper-bound results. [13, Lemma 5.2] observed that if we have a *balanced* instance in which each node touches at most κ triangles, we can solve matrix multiplication in $O(\kappa)$ rounds. However, the key challenge is what to do in *unbalanced* instances: we have at most κn triangles, but some nodes can touch many more than κ triangles. The following lemma shows how to handle also such instances efficiently:

Lemma 2. *Let $\mathcal{T}$ be a set of triangles with $|\mathcal{T}| \leq \kappa n$, and assume that for each pair of nodes $u, v \in V$ there are at most m triangles $T \in \mathcal{T}$ with $\{u, v\} \in T$. Assume that each computer initially holds at most d elements of A and at most d elements of B, and it will need to output at most d elements of X. Then all triangles in $\mathcal{T}$ can be processed (over fields or semirings) in $O(\kappa + d + \log m)$ rounds of the supported low-bandwidth model.*

We emphasize that this strictly improves over [13, Lemma 5.1] in two different ways: our result is applicable in a more general setting, and we avoid the factor-2 loss in the exponent of the running time. In the remainder of this section, we provide a proof for Lemma 2.

High-Level Plan. Our high-level plan for proving Lemma 2 goes as follows:

1. We start with an arbitrary collection of κn triangles $\mathcal{T}$ over the original set of $3n$ nodes $V = I \cup J \cup K$.
2. We construct a new *virtual* collection of κn triangles $\mathcal{T}'$ over a new set of $O(n)$ virtual nodes V'.
3. The new virtual collection of triangles is *balanced* in the sense that each virtual node touches only κ triangles.
4. The virtual nodes are assigned to real computers so that each real computer is responsible for only $O(1)$ virtual nodes.
5. We show how we can route original data to the virtual nodes, so that each virtual node only needs to do work for each triangle that it touches, and we can also aggregate the results back to the original nodes that need it.

The most challenging part is the final step, routing.

Virtual Instance. For each node $v \in V$, let $\mathcal{T}(v)$ be the set of triangles $T \in \mathcal{T}$ with $v \in T$. Let $t(v) = |\mathcal{T}(v)|$ and $\ell(v) = \lceil t(v)/\kappa \rceil$. Our set of virtual nodes is

$$\begin{aligned}
V' &= I' \cup J' \cup K', \\
I' &= \big\{(i, x) \bigm| i \in I, x \in \{1, 2, \dots, \ell(i)\}\big\}, \\
J' &= \big\{(k, x) \bigm| j \in J, x \in \{1, 2, \dots, \ell(j)\}\big\}, \\
K' &= \big\{(j, x) \bigm| k \in K, x \in \{1, 2, \dots, \ell(k)\}\big\}.
\end{aligned}$$

The key observation is that the set of virtual nodes is not too large:

$$|V'| = \sum_v \ell(v) \leq \sum_v (t(v)/\kappa + 1) \leq 3|\mathcal{T}|/\kappa + n = 4n.$$

The last step follows from the fact that each triangle is counted 3 times, once by each of its nodes. Now we can construct a new balanced collection of triangles $\mathcal{T}'$ over V' such that each virtual node $v' \in V'$ touches at most κ triangles: let $\mathcal{T}(v) = \{t_1, \ldots, t_{t(v)}\}$, for the first κ triangles in $\mathcal{T}(v)$ we replace v with the virtual node $(v, 1)$ (e.g., triangle $t_1 = \{v, x, y\}$ is replaced by $t_1' = \{(v, 1), x, y\}$), for the next κ triangles we replace v with $(v, 2)$, etc. Note that after repeating the process for all vertices $v \in V$, all three endpoints of a triangle will be mapped to a virtual node.

We assign the virtual nodes to real computers so that each computer is responsible for at most 4 virtual nodes. Therefore, in what follows, we can assume that the virtual nodes are computational entities, and we can simulate their work in the real computer network with constant overhead. Note that computers know this virtual node assignment at the start of the algorithm without any communication since we only used structural property of the triangles which is available as support to all computers. However, computers do not know the values associated with the edges of the triangles, and therefore a routing protocol is required.

Routing. Now let us see how we will process the triangles. Consider a triangle $\{i, j, k\} \in \mathcal{T}$, and let its unique virtual copy be $\{i', j', k'\} \in \mathcal{T}'$. Conceptually, we proceed as follows:

1. The real computer $p(i, j)$ that initially holds A_{ij} transmits it to virtual computer i'.
2. The real computer $p(j, k)$ that initially holds B_{jk} transmits it to virtual computer i'.
3. Virtual computer i' computes the product $A_{ij}B_{jk}$ and transmits it to the real computer $p(i, k)$ that is responsible for storing X_{ik}.

If we did this in a naive manner, it would be prohibitively expensive, e.g., a real computer might need to send a single value to all other computers; a more careful routing scheme is needed. The key idea is to do intermediate-point routing. We present a schematic overview of our routing scheme in Fig. 1; we refer to the full version of this work [14] for the details.

4 Algorithm for [US : US : AS]

In our terminology, [13] proved the following statement:

Theorem 1 ([13]). *In the supported low-bandwidth model, sparse matrix multiplication of the form* [US : US : US] *can be computed in* $O(d^{1.927})$ *rounds over semirings and in* $O(d^{1.907})$ *rounds over fields.*

In this section, we both improve the running time and widen the scope of applicability:

Theorem 2. *In the supported low-bandwidth model, sparse matrix multiplication of the form* [US : US : AS] *can be computed in* $O(d^{1.867})$ *rounds over semirings and in* $O(d^{1.832})$ *rounds over fields.*

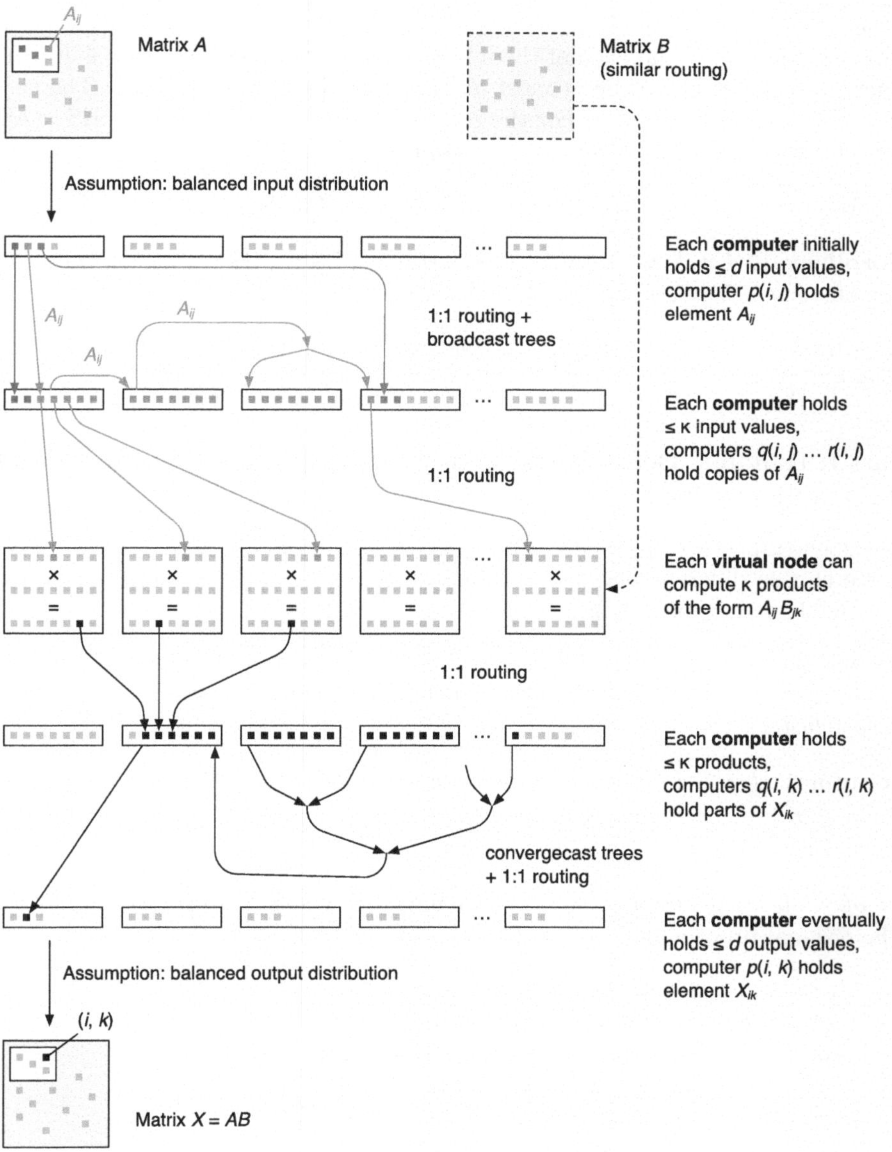

Fig. 1. Routing scheme from the proof of Lemma 2

Preliminaries. We start with the following technical lemma:

Lemma 3. *Let* $\hat{\mathcal{T}}$ *be defined by an instance of* [US : US : AS]. *For each node* $x \in V$ *there are at most* d^2 *triangles* $T \in \hat{\mathcal{T}}$ *with* $x \in T$.

Proof. Consider first the case of US × AS = US. If $x \in I$, then x is incident to at most d nodes $j \in J$ and at most d nodes $k \in K$, and hence in total there can be at most d^2 triangles of the form $\{x, j, k\}$. If $x \in J$, then x is incident to at most d nodes $i \in I$, and each of them is incident to at most d nodes $k \in K$, and hence in total there can be at most d^2 triangles of the form $\{i, x, k\}$. The case of $x \in K$ is similar to $x \in J$. For the cases of $\mathsf{AS} \times \mathsf{US} = \mathsf{US}$ and $\mathsf{US} \times \mathsf{US} = \mathsf{AS}$, permute the roles of I, J, and K as appropriate.

We have two simple corollaries that will be useful shortly:

Corollary 1. *Let $\hat{\mathcal{T}}$ be defined by an instance of* $[\mathsf{US} : \mathsf{US} : \mathsf{AS}]$. *For each pair of nodes $u, v \in V$ there are at most d^2 triangles $T \in \hat{\mathcal{T}}$ with $\{u, v\} \in T$.*

Corollary 2. *The total number of triangles in any* $[\mathsf{US} : \mathsf{US} : \mathsf{AS}]$ *instance is bounded by $d^2 n$.*

Part 1: Handling Dense Parts. In the first part we follow the basic structure of [13], with two changes:

1. we generalize it to $[\mathsf{US} : \mathsf{US} : \mathsf{AS}]$,
2. we choose different parameter values so that we can fully benefit from the new algorithm for part 2.

We will then diverge from [13] in part 2, where we will apply Lemma 2.

Similar to [13], we first show that if there are many triangles, we can find a *dense cluster*:

Lemma 4. *Let $\hat{\mathcal{T}}$ be defined by an instance of* $[\mathsf{US} : \mathsf{US} : \mathsf{AS}]$, *and let $\mathcal{T} \subseteq \hat{\mathcal{T}}$. If $|\mathcal{T}| \geq d^{2-\varepsilon} n$ for some $\varepsilon > 0$, then there exists a cluster $U \subseteq V$ with $|\mathcal{T}[U]| \geq d^{3-4\varepsilon}/24$.*

Proof. This is a generalization of [13, Lemma 3.1], which holds for $[\mathsf{US} : \mathsf{US} : \mathsf{US}]$. The original proof makes use of the following facts about $G(\mathcal{T})$:

- there are at most dn edges between J and K,
- each $i \in I$ is incident to at most d nodes of J,
- each $i \in I$ is incident to at most d nodes of K,
- each $j \in J$ is incident to at most d nodes of I,
- each $k \in K$ is incident to at most d nodes of I.

All of these also hold for $\mathsf{US} \times \mathsf{AS} = \mathsf{US}$. For the cases of $\mathsf{AS} \times \mathsf{US} = \mathsf{US}$ and $\mathsf{US} \times \mathsf{US} = \mathsf{AS}$, permute the roles of I, J, and K as appropriate.

Then by plugging in Lemma 4 instead of [13, Lemma 3.1] in the proof of [13, Lemma 4.1], we obtain the following corollary:

Lemma 5. *Let $\delta > 0$ and $\varepsilon > 0$, and assume that d is sufficiently large. Let $\hat{\mathcal{T}}$ be defined by an instance of* $[\mathsf{US} : \mathsf{US} : \mathsf{AS}]$, *and let $\mathcal{T} \subseteq \hat{\mathcal{T}}$. Assume $|\mathcal{T}| \geq d^{2-\varepsilon} n$. Then we can partition $\mathcal{T}$ into disjoint sets $\mathcal{P}$ and $\mathcal{T}'$ such that $\mathcal{P}$ is clustered and $|\mathcal{P}| \geq d^{2-5\varepsilon-4\delta} n$.*

Proof. We follow the strategy in the proof of [13, Lemma 4.1]: Apply Lemma 4 to find a cluster U. Put the triangles $T \in \mathcal{T}[U]$ that are fully contained in U to $\mathcal{P}$, and put the triangles $T \in \mathcal{T}$ that only partially touch U to $\mathcal{T}'$. Repeat until there are sufficiently few triangles left; put all remaining triangles to $\mathcal{T}'$.

The original proof makes use of the fact that each node $x \in V$ is contained in at most d^2 triangles. This is trivial for [US : US : US]. For [US : US : AS] we apply Lemma 3, and then the original analysis holds verbatim.

Finally, by plugging in Lemma 5 instead of [13, Lemma 4.1] in the proof of [13, Lemma 4.2], we obtain the following lemma:

Lemma 6. *Let $\delta > 0$ and $0 \leq \gamma < \varepsilon$, and assume that d is sufficiently large. Let $\hat{\mathcal{T}}$ be defined by an instance of* [US : US : AS], *and let $\mathcal{T} \subseteq \hat{\mathcal{T}}$. Assume $|\mathcal{T}| \leq d^{2-\gamma}n$. Then we can partition $\mathcal{T}$ into disjoint sets $\mathcal{T} = \mathcal{P}_1 \cup \cdots \cup \mathcal{P}_L \cup \mathcal{T}'$ such that each $\mathcal{P}_i$ is clustered, $L \leq 144d^{5\varepsilon-\gamma+4\delta}$ and $|\mathcal{T}'| \leq d^{2-\varepsilon}n$.*

Proof. We follow the proof of [13, Lemma 4.2]: Apply Lemma 5 repeatedly to construct clusterings $\mathcal{P}_1, \mathcal{P}_2, \ldots$. Stop once the residual set of triangles $\mathcal{T}'$ is sufficiently small. The original analysis holds verbatim.

Now we are ready to prove the following result. Note that unlike the analogous result from [13], we have got here a convenient formulation in which the exponent in the time complexity matches the exponent in the size of the residual part:

Lemma 7. *Let $\hat{\mathcal{T}}$ be defined by an instance of* [US : US : AS]. *We can partition $\hat{\mathcal{T}}$ into $\mathcal{T}_1 \cup \mathcal{T}_2$ such that:*

1. *For matrix multiplication over semirings, $\mathcal{T}_1$ can be solved in $O(d^{1.867})$ rounds and $|\mathcal{T}_2| \leq d^{1.867}n$.*
2. *For matrix multiplication over fields, $\mathcal{T}_1$ can be solved in $O(d^{1.832})$ rounds and $|\mathcal{T}_2| \leq d^{1.832}n$.*

Proof. For the case of semirings, we apply Lemma 6 with the parameters δ, γ, and ε shown in Table 3. In the first step, we start with $\gamma = 0$; note that by Corollary 2 we satisfy the assumption that our initial set of triangles has size at most $d^{2-\gamma}n = d^2n$. We pick a very small δ and choose a suitable ε. Then we use Lemma 1 to process each of $\mathcal{P}_i$. The total processing time will be $O(d^\alpha)$, where $\alpha = 5\varepsilon - \gamma + 4\delta + 4/3$. Our choice of ε is optimized so that $\alpha \leq 1.867$ and hence we do not exceed our time budget. After processing all of $\mathcal{P}_i$, we are left with the residual set of triangles of size $|\mathcal{T}'| \leq d^\beta n$, where $\beta = 2 - \varepsilon$. We then repeat the same process, starting with $\gamma = 2 - \beta$, and again optimizing ε. We continue until $\beta \leq 1.867$.

For the case of fields, the idea is the same, but we make use of different parameter values; we refer to the full version [14] for the details.

Table 3. Parameters for the proof of Lemma 7 (semirings)

Step	δ	γ	ε	α	β
1	0.00001	0.00000	0.10672	1.86698	1.89328
2	0.00001	0.10672	0.12806	1.86696	1.87194
3	0.00001	0.12806	0.13233	1.86697	1.86767
4	0.00001	0.13233	0.13319	1.86700	1.86681

Part 2: Handling Few Triangles. Let us recap: We started with an arbitrary instance, with possibly up to d^2n triangles. Then we have used $O(d^\alpha)$ rounds to process "dense" parts, and we are left with only $d^\alpha n$ triangles; here $\alpha = 1.867$ for semirings and $\alpha = 1.832$ for fields.

Now we can apply Lemma 2 to handle all these remaining $d^\alpha n$ triangles in $O(d^\alpha)$ rounds. When we apply the lemma, we set $\kappa = d^\alpha$, and by Corollary 1 we have $m = d^2$. The overall running time is $O(d^\alpha + d + \log d^2) = O(d^\alpha)$. This completes the proof of Theorem 2.

5 More General Algorithms

We will now step beyond [US : US : AS] and consider more general settings. We will show that we can handle much more general notions of sparsity in $O(d^2 + \log n)$ rounds. For each case we prove a bound on the total number of triangles, and then apply Lemma 2.

5.1 Algorithm for [US : AS : GM]

Lemma 8. *Let $\hat{\mathcal{T}}$ be defined by an instance of* [US : AS : GM]. *Then* $|\hat{\mathcal{T}}| \le d^2n$.

Proof. We give a proof for US × AS = GM; the remaining cases can be proved similarly. We can bound the number of triangles of the form $\{i, j, k\}$ with $i \in I$, $j \in J$, and $k \in K$ as follows: there are at most dn edges of the form $\{j, k\}$, since $B \in$ AS. For each such edge $\{j, k\}$ there are at most d edges of the form $\{i, j\}$, since $A \in$ US. It follows that the number of triangles is at most d^2n.

Theorem 3. *In the supported low-bandwidth model, sparse matrix multiplication of the form* [US : AS : GM] *can be computed in* $O(d^2 + \log n)$ *rounds over semirings and fields.*

Proof. Follows from Lemma 8 and Lemma 2 by setting $\kappa = d^2$ and $m = n$.

5.2 Algorithm for [BD : AS : AS]

When we consider [BD : AS : AS], it will be convenient to decompose BD into RS and CS.

Lemma 9. *Let* $\hat{\mathcal{T}}$ *be defined by an instance of* $[\mathsf{RS} : \mathsf{AS} : \mathsf{AS}]$. *Then* $|\hat{\mathcal{T}}| \leq d^2 n$.

Proof. Let us consider the case $\mathsf{RS}\times\mathsf{AS} = \mathsf{AS}$; the other permutations are similar. We can bound the number of triangles of the form $\{i, j, k\}$ with $i \in I$, $j \in J$, and $k \in K$ as follows: there are at most dn edges of the form $\{i, k\}$, since $X \in \mathsf{AS}$. For each such edge $\{i, k\}$ there are at most d edges of the form $\{i, j\}$, since $A \in \mathsf{RS}$. It follows that the number of triangles is at most $d^2 n$.

Lemma 10. *Let* $\hat{\mathcal{T}}$ *be defined by an instance of* $[\mathsf{CS} : \mathsf{AS} : \mathsf{AS}]$. *Then* $|\hat{\mathcal{T}}| \leq d^2 n$.

Proof. Transpose the matrices in Lemma 9.

Lemma 11. *Let* $\hat{\mathcal{T}}$ *be defined by an instance of* $[\mathsf{BD} : \mathsf{AS} : \mathsf{AS}]$. *Then* $|\hat{\mathcal{T}}| \leq 2d^2 n$.

Proof. Let us consider the case $\mathsf{BD} \times \mathsf{AS} = \mathsf{AS}$; the other permutations are similar. As we discussed in Sect. 1, we can decompose $X = AB$ into $X = A_1 B + A_2 B$, where $A_1 \in \mathsf{RS}$ and $A_2 \in \mathsf{CS}$. The claim follows by applying Lemma 9 to $X_1 = A_1 B$ and Lemma 10 to $X_2 = A_2 B$.

Theorem 4. *In the supported low-bandwidth model, sparse matrix multiplication of the form* $[\mathsf{BD} : \mathsf{AS} : \mathsf{AS}]$ *can be computed in* $O(d^2 + \log n)$ *rounds over semirings and fields.*

Proof. Follows from Lemma 11 and Lemma 2 by setting $\kappa = 2d^2$ and $m = n$.

6 Lower Bounds

In our full version [14], we show lower bounds that are summarized as follows:

- We present lower bounds that build on the hardness of broadcasting and aggregation. For example, an extreme case of $\mathsf{BD} \times \mathsf{BD} = \mathsf{US}$ is the multiplication of one dense row with one dense column. This enables us to compute arbitrary dot products, and in particular arbitrary sums. Hence matrix multiplication is at least as hard as computing the sum of n values. Similar arguments show that matrix multiplication is at least as hard as broadcasting one value to n computers. We show that both of these tasks require $\Omega(\log n)$ rounds.
- We present lower bounds that arise from communication complexity arguments. The key observation is that sparse matrix multiplication can be used to transmit information between two parties, Alice holding a subset of computers (and hence is able to manipulate their input) while Bob holds the rest of the computers (and hence is able to see their output). If there is a fast algorithm for matrix multiplication, Alice and Bob can simulate it to exchange information with each others too efficiently, leading to a contradiction.

- We present lower bounds that connect sparse matrix multiplication with dense matrix multiplication. If we have an algorithm for computing $\mathsf{AS} \times \mathsf{AS} = \mathsf{AS}$, we can pack a small dense matrix into one corner, and this way solve an arbitrary dense matrix multiplication. Particular care is needed here to account for the dual role that n plays in our setting: it is both the matrix dimension and the number of computers.
- Similarly, we show a lower bound using communication complexity argument. However, this time it connects matrix multiplication to the task of computing dot products. The key observation is that no matter how the input and output responsibility are distributed, there are separate computers (Alice and Bob) tasked to compute the dot product of long vectors.

Here it is good to note that the lower bounds in first group above are fundamentally related to the *number* of messages (they hold even if the computers can send in each round one message that is arbitrarily large), while the lower bounds in the second group are fundamentally related to the *total size* of messages.

Acknowledgments. We are grateful to Juho Hirvonen and Jara Uitto for several fruitful discussions, and to the anonymous reviewers for their helpful feedback on prior version of this work. This work was supported in part by the Research Council of Finland, Grants 321901 (Studený), and 333837 (Studený and Vahidi).

Disclosure of Interests. The authors have no competing interests to declare that are relevant to the content of this article.

References

1. Augustine, J., et al.: Distributed computation in node-capacitated networks. In: SPAA 2019, Phoenix, AZ, USA, pp. 69–79. ACM (2019). https://doi.org/10.1145/3323165.3323195
2. Censor-Hillel, K., Dory, M., Korhonen, J.H., Leitersdorf, D.: Fast approximate shortest paths in the congested clique. Distrib. Comput. **34**(6), 463–487 (2021). https://doi.org/10.1007/S00446-020-00380-5
3. Censor-Hillel, K., Kaski, P., Korhonen, J.H., Lenzen, C., Paz, A., Suomela, J.: Algebraic methods in the congested clique. Distrib. Comput. **32**(6), 461–478 (2019). https://doi.org/10.1007/S00446-016-0270-2
4. Censor-Hillel, K., Le Gall, F., Leitersdorf, D.: On distributed listing of cliques. In: PODC 2020, Italy, pp. 474–482. ACM (2020). https://doi.org/10.1145/3382734.3405742
5. Censor-Hillel, K., Leitersdorf, D., Turner, E.: Sparse matrix multiplication and triangle listing in the congested clique model. In: OPODIS 2018, Hong Kong, China. LIPIcs, vol. 125, pp. 4:1–4:17 (2018). https://doi.org/10.4230/LIPICS.OPODIS.2018.4
6. Chang, Y.J., Pettie, S., Saranurak, T., Zhang, H.: Near-optimal distributed triangle enumeration via expander decompositions. J. ACM **68**(3), 21:1–21:36 (2021). https://doi.org/10.1145/3446330
7. Chang, Y.J., Saranurak, T.: Improved distributed expander decomposition and nearly optimal triangle enumeration. In: Proceedings PODC 2019, Toronto, ON, Canada, pp. 66–73. ACM (2019). https://doi.org/10.1145/3293611.3331618

8. Cook, S.A., Dwork, C., Reischuk, R.: Upper and lower time bounds for parallel random access machines without simultaneous writes. SIAM J. Comput. **15**(1), 87–97 (1986). https://doi.org/10.1137/0215006
9. Dietzfelbinger, M., Kutylowski, M., Reischuk, R.: Exact lower time bounds for computing Boolean functions on CREW PRAMs. J. Comput. Syst. Sci. **48**(2), 231–254 (1994). https://doi.org/10.1016/S0022-0000(05)80003-0
10. Dolev, D., Lenzen, C., Peled, S.: "Tri, Tri Again": finding triangles and small subgraphs in a distributed setting. In: Aguilera, M.K. (ed.) DISC 2012. LNCS, vol. 7611, pp. 195–209. Springer, Heidelberg (2012). https://doi.org/10.1007/978-3-642-33651-5_14
11. Foerster, K.T., Hirvonen, J., Schmid, S., Suomela, J.: On the power of preprocessing in decentralized network optimization. In: IEEE Conference on Computer Communications, INFOCOM 2019, Paris, France, pp. 1450–1458. IEEE (2019). https://doi.org/10.1109/INFOCOM.2019.8737382
12. Foerster, K.T., Korhonen, J.H., Rybicki, J., Schmid, S.: Does preprocessing help under congestion? In: Proceedings PODC 2019, Toronto, ON, Canada, pp. 259–261. ACM (2019). https://doi.org/10.1145/3293611.3331581
13. Gupta, C., Hirvonen, J., Korhonen, J.H., Studený, J., Suomela, J.: Sparse matrix multiplication in the low-bandwidth model. In: SPAA 2022 Philadelphia, PA, USA, pp. 435–444. ACM (2022). https://doi.org/10.1145/3490148.3538575
14. Gupta, C., Korhonen, J.H., Studený, J., Suomela, J., Vahidi, H.: Low-bandwidth matrix multiplication: faster algorithms and more general forms of sparsity (2024). https://arxiv.org/abs/2404.15559
15. Izumi, T., Le Gall, F.: Triangle finding and listing in CONGEST networks. In: Proceedings of PODC 2017, Washington, DC, USA, pp. 381–389. ACM (2017). https://doi.org/10.1145/3087801.3087811
16. Korhonen, J.H., Rybicki, J.: Deterministic subgraph detection in broadcast CONGEST. In: OPODIS 2017, Lisbon, Portugal. LIPIcs, vol. 95, pp. 4:1–4:16. Schloss Dagstuhl - Leibniz-Zentrum für Informatik (2017). https://doi.org/10.4230/LIPICS.OPODIS.2017.4
17. Le Gall, F.: Further algebraic algorithms in the congested clique model and applications to graph-theoretic problems. In: Gavoille, C., Ilcinkas, D. (eds.) DISC 2016. LNCS, vol. 9888, pp. 57–70. Springer, Heidelberg (2016). https://doi.org/10.1007/978-3-662-53426-7_5
18. Lotker, Z., Pavlov, E., Patt-Shamir, B., Peleg, D.: MST construction in o(log log n) communication rounds. In: SPAA 2003, San Diego, California, USA, pp. 94–100. ACM (2003). https://doi.org/10.1145/777412.777428
19. Pandurangan, G., Robinson, P., Scquizzato, M.: On the distributed complexity of large-scale graph computations. ACM Trans. Parallel Comput. **8**(2), 7:1–7:28 (2021). https://doi.org/10.1145/3460900
20. Roughgarden, T., Vassilvitskii, S., Wang, J.R.: Shuffles and circuits (on lower bounds for modern parallel computation). J. ACM **65**(6), 41:1–41:24 (2018). https://doi.org/10.1145/3232536
21. Schmid, S., Suomela, J.: Exploiting locality in distributed SDN control. In: Proceedings of SIGCOMM HotSDN 2013, pp. 121–126. ACM (2013). https://doi.org/10.1145/2491185.2491198
22. Valiant, L.G.: A bridging model for parallel computation. Commun. ACM **33**(8), 103–111 (1990). https://doi.org/10.1145/79173.79181
23. Williams, V.V., Xu, Y., Xu, Z., Zhou, R.: New bounds for matrix multiplication: from alpha to omega. In: Proceedings of SODA 2024, Alexandria, VA, USA, pp. 3792–3835. SIAM (2024). https://doi.org/10.1137/1.9781611977912.134

Lower Bounds for Uniform Read-Once Threshold Formulae in the Randomized Decision Tree Model

Nikos Leonardos(✉)

School of Electrical and Computer Engineering, National Technical University of Athens, Athens, Greece
nikos.leonardos@gmail.com

Abstract. We investigate the randomized decision tree complexity of a specific class of read-once threshold formulae. A read-once threshold formula can be defined by a rooted tree, every internal node of which is labeled by a threshold function T_k^n (with output 1 only when at least k out of n input bits are 1) and each leaf by a distinct variable. Such a tree defines a Boolean function in a natural way. We focus on the randomized decision tree complexity of such functions, when the underlying tree is a uniform tree with all its internal nodes labeled by the same threshold function and all leaves at the same level. We prove lower bounds of the form $c(k, n)^d$, where d is the depth of the tree. We also treat trees with alternating levels of AND and OR gates separately and show asymptotically optimal bounds, extending the known bounds for the binary case.

Keywords: Decision trees · read-once threshold formulae

1 Introduction

Boolean decision trees constitute one of the simplest computational models. It is therefore intriguing when the complexity of a function is still unknown. A notable example is the recursive majority-of-three function. This function can be represented by a uniform ternary tree of depth d, such that every internal node has three children and all leaves are on the same level. The function computed by interpreting the tree as a circuit with internal nodes labeled by majority gates (with output 1 only when at least two of the three inputs are 1) is maj_d, the recursive majority-of-three of depth d.

This function seems to have been given by Ravi Boppana (see Example 1.2 in [15]) as an example of a function that has deterministic complexity 3^d, while

This research is co-financed by Greece and the European Union (European Social Fund—ESF) through the Operational Programme "Human Resources Development, Education and Lifelong Learning" in the context of the project "Reinforcement of Postdoctoral Researchers—2nd Cycle" (MIS-5033021), implemented by the State Scholarships Foundation (IKY).

U. Schmid and R. Kuznets (Eds.): SIROCCO 2025, LNCS 15671, pp. 350–365, 2025.
https://doi.org/10.1007/978-3-031-91736-3_21

its randomized complexity is asymptotically smaller. Other functions with this property are known. Another notable example is the function nand_d, first analyzed by Snir [17]. This is the function represented by a uniform binary tree of depth d, with the internal nodes labeled by nand gates. Equivalently, the internal nodes can be labeled by AND and OR gates, alternately at each level.

A simple randomized framework that can be used to compute both maj_d and nand_d is the following. Start at the root and, as long as the output is not known, choose a child at random and evaluate it recursively. Algorithms of this type are called *directional.* For maj_d the directional algorithm computes the output in $(8/3)^d$ queries. It was noted by Boppana and also in [15] that better algorithms exist for maj_d. (See [1] for a more recent study of directional algorithms.) Interestingly, Saks and Wigderson show that the directional algorithm is optimal for the nand_d function, and show that its zero-error randomized decision tree complexity is $\Theta\big((\frac{1+\sqrt{33}}{4})^d\big)$. Their proof uses a bottom up induction and generalized costs. Their method of generalized costs allows them to charge for a query according to the value of the variable. In this work we use this method to show that the directional algorithm is optimal for uniform AND-OR trees where each gate receives n bits.

More generally, there have been studied functions that can be represented by formulae involving threshold functions as connectives. A threshold k-out-of-n function, denoted T_k^n, is a Boolean function of n arguments that has value 1 if at least k of the n Boolean input values are 1. A threshold formula can be defined as a rooted tree with labeled nodes; each internal node is labeled by a threshold function and each leaf by a variable. If each variable appears exactly once the formula is called *read-once*; if only AND and OR connectives appear, it is called a *Boolean* read-once formula. A formula represents a Boolean function in a natural way: the tree is evaluated as a circuit where each internal node is the corresponding threshold gate. If the formula is read-once, the function it represents is also called read-once. If no OR gate is an input to another OR gate, and the same for AND gates, then the formula is non-degenerate (see Theorem 2.2 in [6]) and uniquely represents the corresponding function. Thus, we may define the depth of f as the maximum depth of a leaf in the unique tree-representation. Note also that, for read-once formulae, the set of variables coincides with the set of leaves in their representation as a tree. We prove lower bounds for the subclass that is represented by uniform trees (full and complete trees with all their leaves at the same level and with each internal node having the same number of children).

Related Work and Our Results. Heiman and Wigderson [7] managed to show that for every Boolean read-once function f we have $R(f) \in \Omega(D(f)^{0.51})$, where $R(f)$ and $D(f)$ are the randomized and deterministic complexity of f respectively. For the subclass of such functions that their representing tree is binary, the question was investigated more recently in [2]. Heiman, Newman, and Wigderson [6] showed that read-once formulae with threshold gates have zero-error randomized complexity $\Omega(N/2^d)$ (here N is the number of variables and d the depth of a canonical tree-representation of the read-once function).

Santha showed in [16] that when considering a notion of balanced read-once formulae with AND and OR gates, the zero-error and bounded-error complexities are intimately related. This holds, in particular, for the class of uniform formulae that we consider. Furthermore, it is not hard to verify that our lower bounds for uniform threshold read-once formulae also hold for bounded-error algorithms if multiplied by $(1 - 2\delta)$, where δ is the error allowed. Consequently, we focus on zero-error complexity in this work.

With respect to specific read-once functions, in contrast to the exact asymptotic bounds we have for nand_d (also in the quantum model [3]) there had been no progress on the randomized decision tree complexity of maj_d for several years. However, recent papers have narrowed the gap between the upper and lower bounds for recursive majority. An $\Omega\big((7/3)^d\big)$ lower bound was showed in [9]. Jayram, Kumar, and Sivakumar, proved their bound using tools from information theory and a top down induction. Furthermore, they presented a non-directional algorithm that improves the $O\big((8/3)^d\big)$ upper bound. Landau, Nachmias, Peres, and Vanniasegaram [10], showed how to remove the information theoretic notions from the proof in [9], keeping its underlying structure the same. Magniez, Nayak, Santha, and Xiao [12], significantly improved the lower bound to $\Omega\big((5/2)^d\big)$ and the upper bound to $O(2.64946^d)$. Subsequently, in [11], the lower bound was further improved to $\Omega(2.55^d)$, building upon the techniques of [15]. The bound was further improved with a computer-assisted proof in [13]. The currently known best lower bound is $\Omega(2.59^d)$ from [5].

The above methods work by reducing the complexity of maj_d to a problem of constant size. It is interesting to note that when this constant-size problem relates to maj_1 (i.e., the majority of three bits), all these methods obtain the 2.5^d lower bound. Furthermore, in order to make further progress, analysis of analogous problems of greater depth is required. In particular, in [11], an intermediate problem between maj_1 and maj_2 is constructed and analyzed with a reasonable case analysis (without the need of computer assistance) to obtain a 2.55^d lower bound; in [13], the 2.55^d bound is surpassed by solving depth-3 and depth-4 problems using a computer; finally, [5] obtains the currently best bound via the solution of a linear program (obtained using a computer) which relates to maj_3. Pushing any of the above further seems computationally intractable and will yield possibly modest improvements. We believe a human-readable solution can shed more light into the problem.

In this work we focus our attention to threshold read-once functions which are represented by uniform trees and each internal node is labeled by the same threshold gate. Let us denote by $F^d_{\wedge,n}$ (resp. $F^d_{\vee,n}$) the function represented by a uniform tree of depth d with AND and OR gates of fan-in n alternating with each level and the root labeled by an AND gate (resp. an OR gate). Denote by $F^d_{k,n}$ the function represented by a uniform tree of depth d with each gate being T^n_k. With respect to these classes of read-once functions we prove the following theorems in the model of randomized decision trees. (See Sect. 2 for relevant definitions).

Theorem 1. *The directional algorithm is optimal for* $F^d_{\wedge,n}$ *and* $F^d_{\vee,n}$*. In particular, for even d,*

$$R(F^d_{\wedge,n}), R(F^d_{\vee,n}) = \Theta\left(\left[\frac{n-1}{4}\left(1+\sqrt{1+\frac{16n}{(n-1)^2}}\right)\right]^d\right).$$

Theorem 2. *For* $1 < k < n$*,*

$$R(F^d_{k,n}) = O\left(\left[\frac{n+1}{2}\left(1+\sqrt{1-\frac{8k(n-k+1)}{(n-k+2)(k+1)(n+1)}}\right)\right]^d\right).$$

Theorem 3. *For* $1 < k < n$*,*

$$R(F^d_{k,n}) = \Omega\left(\left[\frac{n+1}{2}\left(1+\sqrt{1-\frac{3k(n-k+1)+n}{(n+1)^2}}\right)\right]^d\right).$$

Note that the last two expressions are symmetric in k around $(n+1)/2$, reflecting the fact that $R(F^d_{k,n}) = R(F^d_{n+1-k,n})$ (as can be seen by negating and pushing the negations down to the leaves). Setting $n = 2$ in Theorem 1 we retrieve the bound of Saks and Wigderson for nand_d. Similarly, for $k = 2$ and $n = 3$, Theorems 2 and 3 give known bounds for maj_d: the $(8/3)^d$ upper bound of the directional algorithm and the 2.5^d lower bound respectively. We expect neither the upper bound nor the lower bound to be optimal. Our interest is mostly in Theorem 3 and also Theorem 1. Theorem 2 is obtained by a straightforward analysis of the directional algorithm and is stated mostly for reference and comparison with the bound of Theorem 3.

We note that one could obtain similar bounds using the methodology of [9,13] or [5]. In particular, formal descriptions of feasible solutions to the corresponding linear programs defined in [5], can be used along with their Composition Theorem to obtain lower bounds for the class of functions we consider.

2 Definitions and Notation

In this section we introduce basic concepts related to decision tree complexity and the tree-functions that we consider. The reader can find a more complete exposition of decision tree complexity in the survey of Buhrman and de Wolf [4].

2.1 Definitions Pertaining to Trees

For a rooted tree T, the *depth* of a node is the number of edges on the path to the root; the *depth* of the tree is the maximum depth of a leaf; the *level* i of a tree consists of all nodes of depth i; the *height* of a node is the number of edges on the longest path between the node and any descendant leaf. We call a tree *uniform* if all the leaves are on the same level and all internal nodes have the same number of children.

Consider the uniform tree of depth d in which every internal node has n children and is labeled by T_k^n (the k-out-of-n threshold gate) with $1 < k < n$. We denote both the tree and the corresponding read-once threshold formula by the same symbol $F_{k,n}^d$ as it will be clear from the context what it refers to.

For the case of AND or OR gates, we consider uniform trees with the root labeled by an AND (resp. OR) gate and subsequent levels of the tree are labeled by OR (resp. AND) and AND (resp. OR) gates alternately. When each gate receives n inputs, we denote these trees by $F_{\wedge,n}^d$ and $F_{\vee,n}^d$, when the root is labeled by an AND and an OR respectively.

The inputs considered hard for the functions we consider are the *reluctant* inputs [15]. Call an input to a T_k^n-gate *reluctant*, if it has exactly k or $k-1$ ones. Call an input to a threshold read-once formula *reluctant*, if it is such that the input to every gate is reluctant. A *reluctant distribution* has only reluctant inputs in its support.

We will consider trees that have all leaves in the two greatest levels. We now define the reluctant distribution over such trees that we will work with. We focus on *positive inputs* only; i.e., inputs that evaluate to true. The definition is recursive. If the tree consists of a single variable, then all the mass in on 1. Otherwise, consider a tree T of depth d with all leaves in levels $d-1$ and d (level $d-1$ might not have any leaves if T is uniform). Obtain a tree T' by removing all the children of an internal node u with label T_k^n in level $d-1$. Let μ' denote the reluctant distribution for T'. Consider any reluctant input x to T and let b be the value that u (as a gate) outputs. Let x' be the corresponding (reluctant) input to T'; i.e., x' assigns b to u and agrees with x everywhere else. Define $\mu(x) = \mu'(x')/\binom{n}{k}$ if $b = 1$ and $\mu(x) = \mu'(x')/\binom{n}{k-1}$ if $b = 0$. The reluctant distribution for $F_{k,n}^d$, $F_{\wedge,n}^d$, and $F_{\vee,n}^d$, is denoted $\mu_{k,n}^d$, $\mu_{\wedge,n}^d$, and $\mu_{\vee,n}^d$, respectively. Note that although we will restrict the analysis over positive inputs, the algorithms are still required to be zero-error over any input.

2.2 Definitions Pertaining to Decision Trees

A *deterministic Boolean decision tree* Q over $\{0,1\}^n$ is a rooted and ordered binary tree. Each internal node is labeled by $i \in [n] = \{1, 2, \ldots, n\}$ and each leaf by a value from $\{0,1\}$. An *assignment* or an *input* is a member of $\{0,1\}^n$. The output $Q(x)$ of Q on an input x is defined recursively as follows. Start at the root and let its label be i. If $x_i = 0$, we continue with the left child of the root; if $x_i = 1$, we continue with the right child of the root. We continue recursively until we reach a leaf. We define $Q(x)$ to be the label of that leaf. When we reach an internal node, we say that Q *queries* the corresponding variable and *reads* its value. We say that Q *computes* a Boolean function $f : \{0,1\}^n \to \{0,1\}$, if for all $x \in \{0,1\}^n$, $Q(x) = f(x)$. The *cost of* Q *on input* x, $\mathrm{cost}(Q; x)$, is the number of variables queried when the input is x. The *cost of* Q, $\mathrm{cost}(Q)$, is its *depth*, the maximum distance of a leaf from the root. The *deterministic complexity*, $D(f)$, of a Boolean function f is the minimum cost over all Boolean decision trees that compute f.

A *randomized Boolean decision tree* Q_{R} is a distribution p over deterministic decision trees. On input x, a deterministic decision tree is chosen according to p and evaluated. The *cost of* Q_{R} *on input* x is $\mathrm{cost}(Q_{\mathrm{R}}; x) = \sum_Q p(Q)\mathrm{cost}(Q; x)$. The *cost* of Q_{R}, $\mathrm{cost}(Q_{\mathrm{R}})$, is $\max_x \mathrm{cost}(Q_{\mathrm{R}}; x)$. A randomized decision tree Q_{R} *computes* a Boolean function f (with zero error), if $p(Q) > 0$ only when Q computes f. The *randomized complexity*, $R(f)$, of a Boolean function f is the minimum cost over all randomized Boolean decision trees that compute f.

We are going to take a distributional view on randomized algorithms. Let μ be a distribution over $\{0,1\}^n$ and Q_{R} a randomized decision tree. The *expected cost of* Q_{R} *under* μ is $\mathrm{cost}_\mu(Q_{\mathrm{R}}) = \sum_x \mu(x)\mathrm{cost}(Q_{\mathrm{R}}; x)$. The *expected complexity under* μ, $R_\mu(f)$, of a Boolean function f is the minimum expected cost under μ of any randomized Boolean decision tree that computes f. Clearly, $R(f) \geq R_\mu(f)$, for any μ, and thus we can prove lower bounds on randomized complexity by providing lower bounds for the expected cost under any chosen distribution.

3 The Method of Generalized Costs for Read-Once Threshold Formulae

Our goal is to prove a lower bound on the expected cost (under the reluctant distribution we defined in the previous section) of any randomized decision tree Q_{R} that computes a uniform threshold read-once function. The high-level outline of our proof is as follows. Given any decision tree Q_{R} computing the function $F^d_{k,n}$, we define a randomized decision tree Q'_{R} that computes $F^{d-1}_{k,n}$. The crucial part is to show that we may charge Q'_{R} more for each query, while maintaining that the cost of Q'_{R} is upper bounded by that of Q_{R}. Applying this step repeatedly reduces our task to determining the complexity of a function on a single variable with a high cost to query it. To implement this plan we utilize the method of generalized costs of Saks and Wigderson [15].

The reduction from $F^d_{k,n}$ to $F^{d-1}_{k,n}$ proceeds inductively and in several steps. In each step we shrink to a leaf the parent of n leaves that are siblings. That is, we start with a function $F(x, y)$, where $x \in \{0,1\}^n$ (recall that n is the fan-in of T^k_n), and reduce its complexity to a function $F'(u, y)$, where $u \in \{0,1\}$, such that $F(x, y) = F'(T^n_k(x), y)$. Since we only need one query to discover the value of u in F', but at least k in F, we should be able to charge more an algorithm computing F' for querying u, than an algorithm computing F for querying any of the bits in x. This is captured by the introduction of cost functions, discussed next. Continuing this way, all nodes in level $d-1$ will become leaves and the uniform tree of depth d that we started with will be transformed to one of depth $d-1$.

3.1 Cost Functions

Define a *cost-function* over $\{0,1\}^N$ to be a function $\phi : \{0,1\}^N \times [N] \to \mathbb{R}$. We extend the previous cost-related definitions as follows. The cost of a decision tree

Q under cost-function ϕ on input $x \in \{0,1\}^N$ is

$$\text{cost}(Q;\phi;x) = \sum\nolimits_{i \in S} \phi(x;i), \quad \text{where } S = \{i \,|\, i \text{ is queried by } Q \text{ on input } x\}.$$

The cost of a randomized decision tree Q_R on input x under cost-function ϕ is

$$\text{cost}(Q_\text{R};\phi;x) = \sum\nolimits_Q p(Q)\text{cost}(Q;\phi;x),$$

where p is the corresponding distribution over deterministic decision trees. Finally, the expected cost of a randomized decision tree Q_R under cost-function ϕ and distribution μ is

$$\text{cost}_\mu(Q_\text{R};\phi) = \sum\nolimits_x \mu(x)\text{cost}(Q_\text{R};\phi;x).$$

We will mostly work with simple cost functions (as was also the case in [15]), in which $\phi(x,i)$ depends only on x_i and i. In particular, we may express such a function with a pair $c = (c_0, c_1)$, where c_0, c_1 are functions mapping each variable to a real number and $c_{x_i}(i) = \phi(x,i)$. We call such cost functions *local.* Defining $\text{cost}(Q;c;x) = \sum_{i \in S} c_{x_i}(i)$, for a local cost-function c, we use both expressions interchangeably in the above quantities.

3.2 The Method and Some Preliminary Definitions

Consider a tree T of depth $d+1$ such that all internal nodes have degree n and all leaves are on levels d and $d+1$. If we treat every internal node as a T^n_k-gate, this tree represents a function F and has an associated reluctant distribution μ. Suppose Q is a randomized decision tree that computes F and $c = (c_0, c_1)$ a local cost-function. We define a process that shrinks T to a smaller tree T' (of depth d or $d+1$) and also a corresponding randomized decision tree Q' that computes the function F' represented by T'. The crucial part is to show that for a "more expensive" local cost-function c' and the reluctant distribution μ' for F',

$$\text{cost}_\mu(Q;c) \geq \text{cost}_{\mu'}(Q';c'). \tag{1}$$

The main ingredient in this framework is the shrinking process. It entails removing n leaves, which we identify with $[n]$, so that their parent u would become a leaf in T'. Given an algorithm Q for F we obtain an algorithm Q' for F' as follows. On input (x_u, y), Q' first chooses uniformly at random $z \in \binom{[n-1]}{k-1}$ and $i \in [n]$ and then simulates Q on $x = (z_1, \ldots, z_{i-1}, x_u, z_i, \ldots, z_{n-1}, y)$. Clearly, if Q computes F, then Q' is a randomized decision tree that computes F'.

Our goal is to determine the "most expensive" cost function c' for T' for which we can argue (1). To that end, it will be useful to express $\text{cost}_{\mu'}(Q';c')$ in terms of Q and μ. For an input x for F and a leaf i of T, let $b = T^n_k(x_1, \ldots, x_n)$ and define a cost function ψ for F as follows.

$$\psi(x;i) = \begin{cases} c'_{x_i}(i), & \text{if } i \notin [n]; \\ c'_1(u)/k, & \text{if } i \in [n] \text{ and } x_i = b = 1; \\ c'_0(u)/(n-k+1), & \text{if } i \in [n] \text{ and } x_i = b = 0; \\ 0, & \text{if } i \in [n] \text{ and } x_i \neq b. \end{cases} \tag{2}$$

Informally, ψ agrees with c' outside $[n]$, is zero when Q' will never query i, and in the rest of the cases has value so that the following proposition holds.

Proposition 1. $\mathrm{cost}_{\mu'}(Q';c') = \mathrm{cost}_{\mu}(Q;\psi)$.

Proof. It suffices to prove the equality for a deterministic Q. The expected cost of any variable outside $[n]$ is clearly the same for both algorithms. Thus, it suffices to show that the contribution of u to the cost of Q' equals that of the variables in $[n]$ to the cost of Q. Intuitively, a query of Q to a variable $i \in [n]$ such that $x_i = b$, leads to a query of Q' to u with probability $1/k$ or $1/(n-k+1)$, according to $b = 1$ or $b = 0$ respectively. More formally, let $\chi_i(x)$ be the indicator function of the event that Q queries i when the input is x. We consider the inputs with $x_u = 1$.

Recalling the recursive distribution of the reluctant distribution μ, the contribution of the leaves in $[n]$ to the cost of Q can be written

$$\sum_y \sum_{z,i} \frac{\mu'(x_u, y)}{\binom{n}{k}} \cdot \chi_i(x) \cdot \frac{c'_1(u)}{k}.$$

In the expression above, $x = (z_1, \ldots, z_{i-1}, x_u, z_i, \ldots, z_{n-1}, y)$, as in the description of Q'. Note that although each x is encountered k times, each $i \in [n]$ such that $z_i = 1$ contributes exactly once (those with $z_i = 0$ contribute 0 anyway). The contribution of u to the cost of Q' is

$$\sum_y \mu(x_u, y) \sum_{z,i} \frac{1}{n\binom{n-1}{k-1}} \cdot \chi_i(x) \cdot c'_1(u),$$

since there are n choices for i and $\binom{n-1}{k-1}$ choices for z. Recalling $\binom{n}{k} = \frac{n}{k}\binom{n-1}{k-1}$, the equality of the two expressions follows.

The inputs with $x_u = 0$ can be analysed in the same way, substituting c'_0 for c'_1 in both expressions and $n - k + 1$ for k in the first one.

Let us go back to our plan: given a local cost-function c for F, determine the most expensive local cost-function c' for F' for which we can argue (1). We will work with local cost-functions that charge the same value for discovering a 0 and the same for discovering a 1 at a given level. That is, if i and j are leaves at the same level, then $c_0(i) = c_0(j)$ and $c_1(i) = c_1(j)$.

We shall define c' so that it agrees with c for all $i \notin [n]$. Defining a cost-function $\phi(x;i) = c_{x_i}(i) - \psi(x;i)$ for $i \in [n]$ and $\phi(x;i) = 0$ otherwise (where ψ is the one obtained from Proposition 1), we have

$$\mathrm{cost}_{\mu}(Q;c) \geq \mathrm{cost}_{\mu'}(Q';c') \iff \mathrm{cost}_{\mu}(Q;\phi) \geq 0. \tag{3}$$

Clearly, we only need to verify $\mathrm{cost}_{\mu}(Q;\phi) \geq 0$ with respect to the leaves in $[n]$. Thus, we are led to study decision trees over $\binom{[n]}{k}$ that are non-empty (i.e., query at least one variable). Note also that we don't care about what these algorithms output. Their only task is to achieve $\mathrm{cost}_{\mu}(Q;\phi) < 0$ for our choice of ϕ. Given a cost function c for F, what is the most expensive c' for F' so that $\mathrm{cost}_{\mu}(Q;\phi) \geq 0$ for any such decision tree?

The Quantity $P(k,n)$. Consider the T_k^n function. Looking into the definition of cost function ϕ above, note that it charges an algorithm a positive value for reading a 0 and a negative value for a 1. Furthermore, the greater the negative value is, the more expensive c' is. We are interested in the values defined below.

Definition 1. *Let* $\nu_{k,n}$ *be the uniform distribution over* $\binom{[n]}{k}$. *For* $\eta \in \mathbb{R}_{\geq}$ *and* $0 < k \leq n$, *define a local cost-function* $c_\eta = (1, -\eta)$,

$$P_\eta(k,n) = \min_Q\{\text{cost}_{\nu_{k,n}}(Q; c_\eta))\} \quad \textit{and} \quad P(k,n) = \max\{\eta : P_\eta(k,n) \geq 0\},$$

where Q *ranges over all non-empty decision trees.*

The relevance of this definition is demonstrated by Proposition 2 below.

Remark 1. We note that, for any given algorithm Q, its cost is a linear function of η: $\text{cost}_{\nu_{k,n}}(Q; c_\eta) = A - \eta B$, where A is the expected number of 0s and B the expected number of 1s that Q reads under distribution $\nu_{k,n}$.

It follows that $P_\eta(k,n)$ is continuous as a function of η and so for $\eta = P(k,n)$ we must have $P_\eta(k,n) = 0$. Furthermore, since we consider non-empty algorithms only, $P_\eta(k,n)$ is decreasing as a function of η.

We call an algorithm Q with $\text{cost}_{\nu_{k,n}}(Q; c_\eta) = P_\eta(k,n)$, an *optimal algorithm for* $P_\eta(k,n)$; if $\eta = P(k,n)$, an *optimal algorithm for* $P(k,n)$.

The normalized local cost-function c_η charges an algorithm 1 for reading a 0 and pays the algorithm η when it discovers a 1. What is the greatest value of η for which $\text{cost}_{\nu_{k,n}}(Q; c_\eta) \geq 0$ for any Q?

We now discuss how lower bounds for $P(k,n)$ and $P(n-k+1,n)$ lead to lower bounds for threshold read-once functions. Suppose $c_0(i) = c_0$ and $c_1(i) = c_1$, for all $i \in [n]$, and write $c'_1 = c'_1(u)$ and $c'_0 = c'_0(u)$. If $\alpha \leq P(k,n)$ and $\beta \leq P(n-k+1,n)$, then we may set

$$\begin{pmatrix} c'_1 \\ c'_0 \end{pmatrix} = \begin{pmatrix} k & \alpha k \\ \beta(n-k+1) & n-k+1 \end{pmatrix} \begin{pmatrix} c_1 \\ c_0 \end{pmatrix}. \tag{4}$$

Proposition 2. $\text{cost}_\mu(Q; c) \geq \text{cost}_{\mu'}(Q'; c')$.

Proof. Recall the definition of cost function ϕ and the related expression (3). Note that ϕ charges 0 for $i \notin [n]$. For $i \in [n]$, ϕ charges c_0 for reading a 0 and $-\alpha c_0$ for reading a 1 over the support of $\nu_{k,n}$ and, respectively, $-\beta c_1$ and c_1 over the support of $\nu_{k-1,n}$. An assignment on the variables outside $[n]$, determines an algorithm over $\{0,1\}^n$. The conditional distribution of the variables in $[n]$, given such an assignment and a value for u, is either $\nu_{k,n}$ or $\nu_{k-1,n}$. It follows that $\text{cost}_\mu(Q; \phi)$ is lower-bounded by a convex combination of $c_0 P_\alpha(k,n) \geq 0$ and $c_1 P_\beta(n-k+1,n) \geq 0$.

Applying this process repeatedly, shrinking all sibling leaves to their parent, we reduce a tree of depth d to one of depth $d-1$. After d such steps and starting with a local cost-function that charges c_1 and c_0 for reading 1 and 0 respectively, we are left with a single node and the cost function defined by

$$\begin{pmatrix} c_1' \\ c_0' \end{pmatrix} = \Gamma_{k,n}^d \begin{pmatrix} c_1 \\ c_0 \end{pmatrix}, \quad \text{where } \Gamma_{k,n} = \begin{pmatrix} k & \alpha k \\ \beta(n-k+1) & n-k+1 \end{pmatrix}. \tag{5}$$

We will obtain a lower bound in the order of λ^d, where λ is the largest eigenvalue. Denoting the trace of $\Gamma_{k,n}$ by T and its determinant by D,

$$\lambda = \frac{T}{2} + \sqrt{\frac{T^2}{4} - D}, \quad \text{where } T = n+1 \text{ and } D = (1-\alpha\beta)k(n-k+1). \tag{6}$$

To accomplish this, we will set $c_1 = 1$ and c_0 so that $\binom{c_1}{c_0}$ is an eigenvector associated with λ. The matrices considered are positive and it can be shown using a theorem of Perron (see Lemma 8.2.1 in [8]) that c_0 is positive. As long as c_0 is also $\Theta(1)$, our asymptotic bounds are not affected. We note for reference that for a matrix $\left(\begin{smallmatrix} a & b \\ c & d \end{smallmatrix}\right)$ with $b \neq 0$, $\left(\begin{smallmatrix} 1 \\ (\lambda-a)/b \end{smallmatrix}\right)$ is an eigenvector associated with its greatest eigenvalue λ.

3.3 Properties and Estimations of $P(k,n)$

Given (6), we seek a lower bound of $P(k,n)$. We start with some monotonicity properties of $P(k,n)$. Note that, by symmetry, we may consider algorithms that read the variables in the order $x_1, x_2, \ldots, x_n$.

Intuitively, if we increase n, we should be able to increase η as it becomes harder for an algorithm to discover a 1. This is captured in the next statement.

Theorem 4. *For any $\eta > 0$ and $0 < k \leq n$, $P_\eta(k,n) < P_\eta(k,n+1)$.*

Proof. Consider any algorithm Q over $\{0,1\}^{n+1}$ and define a randomized algorithm Q' over $\{0,1\}^n$ as follows. Algorithm Q' on input $(x_1,\ldots,x_n)$, chooses $i \in [n+1]$ and simulates Q on input $(x_1,\ldots,x_{i-1},0,x_i,\ldots,x_n)$. We claim that the cost of Q' is less than the cost of Q. This is not hard to see as Q' will gain $-\eta$ whenever Q queries a 1 and also gains $-\eta$, but with positive probability (when $i=1$ for example) will pay 0 when Q queries i and pays 1.

Similarly, increasing both k and n by 1, and since $k/n < (k+1)/(n+1)$, one might expect that we would need to decrease η.

Theorem 5. *For any $\eta > 0$ and $0 < k \leq n$, $P_\eta(k,n) > P_\eta(k+1,n+1)$.*

Proof. Consider any algorithm Q over $\{0,1\}^n$ and define a randomized algorithm Q' over $\{0,1\}^{n+1}$ as follows. Algorithm Q' chooses $i \in [k+1]$ and simulates Q, but when it encounters the i-th 1 while querying a variable j, Q' continues with the rest of the variables as if x_j was skipped or missing.

More formally, for $i \in [k+1]$, let Q_i be the algorithm which results from Q as follows. Consider a path $(u_1, x_1 = b_1, u_2, x_2 = b_2, \ldots, u_j, x_j = 1, u_{j+1})$ in Q, such that $x_j = 1$ is the i-th query that returned 1 (i.e., exactly i of the bits $x_1, \ldots, x_j$ are 1 and $x_j = 1$), and let T denote the sub-tree rooted at node u_j of Q that queries x_j. Let T' be identical to T, but with every query shifted by 1; that is, every node of T labeled by $\ell \in [n]$ is replaced with $\ell + 1 \in [n+1]$. Let Q_i be the tree which results from Q by replacing the (possibly empty) sub-tree rooted at node u_{j+1} with T' and applying this transformation for every such path in Q. Call the edge $x_j = 1$ a *skipped* edge. Finally, Q' is the uniform distribution over the Q_i's.

It suffices to show that Q' has strictly smaller cost than Q. Consider a path P in Q of length m with ℓ 1's. The probability P is executed in Q is $\binom{n-m}{k-\ell}/\binom{n}{k}$. There is a natural map of a path in Q_i to a path in Q: ignore a skipped edge (if it exists) and shift back by one any shifted label.

We compute the probability of the paths in Q' that map to P by counting the number of pairs (i, x), such that the execution path of Q_i on input x is mapped on P.

$$\left[m\binom{n-m}{k-\ell} + (k+1-\ell)\binom{n+1-m}{k+1-\ell}\right] \Big/ \left[(k+1)\binom{n+1}{k+1}\right].$$

The second term of the sum is the number of such pairs without skipped edges; equivalently, with $i > \ell$ and the first m bits of x consistent with P. The first term counts the rest of the pairs (i, x), since there are m choices for the skipped 1 and this choice determines i. This probability can be verified to equal $\binom{n-m}{k-\ell}/\binom{n}{k}$.

Furthermore, since Q is non-empty, there is at least one skipped edge in Q_1 taken with positive probability, implying Q' has strictly smaller cost.

Corollary 1. *An optimal algorithm for $P(k, n)$ stops if the first query reads 1.*

Proof. Consider an optimal tree for $P(k, n)$. By Theorem 5, for $\eta = P(k, n)$, we have $P_\eta(k-1, n-1) > P_\eta(k, n) = 0$. Thus, descending further after reading a 1 will add a positive expected cost.

Using the above property of optimal algorithms, we can determine $P(1, n)$ and $P(n, n+1)$.

Lemma 1. *For all $n > 1$, $P(1, n) = (n-1)/2$.*

Proof. Theorem 4 determines the optimal decision tree Q for $P(1, n)$: it queries the variables one by one until a 1 is read. This is because, for $\eta = P(1, n)$, by the theorem $P_\eta(1, n-1) < 0$; thus, stopping after a 0 is read is not optimal. For $\eta > 0$,

$$\mathrm{cost}_{\nu_{1,n}}(Q; c_\eta) = \sum_{i=1}^{n} \frac{1}{n} \cdot (i - 1 - \eta) = \frac{n-1}{2} - \eta.$$

It follows that $\mathrm{cost}_{\nu_{1,n}}(Q; c_\eta) = 0$ if and only if $\eta = (n-1)/2$.

Lemma 2. *For all $n > 1$, $P(n, n+1) = 1/(2n)$.*

Proof. By Corollary 1, the optimal (non-empty) tree stops if $x_1 = 1$ and if $x_1 = 0$ it proceeds to discover all 1s. Its cost, for $\eta > 0$, is $-\eta n/(n+1)+(1-\eta n)/(n+1)$. This equals zero if and only if $\eta = 1/(2n)$.

We can now prove the following lower bound on $P(k, n)$.

Theorem 6. *For* $0 < k < n$, $\frac{n-k}{2k} \leq P(k, n)$.

Proof. The intuition of the inductive proof below is that if $P(k, n)$ is smaller than $(n-k)/(2k)$, then $P_{\frac{n-k}{2k}}(k, n-1)$ should be absurdly small.

Note first that for any $k > 0$ and $n = k+1$ the statement follows (with equality) from Lemma 2. Suppose—towards a contradiction—that for a lexicographically least pair (k, n) with $n > k+1$, $P(k, n) < (n-k)/(2k)$. Thus,

$$P_{\frac{n-k}{2k}}(k, n-1) < P_{\frac{n-k}{2k}}(k, n) < 0, \tag{7}$$

where the first inequality follows from Theorem 4 and the second from our assumption that $P(k, n) < (n-k)/(2k)$. Furthermore, by the minimality of (k, n),

$$P(k-1, n-1) \geq \frac{n-k}{2(k-1)} > \frac{n-k}{2k},$$

which in turn implies

$$0 < P_{\frac{n-k}{2k}}(k-1, n-1).$$

This last inequality implies that an optimal decision tree for $P_{\frac{n-k}{2k}}(k, n)$ will not descent further after reading $x_1 = 1$. We conclude that

$$P_{\frac{n-k}{2k}}(k, n) = \frac{n-k}{n} \cdot \left[1 + P_{\frac{n-k}{2k}}(k, n-1)\right] - \frac{n-k}{2k} \cdot \frac{k}{n}. \tag{8}$$

Suppose Q is an optimal algorithm for $P_{\frac{n-k}{2k}}(k, n-1)$. There are constants u and v such that, for any $\eta > 0$,

$$\text{cost}_{\nu_{k,n-1}}(Q; c_\eta) = u - \eta \cdot \text{v}.$$

Combining (7) and (8) and substituting the above expression with $\eta = (n-k)/(2k)$, we obtain after simplification

$$u - \frac{n-k}{2k} \cdot v < -\frac{1}{2}.$$

On the other hand, by the minimality of (k, n), $P(k, n-1) \geq (n-1-k)/(2k)$. This implies

$$u - \frac{n-1-k}{2k} \cdot v \geq 0.$$

The above two inequalities are satisfiable only if $v > k$. However, v is the expected number of 1s read by Q, which is at most k; a contradiction.

4 Bounds for Uniform Read-Once Functions

We begin with an analysis of the (straightforward) directional algorithm. We then show that this algorithm is optimal for uniform AND-OR trees. Finally, we obtain bounds for uniform threshold read-once functions.

4.1 Analysis of the Directional Algorithm

A simple directional algorithm evaluates a node by evaluating its children in a randomly chosen order and stops if the value is already determined. To analyze this algorithm on a reluctant input, let Φ_d denote the expected cost to evaluate a node of height d when the output is 1 and Ψ_d when it is 0. Consider a node with exactly k of its children evaluating to 1. The number of its children that will be evaluated until its value is determined by the directional algorithm follows a negative hypergeometric distribution. The expected number of children that will be evaluated to 0 is $(n-k)k/(k+1)$. (An elementary exposition of this distribution and calculation of its mean can be found in [14, Chapter 7].) Similarly, when dealing with a node with exactly $n-k+1$ of its children evaluating to 0, the expected number of children that will be evaluated to 1 is $(k-1)(n-k+1)/(n-k+2)$. Thus, along with $\Phi_0 = c_0$ and $\Psi_0 = c_1$, we obtain the recurrence

$$\begin{pmatrix}\Phi_d\\ \Psi_d\end{pmatrix} = \Delta_{k,n}\begin{pmatrix}\Phi_{d-1}\\ \Psi_{d-1}\end{pmatrix}, \quad \text{where } \Delta_{k,n} = \begin{pmatrix} k & (n-k)\cdot\frac{k}{k+1} \\ (k-1)\cdot\frac{n-k+1}{n-k+2} & n-k+1\end{pmatrix}. \tag{9}$$

4.2 Uniform Alternating AND-OR Trees

In this subsection we prove Theorem 1, by showing that the directional algorithm for the alternating AND-OR tree satisfies the same recurrence as the one we obtain with the method of generalized costs. In particular, we show that the recurrences (4) and (9) coincide for uniform alternating AND-OR trees.

Let us define the analogous quantities $\Phi_d^\wedge$ and $\Psi_d^\wedge$ (resp. $\Phi_d^\vee$, $\Psi_d^\vee$) for the expected cost when evaluating a node of height d and labeled by an AND (resp. OR) gate to 1 and 0 respectively. Recall Eq. (5) and note that $\Delta_{n,n} = \Gamma_{n,n}$ and $\Delta_{1,n} = \Gamma_{1,n}$, since $P(n,n) = 0$ and $P(1,n) = (n-1)/2$ (Lemma 1). Thus, Eq. (9) instantiates to

$$\begin{pmatrix}\Phi_d^\wedge\\ \Psi_d^\wedge\end{pmatrix} = \Gamma_{n,n}\begin{pmatrix}\Phi_{d-1}^\vee\\ \Psi_{d-1}^\vee\end{pmatrix} \quad\text{and}\quad \begin{pmatrix}\Phi_d^\vee\\ \Psi_d^\vee\end{pmatrix} = \Gamma_{1,n}\begin{pmatrix}\Phi_{d-1}^\wedge\\ \Psi_{d-1}^\wedge\end{pmatrix}. \tag{10}$$

Defining $A = \Gamma_{n,n}$ and $B = \Gamma_{1,n}$ we have

$$\begin{pmatrix}\Phi_{2d}^\wedge\\ \Psi_{2d}^\wedge\end{pmatrix} = (AB)^d\begin{pmatrix}c_1\\ c_0\end{pmatrix} \quad\text{and}\quad \begin{pmatrix}\Phi_{2d}^\vee\\ \Psi_{2d}^\vee\end{pmatrix} = (BA)^d\begin{pmatrix}c_1\\ c_0\end{pmatrix}. \tag{11}$$

With respect to the lower bound, in view of Eqs. (4), (5), and (6), we obtain that if there is a decision tree Q for $F_{\wedge,n}^{2d}$ of cost C, then there is a decision tree for a single variable of cost C under the cost-function

$$\begin{pmatrix} c_1' \\ c_0' \end{pmatrix} = (AB)^d \begin{pmatrix} c_1 \\ c_0 \end{pmatrix}, \quad \text{where } AB = \begin{pmatrix} n & \frac{n(n-1)}{2} \\ \frac{n-1}{2} & \frac{(n+1)^2}{4} \end{pmatrix}.$$

Comparing the above with Eq. (11), the first part of Theorem 1 follows.

To obtain the asymptotic bound, observe that the trace T and the determinant D of AB are $T = n + (n+1)^2/4$ and $D = n^2$. Let $c_1 = 1$ and $c_0 = (\lambda - n)/(n(n-1)/2)$, so that $\binom{c_1}{c_0}$ is an eigenvector associated with the greatest eigenvalue λ of AB, where

$$\lambda = n + \frac{(n-1)^2}{8} + \frac{(n-1)^2}{8}\sqrt{1 + \frac{16n}{(n-1)^2}} = \left[\frac{n-1}{4}\left(1 + \sqrt{1 + \frac{16n}{(n-1)^2}}\right)\right]^2.$$

Since c_0 is $\Theta(1)$ and positive, we obtain $R(F^{2d}_{\wedge,n}) = \Theta(\lambda^d)$. Bounds for trees of odd depth, if desired, can be obtained via the recurrence relations (10).

4.3 Uniform Threshold Read-Once Functions

With respect to the upper bound for $F^d_{k,n}$, we obtain directly from (9)

$$\begin{pmatrix} \Phi_d \\ \Psi_d \end{pmatrix} = \Delta^d_{k,n} \begin{pmatrix} c_1 \\ c_0 \end{pmatrix}.$$

We set $c_1 = 1$ and $c_0 = (\lambda - k)/\left[(n-k)\frac{k}{k+1}\right]$, so that $\binom{c_1}{c_0}$ is an eigenvector associated with the greatest eigenvalue λ of $\Delta_{k,n}$, where

$$\lambda = \frac{n+1}{2}\left(1 + \sqrt{1 - \frac{8k(n-k+1)}{(n-k+2)(k+1)(n+1)}}\right).$$

Since $c_0 = \Omega(1)$, Theorem 2 follows.

We now turn to the lower bound. From the bounds on $P(k,n)$ we obtained in the preceding section it follows we can apply the method of generalized costs with

$$\Gamma_{k,n} = \begin{pmatrix} k & \frac{n-k}{2} \\ \frac{k-1}{2} & n-k+1 \end{pmatrix}.$$

Its trace and determinant are $T = n+1$ and $D = k(n-k+1) - (k-1)(n-k)/4$ respectively. Again, with λ the largest eigenvalue,

$$\lambda = \frac{n+1}{2} + \frac{n+1}{2}\sqrt{1 - \frac{3k(n-k+1)+n}{(n+1)^2}},$$

we set $c_1 = 1$ and $c_0 = (\lambda - k)/\frac{n-k}{2}$. Since (c_1, c_0) is an eigenvector associated with λ and $c_0 = O(1)$, $R(F^d_{k,n}) = \Omega(\lambda^d)$ and we have obtained Theorem 3.

5 Future Directions

The main motivation for this work has been the recursive majority-of-three function. Although the value $P(2,3)$ leads to the known 2.5^d lower bound, one can define an analogous problem on trees of some height t. Informally, $P^t(2,3)$ could ask for the greatest value η so that any algorithm for maj_t would have a non-negative cost under c_η. Such a problem can be studied either rigorously or computationally (say using an LP-based approach as in [5]).

Another direction is to generalize the current work to unbalanced functions, in the spirit of [7] or [6]. A first step in this direction would be to obtain the results of [2] using the generalized-costs method.

Acknowledgments. This work has benefit greatly from the constructive comments of several reviewers. I am grateful to them for correcting errors and improving the exposition.

References

1. Amano, K.: On directional vs. undirectional randomized decision tree complexity for read-once formulas. In: Proceedings of the Sixteenth Symposium on Computing: The Australasian Theory, vol. 109, pp. 25–30 (2010)
2. Amano, K.: Bounding the randomized decision tree complexity of read-once Boolean functions. In: Proceedings of the Twenty-Second Annual ACM-SIAM Symposium on Discrete Algorithms, pp. 1729–1744. SIAM (2011)
3. Barnum, H., Saks, M.: A lower bound on the quantum query complexity of read-once functions. J. Comput. Syst. Sci. **69**(2), 244–258 (2004)
4. Buhrman, H., de Wolf, R.: Complexity measures and decision tree complexity: a survey. Theor. Comput. Sci. **288**(1), 21–43 (2002)
5. Göös, M., Jayram, T.S.: A composition theorem for conical juntas. In: Raz, R. (ed.) 31st Conference on Computational Complexity (CCC 2016). Leibniz International Proceedings in Informatics (LIPIcs), vol. 50, pp. 5:1–5:16. Schloss Dagstuhl–Leibniz-Zentrum fuer Informatik, Dagstuhl, Germany (2016). http://drops.dagstuhl.de/opus/volltexte/2016/5849
6. Heiman, R., Newman, I., Wigderson, A.: On read-once threshold formulae and their randomized decision tree complexity. Theor. Comput. Sci. **107**(1), 63–76 (1993)
7. Heiman, R., Wigderson, A.: Randomized vs. deterministic decision tree complexity for read-once Boolean functions. Comput. Complex. **1**, 311–329 (1991)
8. Horn, R.A., Johnson, C.R.: Matrix Analysis, 2nd edn. Cambridge University Press (2012)
9. Jayram, T.S., Kumar, R., Sivakumar, D.: Two applications of information complexity. In: STOC, pp. 673–682. ACM (2003)
10. Landau, I., Nachmias, A., Peres, Y., Vanniasegaram, S.: The lower bound for evaluating a recursive ternary majority function and an entropy-free proof. Undergraduate Research Reports, Department of Statistics, University of California, Berkeley (2006)
11. Leonardos, N.: An improved lower bound for the randomized decision tree complexity of recursive majority. In: Fomin, F.V., Freivalds, R., Kwiatkowska, M., Peleg, D. (eds.) Automata, Languages, and Programming, pp. 696–708. Springer, Heidelberg (2013)

12. Magniez, F., Nayak, A., Santha, M., Xiao, D.: Improved bounds for the randomized decision tree complexity of recursive majority. In: Aceto, L., Henzinger, M., Sgall, J. (eds.) ICALP 2011. LNCS, vol. 6755, pp. 317–329. Springer, Heidelberg (2011). https://doi.org/10.1007/978-3-642-22006-7_27
13. Magniez, F., Nayak, A., Santha, M., Sherman, J., Tardos, G., Xiao, D.: Improved bounds for the randomized decision tree complexity of recursive majority. Random Struct. Algorithms **48**(3), 612–638 (2016)
14. Ross, S.: A First Course in Probability. Pearson Education (2015)
15. Saks, M.E., Wigderson, A.: Probabilistic Boolean decision trees and the complexity of evaluating game trees. In: FOCS, pp. 29–38. IEEE Computer Society (1986)
16. Santha, M.: On the Monte Carlo Boolean decision tree complexity of read-once formulae. Random Struct. Algorithms **6**(1), 75–88 (1995)
17. Snir, M.: Lower bounds on probabilistic linear decision trees. Theor. Comput. Sci. **38**, 69–82 (1985)

Distributed Distance Sensitivity Oracles

Vignesh Manoharan(✉) and Vijaya Ramachandran

The University of Texas at Austin, Austin, TX, USA
{vigneshm,vlr}@cs.utexas.edu

Abstract. We present results for the distance sensitivity oracle (DSO) problem, where one needs to preprocess a given directed weighted graph $G = (V, E)$ in order to answer queries about the shortest path distance from s to t in G that avoids edge e, for any $s, t \in V, e \in E$. No non-trivial results are known for DSO in the distributed CONGEST model even though it is of importance to maintain efficient communication under an edge failure. We present DSO algorithms with different tradeoffs between preprocessing and query cost – one that optimizes query response rounds, and another that prioritizes preprocessing rounds. We complement these algorithms with unconditional CONGEST lower bounds. Additionally, we present almost-optimal upper and lower bounds for the related all pairs second simple shortest path (2-APSiSP) problem.

Keywords: Graph Algorithms · Shortest Paths

1 Introduction

In distributed networks, maintaining communication in the event of a link failure is an important problem. In a communication network modeled by a graph $G = (V, E)$, consider the failure of some edge $e \in E$. We investigate the problem of computing shortest path distance when such an edge fault occurs. This is a fundamental problem in networks for routing and communication. Specifically, we need to answer queries of the form $d(s, t, e)$ for $s, t \in V, e \in E$, where $d(s, t, e)$ is the shortest path distance from s to t when edge e is removed from G. In the special case when s and t are fixed vertices this is known as the replacement paths (RPaths) problem.

Let $|V| = n, |E| = m$. A naive algorithm could compute all n^2m distances $d(s, t, e)$ (this can be reduced to $O(n^3)$ distances since we are only interested in edges on the s-t shortest path), but such an algorithm would have high complexity ($\Omega(n^2)$ rounds in CONGEST) simply due to the output size. So, we instead consider Distance Sensitivity Oracles (DSO), which have been studied in sequential setting. In DSO we first preprocess the network G and store certain information about distances. Then, we answer query $d(s, t, e)$ using the stored information. Without any preprocessing, we can answer a query using a single source shortest path (SSSP) computation.

This work was supported in part by NSF grant CCF-2008241.

U. Schmid and R. Kuznets (Eds.): SIROCCO 2025, LNCS 15671, pp. 366–383, 2025.
https://doi.org/10.1007/978-3-031-91736-3_22

In the sequential setting, the first algorithm for constructing a DSO was given in [8], which was improved in [4] to obtain an algorithm with $O(1)$ query time and $\tilde{O}(mn)$ preprocessing time, which matches the sequential APSP running time. Despite the importance of DSO especially in the distributed setting, no CONGEST results were known.

In this paper, we present the first algorithms to construct a DSO tailored to the CONGEST model. We present two algorithms – one that prioritizes preprocessing rounds, and another that obtains efficient queries at the cost of higher preprocessing. We complement our algorithms with unconditional lower bounds on the tradeoff between preprocessing and query rounds. Adapting this technique, we present a CONGEST lower bound for k-source SSSP – this appears to be the first CONGEST lower bound between n and $\sqrt{n}$ for a shortest path problem.

Another problem we consider is the closely related all pairs second simple shortest paths problem (2-APSiSP), where instead of answering queries $d(s,t,e)$ when a particular edge e fails, we are only interested in a simple s-t path that differs from a shortest path in G by at least one edge. The shortest such path is called a second simple shortest path (2-SiSP) [23], and we denote its distance by $d_2(s,t)$. In the 2-APSiSP problem, we need to compute $d_2(s,t)$ for all pairs $s,t \in V$. This problem was studied in the sequential setting in [1]. In this paper we present a CONGEST algorithm to compute 2-APSiSP in $\tilde{O}(n)$ rounds and we present a matching lower bound that applies even when APSP distances in the original graph are known beforehand. Unlike the DSO problem, there is no need for separate preprocessing and query response rounds.

A core procedure used in sequential DSO and 2-APSiSP algorithms [1,4,8] is that of computing excluded shortest paths distances, where we need to compute shortest path distances when certain types of paths are avoided one at a time. We present a distributed CONGEST procedure for this computation which we use in our 2-APSiSP algorithm and our first DSO algorithm.

Roadmap. The rest of this paper is organized as follows. After presenting background information and problem definitions in Sect. 1.1 we describe our results in Sect. 1.2 and prior work in Sect. 1.3. We present our DSO algorithms and lower bounds in Sect. 2 and our near-optimal 2-APSiSP upper and lower bounds in Sect. 3. Some proof details are deferred to the full version of this paper [16].

1.1 Preliminaries

The CONGEST Model. In the CONGEST model [19], a communication network is represented by a graph $G = (V, E)$ where nodes model processors and edges model bounded-bandwidth communication links between processors. Each node has a unique identifier in $\{0, 1, \dots n-1\}$ where $n = |V|$, and each node only knows the identifiers of itself and its neighbors in the network. Each node has infinite computational power. The nodes perform computation in synchronous rounds, where each node can send a message of up to $\Theta(\log n)$ bits to each neighbor and can receive the messages sent to it by its neighbors. The complexity of an algorithm is measured by the number of rounds until the algorithm terminates.

We mainly consider directed weighted graphs G in this paper, where each edge has an integer weight known to the vertices incident to the edge. Following the convention for CONGEST algorithms [6,18], the communication links are always bi-directional and unweighted.

Notation and Terminology. We will deal primarily with directed weighted graphs $G = (V, E)$. Let $|V| = n$. Each edge $(s, t) \in E$ (for $s, t \in V$) can have non-negative integer weight $w(s, t)$, where $W = poly(n)$. We denote the shortest path distance from s to t by $d(s, t)$, and a shortest path from s to t by P_{st}. The undirected diameter of G, denoted D, is the maximum shortest path distance between two vertices in the underlying undirected unweighted graph of G. We use $G^r = (V, E^r)$ to denote the reversed graph of G, where each edge is oriented in the opposite direction.

We use $d(s, t, e)$, for $s, t \in V, e \in E$, to denote the shortest path distance from s to t in the graph $G - \{e\}$, i.e., the graph G with edge e removed, also called replacement path distance. We generalize this definition for a path P, and $d(s, t, P)$ denotes the shortest path distance from s to t in $G - E(P)$. We use $d_2(s, t)$ to denote the second simple shortest path distance from s to t (2-SiSP distance), which is the minimum distance of a simple s-t path that differs from an s-t shortest path P_{st} by at least one edge. Note that $d_2(s, t) = \min_{e \in P_{st}} d(s, t, e)$. The distance $d_2(s, t)$ is independent of the chosen shortest path P_{st}, and $d_2(s, t) = d(s, t)$ if there are multiple s-t shortest paths.

Shortest Path Trees and Independent Paths: We denote the out-shortest path tree rooted at a vertex $x \in V$ by T_x. For a subpath P of T_x, we define $T_x(P)$ to be the subtree of T_x rooted at the vertex in P *furthest* from source x. We define the notion of a set of *independent* paths $\mathcal{R}$ with respect to a shortest path tree T_x as follows: Each path $P \in \mathcal{R}$ is a subpath of T_x, and for any pair of paths $P, P' \in \mathcal{R}$, for any two vertices $u \in P, u' \in P'$, the subtree of T_x rooted at u and the subtree of T_x rooted at u' are disjoint.

Scheduling with Random Delays: Following [9,12], we use scheduling with random delays in order to efficiently perform multiple computations on the network. Let the *congestion* of a CONGEST algorithm be the maximum number of messages it sends through a single edge, over all rounds of the algorithm. Consider an algorithm that runs in R rounds and has congestion C. Then, we can run k instances of the algorithm in $O(kC + R \log n)$ rounds, with total congestion $O(kC)$ over all instances.

We use basic CONGEST primitives such as breadth-first search (BFS) [13], broadcast and convergecast [19] in our algorithms. We use $SSSP$ and $APSP$ to denote the round complexity in the CONGEST model for weighted single source shortest paths (SSSP) and weighted all pairs shortest paths (APSP) respectively. The current best algorithm for weighted APSP runs in $\tilde{O}(n)$ rounds, randomized [5]. For weighted SSSP, [6] provides an $\tilde{O}(\sqrt{n} + n^{2/5+o(1)}D^{2/5} + D)$ round randomized algorithm. The current best lower bounds are $\tilde{\Omega}(\sqrt{n} + D)$ for weighted SSSP [20,21] and $\tilde{\Omega}(n)$ for (weighted and unweighted) APSP [18]. In our algorithms, we also use a low congestion SSSP procedure presented in [11], which takes $\tilde{O}(n)$ rounds but incurs only $\tilde{O}(1)$ congestion per edge. When computing shortest path distances from a source x, we assume that each vertex v knows the preceding and succeeding vertices on a shortest path from x to v. This can be achieved using $O(1)$ rounds per source, by each vertex sharing its distance from x with its neighbors, e.g. y precedes v on the x-v shortest path if $d(x,v) = d(x,y) + w(y,v)$.

Problem Definitions. In the following definitions, $G = (V, E)$ is a directed weighted graph.

Excluded Shortest Paths: Given a set of sources $X \subseteq V$ and sets of independent paths $\mathcal{R}_x$ for each $x \in X$ (as defined in Sect. 1.1), we need to compute $d(x,y,P)$ for each $x \in X, y \in V$ and $P \in \mathcal{R}_x$, i.e., the x-y shortest path distance when path P is removed. The input set of paths $\mathcal{R}_x$ is given as follows: Each edge $e \in E$ that belongs to some path $P \in \mathcal{R}_x$ knows the start vertex of P. For the output, each distance $d(x,y,P)$ must be known at node y.

Distance Sensitivity Oracles (DSO): We need to compute $d(x,y,e)$, the x-y shortest path distance when e is removed, for any pair of vertices $x, y \in V$ and $e \in E$. A DSO algorithm is allowed to perform some preprocessing on the graph G, and then answer queries of the form $d(x,y,e)$. We assume that at query time the query $d(x,y,e)$ is announced to every vertex in the graph. The round complexities of our algorithms are unchanged if we instead assume the query is known to only one node, since we can broadcast the query to all nodes in D rounds once it is known (and all our query bounds are $\Omega(D)$). After the query is answered, distance $d(x,y,e)$ must be known at node y.

All Pairs Second Simple Shortest Path (2-APSiSP): We need to compute $d_2(x,y)$, the x-y second simple shortest path distance (2-SiSP distance, see Sect. 1.1), for all pairs of vertices $x, y \in V$. For the output, each distance $d_2(x,y)$ $(\forall x \in V)$ must be known at node y. Unlike DSO, there are no separate rounds for preprocessing and answering queries.

1.2 Results

We summarize our results for DSO and 2-APSiSP in a directed weighted graph $G = (V, E)$ in the CONGEST model. Recall that $n = |V|$, and D is the undirected diameter of G. All our algorithms are randomized and are correct w.h.p. in n (with probability $1 - (1/poly(n))$).

Excluded Shortest Paths. Our first DSO algorithm and our 2-APSiSP algorithm rely on a subroutine to efficiently compute shortest path distances from multiple sources when certain parts of shortest path trees are excluded, one at a time. We present an algorithm for the excluded shortest paths problem defined in Sect. 1.1. Our CONGEST implementation is inspired by a sequential algorithm for a single source in [8], and we extend the result to multiple sources by using low congestion SSSP combined with efficient random scheduling. We defer the proof of our result stated in Theorem 1 to the full version of this paper [16].

Theorem 1. *Given a set of sources $X \subseteq V$ and an independent set of paths $\mathcal{R}_x$ for each source $x \in X$, we can compute $d(x, y, P)$ for each $x \in X, y \in V, P \in R_x$ in $\tilde{O}(n)$ rounds. Additionally, the maximum congestion is $\tilde{O}(|X|)$.*

Distance Sensitivity Oracles (DSOs). We present two CONGEST algorithms for DSO, one with fast query responses and one with fast preprocessing. We complement these algorithms with unconditional lower bounds on preprocessing rounds. Our DSO algorithms focus on computing the distance value of a given query $d(s, t, e)$, but we can also readily augment our algorithms so that each vertex on the replacement path from x to y avoiding e knows the next vertex on that path.

DSO with Fast Query Responses. Our first DSO optimizes query time, answering any query $d(s, t, e)$ in $O(D)$ rounds. More specifically, the DSO uses a constant number of broadcasts to answer a query, after which all nodes know $d(s, t, e)$. To answer a batch of k different queries, we pipeline the broadcasts for each query, answering them all in $O(k + D)$ rounds. This DSO uses $\tilde{O}(n^{3/2})$ rounds for preprocessing, and is described in Sect. 2.1.

The algorithm performs $\tilde{O}(\sqrt{n})$ exclude computations from every vertex, such that any query $d(s, t, e)$ can be answered using $O(1)$ excluded shortest paths distances stored at different vertices. We use two types of exclude computations: (i) to directly compute $d(s, t, e)$ when edge e is within $\sqrt{n}$ hops of s or t, and (ii) to exclude independent segments of each out-shortest path tree T_s, and combine appropriate excluded shortest path distances to compute $d(s, t, e)$ if e is not within $\sqrt{n}$ hops of s or t. Once the query is known, we broadcast the relevant precomputed distances and compute $d(s, t, e)$ at t in $O(D)$ rounds.

Theorem 2. *We can construct a DSO for directed weighted graphs that takes $\tilde{O}(n^{3/2})$ preprocessing rounds, and then answers a batch of any k queries of the form $d(s, t, e)$ in $O(k + D)$ rounds.*

DSO with Fast Preprocessing. We present a DSO construction that prioritizes preprocessing time, and takes $\tilde{O}(n)$ rounds. Each query $d(s, t, e)$ takes $\tilde{O}(\sqrt{n} + D)$ rounds to answer. To answer a batch of queries, we pipeline part of the query procedure, answering k queries in $\tilde{O}(k\sqrt{n} + D)$ rounds. Note that answering k queries $d(s, t, e)$ without any preprocessing can be done trivially with k SSSP computations in $\tilde{O}(k \cdot (\sqrt{n} + n^{2/5}D^{2/5} + D))$ rounds [6], and we improve on this bound for all $k \geq 1$. We describe this algorithm in Sect. 2.2.

In order to reduce preprocessing rounds to $\tilde{O}(n)$, we cannot perform too many exclude computations. So, we handle short replacement paths of hop-length $\leq \sqrt{n}$ at query time using a $\sqrt{n}$-hop Bellman-Ford procedure. For longer replacement paths ($> \sqrt{n}$ hops), we construct randomly sampled graphs where each edge is included with certain probability (inspired by a sequential DSO in [22]). In these graphs, we perform SSSP computations only from certain sampled vertices during preprocessing. This allows us to quickly compute long replacement path distances at query time.

Theorem 3. *We can construct a DSO for directed weighted graphs that takes $\tilde{O}(n)$ preprocessing rounds, and then answers a batch of any k queries of the form $d(s,t,e)$ in $\tilde{O}(k\sqrt{n}+D)$ rounds.*

DSO Lower Bounds. We show a lower bound on the preprocessing rounds used by a DSO that achieves a given query response round complexity. We consider algorithms that answer a batch of k queries in $O(k \cdot Q)$ rounds where $Q = o(\sqrt{n}/\log n)$. This covers the salient range of query response complexities, since we can use SSSP to answer queries with no preprocessing and the current best CONGEST lower bound for SSSP is $\Omega(\sqrt{n}+D)$. Our lower bound applies even for directed unweighted graphs with diameter $\log n$.

Theorem 4. *Consider a DSO algorithm for directed unweighted graph $G = (V,E)$ on n nodes with diameter $D = \Omega(\log n)$ that performs P rounds of preprocessing and then answers any batch of $k \leq \frac{n}{Q^2 \log^2 n}$ queries in $O(k \cdot Q + D)$ rounds, where $Q = o\left(\frac{\sqrt{n}}{\log n}\right)$. Then, P is $\tilde{\Omega}\left(\frac{n}{Q}\right)$.*

For the setup in Theorem 2, for a DSO algorithm that answers k queries in $O(k+D)$ rounds, we show a lower bound of $\tilde{\Omega}(n)$ preprocessing rounds. Complementing Theorem 3, we show a lower bound of $\tilde{\Omega}(\sqrt{n})$ preprocessing rounds for a DSO algorithm that answers k queries in $o\left(k\frac{\sqrt{n}}{\log n}+D\right)$ rounds. We present these lower bounds for directed unweighted graphs and they trivially apply to directed weighted graphs as well.

Our main technique to obtain this lower bound is to adapt a CONGEST lower bound for SSSP given in [21]. We adapt their construction to k sources for any $1 \leq k \leq \frac{n}{2}$, obtaining a lower bound of $\tilde{\Omega}(\sqrt{nk})$ for the problem of computing shortest path distances from k sources (k-source SSSP) in directed unweighted graphs, stated in Theorem 5. We note that CONGEST algorithms for k-source SSSP have been studied in [14], and our results show that the round complexity $\tilde{O}(\sqrt{nk}+D)$ in [14] is almost optimal for $k \geq n^{1/3}$.

Theorem 5. *Given a directed unweighted graph $G = (V,E)$ with diameter $D = \Omega(\log n)$, computing k-source SSSP requires $\tilde{\Omega}(\sqrt{nk}+D)$ rounds.*

All Pairs Second Simple Shortest Paths (2-APSiSP). We present an algorithm for 2-APSiSP that takes $\tilde{O}(n)$ rounds to compute the second simple shortest path distance $d_2(s,t)$ for all pairs of vertices $s,t \in V$. Our algorithm builds on a characterization of the second simple shortest path used in a sequential algorithm [1]. We use one exclude computation from each vertex, followed by non-trivial local computation in order to compute all 2-APSiSP distances.

Theorem 6. *We can compute directed weighted 2-APSiSP in $\tilde{O}(n)$ rounds.*

2-APSiSP Lower Bound. We present a lower bound of $\tilde{\Omega}(n)$ for computing 2-APSiSP in undirected unweighted graphs, which proves that our $\tilde{O}(n)$ round algorithm is almost optimal for any class of graphs: directed or undirected, weighted or unweighted. An $\tilde{\Omega}(n)$ CONGEST lower bound for computing directed weighted 2-SiSP for one pair was shown in [15] and trivially applies to 2-APSiSP as well. But our lower bound for 2-APSiSP is stronger as the bound in [15] (for 2-SiSP) applies only to directed weighted graphs and requires the input pair of vertices to have a $\Theta(n)$-hop shortest path. Our lower bound applies even when APSP distances are known, and in constant diameter graphs. We present our result in Sect. 3.

Theorem 7. *Given an undirected unweighted graph $G = (V, E)$, computing 2-APSiSP requires $\tilde{\Omega}(n)$ rounds, even if APSP distances are known (i.e. vertex t knows distances $d(s,t)$ $\forall s \in V$). This lower bound applies even for graphs with constant diameter, and against $(3-\epsilon)$-approximation algorithms (for any constant $\epsilon > 0$). In directed graphs, it holds against any approximation ratio.*

1.3 Prior Work

Distance Sensitivity Oracle. The DSO problem has been studied extensively in the sequential setting (e.g. [4,8,22]). No non-trivial algorithms for DSO have been studied in the distributed CONGEST setting, but some related problems have received attention. For computing replacement path distances between a single pair of vertices (RPaths), CONGEST algorithms and lower bounds were presented in [15] for directed and undirected, weighted and unweighted graphs with almost optimal bounds in most cases: these include a $\tilde{\Theta}(n)$ bound (tight up to polylog factors) for directed weighted graphs, and bounds almost matching $SSSP$ round complexity for undirected graphs. Very recently, matching CONGEST upper and lower bounds for RPaths in directed unweighted graphs were given in [7]. A distributed $O(D \log n)$ algorithm for single source replacement paths in undirected unweighted graphs was given in [10].

All Pairs Second Simple Shortest Paths. For a single pair of vertices, an $\tilde{O}(mn)$ sequential time algorithm for the 2-SiSP problem in directed weighted graphs was given in [23]. For 2-APSiSP, a sequential $\tilde{O}(mn)$ time algorithm matching the 2-SiSP running time was given in [1]. A fine-grained sequential lower bound of $\Omega(mn)$ complementing these upper bounds was shown in [2] for directed weighted 2-SiSP. In the distributed setting, the problem of computing 2-SiSP for a single pair has been studied in [15] with both upper and lower bounds: these include an almost optimal $\tilde{\Theta}(n)$ bound for directed weighted graphs and a $\Theta(SSSP)$ bound for undirected graphs.

2 Distance Sensitivity Oracles

In the DSO problem, we need to preprocess an input graph $G = (V, E)$ so that we can answer query $d(x, y, e)$ for any pair of vertices $x, y \in V$, edge $e \in E$ to be avoided. We present algorithms for directed weighted graphs, though they trivially also apply to undirected or unweighted graphs. Note that with no preprocessing, a query can be answered in $O(SSSP) = \tilde{O}(\sqrt{n} + n^{2/5+o(1)}D^{2/5} + D)$ rounds [6]. We present two algorithms that beat this bound in Sects. 2.1 and 2.2. We complement these algorithms with lower bounds in Sect. 2.3.

In both our algorithms we assume that e is on a x-y shortest path. As both algorithms first compute APSP distances, we can readily check if $e = (z, z')$ is on a x-y shortest path by checking if $d(x, y) = d(x, z) + w(e) + d(z', y)$, and return $d(x, y, e) = d(x, y)$ if not.

2.1 DSO with Fast Query Responses

Our first method for directed weighted DSO with fast query responses is described in Algorithm 1. Our algorithm builds on a sequential DSO algorithm in [8] that runs in time $O(mn^{1.5} + n^{2.5} \log n)$ and answers queries in $O(1)$ time. The main idea is to consider replacement distances $d(x, y, e)$ of two forms in the preprocessing phase. If e is within $\sqrt{n}$ hops of x in its shortest path tree T_x, we perform exclude computations from each node, excluding edges within $\sqrt{n}$ depth in its outgoing shortest path tree. We do the same on the reversed graph.

Tree Cutting Procedure. To compute $d(x, y, e)$ when edge e is at least $\sqrt{n}$ hops away from both x and y, we make use of interval computations. We use a *tree cutting procedure* on each shortest path tree T_x, which designates certain vertices as *level vertices* such that the subpaths of T_x in the intervals between level vertices can be arranged into $O(\sqrt{n})$ sets of independent paths. We ensure that any edge e that is $\sqrt{n}$ hops away from both x and y is flanked by some pair of level vertices w.h.p. in n, and use the distances when these intervals are excluded to answer query $d(x, y, e)$ as shown below.

In order to identify appropriate level vertices, we use the following method from [8]: In shortest path tree T_x, let the size of vertex y be the number of nodes in the subtree of T_x rooted at y. If a vertex y has more than one child with size $\geq \sqrt{n}$, all vertices at the same depth as y in T_x are designated *type 1* level vertices. As noted in [8] there are at most $\sqrt{n}$ such y. We are only concerned with $\sqrt{n}$-length paths, so we do not need to consider nodes of size $< \sqrt{n}$ when computing *type 1* level vertices. Additionally, all vertices at depth $i\sqrt{n}$ in T_x, for $i = 1, 2, \cdots, \sqrt{n}$ are designated *type 2* level vertices so that any path of length $\geq \sqrt{n}$ contains a level vertex. Specifically, if edge e on a x-y shortest path is $\sqrt{n}$ hops away from both x and y, e is flanked by some pair of level vertices on the x-y path. Combining both types, level vertices occur at $\leq 2\sqrt{n}$ depth values, and vertices at these depths partition the tree into $O(\sqrt{n})$ intervals. The set of paths in a given interval with level vertex endpoints on either side are independent

Algorithm 1 DSO with fast query: $\tilde{O}(n^{3/2})$ preprocessing and $\tilde{O}(D)$ query

Input: Graph $G = (V, E)$.
Output: Answer query for replacement path distance $d(x, y, e)$ for any $x, y \in V, e \in E$.

1: ▷ *Preprocessing Algorithm* ◁
2: Compute APSP on G and the reversed graph G^r, remembering parent nodes.
3: Perform excluded shortest path computations from each vertex $x \in V$, with sets of edges $R_x^{(1)}, \cdots, R_x^{(\sqrt{n})}$ where $R_x^{(i)}$ is the set of edges i hops from x in the out-shortest path tree T_x rooted at x. Repeat in G^r.
4: **for** $x \in V$ **do**
5: ▷ *Tree cutting procedure on* T_x ◁
6: Perform an upcast along T_x, computing size and depth of each vertex, and identify level vertices of both types. Perform an additional upcast and downcast, to identify all *type 1* level vertices. (details described in Section 2.1)
7: Perform a downcast along T_x so that each edge $e \in T_x$ knows the closest level vertex (if there is one) on the path from x to e. Repeat in G^r. Each edge identifies the subpath between interval vertices it belongs to.
8: Perform exclude computations from x with the $O(\sqrt{n})$ sets of independent paths identified in line 7.
9: ▷ *Query Algorithm: Given query* $d(x, y, e)$ ◁
10: ▷ *e is within h hops from x or y in T_x: use excluded distances from line 3* ◁
11: If $d(x, y, e)$ has been computed in line 3 at x or y, broadcast $d(x, y, e)$.
12: ▷ *e is not within h hops from x or y in T_x: use excluded distances from line 8* ◁
13: From line 7, e locally determines a as the closest level vertex on the path from x to e, and b as the closest level vertex on the path from e to y. e broadcasts the identities of a and b.
14: Then, a broadcasts $d(a, y, e)$ and x broadcasts $d(x, a), d(x, b, e)$. Distances $d(a, y, e)$ and $d(x, b, e)$ were computed in line 3.
15: Finally, y broadcasts $d(x, y, e) = \min(d(x, a) + d(a, y, e), d(x, b, e) + d(b, y), d(x, y, [a, b]))$. Distance $d(x, y, [a, b])$ was computed in line 8.

(as shown in [8]). So, we have arranged the subpaths of the tree between level vertices into $O(\sqrt{n})$ sets of independent paths.

In lines 6–8 of Algorithm 1, we implement this method as follows: In line 6, we perform an upcast along T_x in $O(n)$ rounds, so that each vertex knows its size and depth in T_x, and use this to identify level vertices of both types described in the paragraph above. With an additional upcast and downcast, all vertices at the same depth as some type 1 level vertex also identifies itself as a level vertex. After the downcast in line 7, each edge knows its closest level vertex in either direction. So, in the exclude computation in line 8, each edge knows which interval subpath it is a part of. We use $O(n)$ rounds and $O(1)$ congestion per tree T_x for the computation in lines 6, 7.

Analysis of Algorithm 1. We utilize the multiple excluded shortest paths algorithm from Theorem 1 for efficiently excluding paths in shortest path trees. We perform a total of $\sqrt{n}$ exclude computations per source x in lines 3, 8, which takes $\tilde{O}(n^{3/2})$ rounds using Theorem 1. In the query algorithm, given query $d(x, y, e)$, we identify the level vertices flanking e if they exist in line 13. We broadcast the excluded shortest paths distances necessary to compute $d(x, y, e)$ in lines 11, 14 in $O(D)$ rounds, and compute the result at y.

Proof (of Theorem 2). **Correctness:** Let P_{xy} be a replacement path from x to y when e is removed. If e is within $\sqrt{n}$ hops from x or y in the shortest path tree, then $d(x, y, e)$ is computed in line 3 and the query is answered in line 11.

Now, consider the other case where e is not within h hops of x or y on the x-y shortest path. After the tree cutting procedure on T_x, there are level vertices a, b such that a is on the shortest path from x to e and b is on the shortest path from e to y. Additionally, the interval subpath from a to b has at most $\sqrt{n}$ hops, so e is within $\sqrt{n}$ hops of both a and b in their respective shortest path trees. Thus, $d(a, y, e)$ is computed at y in line 3 and similarly $d(x, b, e)$ is computed at x in the reversed graph. After the identities of a, b are broadcast, the replacement path distance is correctly computed in line 15—the first term is correct when P_{xy} passes through a and the second is correct when it passes through b. If P_{xy} passes through neither a nor b, it skips the whole interval $[a, b]$, and the distance is correctly computed as $d(x, y, [a, b])$.

Round Complexity: Computing APSP takes $\tilde{O}(n)$ rounds. We perform $\tilde{O}(n^{3/2})$ different exclude computations in lines 3 and 8. Using Theorem 1, this takes $\tilde{O}(n^{3/2})$ rounds. The downcast in line 7 takes $O(n)$ rounds and $O(1)$ congestion per vertex, and can be scheduled for all x in $\tilde{O}(n)$ rounds. The query algorithm broadcasts $O(1)$ values and then broadcasts the correct answer, so it takes $O(D)$ rounds with $O(1)$ congestion per edge. Thus, we can answer a batch of k queries by pipelining all broadcasts in $O(k + D)$ rounds.

2.2 DSO with Fast Preprocessing

In this section, we present Algorithm 2 to construct a DSO with $\tilde{O}(n)$ rounds for preprocessing, which answers a batch of k queries in $\tilde{O}(k \cdot \sqrt{n} + D)$ rounds. The algorithm uses a graph sampling approach to compute hop-limited replacement paths, inspired by a sequential DSO construction [22]. We first describe this graph sampling method and then describe our distributed algorithm which uses it along with vertex sampling to achieve efficient preprocessing.

Hop-Limited Replacement Paths. Given a hop parameter h, and a graph $G = (V, E)$ along with a set of sources $S \subseteq V$, we wish to compute h-hop replacement path distances $d_h(s, x, e)$ for $s \in S, x \in V, e \in E$. Here, $d_h(s, x, e)$ denotes the minimum weight of an s-x path containing up to h edges that does not contain e. The following lemma can be established by adapting proofs in [22], see the full version [16] for details.

Lemma 1. *Sample* $\tilde{h} = 15h \log n$ *graphs* $G_1, \cdots, G_{\tilde{h}}$ *as subgraphs of* G *by including each edge of* G *with probability* $1 - \frac{1}{h}$. *Then,*

A. W.h.p. in n, *there are* $\Theta(\log n)$ *graphs* G_i *such that* $e \notin G_i$.

Algorithm 2 DSO with fast preprocessing: $\tilde{O}(n)$ preprocessing and $\tilde{O}(\sqrt{n}+D)$ query

Input: Graph $G = (V, E)$.
Output: Answer query for replacement path distance $d(x, y, e)$ for any $x, y \in V, e \in E$.
1: $\triangleright$ *Preprocessing Algorithm* $\triangleleft$
2: **for** $j = \log(\sqrt{n}), \cdots, \log n$ **do**
3: $\triangleright$ *Preprocessing for replacement paths of hop-length* $[2^j, 2^{j+1}]$, *where* $2^j \geq \sqrt{n}$ $\triangleleft$
4: Each vertex $s \in V$ is sampled and added to S_j with probability $\frac{c \log n}{2^j}$.
5: Let $h = 2^{j+1}$. Sample $\tilde{O}(h)$ graphs G_i^j as in Lemma 1 by randomly sampling each edge with probability $1 - (1/h)$. $\triangleright$ *Each edge locally samples itself*
6: For each $e \in G$, $I_e^j \leftarrow \{(j, i) \mid e \notin G_i^j\}$ $\triangleright$ *Computed locally at endpoints of* e
7: Compute SSSP from each vertex in S_j, on each G_i^j (and reversed graph). The distance in graph G_i^j is denoted d_i^j. Distances $d_i^j(s, x), d_i^j(x, s)$ are computed at x for each $s \in S_j, x \in V$.
8: $\triangleright$ *Query Algorithm: Given query* $d(x, y, e)$ $\triangleleft$
9: $\triangleright$ *Find best replacement path of hop-length* $\leq \sqrt{n}$ $\triangleleft$
10: Perform a distributed Bellman-Ford computation in $G - \{e\}$ starting from x, up to $\sqrt{n}$ hops. This computes $d_{\sqrt{n}}(x, y, e)$ at y.
11: $\triangleright$ *Find best replacement path of hop-length* $> \sqrt{n}$ $\triangleleft$
12: **for** $j = \log(\sqrt{n}), \cdots, \log n$ **do**
13: $\triangleright$ *Handle paths of hop-length* $[2^j, 2^{j+1}]$ $\triangleleft$
14: Endpoint of e broadcasts I_e^j.
15: x broadcasts $\{d_i^j(x, s) \mid (j, i) \in I_e^j, s \in S_j\}$, i.e. its distance to sampled vertices in S_j in graphs where e is not present.
16: **for** $s \in S_j$ **do** $\triangleright$ *Local computation at* y
17: $\triangleright$ d_s *is the best* x-y *replacement path distance through* s $\triangleleft$
18: Compute $d_s = \min_{(j,i) \in I_e^j} d_i^j(x, s) + d_i^j(s, y)$
19: Compute $d(x, y, e) = \min(d_{\sqrt{n}}(x, y, e), \min_{s \in V} d_s)$ $\triangleright$ *Local computation at* y

B. Let P_{sx}^e be a h-hop replacement path from s to x avoiding e. W.h.p. in n, there is at least one graph G_i such that G_i contains all edges in P_{sx}^e but does not contain e.

CONGEST DSO with Fast Preprocessing. We now describe Algorithm 2. To obtain fast preprocessing, we only deal with replacement paths of hop-length $\geq \sqrt{n}$ during preprocessing. Shorter replacement paths are computed at query time using $\sqrt{n}$ steps of Bellman-Ford with the appropriate edge deleted (line 10).

To compute replacement paths of hop-length $[2^j, 2^{j+1}]$, for $j \geq \frac{1}{2} \log n$, we use the result in Lemma 1 with $h = 2^{j+1}$. After sampling graphs in line 4, we compute shortest paths to and from sampled vertices in line 7. With this information, we can compute replacement path distances at query time (lines 14–18) without needing to compute APSP during preprocessing. Although we could use the graph sampling method for $\leq \sqrt{n}$ hop-length replacement paths and stay within the $\tilde{O}(n)$ preprocessing bound, the number of sampled vertices would exceed $\sqrt{n}$ and answering a query by broadcast (line 15) would take $\omega(\sqrt{n})$ rounds.

Proof (of Theorem 3). **Correctness:** For query $d(x, y, e)$, if the x-y replacement path has hop-length $\leq \sqrt{n}$, the distance is correctly computed in line 10 using the Bellman-Ford computation from x up to $\sqrt{n}$ hops. We now address replacement paths of hop-length $[2^j, 2^{j+1}]$, for $2^j \geq \sqrt{n}$.

Fix one such replacement path P from x to y avoiding edge e, with hop length $2^j \leq h(P) < 2^{j+1}$ and weight $w(P)$. In the preprocessing algorithm, consider the j'th iteration of the loop (lines 2–7). Since we sample vertices into S_j with probability $\tilde{O}(1/2^j)$, w.h.p in n there is at least one vertex s from S_j that is on P. By Lemma 1.B at least one of the sampled subgraphs G_i^j contains all edges of P and does not contain e w.h.p. in n. In this particular G_i^j, the distances $d_i^j(s, y)$ and $d_i^j(x, s)$ are computed in line 7 and we have $w(P) = d_i^j(x, s) + d_i^j(s, y)$.

When answering query $d(x, y, e)$, we use the set I_e^j computed in line 7 and broadcast in line 14. In line 18, we only consider shortest path distances in graphs not containing e, so all distances d_s correspond to valid x-y paths not containing e. For the particular G_i^j that contains P and not e, we correctly compute $d(x, y, e) = d_i^j(x, s) + d_i^j(s, y)$ in line 18.

Round Complexity: We first analyze the preprocessing algorithm. Consider an iteration j: lines 4–6 only involve local computation, and line 7 performs SSSP from $|S_j| = \tilde{O}(n/2^j)$ vertices on $\tilde{O}(2^j)$ graphs. Using the $\tilde{O}(n)$-round low congestion SSSP algorithm from [11], the total congestion over all SSSP computations is at most $\tilde{O}(|S_j| \cdot 2^j) = \tilde{O}(n)$. Thus, using random scheduling, line 7 takes $\tilde{O}(n)$ rounds.

Now, to answer a query $d(x, y, e)$, computing Bellman-Ford up to $\sqrt{n}$ hops in line 10 takes $O(\sqrt{n})$ rounds. The set I_e^j with size $O(\log n)$ is broadcast in $\tilde{O}(D)$ rounds in line 14. Vertex x then broadcasts a set of size $|I_e^j| \cdot |S_j| = \tilde{O}(\frac{n}{2^j}) = \tilde{O}(\sqrt{n})$ (since $2^j \geq \sqrt{n}$) in line 15, which takes $\tilde{O}(\sqrt{n} + D)$ rounds. Lines 18, 19 only involve local computation at y, so the total rounds to answer the query is $\tilde{O}(\sqrt{n} + D)$. To answer a batch of k queries, we can broadcast the $\tilde{O}(k\sqrt{n})$ required values in lines 14, 15 in $\tilde{O}(k\sqrt{n} + D)$ rounds, and perform k Bellman-Ford computations sequentially in $k\sqrt{n}$ rounds – totaling $\tilde{O}(k\sqrt{n} + D)$ rounds.

2.3 DSO Lower Bounds

In this section, we show lower bounds on the preprocessing rounds used by any DSO algorithm with a given query response round complexity. Our proof uses a graph construction to obtain a reduction from set disjointness to the problem of computing k DSO queries in a directed unweighted graph, building on a $\tilde{\Omega}(\sqrt{n})$ CONGEST lower bound for SSSP in [21]. Our construction also gives a lower bound for the problem of computing k-source shortest paths. While similar techniques have been used to show $\sqrt{n}$ lower bounds, our construction appears to give the first CONGEST lower bounds between n and $\sqrt{n}$ for natural shortest path problems.

We start our DSO lower bound with the following preliminary lemma.

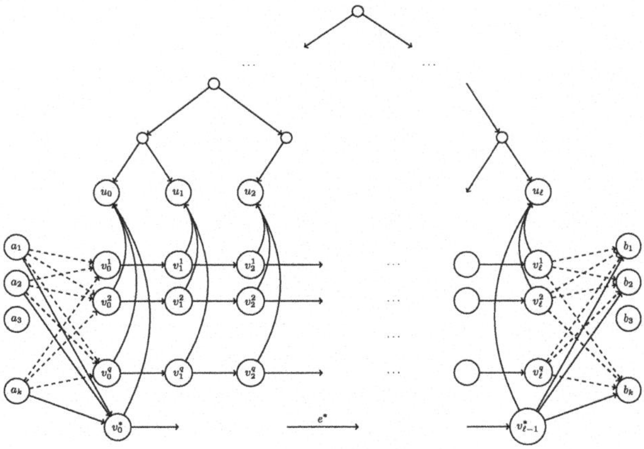

Fig. 1. Lower Bound for DSO Preprocessing Time.

Lemma 2. *Consider an algorithm $\mathcal{A}$ for a directed unweighted graph $G = (V, E)$ on n nodes that computes the answer to any batch of $k \leq n/2$ queries $d(a_i, b_i, e_i)$ for $1 \leq i \leq k$. The answer for query i must be known to b_i after the algorithm terminates. Then, $\mathcal{A}$ must have round complexity $\Omega\left(\frac{\sqrt{nk}}{\log n}\right)$, even on graphs of undirected diameter $\Theta(\log n)$.*

Proof. **Graph Construction:** We construct directed unweighted graph $G = (V, E)$, pictured in Fig. 1, as a balanced binary tree with ℓ leaves, numbered $u_1, \ldots u_\ell$, along with a set of q paths v_0^i-v_ℓ^i of length ℓ, and a path v_0^*-$v_{\ell-1}^*$ of length $\ell-1$. The paths are connected to the leaves of the tree as in the figure, with all edges oriented from path to leaf. Additionally, we have $2k$ vertices a_i, b_i for $1 \leq i \leq k$. We perform a reduction from (kq)-bit set disjointness, and the dashed edges in our construction depend on the inputs S_a, S_b of the set disjointness instance. We index the input such that $S_a[i, j]$ represents the $((i-1) \cdot q + j)$'th bit for $1 \leq i \leq k, 1 \leq j \leq q$ (similarly for S_b). The edge (a_i, v_0^j) is present only if $S_a[i, j] = 1$, similarly edge (v_ℓ^j, b_i) is present only if $S_b[i, j] = 1$. This encoding of bits is similar to a lower bound construction in [15]. All other solid edges in Fig. 1 are always present.

Note that $d(a_i, b_i) = \ell + 1$ due to the v_0^*-$v_{\ell-1}^*$ path. When an edge e^* on the v_0^*-$v_{\ell-1}^*$ path is removed, a replacement path from a_i to b_i with weight $\ell + 2$ exists through v_0^j-v_ℓ^j if $S_a[i, j] = S_b[i, j] = 1$ for some j, i.e., $d(a_i, b_i, e^*) = \ell + 2$. Otherwise, either the (a_i, v_0^j) or (v_ℓ^j, b_i) edge is missing for every j, and no a_i-b_i

replacement path exists which means $d(a_i, b_i, e^*) = \infty$. Thus, computing the queries $d(a_i, b_i, e^*)$ for all i allows us to determine set disjointness of S_a, S_b.

Using an analysis similar to ([21], Theorem 3.1), we can prove that if an algorithm $\mathcal{A}$ can compute answer to k queries $d(a_i, b_i, e_i)$ in $R \leq (\ell/2)$ rounds, then we can solve (kq)-bit set disjointness using $Rp \cdot \Theta(\log n)$ bits, where p is the height of tree. See the full version [16] for a detailed proof.

To choose our parameters, we use $p = \Theta(\log n), q = \frac{n-k}{\ell} \geq \frac{n}{2\ell}$ (since $k < n/2$), ℓ will be chosen later. The number of vertices in the graph is $\Theta(\ell \cdot q + k) = \Theta(n)$. The undirected diameter of the graph is $\leq 2p + 4 = \Theta(\log n)$ (using paths through the tree). The number of bits in the set disjointness instance is $kq \geq \frac{kn}{2\ell}$. Using the randomized communication lower bound for set disjointness [3] along with the above claim, we obtain: If $R(n) \leq (\ell/2)$, then $R(n) \cdot \log^2 n = \Omega(kq) \Rightarrow R(n) = \Omega\left(\frac{kn}{\ell \log^2 n}\right)$. Combining the two bounds, $R(n) = \Omega\left(\min(\ell, \frac{kn}{\ell \log^2 n})\right)$. We balance these terms by choosing $\ell = \frac{\sqrt{nk}}{\log n}$, which gives $R(n) = \Omega\left(\frac{\sqrt{nk}}{\log n}\right)$.

Lemma 2 provides a lower bound for computing k replacement path distances in general, and we use it to prove lower bounds for DSOs in terms of preprocessing and query rounds to prove Theorem 4.

Proof (of Theorem 4). Applying Lemma 2, $P + k \cdot Q = \Omega\left(\frac{\sqrt{nk}}{\log n}\right)$. Choosing k such that $k \cdot Q = o\left(\frac{\sqrt{nk}}{\log n}\right)$, would give a lower bound on P. Thus, $\sqrt{k} = o\left(\frac{\sqrt{n}}{Q \log n}\right) \Rightarrow k = o\left(\frac{n}{Q^2 \log^2 n}\right)$. Then, $k \cdot Q + D = \frac{n}{Q \log^2 n} + \Theta(\log n) = o\left(\frac{\sqrt{nk}}{\log n}\right)$. Choosing $k = \Theta\left(\frac{n}{Q^2 \log^3 n}\right)$ for instance, we get $P = \Omega\left(\frac{\sqrt{nk}}{\log n}\right) = \tilde{\Omega}\left(\frac{n}{Q}\right)$.

We now apply Theorem 4 to obtain lower bounds mirroring the upper bounds in Sect. 2, with query setups similar to Algorithm 1 (Corollary 1.A) and Algorithm 2 (Corollary 1.B). Algorithm 2 has query response rounds $\tilde{O}(k\sqrt{n} + D)$, and the query setup in Corollary 1.B differs from it only by a polylog factor.

Corollary 1. *Consider a DSO algorithm $\mathcal{A}$ for directed unweighted graph $G = (V, E)$ on n nodes that uses P rounds for preprocessing.*

A. If $\mathcal{A}$ answers k queries in $O(k + D)$ rounds then P is $\tilde{\Omega}(n)$.
B. If $\mathcal{A}$ answers k queries in $o\left(\frac{k\sqrt{n}}{\log n} + D\right)$ rounds then P is $\tilde{\Omega}(\sqrt{n})$.

Lower Bound for k-source shortest path. By removing path v_0^*-$v_{\ell-1}^*$ in the construction in Lemma 2 and slightly modifying the proof, we obtain a lower bound of $\tilde{\Omega}(\sqrt{nk} + D)$ for computing distances between k vertex pairs $d(a_i, b_i)$. Thus, we get a $\tilde{\Omega}(\sqrt{nk} + D)$ lower bound for k-source SSSP in directed unweighted graphs, proving Theorem 5. This lower bound is almost tight for $k \geq n^{1/3}$, as the k-source SSSP algorithm of [14] takes $\tilde{O}(\sqrt{nk} + D)$ rounds for k-SSSP in directed unweighted graphs. For $k < n^{1/3}$ the bound is not tight, which is not unexpected as we have a gap between the current best upper and lower bounds even for SSSP ($k = 1$), i.e., $\tilde{O}(\sqrt{n} + n^{2/5+o(1)}D^{2/5} + D)$ [6] and $\tilde{\Omega}(\sqrt{n} + D)$ [21].

3 All Pairs Second Simple Shortest Path

In the 2-APSiSP problem, we need to compute the 2-SiSP distance $d_2(x, y)$ for all pairs of vertices $x, y \in V$. We present a $\tilde{O}(n)$ round algorithm for 2-APSiSP in directed weighted graphs and show that it is almost optimal with a $\tilde{\Omega}(n)$ round lower bound even for undirected unweighted graphs with constant diameter.

2-APSiSP Algorithm. Our algorithm uses a characterization of 2-SiSP proved in [1], and we show that all 2-SiSP distances can be computed using n exclude computations followed by local computation at each vertex. Using the excluded shortest paths result from Theorem 1, we obtain our $\tilde{O}(n)$ round bound. Due to space constraints, we defer the detailed presentation of our 2-APSiSP algorithm to the full version [16].

2-APSiSP Lower Bound. We prove a lower bound of $\tilde{\Omega}(n)$ for 2-APSiSP, proving Theorem 7. A lower bound of $\tilde{\Omega}(n)$ rounds for computing directed weighted 2-SiSP for a single pair of vertices (which trivially applies to 2-APSiSP) was proven in [15]. The result in Lemma 2 of Sect. 2.3 with $k = \Theta(n)$ can also be modified to give a $\tilde{\Omega}(n)$ lower bound for 2-APSiSP. But these results only apply to directed graphs with diameter $\Omega(\log n)$. Our lower bound holds for undirected unweighted graphs with diameter as low as $O(1)$ even when APSP distances in the input graph are known, i.e., vertex $y \in V$ knows distances $d(x, y)$ for all $x \in V$.

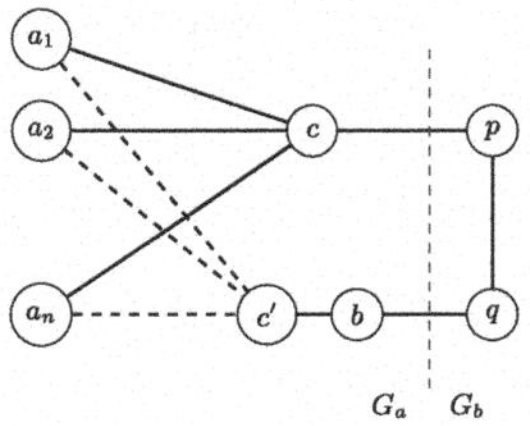

Fig. 2. 2-APSiSP lower bound.

Proof (of Theorem 7). Consider the undirected unweighted graph construction $G = (V, E)$ in Fig. 2, based on a n-bit set disjointness instance S_a, S_b. Solid edges are always present, but for each $1 \leq i \leq n$, edge (a_i, c') is present only if $S_a[i] = 1$. G has $\Theta(n)$ vertices and constant undirected diameter. Alice controls vertices a_i, c, c', b and Bob controls vertices p, q – note that Alice and Bob can construct their sections of the graph with only the bits known to them, and the crossing edges are fixed.

First, we show that set disjointness can be determined by computing 2-APSiSP in G. Note that $d(a_i, q) = 3$, either using the path through vertex c, or using the path through c' when edge (a_i, c') is present is G. If edge (a_i, c') is present, the 2-SiSP distance is the same $d_2(a_i, q) = 3$. If the edge is not present, $d_2(a_i, q) \geq 5$ using a path segment of the form a_i-c-a_j-c'. Since only 2 edges cross the cut separating vertices controlled by Alice and Bob, they can simulate a 2-APSiSP algorithm taking $R(n)$ rounds using $O(R(n) \cdot \log n)$ bits. After computing all 2-APSiSP distances, Bob can identify all bits of S_a using the distances $d_2(a_i, q)$ computed at q. Then, Bob can readily evaluate set disjointness of S_a, S_b and forward the 1-bit result to Alice in one round. So, the communication lower bound for set disjointness [3] gives an $\Omega(n/\log n)$ lower bound for 2-APSiSP. We also claim that APSP distances between vertices controlled by Alice and Bob are the same regardless of bits S_a, S_b. Only distances involving a_i could be affected by the input bits, and regardless of the presence of edge (a_i, c') we have $d(a_i, p) = 2$ and $d(a_i, q) = 3$.

We can generalize this lower bound to hold for $(3 - \epsilon)$-approximation algorithms (for any constant $\epsilon > 0$) by replacing the edges (a_i, c) and (a_i, c') with paths of length k, for a large constant k. The graph still has $\Theta(k)$ vertices, but we have $d_2(a_i, q) = k + 2$. If edge (a_i, c') is not present, then $d_2(a_i, q) \geq 3k + 2$, allowing us to determine set disjointness even if only a $(3 - \epsilon)$-approximation of $d_2(a_i, q)$ is known (by choosing $k = \Theta(1/\epsilon)$). For directed graphs, we can direct all edges from left to right, so there is no 2-SiSP path from a_i to q when the edge (a_i, c') is not present and $d_2(a_i, q) = \infty$. In this case, the lower bound holds for an arbitrarily large approximation factor.

4 Conclusion and Open Problems

In this paper, we present CONGEST upper and lower bounds for DSO and 2-APSiSP. Our bounds for 2-APSiSP are almost optimal, up to polylog factors, for even undirected unweighted graphs. There still remains a gap between upper and lower bounds for directed weighted DSO, leading to the following open problems:

- For a DSO algorithm that can answer a batch of k queries in $O(k+D)$ rounds, we present a lower bound of $\tilde{\Omega}(n)$ preprocessing rounds, and an upper bound that takes $\tilde{O}(n^{3/2})$ rounds. Can we bridge the gap between these bounds?
- For a DSO algorithm that takes $o\left(k\frac{\sqrt{n}}{\log n} + D\right)$ rounds to answer k queries, we show a lower bound of $\tilde{\Omega}(\sqrt{n})$ preprocessing rounds. We show an upper bound that takes $\tilde{O}(n)$ preprocessing rounds and answers k queries in $\tilde{O}(k\sqrt{n}+D)$. It is unlikely the lower bound can be improved for algorithms taking $O(k\sqrt{n}+D)$ rounds, as this would require an improved CONGEST SSSP lower bound, but it is an open problem whether the preprocessing can be reduced to sub-linear rounds.
- Can we achieve a tradeoff between preprocessing rounds and query response rounds, such as $\tilde{O}(n^{3/2-c})$ preprocessing rounds to answer k queries in $\tilde{O}(kn^c + D)$ rounds, for constant $0 < c < 1/2$? We achieve this for $c = 0$ and $c = 1/2$, but our techniques do not immediately extend to arbitrary c.

– Recently, we have obtained the first non-trivial results for DSO on the PRAM including a work-efficient algorithm with $\tilde{O}(mn)$ work and $\tilde{O}(n^{1/2+o(1)})$ parallel time, an algorithm that achieves $\tilde{O}(1)$ parallel time with $\tilde{O}(n^3)$ work, and an algorithm with a work-time tradeoff that can achieve sub-n^3 work and sub-$\sqrt{n}$ time [17]. Improving these results and studying DSO in other parallel models are open problems.

References

1. Agarwal, U., Ramachandran, V.: Finding k simple shortest paths and cycles. In: ISAAC. LIPIcs, vol. 64, pp. 8:1–8:12 (2016)
2. Agarwal, U., Ramachandran, V.: Fine-grained complexity for sparse graphs. In: STOC, pp. 239–252. ACM (2018)
3. Bar-Yossef, Z., Jayram, T.S., Kumar, R., Sivakumar, D.: An information statistics approach to data stream and communication complexity. J. Comput. Syst. Sci. **68**(4), 702–732 (2004)
4. Bernstein, A., Karger, D.R.: A nearly optimal oracle for avoiding failed vertices and edges. In: STOC, pp. 101–110. ACM (2009)
5. Bernstein, A., Nanongkai, D.: Distributed exact weighted all-pairs shortest paths in near-linear time. In: STOC, pp. 334–342. ACM (2019)
6. Cao, N., Fineman, J.T., Russell, K.: Efficient construction of directed hopsets and parallel single-source shortest paths (abstract). In: HOPC@SPAA, pp. 3–4. ACM (2023)
7. Chang, Y.J., Chen, Y., Dey, D., Mishra, G., Nguyen, H.T., Sanchez, B.: Optimal distributed replacement paths (2025). https://arxiv.org/abs/2502.15378
8. Demetrescu, C., Thorup, M., Chowdhury, R.A., Ramachandran, V.: Oracles for distances avoiding a failed node or link. SIAM J. Comput. **37**(5), 1299–1318 (2008)
9. Ghaffari, M.: Near-optimal scheduling of distributed algorithms. In: PODC, pp. 3–12. ACM (2015)
10. Ghaffari, M., Parter, M.: Near-optimal distributed algorithms for fault-tolerant tree structures. In: SPAA, pp. 387–396. ACM (2016)
11. Ghaffari, M., Trygub, A.: A near-optimal low-energy deterministic distributed SSSP with ramifications on congestion and APSP. In: PODC, pp. 401–411. ACM (2024)
12. Leighton, F.T., Maggs, B.M., Richa, A.W.: Fast algorithms for finding O(congestion + dilation) packet routing schedules. Comb. **19**(3), 375–401 (1999)
13. Lenzen, C., Patt-Shamir, B., Peleg, D.: Distributed distance computation and routing with small messages. Distrib. Comput. **32**(2), 133–157 (2019)
14. Manoharan, V., Ramachandran, V.: Computing minimum weight cycle in the CONGEST model. In: PODC, pp. 182–193. ACM (2024)
15. Manoharan, V., Ramachandran, V.: Computing replacement paths in the CONGEST model. In: SIROCCO. LNCS, vol. 14662, pp. 420–437. Springer, Cham (2024)
16. Manoharan, V., Ramachandran, V.: Distributed distance sensitivity oracles (2024). https://arxiv.org/abs/2411.13728
17. Manoharan, V., Ramachandran, V.: Algorithms for distance sensitivity oracles on the PRAM (2025, manuscript)

18. Nanongkai, D.: Distributed approximation algorithms for weighted shortest paths. In: Symposium on Theory of Computing, STOC 2014, pp. 565–573. ACM (2014)
19. Peleg, D.: Distributed Computing: A Locality-Sensitive Approach. SIAM, USA (2000)
20. Peleg, D., Rubinovich, V.: A near-tight lower bound on the time complexity of distributed minimum-weight spanning tree construction. SIAM J. Comput. **30**(5), 1427–1442 (2000)
21. Sarma, A.D., et al.: Distributed verification and hardness of distributed approximation. SIAM J. Comput. **41**(5), 1235–1265 (2012)
22. Weimann, O., Yuster, R.: Replacement paths and distance sensitivity oracles via fast matrix multiplication. ACM Trans. Algorithms **9**(2), 14:1–14:13 (2013)
23. Yen, J.Y.: Finding the k shortest loopless paths in a network. Manag. Sci. **17**(11), 712–716 (1971)

Self-stabilizing Graph Exploration by a Single Agent

Yuichi Sudo[1(✉)], Fukuhito Ooshita[2], and Sayaka Kamei[3]

[1] Hosei University, Tokyo, Japan
sudo@hosei.ac.jp
[2] Fukui University of Technology, Fukui, Japan
[3] Hiroshima University, Hiroshima, Japan

Abstract. In this paper, we present two self-stabilizing algorithms enabling a single (mobile) agent to explore graphs. Starting from any initial configuration, *i.e.,* regardless of the initial states of the agent and all nodes, as well as the initial location of the agent, the algorithms ensure the agent visits all nodes. We evaluate the algorithms based on two metrics: the *cover time*, defined as the number of moves required to visit all nodes, and *memory usage*, defined as the storage needed for maintaining the states of the agent and each node. The first algorithm is randomized. Given an integer $c = \Omega(n)$, its cover time is optimal, *i.e.,* $O(m)$ in expectation, and its memory requirements are $O(\log c)$ bits for the agent and $O(\log(c+\delta_v))$ bits for each node v, where n and m are the number of nodes and edges, respectively, and δ_v is the degree of node v. The second algorithm is deterministic. It requires an input integer $k \geq \max(D, \delta_{\max})$, where D and $\delta_{\max}$ are the diameter and maximum degree of the graph, respectively. The cover time of this algorithm is $O(m + nD)$, and it uses $O(\log k)$ bits of memory for both the agent and each node.

Keywords: self-stabilization · graph exploration · mobile agents

1 Introduction

We focus on the *exploration problem* involving a single mobile entity, referred to as a mobile agent or simply an *agent*, within any undirected, simple, and connected graph $G = (V, E)$. This agent, functioning as a finite state machine, migrates from node to node via edges at each time step. Upon visiting a node, the agent can access and modify the node's local memory, known as a *whiteboard.* The graph is anonymous, *i.e.,* nodes lack unique identifiers. Our objective is to enable the agent to visit every node in the graph in as few steps as possible while minimizing the memory usage of both the agent and the whiteboards. This exploration problem, fundamental in the study of mobile computing entities, has been extensively studied [6,7,10,12,14,17]. Exploration algorithms have frequently served as a foundation for solving other fundamental problems such as rendezvous, gathering, dispersion, and gossip sharing.

U. Schmid and R. Kuznets (Eds.): SIROCCO 2025, LNCS 15671, pp. 384–400, 2025.
https://doi.org/10.1007/978-3-031-91736-3_23

In this paper, we tackle the exploration problem under a more challenging setting: *self-stabilizing exploration* [7,12]. We do not presuppose any specific initial global state (or *configuration*) of the network. This means that at the start of the exploration, (i) the agent's location within G is arbitrary, (ii) the agent's state is arbitrary, and (iii) the content of each whiteboard is arbitrary. The agent is required to visit all nodes in G from any potentially inconsistent configuration. Generally, an algorithm is considered self-stabilizing [4] for problem P if it can solve P starting from any configuration. Self-stabilizing algorithms are capable of handling any type of transient faults, such as temporary memory corruption, making their design both practically and theoretically significant. We evaluate algorithms using two metrics: cover time, which is the number of moves required to visit all nodes, and memory usage, which includes the storage needed for the state of the agent and the state of each node.

Generally speaking, several studies tackle a variety of problems involving mobile agents in the self-stabilizing setting [2,7,9]. In this setting, the number of agents in the graph is fixed. In our case (*i.e.*, self-stabilizing exploration by a single agent), the number of agents is always exactly one: we do not consider configurations where no agent exists, or where two or more agents are present. Therefore, this setting may be particularly suitable for applications where physical robots operate in a field represented by an undirected graph, and the robots can leave information in some way at each intersection in the field.

Throughout this paper, we denote the number of nodes, the number of edges, and the diameter of a graph by n, m, and D, respectively. We denote the degree of a node v by δ_v, and define $\delta_{\min} = \min_{v \in V} \delta_v$ and $\delta_{\max} = \max_{v \in V} \delta_v$.

1.1 Related Work

If we are allowed to use randomization, we can easily solve the self-stabilizing exploration with a well known strategy called *the simple random walk*. When the agent visits a node $v \in V$, it simply chooses a node as the next destination uniformly at random among $N(v)$, where $N(v)$ is the set of all neighbors of v in G. In other words, it moves to any node $u \in N(v)$ with probability $P_{v,u} = 1/\delta_v$. It is well known that the agent running this simple algorithm visits all nodes in G within $O(\min(mn, mD \log n))$ steps in expectation where $n = |V|$, $m = |E|$, and D is the diameter of G. (See [1,8].) Since the agent is oblivious (*i.e.*, the agent does not bring any information at a migration between two nodes) and does not use whiteboards, the simple random walk is obviously a self-stabilizing exploration algorithm.

Ikeda, Kubo, and Yamashita [6] improved the cover time (*i.e.*, the number of steps to visit all nodes) of the simple random walk by setting the transition probability as $P'_{v,u} = \delta_u^{-1/2} / \sum_{w \in N(v)} \delta_w^{-1/2}$ for any $u \in N(v)$. They proved that the cover time of this *biased random walk* is $O(n^2 \log n)$ steps in expectation. However, we cannot use this result directly in our setting because the agent must know the degrees of all neighbors of the current node to compute the next destination. We can implement this random walk, for example, as follows: every time the agent visits node v, it first obtains $(\delta_u)_{u \in N(v)}$ by visiting all v's neighbors in $2\delta_v$ steps, and then decides the next destination according to probability $(P'_{v,u})_{u \in N(v)}$, which is now computable with $(\delta_u)_{u \in N(v)}$. However, this implementation increases the cover time by a factor of at least $\delta_{\min}$ and at most $\delta_{\max}$. Whereas $n^2 \delta_{\max} \log n > mn$ always holds, $n^2 \delta_{\min} \log n < \min(mn, mD \log n)$ may also hold. Thus, we cannot determine which random walk has smaller cover time without detailed analysis. To bound the space complexity, we must know an upper bound Δ on $\delta_{\max}$ to implement this random walk. If the agent stores $(\delta_u)_{u \in N(v)}$ on v's whiteboard, it uses $O(\log \Delta)$ bits in the agent-memory and $O(\delta_v \log \Delta)$ bits in the whiteboard of each node v. If the agent stores $(\delta_u)_{u \in N(v)}$ only on the agent-memory, it uses $O(\Delta \log \Delta)$ bits in the agent-memory.

The algorithm given by Priezzhev, Dhar, Dhar, and Krishnamurthy [12], which is nowadays well known as the *rotor-router*, solves the self-stabilizing exploration deterministically. The agent is oblivious, but it uses only $O(\log \delta_v)$ bits in the whiteboard of each node $v \in V$. The edges $(\{v, u\})_{u \in N(v)}$ are assumed to be locally labeled by $0, 1, \dots, \delta_v - 1$ in a node v. The whiteboard of each node v has one variable $v.\mathtt{last} \in \{0, 1, \dots, \delta_v - 1\}$. Every time the agent visits a node v, it increases $v.\mathtt{last}$ by one modulo δ_v and moves to the next node via the edge labeled by the updated value of $v.\mathtt{last}$. This simple algorithm guarantees that starting from any configuration, the agent visits all nodes within $O(mD)$ steps [17]. Masuzawa and Tixeuil [7] also gave a deterministic self-stabilizing exploration algorithm. This algorithm itself is designed to solve the gossiping problem where two or more agents have to share their given information with each other. However, this algorithm has a mechanism to visit all the nodes starting from any configuration, which can be seen as a self-stabilizing exploration algorithm. The cover time and the space complexity for the whiteboards of this algorithm are asymptotically the same as those of the rotor-router, while it uses a constant space of the agent-memory, unlike oblivious algorithms such as the rotor-router.

In his seminal paper, Reingold [13] proved that given positive integer N, a Universal Exploration Sequence (UXS) with length $poly(N)$ for (possibly non-simple) connected d-regular graphs with a size of at most N can be explicitly constructed in log-space and, hence, in polynomial time. Although we omit the definition of UXS here, from this result, we can immediately derive a self-stabilizing exploration algorithm for arbitrary graphs whose size is at most N, whose cover time is polynomial in N, with memory requirements of $O(\log N)$ bits for the agent and zero for the whiteboards. One might think that Reingold's UXS was designed for regular graphs, thus questioning its applicability to arbitrary graphs. However, this difference is not significant because we can virtually transform any arbitrary graph into a regular graph by adding self-loops (see [16] for details). Later, Ta-shma and Zwick [16] introduced the concept of a Strongly Universal Exploration Sequence (SUXS) and obtained results similar to those of Reingold, which allow us to improve the cover time of the above-mentioned self-stabilizing exploration algorithm from $poly(N)$ to $poly(n)$. Thus, the cover

Table 1. Randomized self-stabilizing graph exploration algorithms for a single agent.

	Expected Cover Time	Agent Memory	Memory on node v
Simple Random Walk	$O(\min(mn, mD\log n))$	0	0
Biased Random Walk [6] (require $\Delta \geq \delta_{\max}$)	$O(n^2\delta_{\max}\log n)$	$O(\log \Delta)$	$O(\delta_v \log \Delta)$
		$O(\Delta \log \Delta)$	0
$\mathcal{R}_c$ (require $c \geq 2$)	$O\left(m \cdot \min\left(D, \frac{n}{c}+1, \frac{D}{c}+\log n\right)\right)$	$O(\log c)$	$O(\log(\delta_v + c))$

Table 2. Deterministic self-stabilizing graph exploration algorithms for a single agent.

	Cover Time	Agent Memory	Memory on node v
Rotor-router [12]	$O(mD)$	0	$O(\log \delta_v)$
$\mathtt{UXS}_N$ [13] (require $N \geq n$)	polynomial in N	$O(\log N)$	0
$\mathtt{SUXS}_N$ [16] (require $N \geq n$)	polynomial in n	$O(\log N)$	0
2-color DFT [7]	$O(mD)$	$O(1)$	$O(\log \delta_v)$
$\mathcal{D}_k$ (require $k \geq \max(D, \delta_{\max})$)	$O(m + nD)$	$O(\log k)$	$O(\log k)$

time no longer depends on a given upper bound N but only on the actual size n.

A few self-stabilizing algorithms were given for mobile agents to solve problems other than exploration. Blin, Potop-Butucaru, and Tixeuil [2] studied the self-stabilizing naming and leader election problem. Masuzawa and Tixeuil [7] gave a self-stabilizing gossiping algorithm. Ooshita, Datta, and Masuzawa [9] gave self-stabilizing rendezvous algorithms.

For additional related work, including studies on non-self-stabilizing graph exploration [10,14] and self-stabilizing depth-first token circulation [3,5,11], see the arXiv version [15].

1.2 Our Contribution

In this paper, we investigate how short a cover time we can achieve in a self-stabilizing setting. One can easily observe that the cover time is lower bounded by $\Omega(m)$: any deterministic algorithm requires $\Omega(m)$ steps and any randomized algorithm requires $\Omega(m)$ steps in expectation before the agent visits all nodes. (For completeness, we will prove this lower bound in the arXiv version [15].) Our goal is to give an algorithm whose cover time is close to this lower bound with as small complexity of agent-memory and whiteboards as possible.

We give two self-stabilizing exploration algorithms $\mathcal{R}_c$ and $\mathcal{D}_k$, where c and k are the design parameters. The cover times and the space complexities of the proposed algorithms and the existing algorithms are summarized in Tables 1 and 2.

Algorithm $\mathcal{R}_c$ is a randomized algorithm that achieves a cover time of $O(m \cdot \min(D, \frac{n}{c}+1, \frac{D}{c}+\log n))$ steps in expectation, utilizes $O(\log c)$ bits of agent memory, and requires $O(\log(\delta_v + c))$ bits for the whiteboard of each node v. Thus, we have trade-off between the cover time and the space complexity. The larger c we use, the smaller cover time we obtain. In particular, the expected cover time is $O(m \log n)$ steps if we set $c = \Omega(D/\log n)$, and it becomes optimal (*i.e.*, $O(m)$ steps) if we set $c = \Omega(n)$. This means that we require the knowledge of $\Omega(n)$ value to make $\mathcal{R}_c$ time-optimal. Fortunately, this assumption can be ignored from a practical point of view: even if c is extremely larger than n, the overhead will be just an additive factor of $\log c$ in the space complexity. Thus, any large $c \in poly(n) \cap \Omega(n)$ is enough to obtain the optimal cover time and the space complexity of $O(\log n)$ bits both in the agent memory and whiteboards. Moreover, irrespective of parameter $c \geq 2$, the cover time is $O(mD)$ steps with probability 1.

Algorithm $\mathcal{D}_k$ is deterministic, and its cover time is $O(m + nD)$ steps, independent of the parameter k. The agent uses $O(\log k)$ bits both for agent memory and the whiteboard at each node. Thus, we do not have a trade-off between cover time and space complexity. Unlike $\mathcal{R}_c$, however, algorithm $\mathcal{D}_k$ requires prior knowledge of an upper bound on the diameter and maximum degree, specifically, it assumes $k \geq \max(D, \delta_{\max})$. If this assumption is not met, the correctness of the algorithm is no longer guaranteed. Nevertheless, this assumption can still be considered weak: since the space complexity grows only logarithmically in k, it is always feasible to choose a sufficiently large polynomial upper bound on n for k without significantly affecting memory usage. Thus, even though the algorithm formally requires knowledge of an upper bound, this requirement does not substantially limit the applicability of this algorithm.

It remains open if there is a deterministic self-stabilizing exploration algorithm with optimal cover time, *i.e.*, $O(m)$ steps.

2 Preliminaries

Let $G = (V, E, p)$ be a simple, undirected, and connected graph where V is the set of nodes and E is the set of edges. The edges are locally labeled at each node: we have a family of functions $p = (p_v)_{v \in V}$ such that each $p_v : \{\{v, u\} \mid u \in N(v)\} \to \{0, 1, \dots, \delta_v - 1\}$ uniquely assigns a *port number* to every edge incident to node v. Two port numbers $p_u(e)$ and $p_v(e)$ are independent of each other for edge $e = \{u, v\} \in E$. The node u neighboring to node v such that $p_v(\{v, u\}) = q$ is called the *q-th neighbor* of v and is denoted by $N(v, q)$. For any two nodes u, v, we define the distance between u and v as the length of a shortest path between them and denote it by $d(u, v)$. We also define the set of the *i-hop neighbors* of v as $N_i(v) = \{u \in V \mid d(u, v) \leq i\}$. Note that $N_0(v) = \{v\}$ and $N_1(v) = N(v) \cup \{v\}$.

An *algorithm* is defined as a 3-tuple $\mathcal{P} = (\phi, \mathcal{M}, \mathcal{W})$, where $\mathcal{M}$ is the set of states for the agent, $\mathcal{W} = (\mathcal{W}_k)_{k \in \mathbb{N}}$ is the family such that $\mathcal{W}_k$ is the set of states for each node with degree k, and ϕ is a function that determines how the agent updates its state (*i.e.*, agent memory) and the state of the current node (*i.e.*, *whiteboard*). At each time step, the agent is located at exactly one node $v \in V$, and moves through an edge incident to v. Every node $v \in V$ has a whiteboard $w(v) \in \mathcal{W}_{\delta_v}$, which the agent can access freely when it visits v. The function ϕ is invoked every time the agent visits a node or when the exploration begins. Suppose that the agent with state $s \in \mathcal{M}$ has moved to node v with state $w(v) = x \in \mathcal{W}_{\delta_v}$ from $u \in N(v)$. Let $p_{\mathrm{in}} = p_v(\{u, v\})$. The function ϕ takes 4-tuple $(\delta_v, p_{\mathrm{in}}, s, x)$ as the input and returns 3-tuple $(p_{\mathrm{out}}, s', x')$ as the output. Then, the agent updates its state to s' and $w(v)$ to x', after which it moves via port p_{out}, that is, it moves to v' such that $p_{\mathrm{out}} = p_v(\{v, v'\})$. At the beginning of an execution, we let p_{in} be an arbitrary integer in $\{0, 1, \ldots, \delta_v - 1\}$ where v is the node that the agent exists on. Note that if algorithm $\mathcal{P}$ is randomized one, function ϕ returns the probabilistic distributions for $(p_{\mathrm{out}}, s', x')$.

Given a graph $G = (V, E)$, a *configuration* (or a global state) consists of the location of the agent, the state of the agent (including p_{in}), and the states of all the nodes in V. Algorithm $\mathcal{P}$ is a self-stabilizing exploration algorithm for a class $\mathcal{G}$ of graphs if for any graph $G = (V, E, p) \in \mathcal{G}$, the agent running $\mathcal{P}$ on G eventually visits all the nodes in V at least once starting from any configuration. Note that, by the above definition, any self-stabilizing exploration algorithm ensures that the single agent visits every node infinitely often.

We measure the efficiency of algorithm $\mathcal{P}$ by three metrics: *the cover time*, *the agent memory space*, and *the whiteboard memory space*. All the above metrics are evaluated in the worst-case manner with respect to graph G and an initial configuration. The cover time is defined as the number of moves that the agent makes before it visits all nodes. If algorithm $\mathcal{P}$ is a randomized one, the cover time is evaluated in expectation. The memory spaces of the agent and the whiteboard on node v are just defined as $\log_2 |\mathcal{M}|$ and $\log_2 |\mathcal{W}_{\delta_v}|$, respectively.

Throughout this paper, we call the node the agent currently exists on *the current node* and denote it by v_{cur}. We call the port number of the current node via which the agent migrates to v_{cur} *in-port* and denote it by p_{in}. As mentioned above, the function ϕ can use p_{in} to compute 3-tuple $(p_{\mathrm{out}}, s', x')$. In other words, the agent can always access p_{in} to update its variables and the whiteboard variables of the current node and compute the destination of the next migration.

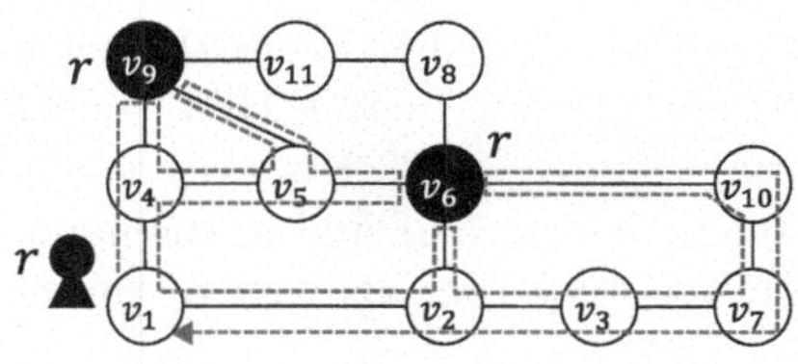

Fig. 1. An example of the trajectory of one DFT.

3 Randomized Exploration

In this section, we give a randomized self-stabilizing exploration algorithm $\mathcal{R}_c$. The cover time and space complexity of $\mathcal{R}_c$ depend on the design parameter $c \geq 2$. Specifically, the cover time is $O(m)$ if $c = \Omega(n)$, while it is $O(m \log n)$ if $c = \Omega(D/\log n)$. The cover time is $O(mD)$ with probability 1, regardless of how small c may be. Algorithm $\mathcal{R}_c$ requires $O(\log c)$ bits of agent memory and $O(\log(\delta_v + c))$ bits for each node $v \in V$. If $c = 2$, the algorithm becomes deterministic and is nearly identical to the exploration algorithm given by [7].

The idea of algorithm $\mathcal{R}_c$ is simple. The agent tries to traverse all the edges in every $2m$ moves according to the Depth-First-Traversal (DFT). However, we have an issue for executing DFT. When the agent visits a node for the first time, it has to mark the node as an *already-visited* node. However, we must deal with an arbitrary initial configuration in a self-stabilizing setting, thus some of the nodes may be marked even at the beginning of an execution. To circumvent this issue, we use *colors* and let the agent make DFTs repeatedly. The agent and all nodes maintain their color in a variable `color` $\in \{1, 2, \ldots, c\}$. Every time the agent begins a new DFT, it chooses a new color ρ uniformly at random from all the colors except for the current color of the agent. In the new DFT, the agent changes the colors of all the visited nodes to ρ. Thus, the agent can distinguish the visited nodes and the unvisited nodes with their colors: it interprets that the nodes colored ρ are already visited and the nodes having other colors have not been visited before in the current DFT. Of course, the agent may make incorrect decision because one or more nodes may be colored ρ when the agent chooses ρ for its color at the beginning of a new DFT. Thus, the agent may not perform a complete DFT. However, the agent can still make a progress for exploration to a certain extent in this DFT: Let r be the node at which the agent is located when it begins the DFT, let S be the set of the nodes colored ρ at that time, and let R be the set of all nodes in the connected component including r of the induced subgraph $G[V \setminus S]$; then it visits all nodes in R and their neighbors. Hence, as we will see later, the agent visits all nodes in V by repeating DFTs a small number of times, which depends on how large c is.

Figure 1 shows an example of one DFT where $r = v_1$ and $S = \{v_6, v_9\}$. The agent tries to make a DFT on the graph starting from node v_1. Unfortunately, whenever the agent visits node v_6 or v_9, which are colored ρ, it immediately *backtracks* to the previous node. This is because it mistakenly *thinks* that it has already visited these nodes even when it visits them for the first time in the DFT. Thus, it never visits v_8 or v_{11} in this DFT, which is unreachable from node v_1 in the induced subgraph $G[V \setminus S]$. However, it visits all the other nodes. Note that the index of the nodes are depicted in the figure only for simplicity of explanation, and we do not assume the existence of any unique identifiers of the nodes.

Following the idea described above, we can design a randomized algorithm $\mathcal{R}_c$ that satisfies the following two lemmas. Due to page limitations, the detailed description of $\mathcal{R}_c$, particularly its pseudocode, is provided in the arXiv version [15].

Lemma 1. *Starting from any configuration, the agent running $\mathcal{R}_c$ changes its color (i.e., begins a new DFT) within $8m + n = O(m)$ rounds with probability 1.*

Lemma 2. *Suppose now that the agent running $\mathcal{R}_c$ changes its color to ρ and begins a new DFT at node r. Let $S \subseteq V$ be the set of the nodes colored ρ at that time and let R be the set of the nodes in the connected component including r in the induced subgraph $G[V \setminus S]$. Then, the agent visits all nodes in R and their neighbors in this DFT with probability 1. Moreover, this DFT finishes (i.e., the agent changes its color) at node r.*

In the DFT mentioned in Lemma 2, the agent changes the colors of all nodes in $R' = R \cup \bigcup_{v \in R} N(v)$ to ρ. In the next DFT, the agent must choose a color different from ρ, ensuring that the agent will visit all nodes in R' and their neighbors (plus possibly many more nodes). Consequently, the number of the nodes visited in the i-th DFT monotonically increases with respect to i. Therefore, we obtain the following corollary.

Corollary 1. *The agent running $\mathcal{R}_c$ visits all nodes before it changes its color $D + 1$ times.*

Theorem 1. *The following propositions hold: (i) Algorithm $\mathcal{R}_c$ is a randomized self-stabilizing exploration algorithm for all simple, undirected, and connected graphs, (ii) irrespective of c, the cover time is $O(mD)$ steps with probability 1, (iii) the expected cover time is $O(m \cdot \min(D, \frac{n}{c} + 1, \frac{D}{c} + \log n))$ steps, and (iv) the agent memory space is $O(\log c)$ and the memory space of each node v is $O(\log(\delta_v + c))$.*

Proof. The first and the second claims of this theorem immediately follow from Lemma 1 and Corollary 1. The fourth claim holds because the agent maintains exactly one color (from the set $\{1, 2, \dots, c\}$) in its memory and on the whiteboard of each node, and stores the parent pointer of each node v (from the set $\{0, 1, \dots, \delta_v\}$) on the whiteboard of each node v, and do not use any other variable with a super-constant size. In the rest of this proof, we prove the third claim of this lemma: the agent visits all nodes within $O\left(m \cdot \min\left(D, \frac{n}{c} + 1, \frac{D}{c} + \log n\right)\right)$ steps in expectation. Let X denote how many times the agent changes its color before it visits all nodes in V. By Lemma 1 and $\Pr[X \le D + 1] = 1$, it suffices to show $\mathbf{E}[X] = O(n/c + 1)$ and $\mathbf{E}[X] = O(D/c + \log n)$. We exclude the case $c = 2$ without loss of generality because these equalities immediately follow from $X \le D + 1$ when $c = 2$.

First, we prove that $\mathbf{E}[X] = O(n/c + 1)$. For each integer $i \ge 1$, we define the i-th DFT as the DFT executed by the agent immediately after it has changed its color exactly i times. Let V_i be the set of nodes visited by the agent during the i-th DFT. Fix some $i \ge 1$ and let $V' \subseteq V$ be an arbitrary superset of V_i satisfying $|V' \setminus V_i| = \lfloor c/2 \rfloor$, with the induced subgraph of V' being connected. If $|V_i| + \lfloor c/2 \rfloor > n$, we define $V' = V$. By Lemma 2, if the color chosen by the agent for the $(i+1)$-th DFT differs from the colors of all nodes in $V' \setminus V_i$, then the agent visits all nodes of V' during the $(i+1)$-th DFT. In that case, we have $|V_{i+1}| \geq \min(|V_i| + \lfloor c/2 \rfloor, n)$. Such an event occurs with probability

$$p \geq 1 - \frac{|V' \setminus V_i|}{c-1} \geq 1 - \frac{\lfloor c/2 \rfloor}{c-1} = \Theta(1),$$

assuming $c \geq 3$. Thus, in expectation, $|V_i|$ increases by at least $\lfloor c/2 \rfloor$ or reaches n within $1/p = O(1)$ DFTs, until $V_i = V$ holds. This yields

$$\mathbf{E}[X] \leq \left\lceil \frac{n}{\lfloor c/2 \rfloor} \right\rceil \cdot \frac{1}{p} = O\left(\frac{n}{c} + 1\right).$$

Next, we prove $\mathbf{E}[X] = O(D/c + \log n)$. By Lemma 2, the agent always begins DFTs at the same node. We denote this node by r. For any node v, consider a shortest path $p = v_0, v_1, \ldots, v_\ell$ from r to v, *i.e.*, $v_0 = r$ and $v_\ell = v$. Let Y be the maximum integer such that v_Y has already been visited. Note that Y is monotonically non-decreasing. For each iteration of DFT, we say that the iteration *succeeds* if by this iteration, Y increases by at least $\lfloor (c-1)/2 \rfloor$ or reaches to ℓ. The agent visits $v = v_\ell$ before it performs $\chi = D/\lceil (c-1)/2 \rceil$ successful iterations of DFT. Since the agent chooses a new color uniformly at random among $c-1$ colors when it begins a new DFT, each iteration of DFT succeeds with a probability of at least $1/2$. Suppose the agent repeats a DFT $2\chi + \lceil 16 \log_e n \rceil$ times, and let Z be the number of successful iterations among those DFTs. By Chernoff bound, $Z \geq \mathbf{E}[Z]/2 \geq \chi$ holds with a probability of at least $1 - \exp(-\mathbf{E}[Z]/8) \geq 1 - \exp(-2 \log_e n) = 1 - n^{-2}$. Thus, the agent visits any node v in $O(D/c + \log n)$ DFTs with probability $1 - n^{-2}$. Thus, $X = O(D/c + \log n)$ with high probability, yielding $\mathbf{E}[X] = O(D/c + \log n)$. □

4 Deterministic Exploration

The randomized algorithm $\mathcal{R}_c$, which is presented in the previous section, achieves self-stabilizing exploration in a small number of steps in expectation. The key idea is to repeat DFTs with changing the color of the agent randomly. Unfortunately, we cannot make use of the same idea if we are not allowed to use randomization and we choose a new color deterministically among c candidates. For example, consider an $(n/2, n/2)$-lollipop graph $G_L = (V_L, E_L)$, which consists of a clique $K_{n/2}$ of $n/2$ nodes and a path $P_{n/2}$ of $n/2$ nodes, connected with a bridge. Let $u_1, u_2, \ldots, u_{n/2}$ be the nodes in $P_{n/2}$ such that one endpoint u_1 neighbors one node of clique $K_{n/2}$. Suppose that the agent begins a new DFT in a configuration where it is located in the clique, and all the nodes in the clique have the same color (Fig. 2). For $i = 1, 2, \ldots, n/2$, let c_i be the colors that the agent (deterministically) chooses for the i-th DFT. Then, the adversary can assign color c_i to each node u_i in the path in the initial configuration. In this case, the deterministic version of $\mathcal{R}_c$ requires $\Omega(n^3)$ steps to visit all the nodes in G_L because the agent visits u_i in the i-th DFT but it immediately backtracks without visiting u_{i+1}, and each DFT involves the move through every edge in the clique, resulting in $\Theta(n^2)$ steps.

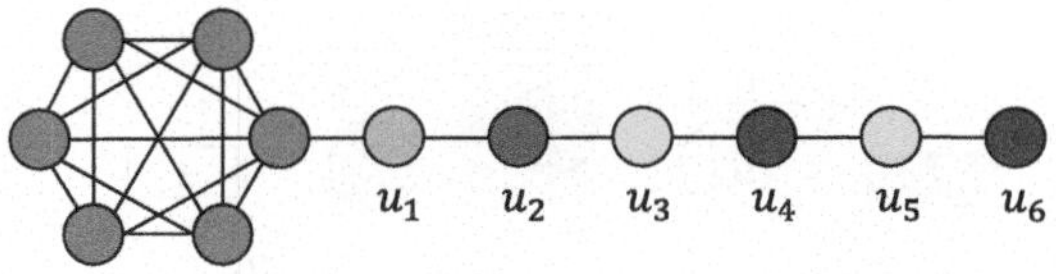

Fig. 2. A (6,6)-lollipop graph.

In this section, we give a deterministic self-stabilizing exploration algorithm $\mathcal{D}_k$ with design parameter k. We must satisfy $k \geq \max(D, \delta_{\max})$, otherwise this algorithm no longer guarantees the correctness. Thus we require the knowledge of an upper bound on $\max(D, \delta_{\max})$. Unlike $\mathcal{R}_c$, the cover time of this algorithm does not depend on parameter k: the agent visits all nodes within $O(m + nD)$ steps starting from any configuration. Algorithm $\mathcal{D}_k$ requires $O(\log k)$ bits both in the agent memory and the whiteboard of each node $v \in V$. Throughout this section, we denote by $dist(u, v)$ the distance between two nodes u and v, *i.e.*, the length of a shortest path from u to v.

Whereas $\mathcal{R}_c$ is based on depth first traversals, $\mathcal{D}_k$ is based on Breadth First Traversals (BFTs). Specifically, the agent attempts to construct a Breadth First Search (BFS) tree rooted at some node $r \in V$. Constructing a BFS tree by a single agent in a self-stabilizing setting poses several challenges, especially when bounding the time complexity by $O(m + nD)$ steps. In the rest of this section, we first introduce some terminologies and then explain the challenges posed by self-stabilization and how we address them.

In the remainder of this section, we present all key ideas of $\mathcal{D}_k$ and the proofs necessary to verify the correctness of our main theorem. The detailed implementation of $\mathcal{D}_k$, including its pseudocode, is provided in the arXiv version [15], which may be helpful for a deeper understanding of $\mathcal{D}_k$.

4.1 Consistent Tree

In $\mathcal{D}_k$, the agent attempts to maintain a BFS tree that it can periodically traverse. Due to memory constraints, explicitly storing this tree structure at the nodes is non-trivial: the set of children of a node v cannot be stored at v, as only $O(\log k)$ bits are available on v's whiteboard. In this subsection, we explain how to embed a tree within the graph while adhering to these memory constraints. The main idea is as follows: since any tree has at most $n - 1$ parent-child relationships, we can store this information in a distributed manner across the n nodes, with each node storing only $O(\log k)$ bits. Our approach is inspired by the well-known "left-child, right-sibling" representation.

Each node $v \in V$ maintains the following whiteboard variables:

$$v.\texttt{parent} \in \{0, 1, 2, \dots, \delta_v - 1\}, \quad v.\texttt{firstChild}, v.\texttt{nextChild} \in \{\bot, 0, 1, \dots, k-1\},$$
$$v.\texttt{level} \in \{0, 1, \dots, k\}.$$

A node v is *valid* if $v.\texttt{firstChild} = \bot$ or there is a sequence of nodes $c_1, c_2, \dots, c_\ell$ belonging to $N(v)$ with $\ell \geq 1$ that satisfies the following conditions:

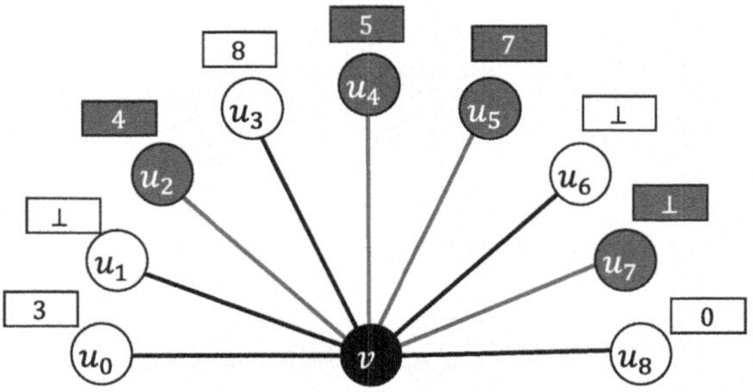

v.**firstChild** = 2

Fig. 3. A valid node v with $N(v) = \{u_0, u_1, \ldots, u_8\}$. The index i of each u_i represents the corresponding number in v, *i.e.*, $u_i = N(v, i)$. The number in each rectangle represents the value of u_i.nextChild. In addition, $v = N(u_i, u_i.\texttt{parent})$ and $u_i.\texttt{level} = v.\texttt{level} = 1$ must hold for all $i \in \{2, 4, 5, 7\}$. Then, u_2, u_4, u_5, and u_7 are the children of v.

- $c_i.\texttt{parent} = v$ and $c_i.\texttt{level} = v.\texttt{level} + 1$ for all $i \in [1, \ell]$,
- $p_v(\{v, c_i\}) < p_v(\{v, c_{i+1}\})$ and $c_i.\texttt{nextChild} = p_v(\{v, c_{i+1}\})$ for all $i \in [1, \ell - 1]$,
- $N(v, v.\texttt{firstChild}) = c_1$, and
- $c_\ell.\texttt{nextChild} = \bot$.

We call those nodes $c_1, c_2, \ldots, c_\ell \in N(v)$ the children of v. A typical example of a valid node is provided in Fig. 3.

A tree $T = (V_T, E_T)$ rooted at $r \in V$ is *consistent* if $r.\texttt{level} = 0$, all nodes in V_T are valid, and for any $u, v \in V$, u is a child of v if and only if $(u, v) \in E_T$ holds. By definition, we observe the following proposition.

Observation 1. *For any configuration C and any node $v \in V$, there is at most one consistent tree that contains v in C.*

We say that a node $v \in V$ is *consistent* if there is a consistent tree $T = (V_T, E_T)$ such that $v \in V_T$. Observation 1 implies that the consistent tree that contains a consistent node v is uniquely determined. We denote such a tree by $T(v)$. Notice $T(v)$ may not be rooted at v.

In the remainder of this paper, for any tree $T = (V_T, E_T)$, we use $v \in T$ and $e \in T$ to denote $v \in V_T$ and $e \in E_T$, respectively. We define $|T|$ as the number of nodes in the tree, i.e., $|T| = |V_T|$.

Note that a consistent tree $T = (V_T, E_T)$ is not necessarily a subgraph of a BFS tree. For example, there may be some node $v \in V_T$ for which $v.\texttt{level} > \mathit{dist}(r, v)$. Additionally, there could be a node $u \notin V_T$ such that $\mathit{dist}(r, u) < \mathit{dist}(r, w)$ for some $w \in V_T$.

4.2 Tree Expansion

Our deterministic algorithm $\mathcal{D}_k$ expands $T(v_{\mathrm{cur}})$ by incorporating new nodes one by one. The first challenge is determining how to expand $T(v_{\mathrm{cur}})$. Consider the

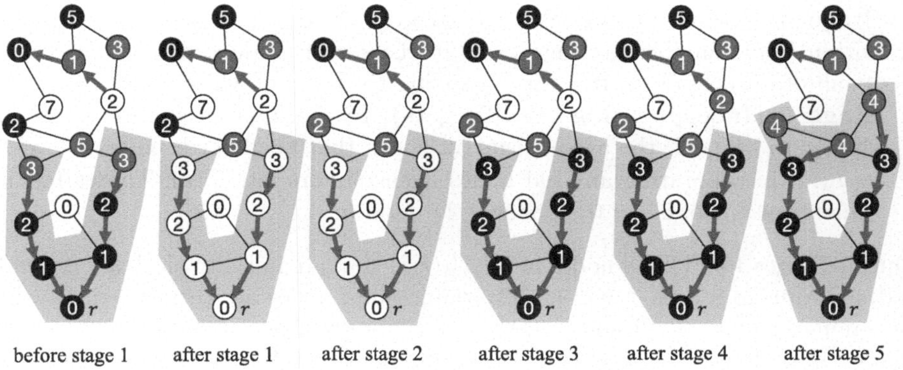

Fig. 4. How to expand tree $T(r)$ in phase 3. A number in a circle represents the level of the corresponding node. An arrow from node u to v indicates that u is a child of v. The orange region represents which nodes join $T(r)$. (Color figure online)

scenario where the agent is located at a consistent node $v \in V$, and explores an edge $e = \{v, u\} \notin T(v)$. If $u \notin T(v)$, the agent should incorporate the destination node u into $T(v)$ by setting $u.\texttt{parent} \leftarrow p_u(e)$ and $u.\texttt{level} = v.\texttt{level} + 1$, initializing $u.\texttt{firstChild} \leftarrow \bot$, and updating $u.\texttt{nextChild}$ and $v.\texttt{firstChild}$ appropriately. However, the agent must refrain from making these changes if u is already part of $T(v)$. A significant challenge arises here because when the agent visits node u, another node $w \in T(v)$ may be in the same state as u, complicating the determination of whether u belongs to $T(v)$.

We resolve this challenge by maintaining two agent variables $\mathit{self}.\texttt{phase} \in \{0, 1, 2, \ldots, k-1\}$ and $\mathit{self}.\texttt{stage} \in \{1, 2, 3, 4, 5\}$, and one whiteboard variable $v.\texttt{color} \in \{B, W, R\}$ for each $v \in V$. Here, B, W, and R stand for *black*, *white*, and *red*, respectively. When $\mathit{self}.\texttt{phase} = i$, the agent is said to be in phase i. Suppose the agent begins phase 0 at node $r \in V$. In this phase, the agent visits all neighbors of r, incorporating them such that $T(r)$ becomes the star centered at r, containing all nodes in $N_1(r)$. For any $i \geq 1$, let $V(r, i)$ be the set of nodes with level i in $T(r)$, and $V_{\text{target}}(r, i)$ be the set of nodes that are neighbors to a node in $V(r, i)$ and are not included in $T(r)$. The objective of phase i is to incorporate all nodes in $V_{\text{target}}(r, i)$ into $T(r)$. Each phase consists of five sub-phases, or *stages*, managed by the variable $\mathit{self}.\texttt{stage} \in \{1, 2, 3, 4, 5\}$, described as follows (Fig. 4 illustrates a typical example).

- **Stage 1 (Coloring the Current Tree White)**: The agent circulates within $T(r)$ (*i.e.,* visits all nodes in $T(r)$ in $2|T(r)| - 2$ steps), coloring all nodes white.
- **Stage 2 (Coloring the Black Targets Red)**: The agent recirculates within $T(r)$. During the circulation, each time it visits a node v at level i, it visits each neighbor $u \in N(v)$ and colors u red if u is black.
- **Stage 3 (Coloring the Current Tree Black)**: The agent recirculates within $T(r)$, this time coloring all nodes in $T(r)$ black.

- **Stage 4 (Coloring the White Targets Red)**: The agent recirculates within $T(r)$. Similar to Stage 2, as it visits each node v at level i, it visits all of v's neighbors, this time changing the color of white nodes to red.
- **Stage 5 (Incorporating the Targets)**: In the final circulation of $T(r)$, similar to Stage 2, the agent visits all nodes that neighbor a node at level i in $T(r)$, incorporating any red nodes it encounters into $T(r)$. Thereafter, it updates $\mathit{self}.\mathtt{phase} \leftarrow (\mathit{self}.\mathtt{phase} + 1) \mod k$.

During Stages 1 and 2, all nodes in $V_{\mathrm{target}}(r, i)$ that are initially black are colored red, and during Stages 3 and 4, all initially white nodes are turned red. By the beginning of Stage 5, all nodes in $V_{\mathrm{target}}(r, i)$ are red, and all nodes in $T(r)$ are black, thus allowing the agent to clearly distinguish and incorporate nodes from $V_{\mathrm{target}}(r, i)$ into $T(r)$.

In every stage, the agent circulates $T(r)$ exactly once. This circulation is easily accomplished as follows: Here, we refer to the movement from a parent to a child as a *forward move* and the movement from a child to a parent as *backtracking*. Each time the agent makes a forward move and visits a node u, it moves to u's first child $N(u, u.\mathtt{firstChild})$ if $u.\mathtt{firstChild} \neq \bot$; otherwise, it simply returns to u's parent. Suppose the agent backtracks from a node u to u's parent, say v, and u is the i-th child of v. Before backtracking, the agent tentatively memorizes $u.\mathtt{nextChild}$ in its own variable $\mathit{self}.\mathtt{nextChild}$. Thus, after backtracking, it can go to the $i+1$-th child of v if one exists. If u is the last child of v, $\mathit{self}.\mathtt{nextChild}$ must be $\bot$ at that time. Then, the agent returns to v's parent, or terminates the circulation if $v = r$. This process clearly circulates $T(r)$, moving through every edge in the tree exactly twice, requiring $2|T(r)| - 2$ steps.

In Stages 2, 4, and 5, each time the agent visits a node v at level $\mathit{self}.\mathtt{phase}$, it must visit all of v's neighbors. At this point, the agent moves to the neighbors in descending order of the port numbers; specifically, it visits $u_{\delta_v - 1}, u_{\delta_v - 2}, \ldots, u_0$ in this order, where $u_i = N(v, i)$ for $i \in [0, \delta_v - 1]$. This ordering helps the agent efficiently incorporate red neighbors of v into $T(r)$ during Stage 5. Initially, the agent sets both $v.\mathtt{firstChild}$ and $\mathit{self}.\mathtt{nextChild}$ to $\bot$. Thereafter, each time the agent visits a red neighbor u_i, it first updates $u_i.\mathtt{nextChild} \leftarrow \mathit{self}.\mathtt{nextChild}$ in addition to $u_i.\mathtt{level} = v.\mathtt{level} + 1$, $u_i.\mathtt{parent} = p_{\mathrm{in}}$, and $u_i.\mathtt{firstChild} \leftarrow \bot$. After returning to v, it updates $\mathit{self}.\mathtt{nextChild} \leftarrow i$ and $v.\mathtt{firstChild} \leftarrow i$. Once this process is completed, all red nodes in $N(v)$ are incorporated into $T(r)$, ensuring that $T(r)$ remains consistent.

Lemma 3. *If the agent begins phase i at a consistent node r with level 0, $V_{\mathrm{target}}(r, i)$ is incorporated into $T(r)$ before the end of phase i. Phase i ends in $O\left(n + \sum_{v \in V(r,i)} \delta_v\right)$ steps.*

Proof. The correctness of the first part of the lemma, concerning the incorporation of $V_{\mathrm{target}}(r, i)$ into $T(r)$ in phase i, follows from the above discussion. We derive the second part from the fact that: (i) each of the five circulations on $T(r)$ requires only $2|T(r)| - 2 = O(n)$ steps, and (ii) in Stages 2, 4, and 5, for each node $v \in T(r, i)$, the agent moves from v to each $u \in N(v)$ and vice versa, which requires exactly $3 \cdot 2 \cdot \sum_{v \in V(r,i)} \delta_v$ steps in total. □

Corollary 2. *Once the agent begins phase* 0*, it visits all nodes within* $O(m+nD)$ *steps.*

Proof. Let r be the node at which the agent begins phase 0. As defined, phase 0 completes within $O(\delta_r)$ steps. During the subsequent phases $1, 2, \ldots, D-1$, by Lemma 3, all nodes are incorporated into $T(r)$ within $O(m+nD)$ steps in total. □

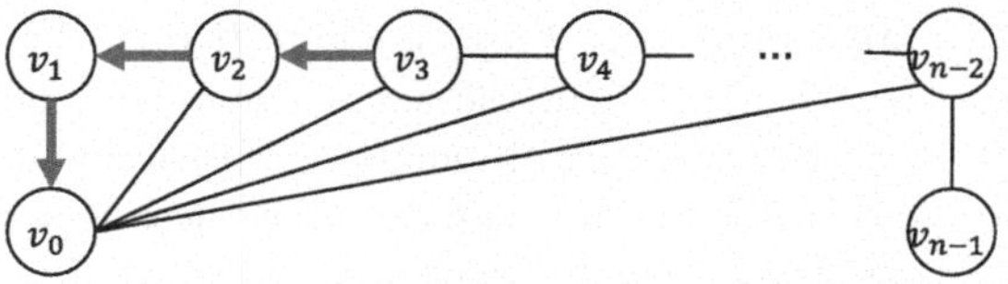

Fig. 5. A consistent but undesirable tree.

4.3 Towards Fast Reset

Since we seek a self-stabilizing solution, we must address arbitrary initial configurations where the current node may not be consistent, or $T(v_{\text{cur}})$ significantly deviates from a BFS tree structure. How can we ensure that the agent visits all nodes within $O(m + nD)$ steps starting from such a configuration? According to Corollary 2, this is feasible if the agent can reset *self*.`phase` to zero and initiate phase 0 within every $\Theta(m + nD)$ steps with a sufficiently large hidden constant. However, the agent does not know the values of n, m, or D; it only knows an upper bound $k \geq \max(D, \delta_{\max})$, which may significantly exceed D, i.e., $k = \omega(D)$. Therefore, we require a sophisticated mechanism to reset *self*.`phase` to zero within $O(m + nD)$ steps from any configuration, without harming tree expansion so that the validity of Corollary 2 is preserved.

It is straightforward to handle configurations where the current node, v_{cur}, is not consistent. As mentioned earlier, the agent attempts to circulate the consistent tree $T(v_{\text{cur}})$ in every stage of each phase, assuming that v_{cur} is consistent. If v_{cur} is not consistent, the agent will encounter some inconsistency within $O(m)$ steps during the circulation process, or v_{cur} becomes consistent. For instance, an inconsistency is detected when the agent moves from a node v to u, one of v's children, but either $u.\texttt{parent} \neq p_u(\{u, v\})$ or $u.\texttt{level} \neq v.\texttt{level} + 1$ holds. The agent begins phase 0 whenever it detects any such *local* inconsistency. (See the arXiv version [15] for detailed rules on detecting inconsistencies). Hence, in the rest of this section, we assume that v_{cur} is consistent without loss of generality.

The main challenge is handling configurations where v_{cur} is consistent, yet $T(v_{\text{cur}})$ significantly deviates from a BFS tree structure. To illustrate the difficulty of this scenario, consider the configuration shown in Fig. 5, where the agent begins phase 3 at a node $r = v_0$ and $T(r) = \{(v_3, v_2), (v_2, v_1), (v_1, v_0)\}$. In this configuration, edges $\{v_0, v_1\}, \{v_0, v_2\}, \ldots, \{v_0, v_{n-2}\}$ are never utilized in subsequent tree circulations, as v_1 is the only child of v_0. Consequently, exactly one node is added to $T(r)$ in each phase, requiring $\Omega(n^2)$ steps to visit all nodes and reset *self.phase* to zero, despite $D = 2$ and $m = O(n)$.

If we ignore time complexity, this challenge can be easily addressed. For example, consider what happens if we require $T(v_{\text{cur}})$ to have some kind of BFS tree structure in addition to being consistent. Specifically, let r be the root of $T(v_{\text{cur}})$ and suppose we require: (i) for any $v \in T(r)$, $v.\texttt{level} = \mathit{dist}(r, v)$, and (ii) all nodes $v \in N_H$ are contained in $T(v_{\text{cur}})$, where H is the height of $T(r)$, *i.e.*, $H = \max\{u.\texttt{level} \mid u \in T(r)\}$. Then, the agent can verify these conditions in each circulation of $T(v_{\text{cur}})$ by visiting all neighbors of each node in $T(v_{\text{cur}})$ and reset *self*.`phase` to zero if it finds any violations of these conditions. However, this may require $\Omega(m)$ steps in each circulation, resulting in $\Omega(mD)$ steps in the entire execution of the algorithm.

We address the second challenge (*i.e.*, deviation from the BFS structure) by adding the following three rules:

- **Rule 1**: In Stage 5 of each phase, the agent checks whether it has incorporated at least one node into $T(v_{\text{cur}})$. If no node has been incorporated, indicating that $T(v_{\text{cur}})$ can no longer be expanded, the agent resets *self*.`phase` to zero.
- **Rule 2**: In Stage 1 of each phase i, the agent resets *self*.`phase` to zero if it finds a node $v \in V(r, i)$ such that $v.\texttt{firstChild} \neq \bot$.
- **Rule 3**: As explained above, in Stage 5 of each phase i, the agent visits all neighbors of each node $v \in V(r, i)$. The agent resets *self*.`phase` to zero if it finds a black neighbor $u \in N(v)$ with $u.\texttt{level} < \mathit{self}.\texttt{phase} - 1 = i - 1$.

In what follows, we first observe that these three rules do not invalidate Corollary 2, and then demonstrate that these rules ensure a high frequency of resetting *self*.`phase`.

Observation 2. *Corollary 2 still holds even with the addition of the above three rules.*

Proof. Suppose the agent begins phase 0 at a node r. The first rule is applicable only after the agent has incorporated all nodes into $T(r)$ and thus does not affect the corollary. The second rule is never triggered because the tree expansion proceeds hop by hop after initiating phase 0. The third rule is also never applicable in the execution subsequent to resetting *self*.`phase` to zero, because at the beginning of each phase i, $T(v_{\text{cur}})$ forms a shortest path (*i.e.*, BFS) tree over $G[N_i(r)]$. This configuration ensures that the difference in levels between any two nodes $u, v \in V'$, where V' is the set of nodes in $T(v_{\text{cur}})$ at the start of phase i, satisfies $|u.\texttt{level} - v.\texttt{level}| \leq 1$. Moreover, all black nodes that the agent visits in Stage 5 must be within V': all neighbors of the nodes in $V(r, i)$ that are not part of V' are colored red in either Stage 2 or 4. Thus, the condition of the third rule is never satisfied. □

Lemma 4. *Starting from any configuration, the agent begins phase 0 in $O(m + nD)$ steps.*

Proof. As mentioned above, the agent begins phase 0 within $O(m)$ steps by finding a local inconsistency if the current node is not consistent. Thus, we assume the current node is consistent. Since each phase requires only $O(m)$ steps, starting from any configuration, the agent initiates some phase i at some node, say r, within $O(m)$ steps. By the second rule, we can assume without loss of generality that at that time all nodes $v \in V(r, i)$ do not have children in $T(r)$, *i.e.*, $v.\mathtt{firstChild} = \bot$; otherwise, the agent begins phase 0 in the next $O(m)$ steps.

In the rest of this proof, we show that the agent resets *self*.phase to zero before it enters phase $i+2D$, thereby supporting the lemma by Lemma 3. Assume for contradiction that it enters phase $i + 2D$ and focus on the configuration C at that time. Let $V_{<i} = \bigcup_{j<i} V(r, j)$, u any node in $V(r, i)$, and v any node in $V(r, i + D + 1)$. Note that the first rule guarantees $V(r, i + D + 1) \neq \emptyset$. After initiating phase i, the agent expands $T(r)$ hop by hop, avoiding the region of $V_{<i}$. This implies that the distance between u and v in the induced subgraph $G[V \setminus V_{<i}]$ is at least $D + 1$; otherwise, $v.\mathtt{level} \leq i + D$ holds, contradicting $v \in V(r, i + D + 1)$. Since $dist(u, v) \leq D$, a shortest path $v_0, v_1, \ldots, v_\ell$ between v and u must contain a node in $V_{<i}$, where $v_0 = v$, $v_\ell = u$, and $\ell \leq D$ holds. Let s be the smallest index such that $v_s \in V_{<i}$, *i.e.*, $v_0, v_1, \ldots, v_{s-1} \notin V_{<i}$. Since $v = v_0$ and v_{s-1} belong to the same connected component in $G[V \setminus V_{<i}]$, v_{s-1} is contained in $T(r)$ and $v_{s-1} \in V(r, i')$ holds for some $i' \in [i + 1, i + D + s]$ in configuration C. Note that we have $i' \neq i$ here because otherwise $v.\mathtt{level} \leq i + s - 1 < i + D + 1$ holds, a contradiction with the fact $v \in V(r, i + D + 1)$. Therefore, in Stage 5 of phase $i' \leq i+D+s < i+D+\ell \leq i+2D$, the agent finds a black node with a level at most $i-1$, *i.e.*, it visits $v_s \in V_{<i}$. Since $i' \geq i+1$, the agent detects $v_s.\mathtt{level} < i' - 1$ and triggers the third rule, resetting *self*.phase to zero, a contradiction. □

The following theorem is derived from Observation 2 and Lemma 4.

Theorem 2. *Algorithm $\mathcal{D}_k$ solves the self-stabilizing exploration problem for all simple, undirected, and connected graphs with a diameter and maximum degree of at most k. The cover time is $O(m + nD)$ steps, regardless of the value of k. The memory requirement is $O(\log k)$ for both the agent and each node.*

Acknowledgments. This work is supported by JST FOREST Program JPMJFR226U and JSPS KAKENHI Grant Numbers JP19K11826, JP20KK0232, JP22K11903, JP23K11059, and JP23K28037.

Disclosure of Interests. The authors have no competing interests to declare that are relevant to the content of this article.

References

1. Aleliunas, R., Karp, R.M., Lipton, R.J., Lovasz, L., Rackoff, C.: Random walks, universal traversal sequences, and the complexity of maze problems. In: 20th Annual Symposium on Foundations of Computer Science (SFCS 1979), pp. 218–223. IEEE (1979)
2. Blin, L., Potop-Butucaru, M.G., Tixeuil, S.: On the self-stabilization of mobile robots in graphs. In: International Conference on Principles of Distributed Systems, pp. 301–314. Springer, Cham (2007)
3. Datta, A.K., Johnen, C., Petit, F., Villain, V.: Self-stabilizing depth-first token circulation in arbitrary rooted networks. Distrib. Comput. **13**(4), 207–218 (2000)
4. Dijkstra, E.W.: Self-stabilization in spite of distributed control. In: Selected Writings on Computing: A Personal Perspective, pp. 41–46. Springer, Cham (1982)
5. Huang, S.T., Chen, N.S.: Self-stabilizing depth-first token circulation on networks. Distrib. Comput. **7**(1), 61–66 (1993)
6. Ikeda, S., Kubo, I., Yamashita, M.: The hitting and cover times of random walks on finite graphs using local degree information. Theor. Comput. Sci. **410**(1), 94–100 (2009)
7. Masuzawa, T., Tixeuil, S.: Quiescence of self-stabilizing gossiping among mobile agents in graphs. Theor. Comput. Sci. **411**(14–15), 1567–1582 (2010)
8. Matthews, P.: Covering problems for Markov chains. Ann. Probab. **16**(3), 1215–1228 (1988)
9. Ooshita, F., Datta, A.K., Masuzawa, T.: Self-stabilizing rendezvous of synchronous mobile agents in graphs. In: International Symposium on Stabilization, Safety, and Security of Distributed Systems, pp. 18–32. Springer, Cham (2017)
10. Panaite, P., Pelc, A.: Exploring unknown undirected graphs. J. Algorithms **33**(2), 281–295 (1999)
11. Petit, F.: Fast self-stabilizing depth-first token circulation. In: International Workshop on Self-stabilizing Systems, pp. 200–215. Springer, Cham (2001)
12. Priezzhev, V.B., Dhar, D., Dhar, A., Krishnamurthy, S.: Eulerian walkers as a model of self-organized criticality. Phys. Rev. Lett. **77**(25), 5079 (1996)
13. Reingold, O.: Undirected connectivity in log-space. J. ACM (JACM) **55**(4), 1–24 (2008)
14. Sudo, Y., Baba, D., Nakamura, J., Ooshita, F., Kakugawa, H., Masuzawa, T.: A single agent exploration in unknown undirected graphs with whiteboards. IEICE Trans. Fundam. Electron. Commun. Comput. Sci. **98**(10), 2117–2128 (2015)
15. Sudo, Y., Ooshita, F., Kamei, S.: Self-stabilizing graph exploration by a single agent (2024). https://arxiv.org/abs/2010.08929
16. Ta-Shma, A., Zwick, U.: Deterministic rendezvous, treasure hunts, and strongly universal exploration sequences. ACM Trans. Algorithms (TALG) **10**(3), 1–15 (2014)
17. Yanovski, V., Wagner, I.A., Bruckstein, A.M.: A distributed ant algorithm for efficiently patrolling a network. Algorithmica **37**(3), 165–186 (2003)

Labeling Embeddings of Planar Graphs for Face-Adjacency

Borna Šimić(✉) and Roger Wattenhofer

ETH Zürich, Zürich, Switzerland
{bosimic,wattenhofer}@ethz.ch

Abstract. We design a $O(\log n)$ labeling scheme for planar graph embeddings. Given the labels of any two vertices we can directly determine whether the two vertices share a face. In the scheme, we use and design a $O(\log n)$ labeling scheme for the property of two vertices sharing a common neighbor in a planar graph, a problem which we show to be equivalent to labeling for face-adjacency. We also prove a lower bound of $(1 + o(1)) \log_2 n$ on label length.

Keywords: Labeling Schemes · Planar Graphs · Graph Embeddings

1 Introduction

Graphs have various different representations, from simple drawings on a piece of paper to adjacency matrices in computer memory. A quirky and often useful representation is a graph labeling scheme. In a labeling scheme, each node is denoted by a label. If chosen carefully, the labels of two nodes of a graph can directly tell us something about the graph, for instance, whether the two nodes are adjacent. Given such a labeling scheme, we do not need to know the entire graph anymore, as just looking at the two labels of the two nodes of interest is enough. These labeling schemes have many unexpected applications. They are for example used in road network routing or social networks, and generally in large scale distributed systems applications.

In this paper we study the mother of all labeling problems: adjacency—however, with a twist. We focus on planar graphs only, i.e., graphs that can be drawn in the plane without any edges of the graph intersecting. Instead of the usual adjacency where two vertices share an edge, we study the case where two vertices share a *face*. Given a drawing of a planar graph, the faces are the tiles of the plane that emerge when we cut along the edges of the graph. Faces are important structural elements of planar graphs. For example, the number of vertices v, the number of edges e and the number of faces f in a planar graph satisfy Euler's formula $v - e + f = 2$.

To be more picturesque, if we think of a graph as a road network, and think of nodes as road crossings or intersections (which inevitably contain road crossings), two nodes are face-adjacent if we can walk from one to the other without crossing

U. Schmid and R. Kuznets (Eds.): SIROCCO 2025, LNCS 15671, pp. 401–415, 2025.
https://doi.org/10.1007/978-3-031-91736-3_24

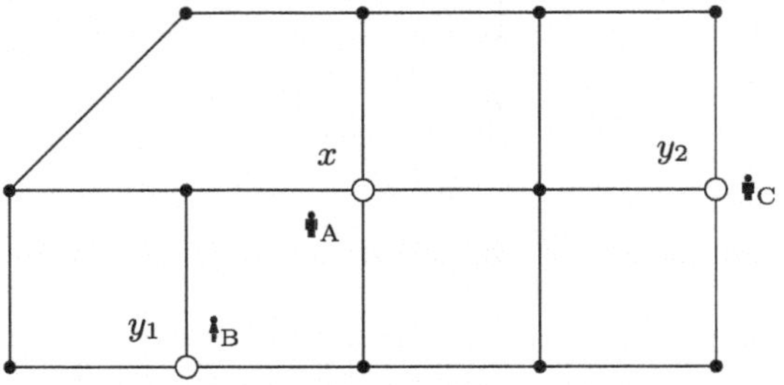

Fig. 1. The child A living in x can visit their friend B in y_1 unsupervised, but would need parental assistance to visit their friend C in y_2.

any other intersections. In this analogy, a concerned parent whose child wants to walk from their home (located at some intersection labeled x) to a friend's home (located at some other intersection labeled y) can check whether the child can go there alone (without needing to cross a road) without knowing the details of the road network, just by knowing the labels x and y. We can see an example in Fig. 1.

In this work, we consider *embeddings* (or *drawings*) of planar graphs, focusing on the property of two vertices in a planar graph sharing the same face in a fixed embedding. As the faces of a planar graph highly depend on the particular embedding we choose, our work is not graph labeling in the strictest sense. Different embeddings will have different vertices be face-adjacent, and will thus require different labelings (see Fig. 2 for an example).

2 Related Work

Adjacency Labeling Schemes. Labeling schemes that can answer, for any pair of vertices, whether the vertices are adjacent are called *adjacency labeling schemes.* They were first considered by Breuer [5] and Breuer and Folkman [6] for general graphs. The work [18] of Moon is equivalent to finding labeling schemes with shorter label lengths for general graphs, but refers to the related notion of universal graphs instead. A labeling scheme giving (almost) optimal label length for general graphs was described somewhat recently by Alstrup et al. [3]. Adjacency labeling schemes for bipartite, directed graphs and tournaments have similar, linear bounds on label size (also see [3]). In sparser settings, graphs tend to have bounds of logarithmic size. Schemes for trees and forests are well-studied, for example by Chung [9], as well as for graphs of bounded degree by Alon et al. [1]. This work does not directly discuss labeling schemes, instead studying the related notion of universal graphs. Esperet [13] and later Alon et al. [2] extend these results to induced-universal graphs and thus labeling schemes. Adjacency labeling of planar graphs was considered by Kannan et al. [15, 16]. Recent results of Gavoille et al. [4] followed up by Dujmović et al. [11] have made the label lengths for planar graphs close to optimal.

Diverse Labeling Schemes. Labeling schemes for properties other than adjacency are also a topic of great interest. Peleg [19] shows labeling schemes for distances (where we wish to be able to compute the distance of any two vertices, given only their labels) and for routing (where we want to find the first edge on a shortest path between two vertices). Further results are outlined in the survey [14] of Gavoille and Peleg as well as the paper [20] of Peleg.

Short Path Queries. We will make use of a connection between face-adjacency and the property of two vertices sharing a common neighbor in (a different) planar graph to design our labeling scheme, as well as showing an equivalence of sorts between the two properties in Sect. 4.

Querying for short paths between pairs of vertices in planar graphs was studied by Eppstein [12] and later by Chen and Xu [7]. A data structure that answers queries of the form "given vertices v, w, are they at a distance of at most k apart" for fixed k in constant time was shown by Kowalik et al. [17] and we use similar ideas to design our common-neighbor labeling scheme.

3 Preliminaries

Graphs and Graph Embeddings. We only deal with simple graphs (graphs without multiple edges and self-loops). An *embedding* (or *drawing*) $\mathfrak{D}$ of a graph G in the plane is considered to be a pair of mappings: one from the set of vertices $V(G)$ to the plane $\mathbb{R}^2$, and one from the edges $E(G)$ to subsets of the plane. The second mapping maps every edge to a continuous, non-self-intersecting curve between the images of its endpoint vertices (under the first mapping) which does not pass through any vertices other than its endpoints. The curves that the edges map to divide the plane (topologically) into regions, which we call *faces* of the embedding. A vertex v is incident to a face if its image under the vertex mapping lies on its boundary.

We will say that two vertices v, w of G are *face-adjacent* (in a fixed embedding) if there exists a face of the graph which they both lie on.

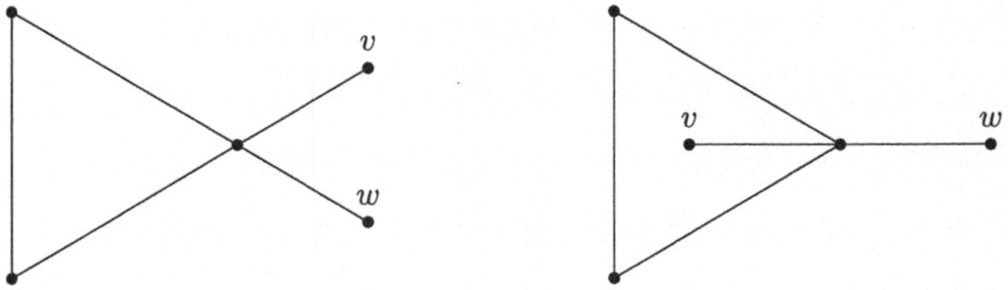

Fig. 2. Two planar embeddings $\mathfrak{D}_1, \mathfrak{D}_2$ of the same planar graph G such that vertices v, w are face-adjacent in $\mathfrak{D}_1$, but not in $\mathfrak{D}_2$.

An embedding is *planar* if no two edges intersect (except in endpoints).

Planar Graphs. A graph G is *planar* if it has a planar embedding. We use some basic facts about planar graphs, stated here without proof (proofs for both can, for example, be found in [10]):

Proposition 1. *Any planar graph G on n vertices has at most $3n - 6$ edges. If G is bipartite, this bound is improved to $2n - 4$.*

Proposition 2. *Any planar embedding $\mathfrak{D}$ of a planar graph G on n vertices has at most $2n - 4$ faces.*

Labeling Schemes. Let $\{0,1\}^*$ be the set of all finite binary strings, and for $s \in \{0,1\}^*$ let $|s|$ be the length of s. The simplest type of labeling schemes are *adjacency labeling schemes.* We give a definition here.

Definition 1. *Let $\mathcal{F}$ be some family of graphs, and let $\mathcal{F}_n$ be the set of n-vertex graphs in $\mathcal{F}$.*

We say that there exists an $f(n)$-bit face-adjacency labeling scheme for $\mathcal{F}$ *if there exists a function $D\colon (\{0,1\}^*)^2 \to \{0,1\}$ such that for any $G \in F_n$ there exists $\ell\colon V(G) \to \{0,1\}^*$ such that $|\ell(v)| \leq f(n)$ for every $v \in V(G)$ and for any two $v, w \in V(G)$, we have $D(\ell(v), \ell(w)) = 1$ if and only if $(v, w) \in E(G)$.*

We give the definition of a labeling scheme for face-adjacency for planar graph embeddings as follows:

Definition 2. *We say that there exists an $f(n)$-bit* face-adjacency labeling scheme for planar graphs *if there exists a function $D\colon (\{0,1\}^*)^2 \to \{0,1\}$ such that for any n-vertex planar graph G and any planar embedding $\mathfrak{D}$ of G there exists $\ell\colon V(G) \to \{0,1\}^*$ such that $|\ell(v)| \leq f(n)$ for every $v \in V(G)$ and for any two $v, w \in V(G)$, we have $D(\ell(v), \ell(w)) = 1$ if and only if there exists a face in $\mathfrak{D}$ which is incident to both v and w.*

We will make use of the notion of a labeling scheme for planar graphs and the property of two vertices sharing a common neighbor, specifically for bipartite graphs.

Definition 3. *We say that there exists an $f(n)$-bit* common-neighbor labeling scheme for (bipartite) planar graphs *if there exists a function $D\colon (\{0,1\}^*)^2 \to \{0,1\}$ such that for any n-vertex (bipartite) planar graph G there exists $\ell\colon V(G) \to \{0,1\}^*$ such that $|\ell(v)| \leq f(n)$ for every $v \in V(G)$ and for any two $v, w \in V(G)$, we have $D(\ell(v), \ell(w)) = 1$ if and only if there exists a vertex $u \in V(G)$ such that u is adjacent both to v and w.*

We assume that f is a nondecreasing function in both definitions.

Induced-Universal Graphs. Adjacency labeling schemes are intimately related to *induced-universal graphs* [21]. We say a graph $\mathcal{G} = (\mathcal{V}, \mathcal{E})$ is an induced-universal graph for some family $\mathcal{F}$ of graphs if for every $G \in \mathcal{F}$, there exists an induced subgraph of $\mathcal{G}$ isomorphic to G.

Adjacency labeling schemes for some family $\mathcal{F}$ of graphs now gives an induced-universal graph for $\mathcal{F}$ on the set of possible vertex labels (connect two labels by an edge if the decoder of the scheme returns 1 for them). This connection shows that the label length ℓ (modulo some details like uniqueness) is equal to the base 2 logarithm of the size of the induced-universal graph. More precisely:

Proposition 3. *Let $\mathcal{F}$ be some infinite family of graphs, and let $\mathcal{F}_n$ be the family of graphs $G \in \mathcal{F}$ on n vertices. There exists a $\lceil \log_2 N(n) \rceil$-bit adjacency labeling scheme for $\mathcal{F}$ if and only there exists an induced-universal graph for each $\mathcal{F}_n$ on $N(n)$ vertices.*

This is a well-known, commonly used property, and we use it to give some lower bounds in Sect. 6.

4 The Common-Neighbor Equivalence

We first show two results that demonstrate the intrinsic connection between the property of two vertices sharing a face in some embedding of a planar graph to the property of sharing a common neighbor in a bipartite planar graph.

In one direction the idea is to construct, given a planar graph G and its embedding $\mathfrak{D}$, a planar bipartite graph G' (on a larger set of vertices) in which two vertices from the vertex set of G share a common neighbor in G' if and only if they are face-adjacent in $\mathfrak{D}$.

Theorem 1. *If there exists a $f(n)$-bit common-neighbor labeling scheme for bipartite planar graphs, there exists a $f(3n-4)$-bit face-adjacency labeling scheme for planar graphs.*

Proof. Assume that there exists a $f(n)$-bit common-neighbor labeling scheme for bipartite planar graphs, denote its decoder as D and its labeling function (on a bipartite planar graph G) as ℓ_G. To construct a $f(3n-4)$-bit face-adjacency labeling scheme for planar graphs, consider some planar graph G and one of its embeddings $\mathfrak{D}$.

We construct an auxiliary planar bipartite graph G' by adding one vertex v' for every vertex v of G and one vertex t_f for each face f of $\mathfrak{D}$, thus adding at most $2n-4$ vertices by Proposition 2. The edge set of the auxiliary graph is then the set of edges (t_f, w) for all $w \in f$. It is easy to see this graph is planar as we can construct a planar embedding by keeping the positions of the vertices v in $\mathfrak{D}$ and adding t_f to the middle of face f, as shown in Fig. 3. To find a bipartition, one can consider the vertices t_f as one set A of the bipartition and the vertices of G as the other set B.

Now, notice that if two vertices $v, w \in A$ in G' have a common neighbor (say t_f), then the corresponding vertices of G must have both been incident to f in the embedding $\mathfrak{D}$ of G.

To construct the labels for G, we now consider the $f(3n-4)$-bit common-neighbor labeling of G'. We fix the label of a vertex $v \in G$ to be $\ell_{G'}(v')$, so exactly the label of its corresponding vertex in G'. This gives labels of length exactly $L = f(3n-4)$.

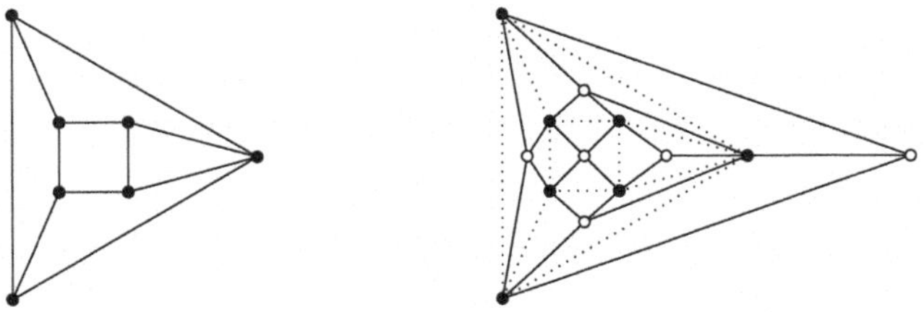

Fig. 3. A planar graph G and its corresponding planar bipartite graph G'

We simply reuse the decoder D of the common-neighbor labeling scheme. By the earlier argument, we will get the result $D(v, w) = 1$ whenever v, w have a common neighbor in G', which is equivalent to them being face-adjacent in G, as desired.

We can, with a bit more effort, prove a similar result in the opposite direction.

Theorem 2. *If there exists a $f(n)$-bit face-adjacency labeling scheme for planar graphs, there exists a $f(5n-8)+2$-bit common-neighbor labeling scheme for bipartite planar graphs.*

Proof. Assume that there exists a $f(n)$-bit face-adjacency labeling scheme for planar graphs, denote its decoder as D and its labeling function (on a bipartite planar graph G) as ℓ_G. Consider a bipartite planar graph G which we wish to label for the common-neighbor property and let $\mathfrak{D}$ be an embedding of G.

We will first construct an auxiliary planar graph G' which we use for the rest of the construction, and we simultaneously construct a planar embedding $\mathfrak{D}'$ of G' (which shows that G' is also planar). Initially, $\mathfrak{D}'$ is a copy of $\mathfrak{D}$.

To construct G' we first delete any isolated vertices of G, as well as any connected components of G which are paths on two vertices. We now fix a bipartition of the remaining vertices into two sets A and B.

We add, for each vertex v, exactly $\deg(v)$ vertices $r_{v,e,e'}$ where e, e' are consecutive edges in $\mathfrak{D}$ from v (in the case where $\deg(v) = 1$, we only add one vertex $r_{v,e,e}$ for the unique e incident to v). This adds $2|E| \leq 4n-8$ vertices by Proposition 1, so G' has at most $5n-8$ vertices. We denote the set of the added vertices as C. We embed those vertices in $\mathfrak{D}'$ in the obvious way: we find a disk D_v centered in the embedding of v such that it contains no other vertices. D_v is partitioned by the edges incident to v into $\deg(v)$ connected components, and we embed $r_{v,e,e'}$ into the component that borders e and e'.

We also add the edges $(v, r_{v,e,e'})$ for every v and any pair e, e' of consecutive (in $\mathfrak{D}$) edges incident to v, which are easy to embed. Additionally, we add edges of the form $(r_{v,b,e}, r_{w,b,f})$ whenever the following two conditions are fulfilled:

- $b = (v, w)$ is an edge of G
- if we order the edges incident to v and w in clockwise (according to their embeddings in $\mathfrak{D}$) orderings, we have that b comes before e in the ordering for v and after f in the ordering for w or vice versa.

An example of this is shown in Fig. 4. Notice that we have $\deg(r_{v,e,e'}) = 3$ for all v, e, e', as it is adjacent to v as well as two other vertices in C that correspond to the edges e, e'.

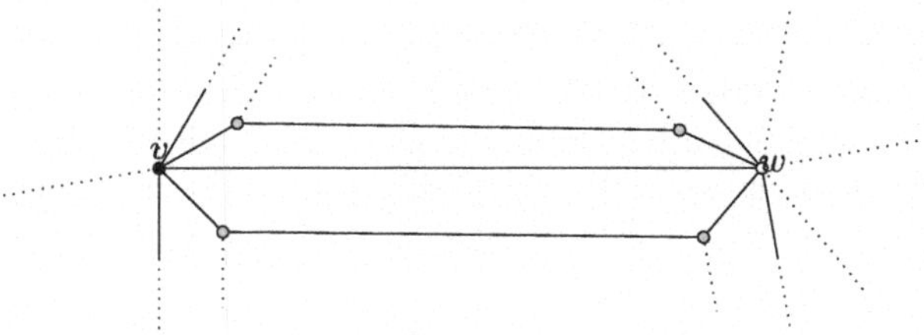

Fig. 4. Vertices v, w that form an edge b, which induces four vertices from the set C which connect to each other as described above.

We embed these edges "close" to the embedding of b in $\mathfrak{D}'$, as we embedded their endpoints close to the vertices they were defined from. We show an example of this construction, for an entire graph G, in Fig. 5.

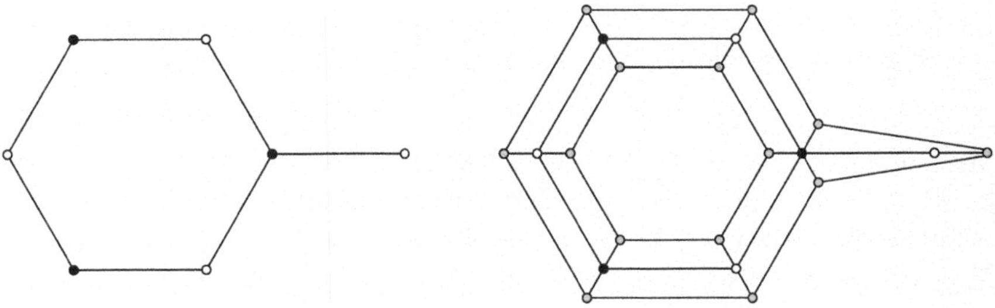

Fig. 5. A bipartite planar graph G and the corresponding planar graph G'

We now wish to show that, in the embedding $\mathfrak{D}'$, two vertices $v, w \in A \cup B$ are face-adjacent if any only if they are connected by an edge in G.

Consider any vertex $v \in A \cup B$ in the embedding $\mathfrak{D}'$, and consider some face F of $\mathfrak{D}'$ such that $v \in F$.

We first deal with the case where $\deg(v) = 1$. In this case, v has a neighbor w of degree at least 2. The single vertex $r_{v,e,e} \in C$ is connected to two vertices r_{w,e,f_1}, r_{w,e,f_2} which are both connected to w, so F can be one of two faces $\{v, w, r_{v,e,e}, r_{w,e,f_i}\}$ for $i \in \{1, 2\}$. Both have exactly four vertices on their boundary, and the only vertex of $A \cup B$ they contain is w, as desired.

Otherwise, if we let e, e' be two consecutive edges incident to v such that both lie on the boundary of F, we see by the construction above that one of them leads to a vertex of C. Assume without loss of generality that it's e', and the vertex is $r_{v,e,f}$ for some edge f incident to v, and that $e = (v, w)$ for some other

$w \in A \cup B$. Now, by construction, there exists some $r_{w,e,t}$ which is connected to $r_{v,e,f}$ by an edge, so the face F has exactly four vertices on its boundary: $v, w, r_{v,e,f}, r_{w,e,t}$, out of which only v, w are not in C and w is a neighbor of v.

Therefore, the only vertices in $A \cup B$ that v can be face-adjacent to are its neighbors, as desired, and all the faces that v is incident to are of size 4 and contain exactly two vertices of $A \cup B$ and exactly two vertices of C.

We are now almost done. We define two auxiliary graphs G'_A, G'_B as induced subgraphs of G' on the vertex sets $A \cup C$, $B \cup C$ with embeddings $\mathfrak{D}'_A, \mathfrak{D}'_B$ obtained in the obvious way. We show this construction in Fig. 6. Now, note that two vertices $v, w \in A$ are face-adjacent in $\mathfrak{D}'_A$ if and only if they share a common neighbor from B in G'.

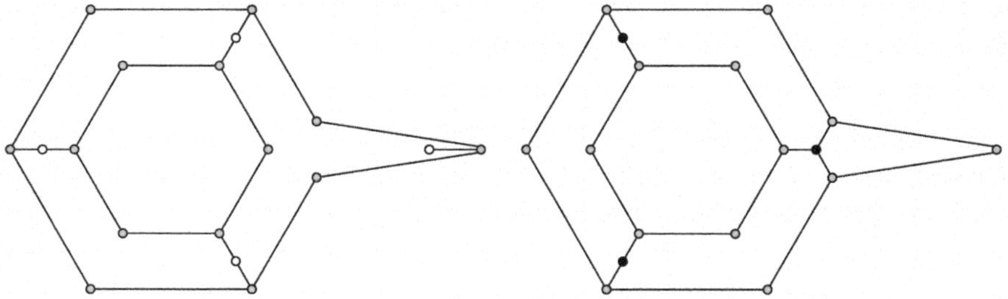

Fig. 6. The graphs G'_A and G'_B for the same G, G' depicted in Fig. 5

We now label each of G'_A, G'_B for face-adjacency with respect to the embeddings $\mathfrak{D}'_A, \mathfrak{D}'_B$. This gives vertex labels of length at most $f(5n-8)$ bits, as G' has at most $5n-8$ vertices by earlier considerations.

We can now label the vertices of G for the common neighbor property in the following way, given a vertex v:

- if $v \notin V(G')$, then v was removed from G to obtain G' and so is either isolated in G' or is part of a connected component which is a path on two vertices. We assign it a special label s which is a string of exactly $f(5n-8)+2$ zeroes.
- if $v \in A$, construct its label by concatenating the string 10 followed by the $f(5n-8)$-bit label given to v by the labeling scheme for face-adjacency for the graph G'_A and the embedding $\mathfrak{D}'_A$.
- if $v \in B$, construct its label by concatenating the string 11 followed by its $f(5n-8)$-bit label given to v by the labeling scheme for face-adjacency for the graph G'_B and the embedding $\mathfrak{D}'_B$.

The label length is exactly $L = f(5n-8)+2$ in all three cases.

Our decoder D' now works in the following way. Given two vertex labels ℓ_1, ℓ_2, we first check whether either $\ell_1 = s$ or $\ell_2 = s$. In that case, we have $D'(\ell_1, \ell_2) = 0$. Otherwise, we look at the second bit of the labels. If it differs, the vertices the labels correspond to are in different parts of the bipartition so they cannot have a common neighbor. Finally, if this bit is the same, we check the final section of $f(5n-8)$ using the decoder D for the face-adjacency labeling scheme.

Because Theorem 1 obtains a bipartite graph in its construction and because we wanted to obtain the tightest possible constant factor relating the common-neighbor and face-adjacency schemes, we focused on common-neighbor schemes exclusively for bipartite planar graphs in this section. Nevertheless, the proof of Theorem 2 can be adapted to show that the existence of a labeling scheme for face-adjacency in planar graphs with a certain label length implies the existence of a labeling scheme for the common-neighbor property with a similar (up to a constant factor) label length in planar graphs in general, by exploiting the 4-colorability of planar graphs (similarly to Theorem 3). We omit the details, but the idea is to have the same construction G' which we then refine into 4 graphs based on the color class that is removed (instead of 2 in the bipartite case). Each vertex will then contain a label component for every color class other than its own.

5 Common-Neighbor Scheme

In this section, we describe our common-neighbor labeling scheme for k-partite planar graphs. The face-adjacency result only requires $k = 2$ which will, due to the dependency of the label lengths on k, give a slightly better bound, but we do prove a more general result as a common-neighbor labeling scheme is also of independent interest. Fortunately, all planar graphs are at most 4-partite, so we can consider k to be bounded.

We note that the scheme we describe uses an adjacency labeling scheme for planar graphs as a building block.

Theorem 3. *If there exists a $f(n)$-bit adjacency labeling scheme for planar graphs, there exists a $13\log_2 n + (k-1)f(n) + O(1)$-bit common-neighbor labeling scheme for k-partite planar graphs.*

Proof. Fix a k-partite planar graph G on n vertices with a vertex k-partition into sets $V_1, V_2, \ldots, V_k$. Assign every vertex an unique and nonzero $\log_2 n + O(1)$-bit ID, which we put at the start of their vertex label.

We first orient the edges of the graph G to obtain a directed graph G' on the same vertex set in such a way that the outdegree $\deg^+(v)$ of every vertex $v \in G'$ is at most 3. We can do this in polynomial time (see [8]), and we will use this orientation to construct a labeling.

Consider some pair of vertices u, v that share a common neighbor t. In G', one of three fundamentally distinct situations may arise:

Each vertex v in G' has at most 3 out-neighbors, and each of them has at most 3 out-neighbours of their own. To assemble the label of some vertex, we will then reserve $12\log_2 n + O(1)$ bits in the label of a vertex to put first the labels of its out-neighbors, and then those of its out-neighbors' out-neighbors.

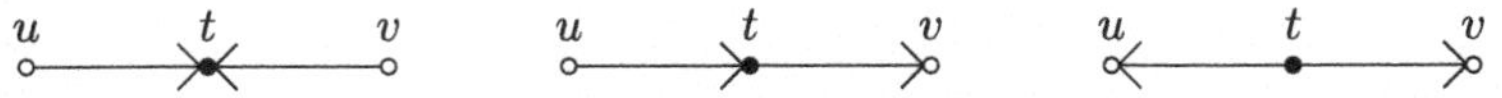

Fig. 7. Three possible cases of a common neighbor t to u and v in the orientation.

We will use this part of the label to check for common neighbours in the first two cases described in Fig. 7.

Dealing with the third case is more involved. We will introduce, for each $i \in [k]$, an auxiliary graph G_i on the vertex set $S_i = V \setminus V_i$, where the edge set of G_i is the set of pairs of distinct vertices (u, v) in S_i such that there exists a vertex $t \in V_i$ so that $t \to u$ and $t \to v$ are directed edges in G'. We say that t *induces* the edge (u, v) of G_i.

We claim that each of the graphs $G_1, G_2, \ldots, G_k$ is in fact planar.

To show this, fix some planar embedding $\mathfrak{D}'$ of G'. We can construct a planar embedding of G_i by first embedding all of its vertices as in $\mathfrak{D}'$. Now, consider some vertex $t \in V_i$. It induces at most three edges of G_i, say e_1, e_2, e_3. We then embed the edges e_1, e_2, e_3 as close as necessary to the embeddings in $\mathfrak{D}'$ of the directed edges from t to ensure there is no overlap with other edges. We present a simplified sketch of how this is done (for t having outdegree exactly 3, as that is the most complex case) in Fig. 8, and this is only possible because t has at most 3 outgoing edges and thus induces at most three edges of G_i.

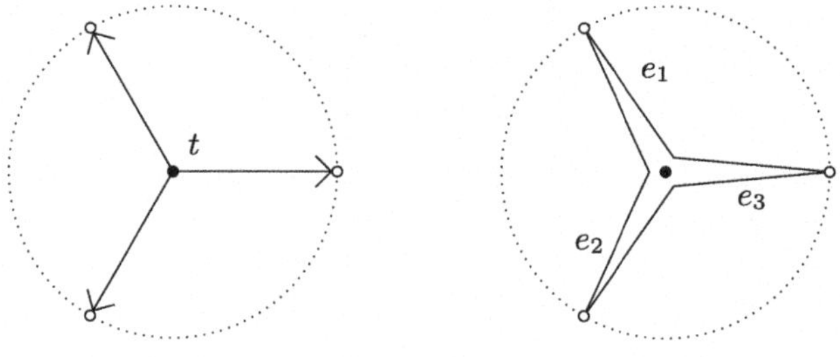

Fig. 8. Converting an outdegree 3 vertex from V_i into three edges in G_i.

As $\mathfrak{D}'$ was a planar embedding, this is always possible.

To finish the label of a vertex v, we first add a section of $\log_2 k + O(1) \leq 10 = O(1)$ bits that marks which set S_i the vertex belongs to by just adding the binary representation of i. We then add $(k-1)$ adjacency labels, each of length $f(n)$ for the planar graphs G_i where $v \in S_i$ in ascending order of i. The total label length is now at most

$$L = \log_2 n + 12 \log_2 n + \log_2 k + (k-1) f(n) + O(1) = 13 \log_2 n + (k-1) f(n) + O(1)$$

as desired.

Our decoder now works as follows: given two labels ℓ_1, ℓ_2, it first checks whether there is a match in the sections corresponding to the outneighbors of vertices in G', which checks for the first case in Fig. 7, where the two vertices have a common outneighbor.

If this is not the case, it then checks the section with labels of the outneighbors of the outneighbors to see whether the second case from the same figure has occured and the section of ℓ_1 containing the ID of the corresponding vertex is anywhere in the outneighbor-of-outneighbor section of ℓ_2, or vice versa, which checks the second case of the same figure where there is a directed path of length 2 from one vertex to the other.

Finally, we need to check for the third case described in the figure, where both of the vertices corresponding to the labels are outneighbors of some third vertex. We first check the sections of the labels of length $\log_2 k + O(1)$ to see which G_i the vertices labels ℓ_1, ℓ_2 refer to are from. Then, we check the corresponding sections in the next $(k-1)f(n)$ bits with the decoder of the $f(n)$-

Together with the $(1 + o(1)) \log_2 n$-bit adjacency labeling for planar graphs of Dujmović et al. [11], we immediately obtain:

Theorem 4. *There exists a $(12 + k + o(1)) \log_2 n$-bit common-neighbor labeling scheme for k-partite planar graphs. Moreover, there exists a $16 \log_2 n$-bit common-neighbor labeling scheme for planar graphs.*

Finally, Theorem 1 combined with Theorem 4 gives our main result:

Theorem 5. *There exists a $(14+o(1)) \log_2 n$-bit face-adjacency labeling scheme for planar graphs.*

6 Lower Bounds

If we assume the labels of vertices are unique, we always have the trivial lower bound of $\lceil \log_2 n \rceil$ on the sizes of labels. The same bound can be achieved without any uniqueness assumption, as we can exhibit a graph on an arbitrarily large number of vertices such that all vertices have distinct "face-adjacency neighborhoods" and therefore no two vertices can have the same label. One such example is a grid:

Proposition 4. *For integers $m, n \geq 5$ let $G_{m,n}$ be the $m \times n$ grid graph, more precisely, the graph on the set of vertices $[m] \times [n]$ where (i_1, j_1) is connected to (i_2, j_2) if and only if their Euclidean distance is 1. Let $\mathfrak{G}$ be the embedding of $G_{m,n}$ mapping each vertex to itself (as a point in $\mathbb{R}^2$) and every edge between two vertices to the shortest path between them. Each vertex of $G_{m,n}$ has a distinct subset of vertices it is face-adjacent to in $\mathfrak{G}$.*

Proof. Any internal vertex v of the grid has a different, unique set of exactly 8 vertices that it's face-adjacent to; those vertices whose coordinates differ from the coordinates of v by at most one. External vertices are face-adjacent to all other external vertices and either 3, 2, or 1 unique (due to $m, n \geq 5$) internal vertices.

We obtain a slightly better lower bound by finding a family of graphs whose adjacency-labeling is equivalent to labeling embeddings of planar graphs for face-adjacency, and exploiting the connection between adjacency labeling and induced-universal graphs.

Definition 4. *Let G be a planar graph, and $\mathfrak{D}$ an embedding of G. The* face-join *of $\mathfrak{D}$ is the graph $J(G, \mathfrak{D})$ on the vertex set of G whose edge set is all pairs of vertices (v, w) such that v and w are face-adjacent.*

Let $\mathcal{J}_n$ be the family of graphs that can be obtained as $J(G, \mathfrak{D})$ for some graph G on n vertices. It is easy to see that labeling the graphs in $\mathcal{J}_n$ for adjacency is equivalent to labeling the underlying set of planar graphs on n vertices for face-adjacency.

This family has to contain many different, diverse graphs, and we can use this to obtain a lower bound on the size of an induced-universal graph for it by find substructures inside it that are necessarily disjoint. To this end, we define the following graphs:

Definition 5. *Let m, k be positive integers. Denote by $T_{m,k}$ the planar graph obtained in the following way: we take a regular polygon with $m+2$ vertices and triangulate it (by connecting one vertex to all others), and then add exactly k points inside each triangle. The vertices of $T_{m,k}$ are then the $m+2$ vertices of the polygon and the set of mk additional interior points, whereas the edges are pairs of endpoints of all the lines of the triangulation. Denote the embedding obtained in this way $\mathfrak{T}_{m,k}$.*

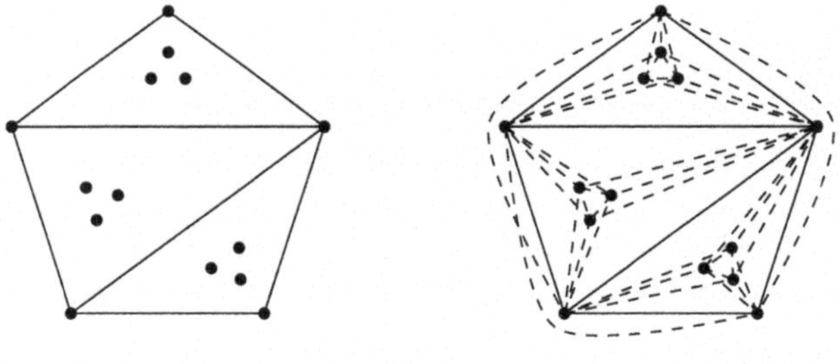

Fig. 9. The embedding $\mathfrak{T}_{3,3}$ of the graph $T_{3,3}$, as well as an embedding of the face-join $J(T_{3,3}, \mathfrak{T}_{3,3})$ with its additional (compared to $T_{3,3}$) edges in dashed lines.

By considering these graphs and the equivalency between adjacency labeling schemes and induced-universal graphs, we can obtain a lower bound on label length (Fig. 9).

We first prove a result about induced-universal graphs for $\mathcal{J}_n$.

Proposition 5. *If G is an induced-universal graph for the family $\mathcal{J}_n$, it must have at least $\frac{1}{2} n \ln n$ vertices.*

Proof. Notice that $J(T_{m,k}, \mathfrak{T}_{m,k})$ contains m disjoint cliques of size k which are not connected by any edges. If we have $m + 2 + mk \leq n$ (which is equivalent to $k \leq \frac{n-2}{m+1}$), we can then find such a configuration in any induced-universal graph for $\mathcal{J}_n$, as we can add $n - 2 - mk - m$ isolated vertices to $T_{m,k}$ to find a graph in $\mathcal{J}_n$ containing this configuration.

By the above consideration, for every $m \geq 2$, we have that G contains m disjoint, disconnected cliques of size $\lfloor \frac{n-2}{m+1} \rfloor$, which we denote by $T_{m,1}, T_{m,2}, \ldots, T_{m,m}$.

Any embedding of the empty graph shows that G must also contain the clique on n vertices, which we denote as $T_{1,1}$.

Due to the disconnectedness condition, for any $m \geq 2$, any clique $T_{m-1,j}$ can intersect at most one clique $T_{m,j'}$. We then have that for each m, at least one $T_{m,j}$ is disjoint from all the cliques $T_{k,j'}$ for $k < m$, so we obtain a sequence of cliques

$$T_{1,i_1}, T_{2,i_2}, \ldots, T_{n-3,i_{n-3}}$$

where $|T_{m,i_m}| = \lfloor \frac{n-2}{m+1} \rfloor$ and they are all pairwise disjoint.

This immediately implies

$$|V(G)| \geq n + \sum_{m=2}^{n-3} \left\lfloor \frac{n-2}{m+1} \right\rfloor > \frac{1}{2} n \ln n$$

where we use standard bounds on the harmonic numbers in the last step, and we are done.

The above result now implies:

Theorem 6. *Any face-adjacency labeling scheme for planar graphs with distinct labels needs to have labels of length at least* $\log_2 n + \log_2 \ln n - 1$.

Proof. We know by Proposition 3 that any induced-universal graph for $\mathcal{J}_n$ with N vertices gives an adjacency labeling scheme (with distinct labels) for J_n with label length $\log_2 N$, and vice versa.

Therefore, Proposition 5 implies that any adjacency labeling scheme for $\mathcal{J}_n$ with distinct labels needs to have labels of length at least

$$\log_2 \left(\frac{1}{2} n \ln n \right) = \log_2 n + \log_2 \ln n - 1$$

bits, so the same holds for any face-adjacency labeling scheme for embeddings of planar graph with at most n vertices.

7 Conclusion

We demonstrate an efficient labeling scheme with label length $(14 + o(1)) \log_2 n$ for the property of face-adjacency on planar graph embeddings. To do this, we show it is equivalent (up to some constant factors) to finding a labeling scheme for the property of two vertices sharing a common neighbor on bipartite planar graphs. We design a labeling scheme for this property in bipartite planar graphs, as well as in planar graphs in general, with slight differences in label length depending on the chromatic number of the graph. We show some lower bounds on the label length, showing in particular that $\log_2 n + O(1)$ is not enough.

The upper and lower bounds are of similar order, but with differing constants, and it would be interesting to see whether they can be made to match or at least brought closer together.

Disclosure of Interests. The authors have no competing interests.

References

1. Alon, N., Capalbo, M.: Sparse universal graphs for bounded-degree graphs. Random Struct. Algorithms **31**(2), 123–133 (2007). https://doi.org/10.1002/rsa.20143
2. Alon, N., Nenadov, R.: Optimal induced universal graphs for bounded-degree graphs. In: Proceedings of the 2017 Annual ACM-SIAM Symposium on Discrete Algorithms (SODA), pp. 1149–1157 (2021). https://doi.org/10.1137/1.9781611974782.74
3. Alstrup, S., Kaplan, H., Thorup, M., Zwick, U.: Adjacency labeling schemes and induced-universal graphs. SIAM J. Discret. Math. **33**(1), 116–137 (2019). https://doi.org/10.1137/16M1105967
4. Bonamy, M., Gavoille, C., Pilipczuk, M.: Shorter labeling schemes for planar graphs. SIAM J. Discret. Math. **36**(3), 2082–2099 (2022). https://doi.org/10.1137/20M1330464
5. Breuer, M.A.: Coding the vertexes of a graph. IEEE Trans. Inf. Theory **12**, 148–153 (1966). https://doi.org/10.1109/TIT.1966.1053860
6. Breuer, M.A., Folkman, J.: An unexpected result in coding the vertices of a graph. J. Math. Anal. Appl. **20**(3), 583–600 (1967). https://doi.org/10.1016/0022-247X(67)90082-0
7. Chen, D.Z., Xu, J.: Shortest path queries in planar graphs. In: Proceedings of the Thirty-Second Annual ACM Symposium on Theory of Computing, STOC 2000, pp. 469–478. Association for Computing Machinery, New York (2000). https://doi.org/10.1145/335305.335359
8. Chrobak, M., Eppstein, D.: Planar orientations with low out-degree and compaction of adjacency matrices. Theor. Comput. Sci. **86**, 243–266 (1991)
9. Chung, F.: Universal graphs and induced-universal graphs. J. Graph Theory **14**(4), 443–454 (1990). https://doi.org/10.1002/jgt.3190140408
10. Diestel, R.: Graph Theory, 5th edn. Springer Graduate Texts in Mathematics (2017)
11. Dujmović, V., Esperet, L., Gavoille, C., Joret, G., Micek, P., Morin, P.: Adjacency labelling for planar graphs (and beyond). J. ACM **68**(6) (2021). https://doi.org/10.1145/3477542
12. Eppstein, D.: Subgraph isomorphism in planar graphs and related problems. J. Graph Algorithms Appl. **3**(3), 1–27 (1999). https://doi.org/10.7155/jgaa.00014
13. Esperet, L., Labourel, A., Ochem, P.: On induced-universal graphs for the class of bounded-degree graphs. Inf. Process. Lett. **108**(5), 255–260 (2008). https://doi.org/10.1016/j.ipl.2008.04.020

14. Gavoille, C., Peleg, D.: Compact and localized distributed data structures. Distrib. Comput. **16** (2002). https://doi.org/10.1007/s00446-002-0073-5
15. Kannan, S., Naor, M., Rudich, S.: Implicit representation of graphs. In: STOC 1988, pp. 334–343. Association for Computing Machinery, New York (1988). https://doi.org/10.1145/62212.62244
16. Kannan, S., Naor, M., Rudich, S.: Implicit representation of graphs. In: Symposium on the Theory of Computing (1992)
17. Kowalik, L., Kurowski, M.: Short path queries in planar graphs in constant time. In: Proceedings of the Thirty-Fifth Annual ACM Symposium on Theory of Computing, STOC 2003, pp. 143–148. Association for Computing Machinery, New York (2003). https://doi.org/10.1145/780542.780565
18. Moon, J.W.: On minimal n-universal graphs. Proc. Glasgow Math. Assoc. **7**(1), 32–33 (1965). https://doi.org/10.1017/S2040618500035139
19. Peleg, D.: Proximity-preserving labeling schemes and their applications. In: Widmayer, P., Neyer, G., Eidenbenz, S. (eds.) Graph-Theoretic Concepts in Computer Science, pp. 30–41. Springer, Heidelberg (1999)
20. Peleg, D.: Informative labeling schemes for graphs. Theor. Comput. Sci. **340**(3), 577–593 (2005). https://doi.org/10.1016/j.tcs.2005.03.015. Mathematical Foundations of Computer Science 2000
21. Rado, R.: Universal graphs and universal functions. Acta Arith **9**(4), 331–340 (1964). https://doi.org/10.4064/aa-9-4-331-340

Brief Announcements

Brief Announcement: Relaxation for Efficient Asynchronous Queues

Samuel Baldwin, Cole Hausman, Mohamed Bakr, and Edward Talmage(✉)

Bucknell University, Lewisburg, PA, USA
elt006@bucknell.edu

Abstract. We explore efficient implementations of shared data structures in an asynchronous message-passing system. Considering the Queue data type, we know that *Dequeue* is inherently expensive. We consider ways to circumvent this performance limitation. Though we cannot improve the worst-case time per operation instance, we show that *relaxation*, weakening the ordering guarantees of the Queue data type, allows most *Dequeue* instances to return after only local computation, giving a low amortized cost per instance. Adjusting the degree of relaxation gives a customizable tradeoff between the ordering of data and the speed of access.

Keywords: Distributed Data Structures · Asynchronous Algorithms · Relaxed Data Types

1 Introduction and Related Work

Shared data is central to modern computing and our increasingly large datasets, but leads to concerns of concurrent data access. Without careful coordination, we can get stale data, overwrite other users' work, or corrupt the data. Distributed data structures allow computing to spread across machines by specifying an interface for users to interact with data and the effects of those operations. This abstraction layer allows a programmer to focus on their larger system, rather than the intricacies of managing distributed data. In co-located computation, shared hardware provides these structures, but in geographically-distributed systems we must build them.

We present an efficient distributed, message-passing algorithm implementing a relaxed queue. Our efficiency measure is the time from a user program invoking an operation on the data structure until getting a response. Past work gave bounds on implementations of various data types in partially-synchronous systems, where processes know bounds on the real time messages may take between participating processes [7,12]. However, these timing assumptions do not match realistic systems. We take the next step towards practical deployment with algorithms for *asynchronous* systems, which have no means to measure time or

A full version of this paper is available [2].

U. Schmid and R. Kuznets (Eds.): SIROCCO 2025, LNCS 15671, pp. 419–424, 2025.
https://doi.org/10.1007/978-3-031-91736-3_25

bounds on message delays. We overcome this by using logical timestamps to agree on the order in which operations take effect [10].

The relaxation we consider weakens the ordering of the Queue data type, allowing better amortized performance than possible for a standard Queue. We consider a version of a Queue where each *Dequeue* may return not strictly the oldest element in the Queue, but any of the k oldest elements, for a parameter k. This allows us to determine most *Dequeue* return values locally and return instantly. Less frequently, we must coordinate with other processes to choose a return value, which takes no longer than an unrelaxed *Dequeue*. In the full version of this paper [2], we also give an unrelaxed Queue implementation with optimal *Dequeue* as a stepping stone to the relaxed algorithm we give here.

There is much work on message-passing data structure implementations in partially-synchronous models [4,7,12]. In these models, we know complex operations like *Dequeue* are inherently expensive, having to wait for communication. In asynchronous models, the presence or absence of failures tends to dictate the possibility of implementations. Lynch's textbook (Sect. 17.1.2) [9] gives a failure-free unrelaxed Queue implementation, which stores all data at a central node, and Lubowich and Taubenfeld perform experiments on another such algorithm which uses partial replication [8], but give no formal complexity analysis. Our algorithms use full replication of data, to enable future failure-tolerance.

2 Algorithm

We work in a standard asynchronous, failure-free, message-passing model. There are n processes $p_0, \ldots, p_{n-1}$, modeled as state machines, each with a user. A process' inputs are user-invoked operations and messages arriving from other processes. A process' outputs are sending messages and returning to its user. The message channel between each pair of processes is reliable and delivers messages in sending order. Processes have no notion of time, nor bounds on message delays and local computation. We can still measure algorithmic time by counting sequential, causally-related messages. In a particular run, d denotes the maximum real-time delay of any message, and we express operations' time complexity in terms of d. See Sect. 2.1.2 in [1] or Sect. 3.2.1 in [3] for details.

We present an implementation of a Queue with k-Out-of-Order (k-OoO) *Dequeue* [5]. This type provides *Enqueue* and *Dequeue* operations, as a basic Queue, but a sequence of operation instances is legal if every *Dequeue* instance returns the argument of one of the k oldest *Enqueue* instances whose argument has not already been *Dequeue*d. We give a *linearizable* implementation [6], which requires there be a legal permutation of all operation instances respecting the real-time order of non-concurrent instances.

We extend Talmage and Welch's partially-synchronous algorithm for a Queue with k-OoO *Dequeue* [11]. The algorithm consists of event handlers, responding to state machine inputs. When a user invokes an operation, that process reads its vector clock, broadcasts an operation request, and waits for acknowledgments (lines 1–4, 12–17). When a process receives an *Enqueue* request, it puts the

Algorithm 1. for each process p_i to implement a queue with k-OoO *Dequeue*.

```
1: function ENQUEUE(val)
2:     EnqResponseCount = 0                                   ▷ Acknowledgment count
3:     Enq_ts = updateTS()                                    ▷ Timestamp from vector clock
4:     send (EnqReq, val, Enq_ts, i) to all processes
5: function RECEIVE(EnqReq, val, Enq_ts, inv) from p_inv
6:     updateTS(Enq_ts)                                       ▷ Update local vector clock
7:     lQueue.insertByTS(val, inv, Enq_ts)                    ▷ Locally execute
8:     send (EnqAck, i) to p_inv                              ▷ Acknowledge receipt
9: function RECEIVE(EnqAck, j) from p_j
10:    EnqResponseCount += 1                                  ▷ Count acknowledgments
11:    if EnqResponseCount == n then return EnqResponse to user
12: function DEQUEUE
13:    Deq_ts = updateTS()                                    ▷ Timestamp from vector clock
14:    if (ret = lQueue.deqByLabel(p_i)) ≠ ⊥ then
15:        send (Deq_f, ret, Deq_ts, i) to all processes      ▷ Deq_f: Fast Dequeue
16:        return ret to user
17:    else send (Deq_s, null, Deq_ts, i) to all processes    ▷ Deq_s: Slow Dequeue
18: function RECEIVE(op, val, Deq_ts, inv) from p_inv
19:    updateTS(Deq_ts)
20:    if Deq_ts is not in PendingDeqs then
21:        PendingDeqs.insertByTS(createList(op, val, Deq_ts, inv))
22:    send (op, val, DeqAck, Deq_ts, inv, i) to all processes
23: function RECEIVE(op, val, DeqAck, Deq_ts, inv, j) from p_j
24:    Let currentList be the list in PendingDeqs with ts == Deq_ts
25:    if currentList == ⊥ then
26:        currentList = PendingDeqs.insertByTS(createList(op, val, Deq_ts, inv))
27:    currentList.responses[j] = True                        ▷ Count acknowledgments
28:    propagateEarlierResponses(PendingDeqs, Deq_ts)
29:    for confList in PendingDeqs, in increasing lexicographic order do
30:        if all responses in confList are True then         ▷ Locally execute
31:            if confList.op == Deq_f, p_i ≠ p_inv then lQueue.remove(confList.val)
32:            else if confList.op == Deq_s then
33:                ret = lQueue.deqUnlabeled(confList.ts)
34:                labelElements(confList.invoker)
35:                if i == confList.invoker then return ret to user
36:            remove confList from PendingDeqs
```

argument in its local replica of the queue and responds with an acknowledgment (lines 5–8). When a process receives a *Dequeue* request, it saves the invocation for later application to its local replica and sends acknowledgments to all processes. When any process receives all processes' acknowledgments for a *Dequeue* instance, it applies that instance to its local replica (lines 30–36).

To utilize relaxation, we use *labels* to assign up to the k oldest elements currently in the queue to processes. There are then two types of *Dequeue* instances. If the invoking process owns an element at invocation, it immediately returns

Algorithm 2. Helper functions for Algorithm 1

function LABELELEMENTS(j) from [11]
  $lQueue.labelOldest(j, x)$, where $x = \min\{\lfloor k/n \rfloor, lQueue.unlabeledSize()\}$
function PROPAGATEEARLIERRESPONSES($PendingDeqs$, Deq_{ts})
  for $confList$ in $PendingDeqs$ with timestamp lexicographically less than or equal to Deq_{ts}, in decreasing lexicographic order **do**
    for $confList' \in PendingDeqs$ with $confList'.ts << confList.ts$ **do**
      for each process id j **do**
        if $confList.responses[j]$ **then** $confList'.responses[j] = True$
function UPDATETS(v_j) ▷ If no argument given, $v_j = [0, \ldots, 0]$
  $v_i[i] \mathrel{+}= 1$
  for $k = 0$ to $n-1$ **do** $v_i[k] = max(v_i[k], v_j[k])$
  return v_i
function CREATELIST(op, val, Deq_{ts}, inv)
  create a new confirmation list $confList$ with $confList.op = op$, $confList.val = val$, $confList.ts = Deq_{ts}$, $confList.invoker = inv$
  $confList.responses = [False, \ldots, False]$, length n, only $True$ at index inv
  return $confList$

that value to the user, and we call this a *fast Dequeue* (line 14). Otherwise, it is a *slow Dequeue* and the invoking process sends a *Dequeue* request without returning to the user (line 17.). When the invoking process receives acknowledgments from all processes and locally executes the *Dequeue* instance, it also chooses a return value and returns to the user (line 35). At this point, we must be careful to only consider elements in the local replica with earlier timestamps, as elements enqueued concurrently, but with later timestamps, may already be in the local replica. During execution of slow *Dequeues*, all processes also label more elements, since slow *Dequeues* only happen when a process owns no elements, implying fewer than k are currently labeled (line 34).

To handle asynchrony, we use *vector clocks* [10]. Each process maintains a list of how many invocations it knows each process has had. When sending invocations, processes attach their vector clock value (lines 3, 13) and on receiving a clock value, they update each component of their own vector to the larger of its previous value and the corresponding value in the received vector (lines 6, 19). To compare timestamps, if k is the first index where two vectors differ and $v_i[k] < v_j[k]$, we say v_i is *lexicographically smaller*, $v_i << v_j$.

When receiving acknowledgments for a particular *Dequeue* instance, we use a structure we call a *confirmation list* to track acknowledgments (lines 21, 26). We store these lists in *PendingDeqs*, sorted in lexicographic timestamp order. When we receive an acknowledgment for one instance, we also count that as an acknowledgment for all instances with smaller timestamps, since it implies that the acknowledging process has updated its vector clock to be later than those instances' timestamps (line 28). Once a confirmation list is full, we locally execute that and all preceding instances in increasing lexicographic timestamp order, which gives agreement between processes and correct return values (line 30).

Pseudocode is in Algorithms 1 and 2, deferring the correctness proof to the full version [2]. In the pseudocode, *lQueue* is the local replica of the queue, *PendingDeqs* stores confirmation lists for *Dequeue* instances, and v_i is the process' vector clock. We assume various utility functions on these lists, allowing us to retrieve the oldest element with a particular label or with no label up to a maximum timestamp (*peekByLabel*, *deqByLabel*, *deqUnlabeled*), to insert in lexicographically-sorted order (*insertByTS*), remove specific values, check the number of unlabeled elements, and apply labels (*remove*, *unlabeledSize*, *labelOldest*).

3 Analysis

We measure the time delay between an operation's invocation and return. Each operation instance returns to the user, at latest, after its announcement reaches all processes and their acknowledgments come back. Thus, each operation has worst-case response time $2d$. However, in many typical scenarios, most *Dequeue* instances return with no delay. The analysis is similar to that in prior work [11], but here slow *Dequeue* instances take $2d$ time and fast *Dequeues* take none. Specifically, in a *heavily-loaded* run, which has at least k more *Enqueue* than *Dequeue* instances in the prefix before any *Dequeue*, only one in every $\lfloor k/n \rfloor$ *Dequeue* instances is slow. This gives the total cost of m *Dequeue* instances, with m_i at each process p_i, as at most $\sum_{i=0}^{n-1} \left\lceil \frac{m_i}{\lfloor k/n \rfloor} \right\rceil (2d) \leq \left(\frac{m-n}{\lfloor k/n \rfloor} + n \right) (2d)$.

Each consecutive sequence of $\lfloor k/n \rfloor$ *Dequeue* instances at one process will take $2d$ total time, which gives the first term. The worst case is if there is one final, slow *Dequeue* instance at each process, giving the extra $n(2d)$. As m increases in longer runs, the effect of a final slow *Dequeue* vanishes, giving a practical runtime of $\left(\frac{m}{\lfloor k/n \rfloor} \right) (2d)$. This is inversely proportional to the degree of relaxation k, so by increasing relaxation, we can reduce the total cost at will. At $k = 2n$, we halve the total runtime, yielding better complexity than possible for an unrelaxed queue, and at larger k, the performance is even better. Thus, even in an asynchronous model, relaxation allows us to practically circumvent lower bounds on the performance of shared data structures, via a tunable tradeoff between ordering guarantees and performance. It is also worth noting the counterintuitive fact that fast *Dequeues* are *faster* in our asynchronous algorithm than in the previous, partially-synchronous algorithm, despite the harder model. We used a more complex linearization, allowing instantaneous fast *Dequeues*, which is an independent contribution.

4 Conclusion

We present a fully-replicated, asynchronous algorithm for a relaxed Queue, with efficient *Dequeue*. This shows that relaxation can improve amortized performance in asynchronous systems, as well as partially synchronous, and improves the common-case complexity of *Dequeue* below the best known even for partially-synchronous systems. This is a stepping stone toward our eventual goal of implementations in asynchronous, failure-prone systems, which more closely model the real world. Previous algorithms for unrelaxed, asynchronous Queues did not have full replication, which we expect to need when we add failure tolerance, as unreplicated data at a failing node would be lost. We are also working to implement other relaxations, which may be more applicable for different uses, in this and more realistic models.

References

1. Attiya, H., Welch, J.L.: Sequential consistency versus linearizability. ACM Trans. Comput. Syst. **12**(2), 91–122 (1994). https://doi.org/10.1145/176575.176576
2. Baldwin, S., Hausman, C., Bakr, M., Talmage, E.: Relaxation for efficient asynchronous queues (2025). https://arxiv.org/abs/2503.02164
3. Barbosa, V.C.: An Introduction to Distributed Algorithms. MIT Press, Cambridge (1996)
4. Chandra, T.D., Hadzilacos, V., Toueg, S.: An algorithm for replicated objects with efficient reads. In: Giakkoupis, G. (ed.) Proceedings of the 2016 ACM Symposium on Principles of Distributed Computing, PODC 2016, pp. 325–334. ACM (2016). https://doi.org/10.1145/2933057.2933111
5. Henzinger, T.A., Kirsch, C.M., Payer, H., Sezgin, A., Şokolova, A.: Quantitative relaxation of concurrent data structures. In: Giacobazzi, R., Cousot, R. (eds.) The 40th Annual ACM SIGPLAN-SIGACT Symposium on Principles of Programming Languages, POPL '13, pp. 317–328. ACM (2013). https://doi.org/10.1145/2429069.2429109
6. Herlihy, M., Wing, J.M.: Linearizability: a correctness condition for concurrent objects. ACM Trans. Program. Lang. Syst. **12**(3), 463–492 (1990). https://doi.org/10.1145/78969.78972
7. Kosa, M.J.: Time bounds for strong and hybrid consistency for arbitrary abstract data types. Chicago J. Theor. Comput. Sci. **1999** (1999)
8. Lubowich, Y., Taubenfeld, G.: On the performance of distributed lock-based synchronization. ACM SIGOPS Oper. Syst. Rev. **45**(2), 28–37 (2011). https://doi.org/10.1145/2007183.2007187
9. Lynch, N.A.: Distributed Algorithms. Morgan Kaufmann, Burlington (1996)
10. Mattern, F.: Virtual time and global states of distributed systems. In: Cosnard, M., et al. (eds.) Parallel and Distributed Algorithms: Proceedings of the 10th International Workshop on Parallel and Distributed Algorithms, pp. 215–226. Elsevier Science Publishers B. V. (1988)
11. Talmage, E., Welch, J.L.: Improving average performance by relaxing distributed data structures. In: Kuhn, F. (ed.) DISC 2014. LNCS, vol. 8784, pp. 421–438. Springer, Heidelberg (2014). https://doi.org/10.1007/978-3-662-45174-8_29
12. Wang, J., Talmage, E., Lee, H., Welch, J.L.: Improved time bounds for linearizable implementations of abstract data types. Inf. Comput. **263**, 1–30 (2018). https://doi.org/10.1016/j.ic.2018.08.004

Brief Announcement: A Sheaf-Theoretic Characterization of Tasks in Distributed Systems

Stephan Felber[1], Bernardo Hummes Flores[2(✉)], and Hugo Rincon-Galeana[1,3]

[1] Technische Universität Wien,Vienna, Austria
[2] LIX, CNRS, École polytechnique, Institut Polytechnique de Paris, 91120 Palaiseau, France
bernardo.hummes-flores@lix.polytechnique.fr
[3] Technische Universität Berlin, Berlin, Germany

Abstract. Task solvability lies at the heart of distributed computing, with direct implications for both theoretical understanding and practical system design. A fundamental challenge in distributed computing is constructing global solutions from local computations and information. Sheaf theory addresses this challenge by providing a mathematical framework for assessing globally consistent properties from locally defined data.

In this paper, we introduce a sheaf-theoretic characterization of task solvability by defining the novel construction of a task sheaf, whose sections correspond to valid solutions of a task. Furthermore, we show that the cohomology of a task sheaf may be used to compute solving protocols, thus establishing a connection between distributed computing and sheaf theory for both protocol design and impossibility analysis. A full version can be found in [6].

1 Introduction

Assessing the solvability or impossibility of a given task in a distributed system is at the center of distributed systems research. A variety of theoretical tools have been developed by sourcing from different fields such as combinatorics, topology and epistemic logic. One of the earliest results is the two generals problem impossibility [14], which roughly states that consensus is impossible in networks with message loss and unbounded delays. Numerous results have followed since: the celebrated FLP [7] impossibility shows that consensus cannot be achieved in asynchronous and crash-prone systems, the lossy-link impossibility [21] that shows that consensus is impossible even in a synchronous two-process model, [10] characterized consensus in distributed systems via *epistemic logic* and [17] provides a *topological* characterization of consensus under general message adversaries.

Sheaf theory originated from studying how local constraints give rise to global solutions, paired with cohomology theory as a measure of obstructions to such

U. Schmid and R. Kuznets (Eds.): SIROCCO 2025, LNCS 15671, pp. 425–430, 2025.
https://doi.org/10.1007/978-3-031-91736-3_26

global constructions. These same principles guide our approach here. Cellular sheaves were introduced by Curry [4] as a combinatorial counterpart to the common notion of sheaves found in the literature [15], with the interest of having a theory that is computable. Most importantly to us, the theory of cellular sheaves presents the concept of sheaves in an approachable manner, whence it becomes clear its usefulness to depict global properties of local information, as is the case for distributed computing. There has been a recent growth in applications of sheaf theory, with instances in sensor data integration [11,20], contextuality in quantum mechanics [1], modal logic [12,13], opinion dynamics [19] and machine learning [2,3]. Distributed applications have been considered before to describe concurrent systems [8,16], but with a considerably limited scope in applicability.

We present, to the best of our knowledge, the first sheaf theoretic characterization of task solvability in a distributed system with dynamic message adversaries. This connection creates space for a variety of tools known in sheaf theory to be applied in distributed computing, and we showcase its effectiveness making use of its relation to cohomology. We define a task sheaf as a construction relating its task to the indistinguishability graph capturing the uncertainty[1] of processes throughout runs, such that its sections correspond directly to task-solving solutions. A full version of this paper can be found in [6].

2 System Model and Results

We formalize how processes evolve, communicate, and make decisions via a state-machine model which executes under communication-wise adversarial influence. This approach follows a common modeling choice for representing distributed protocols (e.g., [18]).

We model a local protocol for each process in a set $\Pi = \{p_i\}_{i \in [n]}$ as a state machine $\mathcal{P}_p$ containing the possible **local states**, together with distinguished input and output states. The behavior of the local protocol is defined by **transition function** τ_p, a **communication function** κ_p, and a **decision function** δ_p. The transition function updates the local state, and the communication function models all of the system interactions. The decision function outputs the result of the distributed computation, and is augmented with $\perp$ to represent the lack of a decision. A **configuration** $\langle g \rangle$ is a tuple of local states, one per process. The projections $\pi_p : S^n \to S_p$ retrieves the local state of process p, and $\pi_I : S^n \to I^n$ retrieves the vector of inputs. A configuration is called **terminal** if all processes' decision functions map their respective local states to an output state. If this is true only for some processes, but not all, it is called **partially terminated**. A distributed task is a triple $T = (\mathcal{I}, \mathcal{O}, \Delta)$, where $\mathcal{I}$ is a set of input vectors, $\mathcal{O}$ a set of output vectors and $\Delta : \mathcal{I} \to 2^{\mathcal{O}}$ the specification that associates to each input vector a set of valid outputs.

The messages generated by κ_p are sent to every other process, but only delivered depending on a **communication trace** $\mathcal{C} = (c_t)_{t \in \mathbb{N}}$, a **communica-**

[1] The indistinguishability graph is a structure inspired by Kripke frames and commonly found in epistemic logic [5].

tion adversary $\mathcal{A}$ is then defined as the set of possible communication traces. A protocol that always eventually provides a terminal configuration for a given adversary is called **terminating**. A set of processes Π, a protocol $\mathcal{P}$, and a communication adversary $\mathcal{A}$ define a **message-passing system** $\mathcal{S}$. The **run** of a system $\mathcal{S}$, for a given input, is a sequence of pairs $\sigma = (\langle g \rangle_t, c_t)_{t \in \mathbb{N}}$. If we consider all communication traces permissible for $\mathcal{A}$, we obtain an **execution** for a set of inputs I, the freely generated directed graph $\mathcal{E}_{\mathcal{P},\mathcal{A},I}$. We define a **local indistinguishability relation** over configurations, relative to each process, where $\langle g \rangle, \langle h \rangle$ are indistinguishable for p whenever p has the same local state in both $\langle g \rangle$ and $\langle h \rangle$. Finally, we obtain an **execution frame** $\mathcal{K} = (V, \{\leq_p\}_{p \in \Pi}, \{\sim_p\}_{p \in \Pi})$ by augmenting the execution with the induced family of indistinguishability relations.

Definition 1 (Execution Cut). *Let $U \subset V(\mathcal{E})$ be the set of configurations in an execution $\mathcal{E}$. We say that U is an **execution cut** iff it is a cut set in $\mathcal{E}$, i.e., it intersects every run in $\mathcal{E}$.*

We can define a **causal closure** of the **local star** of U, $ccl(\mathcal{N}(U))$ as containing $\langle h \rangle$ if it is either indistinguishable to some configuration $\langle g \rangle \in U$, or it is the successor of an indistinguishable configuration $\langle h \rangle$.

Definition 2 (Execution Slice). *Let $U \subset V(\mathcal{E})$ be an execution cut, we define a **execution slice**, denoted by $ccl(\mathcal{N}(U))$, as the causal closure ccl over the local star $\mathcal{N}$ over U.*

We ensure that each process p in any partially terminated configuration $\langle g \rangle$ keeps its decided values across any indistinguishable successor configurations $\langle h \rangle$, reflecting that decisions are final, by constructing a family of causal consistency relations $\{\simeq_p\}_{p \in [n]}$ where $\langle g \rangle \simeq_p \langle h \rangle$. An example is depicted in Fig. 1.

Definition 3 (Causal Consistency Relation). *We define the **causal consistency** relation $\{\simeq_{p_i}\}_{i \in [n]}$ as the **symmetric closure** of the binary relation composition $\leq \circ \sim_{p_i}$.*

Given a system $\mathcal{S}$ and a task specification T, we encode in a cellular sheaf[2] over the associated execution frame $\mathcal{K}$ the task constraints. Namely, the construction of a sheaf requires that processes must provide the same decision across configurations that it cannot distinguish.

Definition 4 (Task Sheaf). *Let $T = (\mathcal{I}, \mathcal{O}, \Delta)$ be a task, $\mathcal{S} = (\Pi, \mathcal{P}, \mathcal{A})$ a message-passing system and $\mathcal{E}_{\mathcal{S},\mathcal{I}}$ the execution frame generated by possible runs of $\mathcal{S}$ for all initial states specified by T. A **task sheaf** $\mathcal{F}_{T,\mathcal{S}}$ is a family of functions that associates valid outputs in the image of Δ to configurations $\langle g \rangle \in \mathcal{K}$, with an encoding of the consistency constraints as follows.*

- ***Valid configuration outputs:** For each configuration $\langle g \rangle \in V(\mathcal{K})$, $\mathcal{F}$ maps to the set of valid output values, i.e. $\mathcal{F}(\langle g \rangle) = \Delta(\pi_I(\langle g \rangle))$.*

[2] See [4] for a thorough introduction of cellular sheaves.

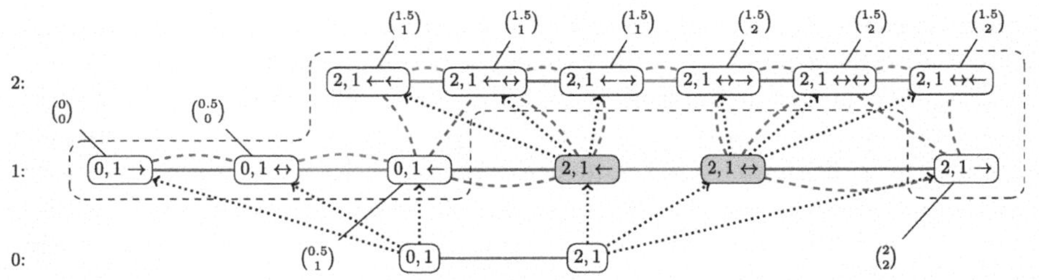

Fig. 1. The execution of the lossy-link synchronous message adversary for two processes $\Pi = \{a, b\}$. Nodes contain the global state, the initial values together with the communication where $\rightarrow$ denotes the message from a arrived at b, but not the other way around. Dotted lines represent causal links, orange and green lines are indistinguishability relations for a, b respectively. Dashed black container is an execution cut U, the solid yellow configurations partially terminated, the dashed colored lines indicate the causal consistency relation. $\binom{0.5}{1}$ denotes an assignment solving ϵ-agreement in that configuration, where $\epsilon = 0.5$. U is intentionally non-trivial for illustrative purposes. (Color figure online)

- ***Valid process outputs:*** *For each process p and each pair of related configurations* $\langle g \rangle \sim_p \langle h \rangle \in \{\sim_p\}$, $\mathcal{F}$ *maps to the set of valid output values for that process, i.e.* $\mathcal{F}(\langle g \rangle \sim_p \langle g \rangle) = \Delta_p(\pi_I(\langle g \rangle)) \cup \Delta_p(\pi_I(\langle g \rangle))$.
- ***Restriction maps:*** *For every pair of related configurations* $\langle g \rangle \sim_p \langle h \rangle$, *for each process p, the maps* $\mathcal{F}_{\langle g \rangle \trianglelefteq (\langle g \rangle, \langle h \rangle)_p} : \mathcal{F}(\langle g \rangle) \rightarrow \mathcal{F}(\langle g \rangle \sim_p \langle h \rangle)$, *that projects the valid outputs vectors of a configuration* $\langle g \rangle$ *its p-th entry.*

A task sheaf $\mathcal{F}$ can be equivalently understood process-wise. It is obtained as the colimit of the sheaf defined for each process $p \in \Pi$, $\mathcal{F}_p : \mathbf{Cell}(\mathcal{K}_p) \rightarrow \mathbf{Cset}_\Pi$, where $\mathbf{Cell}(\mathcal{K}_p)$ is the cellular category associated to $\mathcal{K}_p$, and $\mathbf{Cset}_\Pi$ is the output vector of the task seen as a chromatic semi-simplicial set[3].

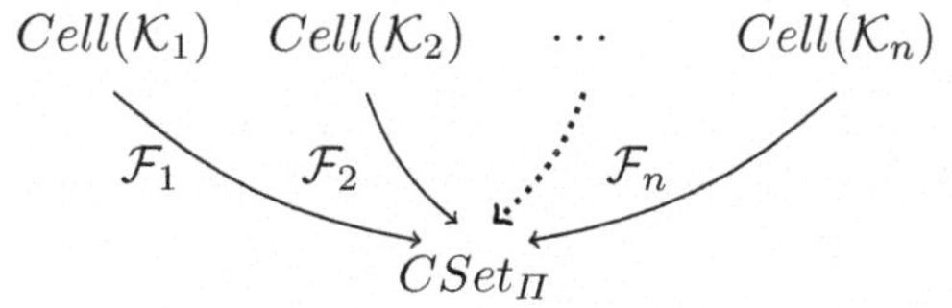

(1)

Theorem 1 (Terminating Task Solvability). *Let* $T = \langle \mathcal{I}, \mathcal{O}, \Delta \rangle$ *be a task, and* $\mathcal{P}$ *a protocol: there exists a **terminating decision map*** δ *solving* T *iff there exists a **execution cut*** A *such that its system slice,* $ccl(\mathcal{N}(A))$, *together with it causal consistency relation* $\{\simeq_{p_i}\}_{i \in [n]}$, *has a section over the task sheaf* $\mathcal{F}_{A,\simeq}$.

[3] Chromatic augmented semi-simplicial sets were introduced in [9] and depict labeled sets with extra structure to be handled algebraically.

Given an execution $\mathcal{E}$ we can enumerate all the possible finite execution cuts A and computationally check for each of them whether or not the induced $ccl(\mathcal{N}(A))$ admits a section by constructing its zeroth cohomology. Thus, we can define a sequential procedure that terminates with a solution iff the task is solvable in a finite execution cut.

Theorem 2 (Computable Decision Maps). *Let $T = \langle \mathcal{I}, \mathcal{O}, \Delta \rangle$ that can be solved by $\mathcal{P}$ under a communication adversary in a bounded number of steps. Then its decision map δ is computable.*

3 Conclusions

Our results and contributions, in particular our task sheaf construction, constitute, to the best of our knowledge, the first sheaf-theoretic characterization of general distributed computing tasks. Moreover, the generality of our model allows us to describe a wide range of systems that only need to satisfy minimal assumptions, namely that the set of processes is finite and that the communication is produced via messages.

Furthermore, by expressing tasks as sheaves, we are able to incorporate cohomology theory as a powerful tool for distributed systems. For instance, the cohomology of a task sheaf is a group that represents the "obstructions" or "limitations" that prevent a specific task from being solved by a specific system. One of the advantages of the sheaf-theoretic perspective is already made explicit in [6, Example 1], where we show an impossibility result by simply looking at the cohomology group of its task sheaf. Note, however, that the applications of task sheaf cohomology go beyond establishing impossibilities, and may also be useful for finding a protocol by highlighting the system limitations.

Finally, the categorical foundations of our approach provides a fertile starting point for further research, such as incorporating failure models and alternative termination conditions.

Acknowledgements. We thank Eric Goubault and Ulrich Schmid for the inspiring discussions and their invaluable input. We extend our thanks to the SIROCCO'25 reviewers for their much valued detailed feedback, which undoubtedly helped to improve this paper. Hugo Rincon-Galeana: Supported by the German Research Foundation (DFG), SPP 2378 (project ReNO), 2023-2027.

References

1. Abramsky, S., Brandenburger, A.: The sheaf-theoretic structure of non-locality and contextuality. New J. Phys. **13**(11), 113036 (2011). https://doi.org/10.1088/1367-2630/13/11/113036
2. Battiloro, C., Wang, Z., Riess, H., Lorenzo, P.D., Ribeiro, A.: Tangent bundle filters and neural networks: from manifolds to cellular sheaves and back. In: ICASSP 2023 - 2023 IEEE International Conference on Acoustics, Speech and Signal Processing (ICASSP), pp. 1–5 (2023). https://doi.org/10.1109/ICASSP49357.2023.10096934

3. Bodnar, C., Di Giovanni, F., Chamberlain, B.P., Liò, P., Bronstein, M.M.: Neural Sheaf Diffusion: A Topological Perspective on Heterophily and Oversmoothing in GNNs. https://doi.org/10.48550/arXiv.2202.04579
4. Curry, J.M.: Sheaves, cosheaves and applications (2014)
5. Fagin, R., Halpern, J.Y., Moses, Y., Vardi, M.Y.: Reasoning About Knowledge. MIT Press (1995). https://doi.org/10.7551/mitpress/5803.001.0001
6. Felber, S., Flores, B.H., Galeana, H.R.: A sheaf-theoretic characterization of tasks in distributed systems (2025). https://arxiv.org/abs/2503.02556
7. Fischer, M.J., Lynch, N.A., Paterson, M.S.: Impossibility of distributed consensus with one faulty process. J. ACM **32**(2), 374–382 (1985). https://doi.org/10.1145/3149.214121
8. Goguen, J.A.: Sheaf semantics for concurrent interacting objects. Math. Struct. Comput. Sci. **2**(2), 159–191 (1992). https://doi.org/10.1017/S0960129500001420
9. Goubault, E., Kniazev, R., Ledent, J., Rajsbaum, S.: Semi-Simplicial Set Models for Distributed Knowledge, pp. 1–13 (2023). https://doi.org/10.1109/LICS56636.2023.10175737
10. Halpern, J.Y., Moses, Y.: Knowledge and common knowledge in a distributed environment. J. ACM **37**(3), 549–587 (1990). https://doi.org/10.1145/79147.79161
11. Joslyn, C.A., et al.: A sheaf theoretical approach to uncertainty quantification of heterogeneous geolocation information. Sensors **20**(12), 3418 (2020). https://doi.org/10.3390/s20123418
12. Kishida, K.: Public announcements under sheaves. In: Motomura, Y., Butler, A., Bekki, D. (eds.) JSAI-isAI 2012. LNCS (LNAI), vol. 7856, pp. 96–108. Springer, Heidelberg (2013). https://doi.org/10.1007/978-3-642-39931-2_8
13. Kishida, K.: Neighborhood-sheaf semantics for first-order modal logic. Electron. Notes Theor. Comput. Sci. **278**, 129–143 (2011). https://doi.org/10.1016/j.entcs.2011.10.011
14. Lamport, L.: The weak byzantine generals problem. J. ACM (JACM) **30**(3), 668–676 (1983)
15. Mac Lane, S., Moerdijk, I.: Sheaves in Geometry and Logic: A First Introduction to Topos Theory. Universitext, Springer, New York (2012). https://doi.org/10.1007/978-1-4612-0927-0
16. Malcolm, G.: Sheaves, objects, and distributed systems. Electron. Notes Theor. Comput. Sci. **225**, 3–19 (2009). https://doi.org/10.1016/j.entcs.2008.12.063
17. Nowak, T., Schmid, U., Winkler, K.: Topological characterization of consensus in distributed systems: dedicated to the 2018 dijkstra prize winners bowen alpern and fred b. schneider. J. ACM (2024). https://doi.org/10.1145/3687302
18. Raynal, M.: Distributed Algorithms for Message-Passing Systems. Springer (2013). https://doi.org/10.1007/978-3-642-38123-2
19. Riess, H., Ghrist, R.: Diffusion of information on networked lattices by gossip. In: 2022 IEEE 61st Conference on Decision and Control (CDC), pp. 5946–5952 (2022). https://doi.org/10.1109/CDC51059.2022.9992539
20. Robinson, M.: Sheaves are the canonical data structure for sensor integration. Inf. Fusion **36**, 208–224 (2017). https://doi.org/10.1016/j.inffus.2016.12.002
21. Santoro, N., Widmayer, P.: Time is not a healer. In: Proceedings of 6th Annual Symposium on Theoretical Aspects of Computer Science (STACS 1989). LNCS, vol. 349, pp. 304–313. Springer (1989)

Brief Announcement: Hardness of Approximate Vertex Ranking by Betweenness Centrality in the CONGEST Model

Yuki Kawashima(✉), Naoki Kitamura, Taisuke Izumi, and Toshimitsu Masuzawa

Osaka University, 1-5 Yamadaoka, Suita-shi, Osaka, Japan
y-kawashima@ist.osaka-u.ac.jp

Abstract. Betweenness centrality is one of the most popular metrics to quantify the importance of vertices within a network. We consider distributed vertex ranking algorithms based on betweenness centrality in the CONGEST model. Hua et al. show an $O(n)$-round algorithm of computing the exact values of betweenness centralities for all vertices in the CONGEST model, where n is the number of vertices. They also show a nearly-tight $\tilde{\Omega}(n)$-round lower bound for computing the exact betweenness centrality of a given single vertex. However, their lower bound does not imply the hardness of ranking based on betweenness centrality. In other words, the hardness of comparing the betweenness centralities of two given vertices is not straightforwardly deduced from the known bound for computing between centralities. It naturally raises the question if there exists a sublinear-time CONGEST algorithm of (approximately) computing the vertex ranking based on betweenness centrality or not. The main contribution of this paper is to provide a negative answer for this question in a very strong sense: Quantifying the preciseness of output rankings by Kendall's τ-distance (i.e., the number of inversion pairs for the correct ranking), we show that there is no $o(n/\log^3 n)$-round algorithm of outputting an approximate ranking within additive error at most $(1-o(1)) \cdot n(n-1)/4$. Since the maximum gap between two rankings is $n(n-1)/2$, it implies that no $o(n/\log^3 n)$-round algorithm can distinguish the instances whose rankings are almost inverted.

Keywords: betweenness centrality · distributed system · lower bound · ranking algorithm · CONGEST model

1 Introduction

The identification of important vertices (in various senses) is one of the key issues in graph data analysis. A considerable number of metrics have been proposed so far for measuring the importance of vertices. The *centrality* of vertices often

U. Schmid and R. Kuznets (Eds.): SIROCCO 2025, LNCS 15671, pp. 431–436, 2025.
https://doi.org/10.1007/978-3-031-91736-3_27

plays an important role in this context, which reflects how influential a vertex is in the network in terms of information spreading and connectivity. The *betweenness centrality* (BC) is one of the major metrics in that class. The betweenness centrality $C_B(v)$ of a vertex v in the graph $G = (V(G), E(G))$ is defined by the following expression:

$$C_B(v) = \sum_{s,t \in V(G) \setminus \{v\}} \frac{\sigma_{st}^G(v)}{\sigma_{st}^G},$$

where σ_{st}^G as the number of distinct shortest paths between vertices s and t in G, and $\sigma_{st}^G(v)$ be the number of the shortest paths between s and t in G which contains v. Note that, we call paths P_1, P_2 distinct if at least one edge of P_1 and P_2 is different. Betweenness centrality has a lot of applications in various areas, such as knowledge network analysis [17], contingency analysis of power grid systems [10], biological graph studies [11], and traffic network monitoring [18]. Many applications of BC in distributed systems are also exhibited: Network load balancing [1], response to attacks on the network [8], reliability in P2P networks [6].

Many applications utilize BC for extracting the nodes with higher importance, and thus much focus on the *importance ranking* by BC rather than the calculation of their values themselves. We refer to the task of vertex ranking based on betweenness centrality as *BC-ranking* hereafter. In this paper, we focus on the round complexity of distributed CONGEST algorithms for the BC-ranking problem. The CONGEST model is one of the standard computational models for distributed graph algorithms, in which each computer (vertex) operates synchronously in time units called rounds, and can exchange at most $O(\log n)$ bits of messages with its neighbors per round. The motivation of considering the BC-ranking task in distributed settings is twofold. First, some of applications of BCs require the *top-k vertex identification* with respect to BC measurements, a weaker form of the BC ranking, over huge graphs which cannot be stored into a single machine. Yet another motivation lies in relation to the context of *graph neural networks (GNNs)*. It has been pointed out that the computation process of conventional GNNs are very close (or equivalent) to distributed message passing systems [13,21], and hardness in the CONGEST model implies some size lower bound for GNNs [13]. More precisely, once we obtain the r-round lower bound for some task T in the CONGEST model, it implies that any GNN with $\tilde{O}(1)$ width (i.e., the dimension of the vectors stored at GNN vertices) outputting the solution of T requires the $\Omega(r)$ depth[1]. The computation/prediction of vertex ranking is one of the popular approaches in machine learning and graph data analysis [7,14,15,19,22,23], and some of those literatures particularly focus on the efficient BC-ranking computation (or weaker top-k vertex identification) uti-

[1] Note that this argument is easily generalized to the B-model, where each edge can transfer $O(B)$-bit information per round. Many r-round lower bounds in the CONGEST model, including the result of this paper, is trivially generalized to $\tilde{\Omega}(r/B)$-round lower bound for the B-model, and those bounds are transformed into $\tilde{\Omega}(r)$ bounds for (depth·width) in GNNs.

lizing GNNs [14,23]. A lower bound in the CONGEST implies some theoretical limit for such an approach.

Looking at the current state-of-the-art for the computation of betweenness centrality in the CONGEST model, there is an algorithm by Hua et al. [9], which computes the BCs of all values in $O(n)$ rounds. Since calculating the ranking of each vertex based on the computed betweenness centrality values is accomplished in $O(n)$ rounds, one can solve the BC-ranking problem in $O(n)$ rounds. Hua et al. [9] also proved that computing the betweenness centrality of a given vertex exactly requires $\Omega(n)$ rounds, which implies that the proposed algorithm attains the optimal time complexity up to polylogarithmic factor.

Intuitively, ranking calculation based on betweenness centrality is as difficult as calculating the betweenness centrality of a single vertex, and it is not difficult to imagine that the problem of computing *exact* rankings has a similar $\tilde{\Omega}(n)$-round lower bound. However, it is not clear whether a similar lower bound holds or not when considering *approximate* rankings. This question entails the issue how we measure the accuracy of rankings. Throughout this paper, we adopt Kendall's τ-distance, which is one of the most common metrics for quantifying the similarity of two rankings. Precisely, τ-distance $\tau(S, S')$ between two rankings S, S' is defined as the total number of pairs which differ among S and S' in order. Obviously the τ-distance $\tau(S, S')$ between the outputted ranking S and the correct ranking S' represents the preciseness of the algorithm's outputs.

In this paper, we show that there is no $o(n/\log^3 n)$-round algorithm of outputting an approximate ranking within additive error at most $(1 - o(1)) \cdot n(n-1)/4$. Since the maximum gap between two rankings is $n(n-1)/2$, it implies that no $o(n/\log^3 n)$-round algorithm can distinguish the instances whose rankings are almost inverted.

1.1 Related Work

In centralized settings, betweenness centrality is obviously computable in polynomial time. Specific exact computation algorithms are known, such as the method of Brandes et al. [4] whose computation time is known to be $O(mn)$. The computation is inherently as expensive as the all-pairs shortest path problem, and the computation time is quite large. It is shown that the lower bound for any algorithm that computes betweenness centrality based on path comparison model is $\Omega(mn)$ time [12]. Bader et al. [2] presented a novel approximation algorithm based on an adaptive sampling technique for computing betweenness centrality of a given vertex, for both weighted and unweighted graphs. Riondato et al. [20] presented two algorithms based on random sampling of shortest paths: one that estimates the betweenness centrality of all vertices, and another that focuses on approximating the betweenness centrality of the top K vertices with the highest betweenness. Pellegrina et al. [16] proposed an algorithm that, with high probability, computes accurate estimates of the betweenness centrality for all nodes in a graph and high-quality approximations of the top k betweenness centralities. Their algorithm follows a progressive sampling approach and builds on novel bounds based on Monte Carlo Empirical Rademacher Averages.

Xiang et al. [24] present an efficient CBCA (Centroids based Betweenness Centrality Approximation) algorithm based on progressive sampling and shortest paths approximation.

Crescenzi et al. [5] designed an efficient algorithm for computing betweenness centrality, which can be implemented by minimal modifications to any distance-vector routing protocol based on Bellman-Ford and they showed that the convergence time of their implementation is proportional to the diameter of the network.

Bergamini et al. [3] is the first to make in-memory computation of a betweenness ranking practical for million-edge semi-dynamic networks. They proposed two incremental approximation algorithms (for weighted and unweighted connected graphs) which provide a provable guarantee on the absolute approximation error. Maurya et al. [14] proposed the first graph neural network (GNN) based model to approximate betweenness and closeness centrality. They demonstrated that their approach significantly outperforms current techniques while taking less amount of time through extensive experiments on a series of synthetic and real-world datasets.

There has also been some researches applying betweenness centrality: AlShammari et al. [1] used a load balancing technique based on betweenness centrality to reduce latency and energy consumption on a network. Holme et al. [8] studied the defense of complex networks subject to attacks on vertices and edges by using strategies based on the value of betweenness centrality.

2 Technical Idea

We explain the brief outline of the proof. The proof follows the standard technique of the reduction from two-party communication complexity of *exact set-disjointness* function, which is a slightly modified version of the standard set-disjointness. The N-bit exact set-disjointness function $\mathrm{edisj}(x, y)$ is defined as follows:

$$\mathrm{edisj}(x, y) = \begin{cases} 0 & \text{if } x[i] = y[i] = 1 \text{ holds for an exactly single } i \in [0, N-1]. \\ 1 & \text{otherwise,} \end{cases}$$

where $x[i]$ and $y[i]$ mean the i-th bits of x and y. It is easy to show that the communication complexity of N-bit exact set-disjointness is $\Omega(N)$ bits as in set-disjointness.

The key idea of our proof is to construct a graph H_j called *digit-j gadget*, which is a graph of $O(N)$ vertices and an instance of N^2-bit exact set-disjointness instance (x, y). A digit-j gadget has two terminal vertices. For each vertex z connected to those terminals, H_j contributes some amount depending on $\mathrm{edisj}(x, y)$ to $C_B(z)$. More precisely, its contribution is β if $\mathrm{edisj}(x, y) = 0$, or $\frac{1}{\alpha 2^{j-1}} + \beta$ otherwise, where α and β are some values not depending on (x, y). Whole lower bound graph consists a target vertex set $Z = \{z_0, z_1, \ldots, z_{M-1}\}$ and $4 \log M$ digit-j gadgets $H_j^k (0 \leq j \leq 2 \log M - 1, k = \{0, 1\})$. Each target vertex is connected to a subset of digit-j gadgets so that it satisfies the following conditions:

- If $\text{edisj}(x, y) = 0$, $C_B(z_i) < C_B(z_{i+1})$ holds for any $0 \leq i < M - 1$.
- If $\text{edisj}(x, y) = 1$, $C_B(z_i) > C_B(z_{i+1})$ holds for any $0 \leq i < M - 1$.

Obviously, (very approximately) solving the BC-ranking problem in the lower bound graph implies the computation of $\text{edisj}(x, y)$. Suppose for contradiction that there exists such an approximation algorithm for the BC-ranking problem. The reduction is to make Alice and Bob simulate the execution of this algorithm in the lower bound graph as constructed above. One can show that one-round simulation costs $O(N \log N)$-bit communication between Alice and Bob. It follows that the worst-case running time of the algorithm must be $\Omega(N/\log N)$ rounds, by the communication complexity lower bound of $\text{edisj}(x, y)$. Setting $M = n - O(n/log^2 n)$ and $N = O(n/\log^2 n)$, we obtain the desired lower bound.

Acknowledgments. The second author (Naoki Kitamura) was supported in part by JSPS KAKENHI Grant Numbers 23K16838 and 22K21277. The third author (Taisuke Izumi) was supported in part by JSPS KAKENHI Grant Numbers JP22H03569 and JP23H04385, JST CRONOS Japan Grant Number JPMJCS24K2, and JST ASPIRE Japan Grant Number JPMJAP2302.

Disclosure of Interests. The authors have no competing interests to declare that are relevant to the content of this article.

References

1. AlShammari, W.M., Alenazi, M.J.: Performance analysis of a graph-theoretic load balancing method for data centers. Int. J. Adv. Comput. Sci. Appl. **11**(8) (2020)
2. Bader, D.A., Kintali, S., Madduri, K., Mihail, M.: Approximating betweenness centrality. In: Algorithms and Models for the Web-Graph: 5th International Workshop, WAW 2007, San Diego, CA, USA, 11–12 December 2007. Proceedings 5, pp. 124–137. Springer, Cham (2007)
3. Bergamini, E., Meyerhenke, H., Staudt, C.L.: Approximating betweenness centrality in large evolving networks. In: 2015 Proceedings of the Seventeenth Workshop on Algorithm Engineering and Experiments (ALENEX), pp. 133–146. SIAM (2014)
4. Brandes, U.: A faster algorithm for betweenness centrality. J. Math. Sociol. **25**(2), 163–177 (2001)
5. Crescenzi, P., Fraigniaud, P., Paz, A.: Simple and fast distributed computation of betweenness centrality. In: IEEE INFOCOM 2020-IEEE Conference on Computer Communications, pp. 337–346. IEEE (2020)
6. Gkorou, D., Pouwelse, J., Epema, D.: Betweenness centrality approximations for an internet deployed p2p reputation system. In: 2011 IEEE International Symposium on Parallel and Distributed Processing Workshops and Phd Forum, pp. 1627–1634. IEEE (2011)
7. He, Y., Gan, Q., Wipf, D., Reinert, G.D., Yan, J., Cucuringu, M.: GNNRank: learning global rankings from pairwise comparisons via directed graph neural networks. In: 39th International Conference on Machine Learning (ICML). Proceedings of Machine Learning Research, vol. 162, pp. 8581–8612 (2022)
8. Holme, P., Kim, B.J., Yoon, C.N., Han, S.K.: Attack vulnerability of complex networks. Phys. Rev. E **65**(5), 056109 (2002)

9. Hua, Q.S., et al.: Nearly optimal distributed algorithm for computing betweenness centrality. In: 2016 IEEE 36th International Conference on Distributed Computing Systems (ICDCS), pp. 271–280. IEEE (2016)
10. Jin, S., Huang, Z., Chen, Y., Chavarría-Miranda, D., Feo, J., Wong, P.C.: A novel application of parallel betweenness centrality to power grid contingency analysis. In: 2010 IEEE International Symposium on Parallel & Distributed Processing (IPDPS), pp. 1–7. IEEE (2010)
11. Joy, M.P., Brock, A., Ingber, D.E., Huang, S.: High-betweenness proteins in the yeast protein interaction network. J. Biomed. Biotechnol. **2005**(2), 96 (2005)
12. Kintali, S.: Betweenness centrality: algorithms and lower bounds. arXiv preprint arXiv:0809.1906 (2008)
13. Loukas, A.: What graph neural networks cannot learn: depth vs width. In: International Conference on Learning Representations (ICLR) (2020)
14. Maurya, S.K., Liu, X., Murata, T.: Graph neural networks for fast node ranking approximation. ACM Trans. Knowl. Discov. Data (TKDD) **15**(5), 1–32 (2021)
15. Park, N., Kan, A., Dong, X.L., Zhao, T., Faloutsos, C.: Estimating node importance in knowledge graphs using graph neural networks. In: KDD 2019, pp. 596–606 (2019)
16. Pellegrina, L., Vandin, F.: Silvan: estimating betweenness centralities with progressive sampling and non-uniform rademacher bounds. ACM Trans. Knowl. Discov. Data **18**(3), 1–55 (2023)
17. Pham, M.C., Klamma, R.: The structure of the computer science knowledge network. In: 2010 International Conference on Advances in Social Networks Analysis and Mining, pp. 17–24. IEEE (2010)
18. Puzis, R., Altshuler, Y., Elovici, Y., Bekhor, S., Shiftan, Y., Pentland, A.: Augmented betweenness centrality for environmentally aware traffic monitoring in transportation networks. J. Intell. Transp. Syst. **17**(1), 91–105 (2013)
19. Rakaraddi, A., Pratama, M.: Unsupervised learning for identifying high eigenvector centrality nodes: a graph neural network approach. In: 2021 IEEE International Conference on Big Data (Big Data), pp. 4945–4954 (2021)
20. Riondato, M., Kornaropoulos, E.M.: Fast approximation of betweenness centrality through sampling. In: Proceedings of the 7th ACM International Conference on Web Search and Data Mining, pp. 413–422 (2014)
21. Sato, R., Yamada, M., Kashima, H.: Approximation Ratios of Graph Neural Networks for Combinatorial Problems (2019)
22. Scarselli, F., Yong, S.L., Gori, M., Hagenbuchner, M., Tsoi, A.C., Maggini, M.: Graph neural networks for ranking web pages. In: The 2005 IEEE/WIC/ACM International Conference on Web Intelligence (WI 2005), pp. 666–672 (2005)
23. Shetty, R.D., Bhattacharjee, S.: Weighted GNN-based betweenness centrality considering stability and connection structure. In: 2023 15th International Conference on COMmunication Systems & NETworkS (COMSNETS), pp. 304–308 (2023)
24. Xiang, N., Wang, Q., You, M.: Estimation and update of betweenness centrality with progressive algorithm and shortest paths approximation. Sci. Rep. **13**(1), 17110 (2023)

Brief Announcement: Perfect Matching with Few Link Activations

Hugo Mirault[1], Peter Robinson[1], Ming Ming Tan[1(✉)], and Xianbin Zhu[2]

[1] School of Computer and Cyber Sciences, Augusta University, Augusta, GA, USA
{hmirault,perobinson,mtan}@augusta.edu
[2] Department of Computer Science, Aalto University, Espoo, Finland

Abstract. We consider the problem of computing a perfect matching problem in a synchronous distributed network, where the network topology corresponds to a complete bipartite graph. The communication between nodes is restricted to activating communication links, which means that instead of sending messages containing a number of bits, each node can only send a pulse over some of its incident links in each round. In the port numbering model, where nodes are unaware of their neighbor's IDs, we give a randomized algorithm that terminates in $O(\log n)$ rounds and has a pulse complexity of $O(n \log n)$, which corresponds to the number of pulses sent over all links. We also show that randomness is crucial in the port numbering model, as any deterministic algorithm must send at least $\Omega(n^2)$ messages in the standard LOCAL model, where the messages can be of unbounded size. Then, we turn our attention to the KT_1 assumption, where each node starts out knowing its neighbors' IDs. We show that this additional knowledge enables significantly improved bounds even for deterministic algorithms. First, we give an $O(\log n)$ time deterministic algorithm that sends only $O(n)$ pulses. Finally, we apply this algorithm recursively to obtain an exponential reduction in the time complexity to $O(\log^* n \log \log n)$, while slightly increasing the pulse complexity to $O(n \log^* n)$. All our bounds also hold in the standard CONGEST model with single-bit messages.

Keywords: distributed graph algorithm · perfect matching

1 Introduction

We study the fundamental problem of computing a perfect matching among the nodes of a network, where the communication capabilities are limited to link activations, which is motivated by simple biological cell networks or sensor networks of resource-restricted devices. Our model is inspired by protein-interaction networks and other biochemical signaling networks, where the alphabet of the messages between nodes are effectively restricted to a binary alphabet (i.e., signal or no signal), as demonstrated by [2].

Concretely, we consider a complete bipartite graph consisting of two vertex sets L and R, each consisting of n nodes, and we assume that each node knows

U. Schmid and R. Kuznets (Eds.): SIROCCO 2025, LNCS 15671, pp. 437–443, 2025.
https://doi.org/10.1007/978-3-031-91736-3_28

whether it is in L or in R. The communication between the nodes follows a *link activation model*, which can be interpreted as a restricted variant of the standard CONGEST model [6]: In each synchronous round, each node may *activate* any subset of its incident edges by sending a *pulse* over these edges. That is, if a node u sends a pulse over the edge $\{u, v\}$ in round r, then node v observes this activation by u at the start of round $r + 1$. We emphasize that these link activations are the only form of communication available in our model, which, in particular, prevents nodes from directly exchanging bit strings of information, as is commonly assumed in other models of distributed computation. In particular, activating the link $\{u, v\}$ does *not* allow u to transmit a sequence of bits to v in round r, but merely corresponds to a pulse that can be observed by v.

A model with similar features is the *beeping model*, introduced in [3], where, in each synchronous round, a node may choose to "beep" across *all* of its incident links, whereas, on the receiving side, each node can only distinguish between the case where none of its neighbors beeped or at least one of them did. We refer the reader to [5] for a detailed survey describing the common aspects and differences of various biological systems and distributed computing models.

To quantify the performance of algorithms in the link activation model, we analyze the *round complexity* (also known as *time complexity*), which is the worst case number of rounds, as well as the *pulse complexity*, whereby the latter corresponds to the worst-case number of pulses over all edges in any execution, for deterministic algorithms. When considering randomized algorithms, we assume that each node has access to a private source of unbiased random bits that it may query during its local computation at the beginning of each round. In that case, we also consider the *expected pulse complexity*, where the expectation is computed over the random bits of the algorithm.

While the pulse complexity is not a standard metric for distributed algorithms, it is closely related to the *message complexity*. In fact, any upper bound on the pulse complexity also implies the same bound on the message complexity. Hence, designing algorithms with low pulse complexity immediately gives rise to message-efficient algorithms that only need to send a small number of single-bit messages in the CONGEST model. Conversely, given an algorithm in CONGEST model that sends only single-bit messages, one can simulate the algorithm in our link activation model with pulse complexity equal to the message complexity.

Our main goal of this work is to quantify the achievable round and pulse complexity for computing a perfect matching in the link activation model. To this end, we distinguish two standard assumptions pertaining to the initial knowledge of the nodes:

1. In the *port numbering* or KT_0 *model* [6,7], we assume that nodes are anonymous and do not have unique IDs. Moreover, the incident edges to each node are numbered $1, \ldots, n$ and, at the start of each round, every node v only learns the subset of its incident port numbers that were activated by the respective other endpoint during the previous round. A crucial difficulty in this setting is that the assignment of port numbers to communication links is not known in advance to the algorithm, and may be chosen adversarially.

2. When considering the KT_1 *assumption* [1], on the other hand, we assume that every node is equipped with a unique integer ID, chosen from some range of polynomial size, and an algorithm needs to satisfy the stated properties for all possible ID assignments to the nodes. Each node knows the IDs of all its neighbors, and, since we consider a complete bipartite graph, this means that every node in L knows the IDs of all nodes in R, and vice versa. In particular, this allows a node in L to directly activate a link to a node v with some specific ID. At the start of each round $r \geqslant 2$ every node u learns the IDs of the nodes that activated a link to u in round $r-1$.

Our Contributions. We present several new algorithms and a lower bound for computing a perfect matching in complete bipartite graphs. As elaborated above, our results also hold in the CONGEST model where the per-link bandwidth restriction is just one bit. The complete proofs of our results can be found in the full version of the paper [4].

- In Sect. 2, we give a simple randomized algorithm that terminates in $O(\log n)$ rounds with a pulse complexity of $O(n \log n)$ with high probability. The algorithm assumes the port numbering model.
- Then, we focus on the impact of the more powerful KT_1 assumption, where nodes know the IDs of their neighbors, and show that this yields significantly improved bounds:
  - We show that the KT_1 assumption is sufficient to circumvent the lower bound of Sect. 4, by giving a deterministic algorithm that computes a perfect matching in $O(\log n)$ rounds and sends only $O(n)$ pulses.
  - Next, we leverage this deterministic algorithm to obtain an exponential improvement in the number of rounds, yielding a time complexity of $O(\log\log n \cdot \log^* n)$, at the cost of a slightly increased pulse complexity of $O(n \log^* n)$.
- Finally, in Sect. 4, we demonstrate that randomization is indeed crucial for achieving a low pulse complexity or message complexity in the port numbering model. We show that any deterministic algorithm must activate at least $\Omega\left(n^2\right)$ links in the port numbering model.

2 An Algorithm for Perfect Matching in KT_0

Throughout this section, we consider a complete bipartite graph of $2n$ nodes, $G=(V,E)$ where the vertices are partitioned into two disjoint subsets L and R. Nodes in L are referred to as *left* nodes, and nodes in R are referred to as *right* nodes. Note that each node knows whether it is a left or a right node. The algorithm is split into two stages. Stage 1 consists of $T=\lceil\log n\rceil-\Theta(\log\log n)$ phases with the goal to reduce the number of unmatched nodes to $O(\log n)$. Stage 2 deals with the remaining $O(\log n)$ unmatched nodes. Each phase consists of a constant number of rounds of communication. In phase 1, each of the left nodes randomly selects one port to send a pulse to the right nodes. If a right node receives a pulse, then it will be matched by randomly selecting one left node from which it has received the pulse to match. We show that with high probability, after the end of phase 1, the number of unmatched nodes reduces by $c := \frac{11}{12}$. Note that the node does not know which port leads to an unmatched neighbor until a pulse is exchanged between the two connected ports. Hence, a pulse sent by an unmatched left node might hit a right node that is already matched. To ensure sufficient unmatched left nodes get matched, in phase 2, each unmatched left node now sends $\lfloor 1/c \rfloor$ pulses. Again, we can show that with high probability, after the end of phase 2, the number of unmatched nodes is reduced further by c. Ultimately, we can show that, for each subsequent phase $i \leqslant T$, the unmatched nodes send $O(\frac{1}{c^{i-1}})$ pulses, and with high probability, the number of unmatched nodes is at most $c^i n$ at the end of each phase. As a result, with high probability, the number of unmatched nodes after T phases is $O(\log n)$.

In Stage 2, a matching for the remaining $O(\log n)$ unmatched nodes is constructed in $O(\log n)$ phases and using $O(n \log n)$ pulses. The key idea to achieve the desired number of pulses and the time complexity bound in Stage 2 is that, by exchanging $O(n \log n)$ pulses in total, each unmatched node learns *all* the ports that lead to its unmatched neighbors in each phase.

Theorem 1. *There exists a randomized algorithm that, with high probability, constructs a perfect matching on a complete bipartite graph with* $2n$ *nodes in* $O(\log n)$ *time and sends* $O(n \log n)$ *pulses.*

3 More Efficient Algorithms Under the KT_1 Assumption

In this section, we present two deterministic algorithms that leverage the knowledge of their neighbors' IDs to not only break the lower bound of Sect. 4, but also improve over the performance of the randomized algorithm in the port numbering model given in Sect. 2.

Here, we assume the standard KT_1 model, where nodes have unique integer IDs, and each node starts out knowing the IDs of its neighbors. We say that node $u \in L$ (resp. R) has *rank* i, if its ID is the i-th smallest among all nodes in L (resp. R). Note that due to KT_1 assumption, each node in L can locally compute the rank of each node in R and vice versa. However, each node is unaware of its own rank. We remark that if each node knew its own rank (which is the case if we assume KT_2 [1], where each node knows the IDs of all nodes at most two hops away from it), then all nodes can output a matching instantly without any communication. Returning to our KT_1 setting, this motivates the algorithmic design strategy of informing each node of its own rank. In the CONGEST model, where each message can carry $O(\log n)$ bit, it takes just two rounds and $O(n)$ messages to inform each left node of its rank. Concretely, in the first round, each left node sends a message to the smallest rank right node r_0. In the second round, r_0 sends one message to each of the left nodes to inform each of them of their rank. On the other hand, in our link activation model, where we can only signal a pulse, a naive implementation of the above approach would take $O(\log n)$

rounds and $O(n \log n)$ pulses, which would not improve over the port numbering algorithm of Sect. 2: When using the convention that a 0 bit sent over a link corresponds to non-activation of the link, and a 1 bit corresponds to a pulse, in round i, node r_0 sends the i-th bit of the binary representation of j to the left node of rank j. This can be completed in $O(\log n)$ rounds.[1] We now describe the algorithm Fast-Interval-Matching, which achieves optimal pulse complexity of $O(n)$ and terminates in $O(\log n)$ rounds.

Interval Construction: We describe a procedure called Interval-Construction that partitions the nodes in L and R into equal size intervals. As we will instantiate this method for subgraphs of different sizes, we state it under the assumption that $|L| = |R| = N$, where N is any positive integer that is greater than 2. We use $l_0, \ldots, l_{N-1}$ to denote the nodes in L ordered in increasing order of their IDs, and analogously define the order $r_0, \ldots, r_{N-1}$ of the nodes in R. Note that this means the rank of l_i as well as r_i is i. Each node in L activates a link to the node r_0, who has the smallest ID in R. We define $s = \lceil \log N \rceil$.

Helper Inform Step: r_0 recruits a set H of s *helpers*, by contacting the nodes with the s-smallest ranks in L, and informing these nodes of their respective rank in L. In more detail, r_0 executes $O(\log s)$ rounds, where in each round i, r_0 sends the i-th bit of rank j to l_j, for each $j = 0, \ldots, s-1$ in parallel.

Partition Step: Our next goal will be to split R into N/s intervals,[2] denoted by $R_0, \ldots, R_{N/s-1}$. For every $i \in [0, N/s-2]$, interval R_i contains exactly s nodes, whereby $R_{N/s-1}$ contains at most $s-1$ nodes. We will obtain the intervals $L_0, \ldots, L_{N/s-1}$ by partitioning L analogously. To implement this partitioning process, we make use of the helper nodes to identify every node $r_{j \cdot s}$, for $j \in \{0, \ldots, n/s-1\}$, as the *leader of interval* R_j. To this end, we need to make sure that each of these interval leaders learns its own rank. Concretely, for every $r_{j \cdot s}$, each helper l_i sends the i-th bit of $r_{j \cdot s}$' rank to $r_{j \cdot s}$ (where 0 and 1 bits are encoded as pulses or absence of pulses, as described above). Note that all helpers perform this operation in parallel, and thus this requires just 1 round. Afterwards, each interval leader $r_{j \cdot s}$ activates a link to $l_{j \cdot s}$. Since the *leader of* L_j, which is $l_{j \cdot s}$, knows the rank of $r_{j \cdot s}$, a pulse is sufficient for it to learn its own rank. Then, we add all edges $\{l_{j \cdot s}, r_{j \cdot s}\}$ to the matching, for each j.

Interval Inform Step: In one additional round, the leader of L_j sends a pulse to each node in interval R_j, and similarly the leader of R_j sends a pulse to the nodes in the L_j, for every j. This completes the Interval-Construction procedure.

Parallel Interval Matching: Finally, we compute a matching in parallel on each interval: Each leader in R_j, i.e., node $r_{j \cdot s}$, sends a pulse to the nodes in L_j sequentially, by activating one link per round. This allows each node in L_j to

[1] An alternate approach for each left node to learn its own rank is by performing the standard binary search. That is, each left node l initially guesses its rank i and sends a pulse to the right node r of rank i. In two rounds, r can inform l if its guess is equal, higher, or lower than the correct value.

[2] To simplify the notation, we assume that N/s is an integer.

deduce its rank relative to the other nodes in L_j. In particular, node $l_{j \cdot s+k}$ will have its link activated after k rounds, prompting it to activate the link to $r_{j \cdot s+k}$, to which it becomes matched. Since there are s nodes per interval this completes in $O(s)$ rounds.

Theorem 2. *Fast-Interval-Matching computes a perfect matching in $O(\log n)$ rounds and with $O(n)$ pulse complexity, assuming the* KT_1 *model.*

We now employ procedure Interval-Construction as a building block for obtaining an exponential improvement in the time complexity, while increasing the message complexity by a factor of $\log^* n$.

Theorem 3. *There is a deterministic algorithm that computes a perfect matching in $O(\log^* n \cdot \log\log n)$ rounds and with a pulse complexity of $O(n \log^* n)$ under the* KT_1 *assumption.*

4 A Lower Bound for Deterministic Algorithms

In this section, we return to the port numbering model and show that randomization is crucial for any perfect matching algorithm that uses a small number of communication links. We make several assumptions that strengthen our lower bound, namely, we consider the standard LOCAL model [6], where nodes may send messages of arbitrary size in each round, and deterministically label the nodes in L with IDs from $\{1, \ldots, n\}$ and the nodes in R with IDs from $\{n+1, \ldots, 2n\}$.

Theorem 4. *Any deterministic algorithm that constructs a perfect matching in the complete bipartite graph of $2n$ nodes has a message complexity of $\Omega(n^2)$ in the* LOCAL *model.*

Acknowledgements. Peter Robinson was supported in part by National Science Foundation (NSF) grant CCF-2402836. Ming Ming Tan was supported in part by National Science Foundation (NSF) grant CCF-2348346. Xianbin Zhu was partially supported by a grant from the Research Grants Council of the Hong Kong Special Administrative Region, China [Project No. CityU 11213620].

References

1. Awerbuch, B., Goldreich, O., Peleg, D., Vainish, R.: A trade-off between information and communication in broadcast protocols. J. ACM **37**(2), 238–256 (1990)
2. Cheong, R., Rhee, A., Wang, C.J., Nemenman, I., Levchenko, A.: Information transduction capacity of noisy biochemical signaling networks. Science **334**(6054), 354–358 (2011)
3. Cornejo, A., Kuhn, F.: Deploying wireless networks with beeps. In: International Symposium on Distributed Computing, pp. 148–162. Springer (2010)
4. Mirault, H., Robinson, P., Tan, M.M., Zhu, X.: Perfect matching with few link activations. arXiv preprint (2025, to appear)

5. Navlakha, S., Bar-Joseph, Z.: Distributed information processing in biological and computational systems. Commun. ACM **58**(1), 94–102 (2014)
6. Peleg, D.: Distributed Computing: A Locality-Sensitive Approach. SIAM, Philadelphia (2000)
7. Suomela, J.: Survey of local algorithms. ACM Comput. Surv. **45**(2), 24:1–24:40 (2013)

Author Index

U. Schmid and R. Kuznets (Eds.): SIROCCO 2025, LNCS 15671, pp. 445–446, 2025.
https://doi.org/10.1007/978-3-031-91736-3

The manufacturer's authorised representative in the EU is Springer Nature Customer Service Centre GmbH, Europaplatz 3, 69115 Heidelberg, Germany. If you have any concerns regarding our products, please contact ProductSafety@springernature.com

Printed and bound by CPI Group (UK) Ltd, Croydon, CR0 4YY
01/10/2026
02225952-0015